PLANT BREEDING METHODOLOGY

PLANT BREEDING METHODOLOGY

NEAL F. JENSEN

Liberty Hyde Bailey Professor of Plant Breeding, Emeritus
Cornell University

A WILEY-INTERSCIENCE PUBLICATION
JOHN WILEY & SONS
New York Chichester Brisbane Toronto Singapore

Library of Congress Cataloging in Publication Data:

Jensen, Neal F.
Plant breeding methodology.

"A Wiley-Interscience publication."
Includes index.
1. Plant breeding—Methodology. I. Title.

SB123.J46 1988 631.5′23 88–2663
ISBN 0–471–60190–X

Printed in the United States of America

10 9 8 7 6 5 4 3 2 1

To my wife, Mary Webb Jensen

PREFACE

Sometimes it seems that I have been a plant breeder all my life, and in fact that is nearly true. As a boy in western North Dakota, I collected native plants of all kinds for my private garden. There were sand cherries, cacti, wild roses, and plenty of garden produce, all surrounded by a magnificent lilac hedge. Then in 1935, when I started college at North Dakota Agricultural College, a most fortunate thing happened: I took a laborer's job in the Agronomy Department, which led to a summer job on the durum wheat breeding project of Dr. Glenn S. Smith. Because I could type, Glenn gave me winter work analyzing data and typing his USDA annual report. I was later to realize that this undergraduate work experience was in no way different from the experience of a graduate research assistant. After graduation and a summer of working for the Soil Conservation Service at the Great Plains Experiment Station at Mandan, North Dakota, I accepted a fellowship in the Department of Plant Breeding at Cornell University. Here I received my "second" graduate training as research assistant to Professor Harry H. Love, who directed the cereals improvement project for New York State. The project embraced all the small grain cereals, wheat, oats, barley, rye, and later, triticale. I have always been grateful for these broader responsibilities and opportunities, for they have given me colleagues who work in all the small grains, and have provided me with the opportunity for variety releases in several crops.

The Ph.D. degree was awarded me January 1943, when I was already on active duty with the U.S. Navy in World War II. In mid-1946, I returned to Cornell University as Assistant Professor in the Plant Breeding Department, working with Dr. Love on the cereal improvement programs. When he retired in 1949, I was placed in charge of the project and administered it until I retired from Cornell University in 1978. For the next five years, I operated a private cereal breeding project, using sites in Arizona, New Mexico, and North Dakota. An account of this period is included in this book.

Thus, subtracting the war years, I have been actively and intimately associated with plant breeding for more than 50 years. I have seen and worked with almost every modification of every method of plant breeding devised—or at least worked with it in my mind. From this I have learned from countless "evolvings" of my own methods that there are still changes in methodology to come. I have learned, too, that progress sometimes is made in the most unlikely ways. We can see this when we survey the whole history of breeding and selection methodology. I have long wanted to respond to this challenge, but practically, I could never find the time until I retired. Then I felt called upon to write this book about plant breeding methodology, a book that I think of as subjective and personal, as well as definitive.

It seems important to say for whom this book is written. Of course I would be pleased if my contemporaries and peers read it; nevertheless, I do have in mind the new, usually younger and less experienced, breeders. It seems a waste to learn a great deal in a career lifetime—things that may not be found in books or obvious to the less experienced—and not pass on some of the lore and art of the plant breeding craft. Therefore, I have attempted to put some things normally not found in textbooks into this book.

This book is about the *methodology* of plant breeding and selection. Each of the approximately 1400 predominantly English language papers reviewed and interpreted were chosen because they had something to say about methodology. The preponderance of papers on the small grains relates to the obvious fact that my experience has been with all of the small grain cereals, and perhaps also to the great wealth of literature on these crops. Nevertheless, the emphasis is on methodology regardless of crop species or mode of pollination, and there are generous contributions from other crops—at least 30—including soybeans, cotton, sorghum, tobacco, and corn.

Finney, in defining his concept of the meaning of "variety" in 1958, made this statement:* "Rather surprisingly, plant breeders appear to have no one word to describe the very much more numerous variants of a species that they first produce and then subject to a process of selection.... These are the 'potential varieties' or 'potential cultivars'...." When I read his paper in 1979, it occurred to me that there was a suitable name that might rest on the same level with "cultivar." Without benefit of authority, I coin it now and I will use it for elite lines throughout the book: it is *candivar*, for *candidate* culti*var* or *var*iety. For the same reason, I coin the name, *multiblend*, to take its place with multiline, uniblend and biblend.

* Finney, D.J. (1958). Plant selection for yield improvement, Euphytica 7:83–106.

Once in a television interview, Eric Hoffer was asked a critical question: how could he justify publishing a book that had but seven or eight short chapters? He vigorously replied that most books, regardless of length, had at most one new idea. His modest book offered the reader a new idea in each chapter. In a sense, I have taken that as my goal, to offer the reader more than one new and helpful idea.

NEAL F. JENSEN

Albuquerque, New Mexico
1988

CONTENTS

PLANT BREEDING METHODOLOGY

SECTION I

INTRODUCTORY TOPICS

1

INTRODUCTION TO METHODOLOGY

In this book I have tried to recognize and describe the various breeding and selection systems, but it is a perverse fact of plant breeding that a system is seldom used in its pure state. George Sprague (1966) put this succinctly:

> A realistic appraisal will often indicate that no one breeding system is completely adequate for a specific situation and that two or more systems should be used simultaneously. When this is true one system may be chosen because of simplicity and the promise of rapid improvement. The second method may be specifically chosen because of long-range possibilities.

Successful strategies for breeding and selection embody three elements: choice of breeding system, operational experience, and style. Of these, operational experience and style contribute more to success than choice of breeding system. Granted, at times a choice of breeding system may be decisive; for example, a backcross program can be powerful. Nevertheless, it is the meld of a breeder's operational experiences and his personal style of administering that dominates a program and shapes the several optional components into an operating breeding system.

"Style" represents a breeder's personal proclivities, a way of doing things, an establishment of a cultural and comfort level for sustained project operation.

The important fundamental in the choice of a breeding system is that it be flexible and open-ended for change. Change is the hallmark of breeding and selection methodology; experienced, successful breeders continually fine-tune their operations. Viewed this way the starting choice of method for a breeding program is less important than the system that gradually but certainly will evolve.

STRATEGY VERSUS TACTICS

The complementary relationship between strategy and tactics is often misunderstood. Tactics implement strategy. Strategies are grander, on a loftier plane, often ponderous and slower to change, and shaped by outside forces. Strategy is analogous to policy. A cultivar may represent a strategic goal; however, its actual creation reflects the tactics of the breeder and his colleagues. There may be changes in tactics—for example, two breeding approaches—without altering strategy. To illustrate, the strategy of the wheat breeding program at Cornell University changed but little and very slowly over decades; that is, it remained a program aimed at producing superior white-kerneled winter wheat cultivars high in yield and milling quality and resistant to the ravages of pests. The strategy admitted new goals of shorter and stiffer straw, changes in needed disease resistance, and so forth. By contrast, the tactics were much more plastic. An original strict pedigree system grew to accommodate the use of the bulk population breeding method, the backcross, mass selection, recurrent selection, male sterility, and various combinations and variations of selection.

Strategy without a grasp of tactics can be sterile—an exercise in futility. A young breeder, for example, a person succeeding to a position of directing a program, usually finds acceptable strategies in place. Of more importance to eventual success is to examine the tactical approaches he or she will use to accomplish the strategic goals.

EXAMPLES OF METHODOLOGY USE

Mac Key (1963) gave an interesting account of the evolution of plant breeding methodology over 75 years in describing procedures used in the Swedish Seed Association. The earliest period, from about 1890 to 1910, was devoted to the exploitation of land varieties, using mass selection and pure line progeny techniques. These activities reached a plateau upon the recognition of a major flaw, which was that although variability was great in the predominant landrace varieties, all genotypes isolated represented already fixed entities. Therefore, it was difficult to obtain types with desired multiple traits. For example, one might find a short genotype, a disease-resistant genotype, and a high-yielding genotype in separate selections, but seldom together in one individual.

This led to the era of combination breeding. The first strategy was an approach that treated a few crosses on a large scale. Later this was seen to be too optimistic an assessment. As problems became more complex and improvement slower than anticipated, changes were made toward a step-by-step, more patient approach that featured a less intensive treatment of more crosses in a recurrent mode. In time the need to introduce foreign germ plasm became urgent and led to a break with the classical two-parent cross. Mac Key described the situation at this juncture thus:

> The successively increasing difficulties to improve a type, which already in essential features has approached the ideal agro-ecotype, is expected as a logical, natural law. It is, however, complicated by the fact that the intention of the breeding becomes more and more restorative in favour of this already gained genic constellation, while the recombination of a cross is equally dependent on the two parents. A delicate genic balance is thus easily broken down, when one of the components is less well adapted. The frequent, intermediate recombinants will here seldom be of potential interest but rather rare extremes with specific characters added to the already given pattern. The chance of finding such segregants, which do not carry too high a load of inferior germ plasm is, however, in practice quite low.

The result was the introduction of the backcross into a planned convergent cross series, referred to by Mac Key as a *discontinuous backcross*. Occasionally, modified backcrosses were used. These were three-way outcrosses to an adapted genotype accepted by the breeder as equivalent in quality to the recurrent parent. In these planned crosses, which spanned time periods of many years, the practice of selected sib crossing was found useful as an aid in concentrating the genes for multiple characters into subsequent progeny individuals. Mac Key gave four examples or variations of the use of backcrosses in convergent breeding.

Selection methods in use when Mac Key wrote reflected some concern about how best to handle the very early generations. An example of the procedure for complex crosses began with F_4-embryo seed selection of 6000 hand-picked seeds. These were space-planted and 200–300 plants saved from the population, about 3–5%. The F_5 was planted in two-rowed plant progeny performance tests; the plants were spaced at commercial crop density, about 1 in. apart, with a 10–15% selection intensity. From this point, selected lines went into optimal plot nurseries, generally embracing varieties and replications in a ratio of 2 : 1.

Even a detailed review is a poor substitute because it cannot impart the flavor and philosophy of an original paper; Mac Key's article has much to recommend its full reading.

Another example of how changes in methodology evolve within a breeding project was given by Qualset and Vogt (1980), who reviewed the breeding and selection methods used in California for wheat. Their climate is described as Mediterranean, a transition zone between rigid spring and winter types. Historically, the California wheat improvement experience encompasses two distinct eras. The first was based on Spanish and Australian germ plasm and lasted into the 1960s, when it was ended by devastating stripe rust epidemics. Resistant germ plasm from the Rockefeller Foundation program in Mexico was introduced and in the ensuing years became the foundation germ plasm of the California program. The abrupt shift in basic germ plasm disrupted the heavy reliance on the backcross program, which has since been resumed, and with time a gradual introgression of the two eras may be expected. An illustration would be the incorporation of Spanish 'Sonora' stripe resistance genes into 'Siete Cerros 66'.

Qualset and Vogt intimated that the range of choice of breeding methods should be wide and available to meet different situations. They had available and used the backcross, multiline, two-parent, and multiparent crosses, diallels, and diallel selective mating (DSM) systems.

The authors felt that in the matter of selection, the breeder must be aware of and direct each program or population toward one of two goals, a commercial variety or parent building. Recurrency, in one form or another, must be the overriding concern in selection. Greater success may be ensured by increases in population size, early generation, and multisite testing.

The breeding and selection methods used, which Qualset and Vogt termed "population management," reflect a mixture of procedures tailored to needs, and were built upon a two-crop-a-year rotation. This latter feature permits release of a variety in as little as seven years. Two population management methods were used, namely, pedigree and bulk hybrid population.

The pedigree method used was procedurally orthodox, starting with the F_2. Selection was more rigid and fruitful when done in the winter nursery, normally positioned for the F_2 and F_4 generations. Truncation proceeded through a three-stage visual evaluation where the phenotypes were viewed successively as families, individual progenies within families, and individual plants within progenies. The continuous selection brought homozygous lines quickly up to a series of yield, regional, and grower trials. They reported, on the basis of tests of strategies, that the use of visual selection enhanced procedures through increased gains, in addition to eliminating the labor of the weighing of individual plant yields.

Qualset and Vogt's discussion of their three-stage visual evaluation of the F_4 is valuable to an understanding of selection procedures that have broad application beyond the pedigree system. They saw each F_3 row (F_2-derived from a single plant) as one unit in the whole range of F_3 rows of a cross. They then viewed the plants within a selected row, and later they saw the F_4 progeny rows of these plants. This procedural routine of theirs illustrates a mental framework that plant breeders use at all stages of selection.

As an illustration, when selecting wheat F_5 head rows in my project at Cornell University I would have a field plan listing the 7000–10,000 rows, arranged sequentially by cross groups. Assuming that the first cross, from either pedigree or a single-cross bulk, might be rows 1–237, I would walk the alley to get a sense of the merit and variance of the whole cross. This preliminary observation can make a breeder sad or excited. In a multiple-cross bulk system the same evaluation philosophy applies, even though one cannot recognize the within-family and between-family relationships. Qualset and Vogt, in exploring several selection strategies on F_3 plants, made the important observation that the use of visual evaluation was a critical element in the success of any strategy.

Concerning bulk population breeding, Qualset and Vogt accepted the concept that bulk population management may dictate different early generation treatment. They recognized two kinds of bulks, selected and random, which may move from one status to the other as circumstances indicate. A selection of desired individuals might be made and bulked for the next generation and thereafter receive no further attention until the year of line selection. They recommended only three generations in bulk for a cross. Lines taken out were evaluated in progeny rows. The overall combination of the pedigree and bulk hybrid population methods, with recurrent selection overtones, makes for a sound, efficient program.

Concluding on theoretical grounds that the pedigree, bulk, and mass pedigree methods are not the most efficient means of processing the segregation generations of crosses, van der Kley (1955) proposed a "gradual selection" procedure with claimed higher efficiency. Van der Kley worked through the well-known formulas for degree and attainment of heterozygosity and homozygosity per generations, culminating in the frequency at F_∞ of best genotypes in populations having certain genetic factors. He concluded from these background considerations that final selections made in F_2 or F_3, or in previously unselected segregating populations, would be inefficient.

Van der Kley emphasized the need for preselection, a euphemism for removal or discard of undesirable characters (genes) accompanied by a corresponding increase in the percentage of other genotypes. An indiscriminate preselection, that is, a massive all-embracing removal of undesirable individuals, beginning in F_2, runs the risk that its imposition upon the most heterozygous generation will lead to a loss of hidden, possibly valuable genetic combinations. Van der Kley thus favored a gradual preselection scheme. The distinction he drew between preselection and gradual selection is that the first method preselects all homozygous recessive deleterious genes (individuals) and continues this in consecutive generations; the latter preselects at a more gentle pace, beginning with one trait and successively adding additional traits as generations proceed. Van der Kley hoped that this gradual attrition would keep in check the expansion of total population numbers. He used disease resistance as an illustration of the gradual method. Recessive susceptibility to an imposed epiphytotic of a single race of a pathogen would be followed the next year by the breeder's use of a different race, or by choosing a new criterion character for selection.

There are so many good ideas in van der Kley's paper that it is difficult to review it adequately. On the one hand breeders accept many of the thoughts expressed, but on the other hand the assumptions and conclusions of other parts are arguable and at times too sweeping. Most breeders practice some form of early grooming at certain times and in selected hybrid populations. Indeed, so variable are the practices that it would be unlikely to find two pedigree, bulk, or mass pedigree breeding programs operating with identical procedures. There would be agreement with van der Kley's statement that reliable selection must be preferred to random selection. The breeder must recognize also the real problem of sampling large harvested seed lots to get the small amount needed for planting the next generation; van der Kley's gradual method addresses this concern.

The contribution of natural selection in van der Kley's assessment of the bulk method was, I thought, unnecessarily negative. When the bulk hybrid or composite breeding and selection method is examined at the practical F_5 line-removing level, the negative aspects of natural selection seldom apply. The bulk breeding system is a benign, efficient, and successful method. The breeder must determine whether the negative cases reported in the literature actually apply to his crop and situation. Natural selection, teleologically speaking, sometimes is halting in its progress and even may reverse its direction, but it also has favorable long-term aspects meriting careful consideration. Suneson has recognized these by applying the term *evolutionary breeding method* to these long-range programs.

These examples illustrate the changes that have taken place in just a few projects. It would be rare to find a breeding program that had not experienced similar changes.

COMBINATION BREEDING METHODS

The logical mating of a pedigree and bulk population system represents a mix of methodology and can be a challenge for breeders. Some variations that attempt this are illustrated below for the small grain cereals:

Variant 1: Combination Pedigree and Composite

This is the simplest variation and uses the pedigree method only in the F_2, where unwanted plants are discarded to improve the later sampling efficiency of the population. This also ensures that plot space will not be taken up by inferior material.

1. Make the cross, grow the F_1, and plant the F_2.
2. Visually select among the F_2 plants.
3. Bulk the selected plants.
4. Grow the bulk populations of F_3 and F_4 (F_5 is optional).
5. "Explode" the population for selection.
6. Proceed to the head row and yield nurseries.

Comments: A useful variant for Step 3 is to form more than one bulk by stratifying for plant characters, such as height, maturity, or disease resistance. The special "explode" procedure for withdrawing lines is mandatory because attempts at random or nonrandom selection of lines from a bulk population at crop density produce disappointing results. These special procedures expand the population through mechanical random space planting in rows, exposing every plant to the breeder's eyes. Any row arrangement will work provided there is row separation for the breeder's path. Details of the system I used may be found in Chapter 9 on the bulk population method.

Variant 2: Combination Pedigree and Composite

This variation is next in simplicity and permits a look at the F_2 progeny in F_3 plant rows:

1. Make the cross and grow the F_1.
2. Space-plant the F_2 and select the best plants.
3. Plant the selected F_2 plants in progeny plant rows and visually select the best rows. Bulk the selected rows.
4. Grow in bulk through the F_4 and F_5.
5. Explode the population and select plants or heads.
6. Proceed through the head row and yield tests.

Variant 3: Combination Pedigree and Composite

This variation uses early pedigree selection, yield testing with visual discrimination, bulking of selections for two generations, and F_6 line selection, followed by nursery testing:

1. Make the cross and grow the F_1.
2. Space-plant the F_2 and select the best plants.
3. Plant a one-replicate F_3 single-row nursery of F_2-derived progenies. Select from visual and yield data.
4. Bulk the selections and grow for the F_4 and F_5.
5. Withdraw line selections using the special explode procedure described under Variant 1.
6. Proceed with nursery plot testing.

Comments: Selection in the F_2 should cause a high discard, which will happen also in the F_3 single-row nursery. The F_3 rows may be any length to accommodate minimum seed amounts and should be mechanically sown if possible for efficiency. Check cultivars should be included. A field selection for harvest only should precede the yield measurements in order to realize the efficiency of discarding. F_4 and F_5 bulk plots should veer in the direction of large size and may be planted mechanically at double seed density. These are a few examples of ways in which plant breeders vary their handling of early generation material, and thus incorporate methodology.

REFERENCES

Mac Key, J. 1963. Autogamous plant breeding based on already highbred material. In E. Akerberg and A. Hagberg (Eds.), Recent plant breeding research, Svalof, 1946–1961, Stockholm.

Qualset, C. O. and H. E. Vogt. 1980. Efficient methods of population management and utilization in breeding wheat for Mediterranean-type climates. Proc. Third Intern. Wheat Conf, Madrid, pp. 166–188.

Sprague, G. F. 1966. The challenge to agriculture. African Soils 11:23–29.

Van der Kley, K. K. 1955. The efficiency of some selection methods in populations resulting from crosses between self-fertilizing plants. Euphytica 4:58–66.

2

GENETIC ENGINEERING AND THE FUTURE

Scientists in crop technology, particularly plant breeders, have no reason to feel complacent about their role in shaping the future. It is true that in many cases the developments in biotechnology will have to pass through the breeders hands to be put into a form suitable for practical agriculture. However, it may be the corporation breeder as well as the institutional breeder who does this. Furthermore, the marketable product, perhaps a new snack food, may be taken through to the consumer with controlled contracts and pricing without ever touching traditional commercial agriculture. This is not necessarily bad, just different. *How* different was brought home to me by a news story (Anonymous,) in which officials of a genetic technology firm talked about their products and objectives. The press reported that one company executive said, "Most of the plant breeding in the past was for the benefit of the farmer, improving things like crop yield. Our work is mostly for the consumer." Another is reported to have said, "It's just a development of the selection process. Farmers have been doing this for years." *Translation*: Consumers have been ignored in the past, benefits going only to the farmers; and farmers have been in charge of crop improvement. So much for the efforts of thousands of agricultural scientists during the past century. Although it is difficult to accept these simplistic views, we in the working end of plant improvement know full well that there has been a "sea change" taking place and that we are in a transition period between two eras.

WHAT HAS BROUGHT US TO THIS TRANSITION PERIOD?

It's been said that all people living at the time remember what they were doing at the time of Pearl Harbor, December 7, 1941. I was playing touch football on the lawn of the Alpha Zeta Fraternity at Cornell University when the news came through. When World War II ended in the Pacific almost four years later in August 1945, it came about because of a stunning technological breakthrough, the atomic bomb. I heard the shouted news while sitting in the rigging of my ship, the USS *Eldorado*, in Manila Bay watching a movie. I well recall the perplexity we felt when we tried to fathom the type of bomb used. This weapon was so different, and it was so difficult to predict its place in future warfare, that for years thereafter the military characterized it simply as a projectile with a "bigger bang."

In an analogous way, a "bomb" has been dropped on the biological sciences—the discovery of the genetic code by Watson and Crick in 1953. And in the same sense, biological science never again will be the same. Today we are living in a transition period of evaluating the future effects of this "bigger bang" on biology. As plant breeders, we must accept the projection that at least the new genetic engineering will provide more "tools" with which to work.

In an historical sense I have written this book in the transition period between these two eras, that of traditional genetics-based methods and that of biotechnological genetic engineering. Although one can believe that conventional plant breeding will not disappear, if for no other reason than that breeders will be needed in many cases to incorporate the new developments into agriculturally acceptable forms, nevertheless the long-range prospect is for a more corporate and less institutional look and a more legalistic and equity position on the exchange and use of information; see, for example, Bradner (1975). Increasingly biotechnology is finding its way into the weekly news and business magazines.

WHAT IS GENETIC ENGINEERING?

I had always thought of biotechnology and genetic engineering as interchangeable terms, with genetic engineering arising out of cytological manipulations and research. Now I am led to believe that the definitions for both have broadened, with "biotechnology" having a more general connotation, and "genetic engineering" expanding to embrace heretofore relatively obscure specializations that had been agreeably accommodated under animal breeding and plant breeding umbrellas. As examples of disciplines subject to genetic engineering we might list several, such as biochemistry, the use of microbes in industrial processes, protoplast fusion and cell-tissue culture, recombinant DNA applications, nitrogen-fixing capability in new crops, and finally, traditional genetics and animal and plant breeding. Recombinant

DNA technology has been the yeast in the genetic engineering dough, in fact redefining the meaning of and our understanding of the gene, from a segment on a chromosome to definite sequences of the four bases that are the building blocks in the double helix structure. Incidentally, the term "near isogenic" becomes more credible when the gene region might be seen as having "sticky edges" or imprecise break points.

When we say that we are in a transition period between two eras, that of traditional plant breeding and that of emerging genetic engineering, I think it fair to say that a hallmark of the first era was the dominance of the plant breeder. The assembly of plant improvements into a final commercial form was the domain of the breeder. Today, and in the era we have begun, this will no longer be a simple truth; already some agricultural consumer products are marketed directly from a corporate research laboratory to the public. But in the main, the breeder will still pursue his craft, most likely with greater opportunities and added "tools." It is also likely that both corporate and institutional plant engineers will need to understand each other's activities. Plant breeding will remain important, and it is my hope that this book will be useful to those corporate scientists now working in genetic engineering who were trained in other disciplines such as physics or chemistry.

For more on genetic engineering, I cannot do better than to recommend the report by the Council for Agricultural Science and Technology (1986).

REFERENCES

Anonymous, 1985. The high-technology carrot: how sweet it is. Associated Press, Albuquerque Journal, front page, July 31.

Bradner, N. R. 1975. Hybrid soybean production. U.S. Patent 3,903,645. Off. Gaz. U.S. Patent Office 938:480.

C.A.S.T. 1986. Genetic engineering in food and agriculture. Council Agric. Sci. Technol. Report No. 110:47 pp.

SECTION II

BREEDING AND SELECTION METHODS

3

PRIMARY METHODS SHAPED BY HISTORICAL USAGE—A PREAMBLE

Modern-day breeding and selection methods are rooted in the past—not a dim past but a comfortable century or so ago. Certainly, many individuals earlier had contemplated plant improvement, but only in the latter half of the nineteenth century do we see the placement of the scientific underpinnings of genetics and plant breeding. It was unfortunate that Mendel's 1865 paper on studies in plant hybridization was not immediately recognized. This left the theory field undisputed to Charles Darwin. According to Akerman, Granhall, Nilsson-Leissner, Muntzing, and Tedin (1938), plant breeding in Sweden started in 1886 using *methodical selection*, a scheme based on a general acceptance of Darwin's theory that character variation could be directed and fixed. It was soon realized that this form of mass selection was rarely successful when practiced on well-maintained varieties. The pedigree selection method developed from a realization that bulking all selected plants obscured the identity of those that might represent genetically different types, but lining out the individual plant progenies by pedigree culture led to purebreeding lines. Pedigree selection was evolving and in use in Europe, Great Britain, and Sweden a decade or so before the turn of the century.

That the early efforts with the pedigree selection method resulted in many successful varieties can be attributed to the fact that existing varieties were the result of mass selection, which, though often producing similar types capable of blending into a seemingly uniform population, nevertheless represented heterogeneous populations in which pedigree selection was effective. An illustration is the story of 'Victory' oats. W. Rimpau in Germany developed the 'Probsteier' variety, known to be heterogeneous. Exported to the United States it became 'Milton' and was later imported into Sweden. In 1892 thirty heads were taken from a field plot. From these selections two became Victory and 'Golden Rain'. Victory went on to become one of the world's most popular and successful oat varieties. Nevertheless, it seems clear that pedigree selection was linked to a slowly declining reservoir of landraces and heterogeneous varieties.

In 1901 Mendel's paper was exhumed, independently, by DeVries, Tschermak, and Correns. Mendel found a ready supporter in Bateson, and DeVries mutation theory complemented the view of heritable units and led people to see, beyond Darwin, the genetics of how existing variation worked, and how only hybridization and mutation could create totally new variations.

To this point we have seen that working plant breeders had interpreted variation from a mind-set conditioned by (1) a belief in an inexhaustible supply of variation reposing in landrace varieties, (2) a population improvement approach only peripherally concerned with varietal uniformity, and (3) a belief that even fixed types could be improved. Enormous effort was applied to the cataloguing of botanical and morphological characters. The predominant breeding approach was, naturally, *continuous mass selection*, a cyclic procedure that attempted to upgrade whole populations by directed selection. With landrace varieties the predominant early crop form, it is understandable that crop uniformity was not a worrisome problem—it only became one when seed export opportunities to other areas were believed to fail because the crop was too impure (for example, Swedish growers at one time anticipated a vigorous seed export market to Europe based on the "uniqueness" of their climate in producing high-quality seeds). Thus crop purity attempts were imposed from the top down (population) rather than from the bottom up (pure line individual).

The metamorphosis from the group to the individual came about as a result of several parallel occurrences. The intense categorization of plants according to botanical and morphological traits resulted in occasional categories containing but one individual, and it was noted that these usually bred true. Also, as might be expected, certain breeders observed this phenomenon of single (homozygous) plants; for example, Vilmorin popularized this system in France, and the Nilsson-Ehles in Sweden. Then, when Johannesen (1903) of Denmark (who gave us the term *genotype*) published his treatise on pure lines showing that continuous selection in pure lines failed to alter the characteristics of successive progenies, all the pieces for the pedigree culture system fell into place. First seen as an efficient way to achieve uniformity, the pedigree method required only to be matched with hybridization to become a powerful breeding tool.

Hybridization also entered the breeder's lexicon or tool chest by the "back door." Long known and in common use before 1900 in Europe, the United States, and Australia, it was originally not seen as of primary importance to plant improvement because of the strongly held belief that old landraces contained all the variations needed for future crop improvement. In Sweden, hybridization was also referred to as *combination breeding*. It was realized that large numbers (of F_2) were required to allow a desired combination to surface—progenies as large as 20,000–50,000 plants were grown and of these only about 200–500, respectively, would be saved. Selected plants would be carried forward by either (1) pedigree, or (2) bulking, with selection and pedigree applied at a later generation, that is, F_6–F_{10}.

Akerman, et al. (1938) expressed a view that today, almost a half century later, seems reasonable:

> Experience has shown that the very best combination aimed at will seldom or never be obtained, even if very extensive material is available. Usually, however, it is possible to approach comparatively close to the ideal and to obtain a new variety of practical value. By means of new crosses further improvements can be made and the desired variety may finally be reached. The principle of cross-breeding is based upon *continuous combination experiments*.

Johannesen's pure line paper also opened the door to an appreciation of the need for and value of statistical inferences from breeding and selection operations; it was seen that the variation observed even in pure lines was due to environmental forces.

Thus by the turn of the century, the groundwork had been laid for all the innovations and refinements of techniques that have brought us to the present. Strangely, nothing has been lost—we still have mass selection, individual selection from bulk populations, and the pedigree method—except for the virtual disappearance of most of the world's landrace populations.

Forty years ago, Hutchinson (1940) made two revealing statements reflecting on the status of plant breeding: "The influence of the science of genetics on the art of plant breeding has been much less profound than was expected by the early geneticists," and "It should be the function of the geneticist to substitute objective evidence for the breeder's intuition."

It is to be hoped that plant breeding is now a recognized science. Increasingly, plant breeders work with the geneticists, the physiologists, the pathologists, and the statisticians.

Harlan and Martini (1937) paid a tribute to the first barley breeders that could stand as a statement for all the early plant breeders in all crops:

> The contributions of the pioneers in breeding are not easily evaluated. They did more than breed plants. They paved the way for the generation that followed. They awakened the interest that led other men into the same fields. They effected contacts with the farming population. And with it all, they were successful breeders. They did not have adequate replications. They could not know the inheritance of many of the qualities for which they were working. They did not have available information on the role of certain diseases. Yet they did produce

good barleys. This is a testimonial to the art of breeding, for it is an art as well as a science. The eye and the judgment of the breeder can never be wholly dispensed with. Judgment can be aided by perfection of technique, but many times in the handling of every extensive program the breeder must choose.

REFERENCES

Akerman, A., I. Granhall, G. Nilsson-Leissner, A. Muntzing, and O. Tedin. 1938. Swedish contributions to the development of plant breeding. 111 pp. New Sweden Tercentenary Publications, Stockholm.

Harlan, H. V. and M. L. Martini. 1937. Problems and results in barley breeding. USDA 1936 Yearbook of Agriculture: 303–346. U.S. Government Printing Office.

Hutchinson, J. B. 1940. The application of genetics to plant breeding. 1. The genetic interpretation of plant breeding problems. J. Genet. 40:271–282.

Johannesen, W. 1903. Ueber Erblichkeit in Populationen und in reinen Linien. G. Fischer, Jena.

4

THE MASS SELECTION METHOD

Mass selection is the formation of a composite propagation stock (usually seed) by the selective harvest of individuals from a heterogeneous population. It is one of the oldest of man's plant improvement practices. Its earliest use was with cross-pollinated crops and with heterogeneous populations composed of predominantly homozygous individuals, for example, landrace varieties. Today it is employed almost exclusively in populations of deliberate hybrid or composite origin. In the cereals its use has been largely as a screening procedure utilizing visual, mechanical, disease control, or other means. It has been shown to be effective with some characters and not with others and, of course, works most successfully with characters having high heritability.

Mass selection may be done from either a positive or negative approach (or both): positive if by selection and rebulking of desired genotypes, negative if by discard or culling from a population. Procedures may be visual, mechanical, or assisted–natural (e.g., an applied screen). They may be simple or sophisticated, using information on heritability, genetic correlations, and other statistics. Composites and/or bulk populations are ideal subjects for mass selection, bearing in mind that homozygosis increases and the nonadditive portion of genetic variance decreases with increased generations.

Mass selection is used in some form by almost all plant breeders. To illustrate, in my cereal breeding projects I clipped plots and pulled plants to lower the extreme height portion, used different seed screening procedures, planted winter crops on poor fields using late sowings to induce winter kill, worked with plant pathologist colleagues to induce rust epiphytotics, and imposed high atrazine applications to acre-sized bulk populations in a search for tolerance. In one long-running project, Professor George Kent of Cornell's Plant Pathology Department and I repeatedly planted bulk populations of oats in fields bordered by buckthorn, the alternate and breeding host for the crown rust organism, in an area 30 miles north of Ithaca, New York. The fields were harvested by combine harvester, and the large volume of seeds was vigorously and variously processed to eliminate all but the heaviest kernels for replanting. Plump heavy seeds are indicators for rust resistance, extended by implication to the plant from the observed lack of damage to the kernel.

EXAMPLES OF THE MASS SELECTION METHOD

The history and use of mass selection in open-pollinated corn is of distinct interest to breeders. [For a history of the evolution of corn breeding methods, passing through mass selection, hybridization, ear-to-row breeding, and pure line breeding—inbreds and hybrid corn—see the paper by Richey (1922).] It is an indubitable fact that open-pollinated varieties of corn contained vast amounts of genetic variance, yet scientists reported inconclusive and conflicting results from mass selection experiments that attempted to improve yields (but positive results were obtained when mass selection was applied to certain morphological and chemical properties). Hull's (1952) hypothesis of overdominance was accepted as one reason for the difficulties in showing progress in yield advance. Here, the favored heterozygote would also be more likely to be chosen because of heterotic advantage, yet because of nonadditivity of the variance it would not be expected to respond positively to mass selection for yield. One is led to believe also that the criteria chosen to represent yield, as embodied in the persistent ear-to-row method, were flawed.

These early experiments with mass selection in corn were largely placed in abeyance with the advent of the commercial acceptance of the double cross hybrid suggested by Jones (1918), but fundamental interest continued. The use of progeny tests showed that yield advances could be documented.

A significant step forward was made when Gardner (1961) instituted gridding of field research areas in order to reduce the environmental variation among plants. This organizing technique, which has profound implications for field experimentation, occupied but a few lines under his "experimental procedure" section:

> Beginning in 1956 the fields have been stratified into small areas of 40 plants each, and seed of the highest yielding 10% of the plants in each stratum has been saved to produce the next generation. The reason for this is to reduce the environmental variation among the plants compared so that the selection would be more effective.

As a result of this technique Gardner was able to show a gain in yield of 3.9%/year.

In the small grain cereals, mass selection is a way of increasing the frequency of favorable genes during the segregation period of inbreeding. In the self-pollinated crops this amounts to a way of concentrating desired genetic variance; however, there are many situations, involving male sterility and intermating, where populational breeding and recurrent selection can be practiced.

Love (1927) compared pedigree and mass selection—the form of mass selection described was the pure, not modified, kind. In this method, the seeds of the F_2 and succeeding generations were bulked each year and sown as mixed seed. The mixtures were grown four or five years before individual selections were made. Love suggested that mass selection extended the time-to-variety release by two years. Advantages cited of mass selection over the pedigree method were less detailed record-keeping and ample seed supplies that enabled the breeder to test in different environments.

McFadden's (1930) early use of the fan mass method to partition and separate out the plump and heavy wheat seeds was a recognition of, and a corrective response to, segregating populations containing disease-injured, hulled (emmer), and low-quality lightweight seeds. The deliberate procedures represented a step beyond the routine use of the fanning mill and a step forward in mechanical mass selection.

Harrington (1937) introduced a variation in mass selection—actually a combination of pedigree and mass selection—which he termed the *mass pedigree method*. This scheme took advantage of environmental situations favorable to selection; in the absence of such situations the populations were handled in a straight mass selection manner. Harrington wrote:

> The method may be illustrated as follows: A given cross, with or without mass selection in F_2, is carried on in mass to F_3. In F_3, owing to a favorable season, very efficient selection reduces the population to a twentieth of its size. The selected plants, instead of being bulked, are separately threshed and sown in progeny rows in F_4. In F_4 selection is carried on as in the pedigree method.

He went on to write that the selection could be implemented or disregarded at any stage but was proposed to take advantage of opportunities presented by changes in the environment. In discussing mass selection later, Harrington (1952) did not mention mass pedigree but introduced a variation, the *fanned mass method*. Probably both should be considered modified mass selection methods.

Ivanoff (1951) achieved the ultimate in mass selection when he developed a laboratory "rag doll" method of testing resistance to Victoria blight (*Helminthosporium victoriae*) in oats. Large volumes of seeds, for example, a hybrid bulk, can be treated by coating seeds with an agar culture of the pathogen, and then incubating or germinating them in gunnysack rolls. The seedling reaction to this pathogen is so decisive that resistant seedlings can be seen immediately among the dead or dying susceptible mass.

Bal, Suneson, and Ramage (1959) examined changes in 21 characters in three F_{30} versions of Composite Cross II (formed by Harlan from a bulk of 378 hybrids of 28 parents). The populations were as follows: without selection, selection for large seed, and selection for small seed. The selection for seed size was conducted over nine generations. The results after 30 generations, expressed as milligrams per 100 kernel weights, were: large 61.5, small 50.6, unselected 57.3, and mean of all parents 48.9. In addition to the large human-selection effect, there was a natural shift to larger seeds, evident from the similar parent mean and small seed weights.

Lonnquist, Cota A., and Gardner (1966) reported on 2 years of mass selection experiments involving both normal and irradiated corn populations drawn from the 'Hays Golden' cultivar. Substantial gains per cycle were made in both populations for all characters, that is, yield and ears per plant, days to flower, and ear height. There was no decline in additive genetic variability. Higher additive genetic variance was observed in the thermal neutron-treated population, being somewhat less noticeable in yield of grain than the other traits.

Sprague (1966) stated, "Genetic theory and the experimental studies currently available clearly indicate that when additive effects are of major importance some type of mass or family selection is one of the most effective breeding procedures that can be employed."

However, Sprague implied that mass selection is better suited to cross-pollinated than to self-pollinated crops for the reason that in the latter, although quick gains are possible, continued progress is slowed by the lack of recombination.

Romero and Frey (1966) mass-selected a 250-cross oat composite of F_2 seeds for height reduction for four years, by clipping all panicles to the arbitrary height of the medium short 'Cherokee' cultivar. Over generations the mass-selected oat population became about 3 in. shorter by the F_6. When other traits in the two populations were examined, it was found that means for other traits had shifted, particularly toward earlier heading and higher grain yields. This may be called *indirect mass selection* (or serendipity!). The genetic variances of height, heading date, and yield were reduced in the mass-selected populations.

Andrus and Bohn (1967) applied mass selection by index to hybrid heterozygous populations of the cantaloupe, *Cucumis melo* L. The cantaloupe, with about 30% interbreeding, falls between the self-pollinated and cross-pollinated categories and may be described as partly cross-pollinated. The research interest centered on index selection efficiency, variability of components, relative stability, merit, and the adaptation range of both populations and inbreds.

The base population was a single F_2 heterozygous plant from which 24 F_3 selections were bulked for the F_4. Seeds from 74 selected fruits of the F_4

produced the F_5 foundation stock (M_1). The population was perpetuated from M_2 through M_9 by index score selections of fruits at alternating sites, taking the best 20% of 300 marketable fruits to seed the next generation.

The index used involved scoring for 18 separate traits on a 1–5 scale (1–3 scores were deemed unacceptable), with total score being the sum of all 4 and 5 ratings. Inbreds were drawn from each cycle and selfed for evaluation.

At the end of nine generations ($F_9 = M_7$) the reserve seed of each mass selection cycle was grown along with an equal number of self-pollinated populations from matching cycles. The observation that variability remained high throughout cycles was attributed to multiple factors controlling fruit characters, leading to slow fixation. It should be mentioned that the deliberate choice of alternating sites was an attempt to provide both high-stress and low-stress environments in the hope that a form of stability might be achieved for both environments. The inbred selections that were not subjected to alternating sites showed lower variability.

Generally, the use of index mass selection was successful; most of the characters were improved without significant loss of genetic variance. The authors were aware that optimum weighting might not have been applied—more information on desired component weights should improve efficiency. Even so, there were significant advances (shifts of means) for several important characters, and the overall judgment of acceptability increased.

Andrus and Bohn cautioned against too rapid and direct an approach to homozygosity; however, as background one must consider the high level of interbreeding that existed, with its consequences *in re* recurrent breeding and genetic variance. Specifically, one would want to keep more generations involved. The possibility of later selection and perhaps some form of selected and assisted intermating to create adjunct populations might be considered.

Lonnquist (1967) tried five generations of selection for increased productivity in corn, using a correlated character, more ears per plant, and found that the gain in yield per cycle of 6.28% approximated the gain made when the direct selection of weight of grain per plant was the criterion.

Reid (1987) applied mass selection in the late 1950s and early 1960s to a male sterile barley population generated by crossing 18 parents in 153 possible crosses. The F_2, F_3, and F_4 generations were tested for winter hardiness in a series that included a bulk of the parents. The F_4, F_5, and F_6 generations were subsequently tested after the years of natural crossing. Winter hardiness increases in the 25–30% range were achieved, a substantial gain induced by natural mass selection forces.

Frey (1967) practiced mass selection for oat seed width over five generations and recorded its effect on 100-seed weight, heading date, and plant height. This exercise in indirect mass selection resulted in 0.48 g gain in 100-seed weight, 5.0 days later in heading, and 8.25 cm increase in average height coupled with a 9% yield increase over the control population.

Rasmusson, Beard, and Johnson (1967) reported on the growing of a composite of 6000 entries from the World Collection of Small Grains (barley) under stress conditions of late planting for 6 years. The objective was to develop commercial cultivars adapted to late planting. Seed aliquots from each year's harvest were replanted for each next crop.

When the sixth-year seed was grown for two years in comparison with remnant basic composite seed and seed from each of the five yearly cycles, there was a stepwise linear increase from a two-year mean of 138 g/plot for the original seed to 217 g/plot for the sixth-year seed. The improvement amounted to 57%, or an average 9.5% per cycle. The yields of two check cultivars, 'Larker' and 'Dickson', were 241 and 347 g/plot, respectively.

Using computer simulation techniques, Bliss and Gates (1968) compared mass selection with single seed descent (SSD) for genetic gain. For mass selection, they specified phenotypic selection of the best 12.5% and 25.0% of the population; SSD was practiced over 20 generations. The population sizes ranged from 16 to 256—low numbers in breeding program terms. They found greater genetic gains from mass selection except in the smaller populations, and because these numbers were smaller than likely to occur in programs, we can take this for a general statement. This was attributed partly to continued selection in every generation (in SSD, selection was practiced only in the final generation). On the other hand, Bliss and Gates noted that mass selection is usually a field operation producing one generation per year and taking several years per cycle, whereas SSD generations may turn over as rapidly as the length of the plant's life cycle. A fair comparison, based on time, is that perhaps two SSD cycles, complete with the hybridizing of the best selections, can be equated with one cycle of mass selection.

Doggett (1968) gave different variations of mass selection systems for sorghum (significant outcrossing) that are directly applicable to the small grain cereals wheat and especially barley, with its male sterile potential. His versions employed the (Coes) genetic male sterile gene, ms_3, and thus embodied not only mass selection but recurrent selection, and had the added merit of using the Gardner grid.

The first system Doggett proposed was "normal mass selection (female choice)." Female choice refers to the male sterile individual, a female phenotype. A composite was formed from $ms_3ms_3 \times Ms_3ms_3$, which in time contained an estimated 40% sterile plants (the expected 50% suffered about a 20% deficiency owing to weather conditions and diseases in Uganda affecting seed set, a situation that also occurs in the cereal steriles). Selection was made in Gardner-gridded areas. Although female choice mass selection is not as efficient in terms of gain as selfed plant mass selection, there are the positive factors of recurrent selection and the opportunity for recombination.

The second version of Doggett's was *alternating female choice and selfed plant mass selection*, which took advantage of the annual bimodal rainfall and two growing seasons in Uganda. In the first half of the season, normal female choice gridded mass selection was practiced. In the second half of the season, gridded mass selection was imposed on the fertile plants only, which made up about two-thirds of the population. Selection pressure may be adjusted to produce a desired number of selections. This system tends to pyramid the efficiency of gains per cycle.

A third variant involved the World Collection of sorghum. Each year a part of the collection was grown, the entries judged visually, and the "poor" group discarded. The retained entries were then crossed to both genetic and cytoplasmic steriles and became gene pool or reservoir composites for use as inputs into the breeding program as described above.

Doggett's proposals find ready acceptance and agreement with my experience. Three decades of working with the *ms* genes in barley (of which there are many) caused me to convert the entire Cornell winter barley project over to procedures using MSFRS (*male sterile facilitated recurrent selection*, a term proposed by Eslick of Montana). My winter barley project contained about 40 composites, all MSFRS-based.

The breeder must be aware that the tangential programs possible from these composites easily could exceed the capacity of a project to absorb. To start a program in any desired direction, for example, toward short straw, or disease or lodging resistance, it is necessary only to take male sterile individuals from any convenient or desired composite base. Hockett, Eslick, Reid, and Wiebe (1968) published a list of 101 spring and 115 winter-type male sterile stocks that were publicly available. The progenies are oriented in the desired direction through controlled pollination.

There are helpful "pushes" that the breeder can give, and Doggett refers to some of these, such as the removal of undesirable fertile plants before they shed pollen, mass selection removal of visually recognizable qualitative traits, and so forth. As an MSFRS composite increases in number of generations, cereal breeders using these systems could graft a harvest index selection procedure onto the annual harvest of new selections.

Matzinger and Wernsman (1968) reported that four cycles of mass selection for increased green leaf production (known to be highly correlated with cured weight in tobacco, an autogamous species) resulted in an average linear increase of 44 g/plant/cycle with no evidence that genetic variability of the populations had been reduced. The method's efficiency was enhanced by measuring and identifying high green leaf weight individuals followed by immediate random hybridizing within this group. The bulked hybrid seeds then produced the next cycle population. This combination of mass and recurrent selection is possible whenever the character can be identified and measured before anthesis of the last tiller.

Smith and Weber (1968) ran two cycles of bidirectional mass selection by specific gravity in two hybrid populations of soybeans. The first cycle effectively created two subpopulations: one (Hi) for high protein and low oil; the other (Lo) for low protein and high oil, as measured against the control. Continued selection (Cycle 2, different environment) showed more inter-hybrid and intrahybrid variation and was not as effective as Cycle 1, perhaps because of decreased variance. No effects were observed on yield, height, and lodging.

Fehr and Weber (1968) reported on mass selection for seed size and specific gravity in soybeans and their effect upon protein and oil contents. They found that the selection combination of large seed and high specific gravity resulted in maximum progress for high protein and low oil. Conversely, maximum progress for low protein and high oil came from selection for small seed and low specific gravity.

Doggett and Eberhart (1967) put forth recurrent and mass selection proposals for use in sorghum that are worth reading in the original.

The use of mass selection does not guarantee genetic gain. Hallauer and Sears (1969) examined a case where, after five and six cycles of selection in the open-pollinated corn varieties 'Iowa Ideal' and 'Krug', respectively, no significant yield advance was made over the original populations. The authors considered five possible reasons for this result: (1) limited genetic variance, (2) imprecise plot techniques, (3) insufficient testing, (4) low intensity of selection, and (5) too high plant density. Points 1, 2, 3, and 5 were dismissed as reasonable causes. It was thought that the exclusion of stalk-lodged plants may have reduced the selection intensity, which moved from 7.5% to 27.4%, but for this there seemed no alternative because stalk lodging could not be accepted in selected lines.

The authors suggested that earlier selection for yield in these two varieties could have lowered the genetic variance. They knew that additive genetic variance constituted a high proportion of total variance (Krug 77.6%, Iowa Ideal 86%); nevertheless, reduced variance would mean this would be a percentage of a smaller total.

Frey and Browning (1971) observed grain yield differences in near-isogenic oat lines containing specific crown rust resistance gene loci. Investigating this phenomenon, Brinkman and Frey (1977) identified and assigned the yield differences to components of yield. In some isolines, increased yield was traced to increased tiller production. In others a slight increase in spikelets per panicle, an increase in panicle number, and an increase in seed weight were responsible. In two isolines fewer spikelets per panicle resulted in a reduction in yield. These are all examples of indirect or associated selection, often found in mass selection.

Tiyawalee and Frey (1971), aware of the debilitating influence of oat crown rust attack on mature seed weight (Simons, 1965), used mass selection for density (seed weight) of harvested oat seeds of a broad-based hybrid F_3 population to determine changes in frequency of crown rust resistance genes over seven generations. The harvested seeds from the populations exposed to crown rust attack were processed and divided into heavy and light density classes. The gene frequency for resistance, which averaged 20%–21% in the F_3 check over the period, rose to 35% by the F_{10}, most of the 14% increase occurring during the first three generations. An association of tall–susceptible and short–resistant was favorable to selection. The authors discussed the import of being unable to push the population beyond approximately one-third resistant. It could have been related to the breeding processes or mutation rates in the pathogen, or to the interfering properties of several races operating on the population, to the existence of a natural equilibrium conferring practical protection, or to the particular races present in the experiment. The study has particular relevence to the disease protection aspects of multilines.

Derera and Bhatt (1972) reported the results of a mass selection study in wheat wherein homogeneous and heterogeneous populations were stratified for seed size and later tested for yield. Mass selection in the heterogeneous and heterozygous F_2 bulks produced an expected reduction in variance. There was a shift in means, upward for large and downward for small seed size, for kernel weight, grain weight per spike, and grain yield per plot in five of the six tests. There was no detectable shift of these statistics in the homogeneous populations (cultivars).

Of particular interest was the finding that selection for larger seeds increased yields, whereas selection in the opposite direction decreased yields. The yield increase was approximately 33% over the control yield in both cycles, suggesting that one cycle devoted to this trait may be sufficient to prepare a population for line selection. No appreciable effect was noted in the homozygous cultivars.

The reduction in yield from the smaller-seeded selections was modest, about 7%. This suggests that their greater number of seeds planted per area of the drilled plots compensated somewhat for the advantages accompanying the larger plump seeds.

Derera and Bhatt described the use of mass selection as a major but routine project procedure, and proposed its use through different approaches to create an "assembly line" of truncated populations moving toward line selection.

Following their earlier 1972 paper, Bhatt and Derera (1973) examined the changes associated with mass selection for seed size. They found that a series of quality attributes (test weight, milling extract, wheat and flour protein, kernel hardness) were size neutral. As to the association of seed size with agronomic characters, the results appeared to be cross-dependent, but overall, it was judged that selection for seed size posed little negative selection risk. An exception may have been a positive association between seed size and height, but in this case, height is an easily selectable visual character.

Chandhanmutta and Frey (1973) found selection for heavy panicle to provide an associated gain of 5.6%/cycle in yield. Later-heading and taller plants were undesirable indirect responses.

The paper by Derera, Bhatt, and Ellison (1974) marked the advent of mechanical mass selection as a routine and general procedure in the New South Wales wheat improvement program. In effect it was experimentally tested on seed attributes, then expanded to field selection involving the traits, earliness, height, spring–winter growth, disease resistance, and preharvest sprouting resistance. Illustrations were given of mass selection at the different F stages of a cross population.

Using the number of green tillers per plant prior to head emergence in wheat and barley, Redden and Jensen (1974) compared two mating systems under mass and recurrent selection. Similar procedures were applied to a single cross of barley, and one cross of wheat cultivars, and all procedures stemmed from the same F_2 population in each crop. The following populations were formed, based on tiller number at heading time. Maximum selection efficiency resulted because selection preanthesis permitted same-generation hybridization. The series were as follows:

1. *Hybrid Series:* two cycles of truncation mass selection followed by recurrent selection random intermating of selected high-tiller number plants.
2. *Selfed Series:* two cycles of truncation mass selection.
3. *Control Series:* random F_2, and beyond, propagation, unselected.
4. *Special Series:* supplementary populations to provide special information.

The populations were brought together for test in what would normally be the F_5, at two sites (Site 2 of wheat was considered impaired because of poor establishment). One of the special cases mentioned under Series 4 was a study of the effect of hand-pollinated seeds, always smaller in size. It was found to have an effect on tillering, but not enough to affect valid estimates of selection response.

The results, measured against unselected controls, in Series 3, showed that the average selection response for increased tillering at two sites was 10.3%/cycle for wheat, and 6.3%/cycle for barley. Comparing the hybrid and the selfed series, the results, respectively, were for wheat (one site), 22.6 and 18.5%; and for barley, Site 1; 17.1 and 8.7%, and Site 2, 13.9 and 10.3%.

Geadelmann and Frey (1975) imposed one cycle of bidirectional direct and indirect mass selection on seven traits of oats in 1500 lines randomly chosen from a heterogeneous population. The lines were grown in three environments representing standard (highly fertile), competitive (crowded), and noncompetitive (open) environments. The field and laboratory measurements were taken of the seven traits on all of the 1500 lines.

The findings were many: (1) direct mass selection for each trait was effective; (2) heavy primary panicle was the most effective trait for indirect mass selection for grain yield; (3) differences in gains in different environments were generally small and varied, and choice of environment could be made on a least-cost basis; (4) population variance suggested additional cycles might be imposed; (5) selection intensity was influential on realized gains; (6) mass selection on weight of seed gave good responses in grain yield; and (7) selection pressure had to be lowered for traits of moderate to low heritability or a disproportionate number of promising individuals would be lost. On the latter point, it was found that turning selection around and selecting for discard required about the same cutoff point as positive selection, about 50%. The authors did not use the Gardner gridding technique, but suggested its use in conjunction with choice of environment.

Finding that selecting sweet corn lines for earworm resistance produced late-maturing lines unsuited for parents in commercial hybrids, and that loss of resistance accompanied selection for agronomic characters, Wann and Hills (1975) applied tandem mass selection to a hybrid population created by intercrossing several relatively resistant inbred lines (complete resistance was difficult to obtain) with a heterogeneous broad germ plasm composite synthesized from 12 intermating commercial hybrids. In the F_1 a large number of plants were selected for earliness and other desired characters and the ears from these plants later examined for earworm infection. This tandem procedure was followed for four effective cycles. The results were a 5-day gain in earliness, an increase in light-colored silks, and an increase of 1.4 kernel rows per ear. Earworm resistance increased slightly but steadily. Observations indicated the population still contained reservoirs of untapped resistance.

Using a diverse alfalfa population, Heinrichs (1976) used bidirectional mass selection for fast and slow emergence. The selected groups produced polycrossed seed followed by selection in a second cycle. This procedure was repeated in a third cycle. The final fast and slow populations were tested for yield and found to have a differential of about 10% in yield.

The role of mass selection in corn genetics and breeding has been studied in detail. In two areas there are close relationships to small grains cereal breeding. First, the relationship is very close in barley and wheat genetics and breeding where the male sterile genes are used; here the situation is much like that in open-pollinated corn. Second, Gardner's 1961 suggestion of stratifying fields into small areas, which rejuvenated and made more precise the measurements of mass selection advances in corn, has numerous applications to cereal breeding and selection operations.

Mass selection has been applied to the problem of simultaneously raising protein percent of grain and total grain yield. Historically, a negative relationship has been found; that is, attempts to raise protein have resulted in lowered yields (Mertz, 1976). In certain fortunate cases, for example, the 'Atlas 66' genes for higher protein and leaf rust resistance in wheat, specific genes have been able to transmit higher protein without lowering yield.

Foster, Jain, and Smeltzer (1980) successfully used 10 cycles of bidirectional mass selection, which they imposed on the three characters of seed weight, plant height, and days to flower, in six continuously selfed populations of grain sorghum. Tenth cycle means for the three characters, low, nonselected, and high, were as follows:

Seed weight (g/1000) :	20.7	25.3	33.9
Plant height (cm) :	70.2	95.7	125.9
Flowering date (days) :	68.9	70.7	78.3

Sorghum is commonly considered to experience outcrossing in the 10–20% range; however, the base population used, 'Double Dwarf Yellow Milo 38', was known to be remarkably free of outcrossing, as could be determined by the uniformity of its different panicle type and double dwarf height appearance. Consequently, it was a bit of a mystery as to how a closed system could provide continuing variability over 10 cycles. The natural assumption would be that there was a continuing supply of heterozygotes present. Although this matter was not satisfactorily resolved, the authors concluded that the population structure of DD 38 had to be complex. Although the population traced to only 10 plants in 1910, it may well be that isolation and rigid roguing to the distinct double dwarf type may have masked the consequences of the outcrossing that likely had occurred over this long period of time.

The destructive economic effects of rain prior to or during harvest, causing sprouting in the head, were presented in detail by Derera (1982), who chronicled the problems of dormancy and ways of breeding for resistance to sprouting. Preharvest sprouting is a problem for both spring and winter wheat cultivars, as well as other small grain cereals. In wheat, it is a major problem with the white types, although there are red wheats that do not show grain dormancy or resistance to preharvest sprouting.

Derera's breeding approaches involved selected parents (a list of several was included) and backcrossing and mass selection methods. In order to get the desirable resistant genotypes for backcrossing cycles, the segregating populations were exposed to mass selection under simulated rainfall conditions. For best results the ear-sprouting tests were conducted at 10, 20, and 30 days after morphological maturity, and selections were then tested and monitored for three to four years in order for them to experience and respond to genetic × environment interactions. The various expressions of sprouting resistance have been incorporated into a sprouting index.

Ross, Kofoid, Maranville, and Voight (1981) used mass selection over two cycles for high and low protein in S_1 sorghum populations, some of which had been random mated, having the same genetic base. Small but significant changes were found, but in the opposite direction from grain yield. An interesting point was that "populations selected for grain yield produced more protein per ha (protein yield) than did populations selected for protein." Because this was so, they suggested that selection for protein yield be explored, since the grain yield penalty would be slight.

Ross, Maranville, Hookstra, and Kofoid (1985) used mass selection for high and low protein at 15% selection pressure in recombinable sorghum populations, and realized a spread of 1.56 percentage points from the base with increased protein accounting for +0.50 points and lower protein for −1.06 points; nevertheless, protein yield remained essentially constant. The authors stressed the need for protein-specific genes, such as in Atlas 66 wheat, or the use of tandem or index selection, which might bring increased protein percentage and grain yield.

DISCUSSION OF MASS SELECTION

Mass selection has progressed through several phases, all of which are still in use to some degree.

1. Mass selection before hybridization, for example, with landraces (sometimes called *methodical selection*).
2. Modified mass selection, that is, selection after hybridization during the segregating generations.
3. Mass pedigree method (cf. Harrington), that is, selection when warranted.

Love's 1927 description of mass selection is interesting in the light of advances in today's procedures. He showed an appreciation of the wider uses of mass selection by not limiting it to a single-cross population: "It is also possible to mix together the seed of several crosses and grow the lot all together. One need not confine the mixture to one cross only."

Furthermore, he demonstrated an understanding of the influence of different sites and environments on the differential survival of populations. Nevertheless, the description is of a simple generation advancing procedure unconnected with directed changes to the population. His contention that there is an inherent and unavoidable time penalty to using mass selection is not, in my opinion, valid (see Jensen, 1978). In his description, Love equates the terms *mass selection* and *bulk method*.

Earlier it was stated that Gardner's stratification technique had deeper implications to research in the small grains. At Ithaca, New York, except for a few favored sites, soil and environmental variability was unusually great. Many tales are told about our "soil". Fist-size rocks were classed as soil particles. Seba Sloughter, longtime farm foreman from the days of horse-drawn corn planters, told me one day, "When planting corn on Tailby Field I could stop the team and hear the kernels rattling for 15 minutes." Exaggeration aside, Gardner's gridding technique is a boon to researchers dealing with sites of high environmental variability. One example of its use at Cornell was an experiment I conducted years ago wherein the "best" wheat plants were to be chosen from a population of many thousands. The seeds were space-planted in wide rows across a field, back and forth in a serpentine fashion. Instead of picking the best plants from the entire variable area I gridded the area into 100-plant segments, following the serpentine turning at each end of the field, and chose an equal number of "best" plants from each segment.

Harrington's 1937 proposal for the mass pedigree method was predicated on the particular environment in which his breeding program rested, namely, one in which dry seasons were frequent. In such a season, F_2 selection under the pedigree system is ineffective and yet as costly as though the opportunities were favorable. For example, one dry period extended over seven years, permitting little chance of selecting for straw strength, earliness, height, and resistance to shattering and some diseases. Despite this special background, the premise of the mass pedigree is sound for a range of situations.

Frey (1967) and Romero and Frey (1966) have pointed out that under direct mass selection, heritability of the trait is paramount,, whereas under indirect mass selection, heritability is confounded with the genetic correlations between two traits. In both (all) cases population variances on all traits may be changed. It would be interesting to see the results of a series of mass selection pressures for different traits applied in the same time period to population subsets of the same base hybrid population, and then to compare and evaluate the ensuing subset populations for direct and indirect changes from the base or control population.

REFERENCES

Andrus, C. F. and G. W. Bohn. 1967. Cantaloup breeding shifts in population means and variability under mass selection. Proc. Am. Soc. Hort. Sci. 90:209–222.

Bal, B. S., C. A. Suneson, and R. T. Ramage. 1959. Genetic shift during 30 generations of natural selection in barley. Agron. J. 51:555–557.

Bhatt, G. M. and N. F. Derera. 1973. Associated changes in the attributes of wheat populations mass selected for seed size. Aust. J. Agric. Res. 24:179–186.

Bliss, F. A. and C. E. Gates. 1968. Directional selection in simulated populations of self-pollinated plants. Aust. J. Biol. Sci. 21:705–719.

Brinkman, M. A. and K. J. Frey. 1977. Yield-component analysis of oat isolines that produce different grain yields. Crop Sci. 17:165–168.

Chandhanmutta, P. and K. J. Frey. 1973. Indirect mass selection for grain yield in oat populations. Crop Sci. 13:470–473.

Derera, N. F. 1982. The harmful harvest rain. 1981 Farrer Memorial Oration. J. Aust. Inst. Agric. Sci. 48:67–75.

Derera, N. F. and G. M. Bhatt. 1972. Effectiveness of mechanical mass selection in wheat (*Triticum aestivum* L.). Aust. J. Agric. Res. 23:761–768.

Derera, N. F., G. M. Bhatt, and F. W. Ellison. 1974. Population improvement through mechanical mass selection. Cereal Res. Commun. 2:95–104.

Doggett, H. 1968. Mass selection systems for sorghum. Crop Sci. 8:391–392.

Doggett, H. and S. A. Eberhart. 1967. Recurrent selection in sorghum. Crop Sci. 8:119–120.

Fehr, W. R. and C. R. Weber. 1968. Mass selection by seed size and specific gravity on soybean populations. Crop Sci. 8:551–554.

Foster, K. W., S. K. Jain, and D. G. Smeltzer. 1980. Responses to 10 cycles of mass selection in an inbred population of grain sorghum. Crop Sci. 20:1–4.

Frey, K. J. 1967. Mass selection for seed width in oat populations. Euphytica 16:341–349.

Frey, K. J. and J. A. Browning. 1971. Association between genetic factors for crown rust resistance and yield in oats. Crop Sci. 11:757–760.

Gardner, C. O. 1961. An evaluation of effects of mass selection and seed irradiation with thermal neutrons on yield of corn. Crop Sci. 1:241–245.

Geadelmann, J. L. and K. J. Frey. 1975. Direct and indirect mass selection for grain yield in bulk oat populations. Crop Sci. 15:490–494.

Hallauer, A. R. and J. H. Sears. 1969. Mass selection for yield in two varieties of maize. Crop Sci. 9:47–50.

Harrington, J. B. 1937. The mass-pedigree method in the hybridization improvement of cereals. J. Am. Soc. Agron. 29:379–384.

Harrington, J. B. 1952. Cereal Breeding Procedures. FAO-UN. Rome, Italy.

Heinrichs, D. H. 1976. Selection for quick and slow emergence in alfalfa and relationship to yield potential. Can. J. Plant Sci. 56:281–285.

Hockett, E. A., R. F. Eslick, D. A. Reid, and G. A. Wiebe. 1968. Genetic male sterility in barley. II. Available spring and winter stocks. Crop Sci. 8:754–755.

Hull, F. H. 1952. Recurrent selection and overdominance. Heterosis. Iowa State College Press, Ames, IA, pp. 451–473.

Ivanoff, S. S. 1951. Mass screening in oat breeding. J. Hered. 42(5):224–230.

Jensen, N. F. 1978. Composite breeding methods and the DSM system in cereals. Crop Sci. 18:622–626.

Jones, D. F. 1918. The effects of inbreeding and crossbreeding upon development. Conn. Agric. Exp. Stn. Bull. 207.

Lonnquist, J. H. 1967. Mass selection for prolificacy in maize. Der Zuchter 37:185–188.

Lonnquist, J. H., O. Cota A, and C. O. Gardner. 1966. Effect of mass selection and thermal neutron irradiation on genetic variances in a variety of corn (*Zea mays* L.). Crop Sci. 6:330–332.

Love, H. H. 1927. 2. A program for selecting and testing small grains in successive generations following hybridization. J. Am. Soc. Agron. 19:705–712.

McFadden, E. S. 1930. A successful transfer of emmer characters to vulgare wheat. J. Am. Soc. Agron. 22:1020–1034.

Matzinger, D. F. and E. A. Wernsman. 1968. Four cycles of mass selection in a synthetic variety of an autogamous species *Nicotiana tabacum* L. Crop Sci. 8:239–243.

Mertz, E. T. 1976. Interaction between yield, protein, and lysine in cereals. *In* H. L.

Wilcke (ed.), Improving the nutrient quality of cereals. II. USAID, Washington, D. C., pp. 107–112.

Rasmusson, D. C., B. H. Beard, and F. K. Johnson. 1967. Effect of natural selection on performance of a barley population. Crop Sci. 7:543.

Redden, R. J. and N. F. Jensen. 1974. Mass selection and mating systems in cereals. Crop Sci. 14:345–350.

Reid, D. A. Personal communication, March, 1987.

Richey, F. D. 1922. The experimental basis for the present status of corn breeding. J. Am. Soc. Agron. 14:1–17.

Romero, C. A. and K. J. Frey. 1966. Mass selection for plant height in oat populations. Crop Sci. 6:283–287.

Ross, W. M., K. D. Kofoid, J. W. Maranville, and R. L. Voight. 1981. Selecting for grain protein and yield in sorghum random-mating populations. Crop Sci. 21:774–777.

Ross, W. M., J. W. Maranville, G. H. Hookstra, and K. D. Kofoid. 1985. Divergent mass selection for grain protein in sorghum. Crop Sci. 25:279–282.

Simons, M. D. 1965. Relationship between the response of oats to crown rust and kernel density. Phytopathology 55:579–582.

Smith, R. R. and C. R. Weber, 1968. Mass selection by specific gravity for protein and oil in soybean populations. Crop Sci. 8:373–377.

Sprague, G. F. 1966. The challenge to agriculture. African Soils. 11:23–29.

Tiyawalee, D. and K. J. Frey. 1971. Mass selection for crown rust resistance in an oat-M population. Iowa State J. Sci. 45(2):217–231.

Wann, E. V. and W. A. Hills. 1975. Tandem mass selection in a sweet corn composite for earworm resistance and agronomic characters. HortScience 10(2):168–170.

5

PEDIGREE AND F_2-DERIVED FAMILY METHODS

At first, several things prompted me to raise the pure line family method to the status of a recognized breeding and selection method. I finally realized, however, that it was so closely intertwined with the pedigree method that separating the two could not be accomplished without considerable duplication.

The pure line family method projects a feeling of natural affinity to the pedigree method, but it might also be said to fit between the pedigree and the bulk population method because it deals with the question of the role of the F_2.

The F_2 is an elemental, pivotal generation, having properties found in no other generation. The F_2 is the first generation breeders can work with in selection, when the population numbers are smallest, when selection or rejection has the greatest impact, and when the most information is present. It might seem expedient to gloss over certain confounding aspects of the F_2, yet they cannot be ignored. A partial listing of hazards, restrictions, obstacles and pitfalls might include:

1. *Competition:* many kinds are involved, including interplant, intraplot, and interplot, and density influences.
2. *GXE Interactions:* the environment is complex, embracing soils, moisture, weather, diseases and insects, wind, cold, drought, weeds, sites, years, and so forth.
3. *Time Requirements:* plant breeding problems may require a decade or more for a successful solution to be reached.

4. *Genetic Status:* many factors, such as heterosis, heterogeneity, heritability, gene numbers, germ plasm relations, dominance, epistasis, additive, and others must be considered.
5. The interplay of any or all of the above.

The degree of success of the F_2-derived line method of breeding hinges in large part on the rapidity of fixation of genes governing quantitatively controlled traits, of which yield is the best example. The method highlights the distinctions between the within-line variance (i.e., the variance contained and advanced in the progeny of each F_2 individual), and the between-line variance (which is represented by the variance among all the F_2-derived streams). The concern is not so great for simply inherited or visually tracked characters.

EXAMPLES OF THE PEDIGREE AND F_2-DERIVED FAMILY METHODS

Love (1927) presented a detailed description of the line or pedigree selection method. His views are authoritative because of his strict adherance to the pedigree method in all of the breeding programs he supervised at Cornell University. Love stated that the pedigree method required the progeny of each F_2 plant to be kept separate and subsequently handled as a line or family. He suggested two ways of handling these F_2-derived F_3 progenies: (1) space planting in rows, or (2) solid sowing in plant rows. He favored the first alternative and, in fact, used it exclusively.

The two options are in fact quite different: the first leads via rapid truncation to line homozygosity, where the original within-line variance of an F_2 plant is represented by the selections from its progeny; the second becomes an F_2-derived family line containing all its within-line variance. In the first option, the breeder deals directly with advancing and testing the retained lines; in the second option, he deduces the potential line value from the family appearance and mean performance, then chooses among family groups, later making selections of plant lines from these retained family groups. This is, in effect, a partitioned bulk system with a moving discard option.

Under Love's procedures, the F_3 space-planted rows were subjected to a retain/discard decision on a row basis. In the retained rows, desired plants were selected. These plants were threshed and individually sown in F_4 plant rows. Retained lines from these plant rows were advanced the next season into nurseries of larger plots, which included variety checks and replication. Further retention became increasingly dependent upon a line's continued superior performance.

As Akerman, Granhall, Nilsson-Leissner, Muntzing, and Tedin (1938) pointed out, pedigree selection predated the rediscovery of Mendel's paper and thus its origin was more closely associated with selection out of mixed heterogeneous populations such as landraces than with hybrid segregating populations.

Mahmud and Kramer (1951) compared 64 soybean families, each of which included the F_3 line, the bulked progeny of seven plants of the F_3 line (thus F_4), and three individual F_3 plant progenies (F_4) in a replicated yield trial in the same year. The test conditions minimized or reduced genetic shift, year, and multiple site effects. The yield of the bulked F_4 lines showed a high correlation with the F_3 family line progenitor for height and maturity, and for yield when the comparison was made in one season (the predictive relationship was negligible when the comparisons were made in different seasons). The conclusions that F_3 lines could be used to predict the potentialities of F_4 segregates thus carried the requirement that genetic shift and $G \times E$ interactions be controlled. Because the F_3 and F_4 are usually sequential year operations in practice, this would entail a modification of practical field procedures.

Frey (1954) studied the use of F_2-derived families and lines in barley in a search for a modified procedure between the pedigree and bulk hybrid method. The early uniformity of plant type among selections from family progenies had suggested to him that selection within strains in generations subsequent to the F_2 would provide diminishing returns. His modification involved selecting among F_2-derived families through the F_5 generation, handling each family as a line entry in nursery trials. There would be no selection within families through the F_5, thus increasing efficiency over the strict pedigree system by eliminating that labor. The selections remaining after routine nursery and evaluation testing would be subjected to within-family selection—a return to pedigree procedures—beginning in the F_6–F_{10}. The procedures are basically a delayed pedigree approach, the delay resulting from the insertion of the bulk family treatment for three generations. The time spent is not lost in any sense because the bulk handling, the steady discard, and the ability to concentrate testing on reduced numbers of near-homozygous material are all marks of efficiency.

Lupton and Whitehouse (1957) credited the pedigree and mass breeding and selection methods with repeated successes in cereal improvement, but felt that both would be improved if quantitative measurements could be made earlier. They pointed out that the pedigree system is greatly influenced by the competence and experience of the breeder, and the mass method is vulnerable to the loss of valuable segregates because of random sampling in the early generation plots. It is not clear whether this refers to selection of planting seeds or to selection of lines, that is, whether the losses were from seeds not planted or from lines overlooked. Regardless of interpretation, they believed that the selection process would be improved by the addition of precise quantitative measurements.

Lupton and Whitehouse described a series of breeding probes undertaken, prior to and leading to two proposals that will be presented shortly. First, they bypassed the laborious assessment of parents to see whether parent potential might be ascertained indirectly through the performance of unselected bulks of their crosses. The results, from F_1 to F_6 and parents tested of one cross, showed marked heterosis in the F_1, a slight amount in the F_2, and yield essentially stabilizing thereafter at a level above the parents midpoint yield.

The next step was to compare unselected bulks of 10 crosses in F_2, F_3, and F_4. There was high cross × year interaction, but the important finding was that although seven of the crosses outyielded the high-quality parent, only two did so with statistical significance.

Neither of the methods tried provided the critical information needed to assess either parents or crosses. This prompted the authors to design the two different methods of selection described below, while acknowledging the existence of similar schemes tracing, in some cases, to the early 1900s.

Pedigree Progeny Method of Selection

This method, applied to each cross, is shown in Lupton and Whitehouse's Figure 1; the steps may be described as follows:

F_2: An approximately 10% visual selection on qualitative appearances is made from a base population of 2000–3000 plants of each cross.

F_3: Equal seed numbers from each F_2-derived plant are sown. Better progenies are visually selected on the same basis as in the F_2, and selected rows are bulk harvested. The bulk lines are examined for grain type, and the numbers are adjusted to fit F_4 yield nursery capacity.

F_4–F_6: Selection of best bulk lines is made from yield trials of increasing precision.

F_7: Single plant selection in the retained bulk lines establishes new lines.

F_8 and thereafter: Yield and quality tests are conducted for the near-homozygous lines.

Lupton and Whitehouse used the mean yield of the final selected group of lines to assess the value of crosses and their parents retroactively. They made the statement that the mean performance of the elite lines from a cross carries more information to the breeder than might be deduced from earlier comparisons. Note that their procedures represent a complete breeding system.

Pedigree Trial Method of Selection

Lupton and Whitehouse's second and alternative proposal involved parallel yield and selection procedures, as follows:

F_2: Same as previous method.

F_3: The same visual selection of best rows. Within progeny rows the best plants are selected using pedigree and visual procedures.

F_4: Lines are grown in plant rows. One or more rows are selected for pedigree procedures; the remainder are bulked to form a check for F_5 yield tests.

F_5–F_7: Yield trials. No subsequent selection necessary.

The authors believed these procedures would save two years of program time. The early isolation of lines attributable to the pedigree features has the effect of reducing the competition associated with the bulk hybrid method.

The F_2-derived family and line method could be a powerful screen for a highly efficient bulk hybrid population system by providing a necessary early discard of F_2 (plants) and F_3 (families) so that the final bulking represents the breeder's choice of superior material.

Voight and Weber (1960) reported on F_5 results of a study comparing pedigree (P), bulk (B), and a form of F_2-derived family testing (F) in soybeans. The procedures were applied separately to five different crosses involving five cultivars. Maturity was arbitrarily maintained to correspond to that of 'Hawkeye'. It was the intent of the authors to apply the three methods to the same biological material, so that each method had equal access for evaluation, that is, the F_2 (F_3-embryo) seeds were divided into aliquot samples to serve each method. The organization of the three methods was as follows:

Prearrangements: Seventy-five F_2 plants were selected from each cross in 1955 and their F_3-embryo seed amounts divided into portions.

Method B: A composite was formed from two seeds from each of 75 F_2 plants for each cross, and was grown during F_3 and F_4. However, the F_4 rows were thinned twice in order to increase seed set and remove non-Hawkeye maturity plants. Visual selection in F_4 singled out the 20 best plants per cross; thus there were 20 from a bulk of 75.

The second, remaining, portion of seed from each of the 75 F_2 plants per cross was planted in a 12-ft F_3 plant row and was used to generate the P and F parts of the study.

Method P: Three good plants were visually selected from the first half of each F_3 row. Each of these F_4-embryo selections was planted in a split row, drilled and spaced corresponding to anticipated line and plant selection, respectively. The drilled portion was used to evaluate line appearance and the spaced portion facilitated easy plant selection from the chosen lines. Twenty F_4 lines per cross were chosen from the 225 rows, each one tracing to a different F_2 plant.

Method F: Three good plants were visually selected from the second half of the F_3 plant row and termed a family, F_2-derived. The three family lines separately were entered in an F_4 yield performance trial and considered as three replicates of the family line. The results of this test enabled the 20 highest-yielding families per cross to be identified (by mean of three replicates), and then the single highest-yielding replicate line was chosen to represent that family in the forthcoming F_5 test.

The F_5 test included the five crosses, each contributing 20 F_4 entries for each of the three methods plus four Hawkeye controls, to a randomized complete block design. Yield, lodging, height, and maturity were measured. Based on combined results of the five crosses (100 F_5 lines from each method) the yields were B 44.5, P 44.2, and F 45.1 bu/acre. Yields of F were significantly higher (5% probability level) than those of P and of the mean of P and B. Apparently F and B were not significantly different in yield. Differences in height, maturity, and lodging were small and unimportant save for higher lodging in B. The authors concluded that F was generally superior to the B and P methods, but acknowledged that efficiency was not a consideration of the study.

As in all research, it is difficult to escape some of the restraints imposed so as to achieve clear, unconfounded results. Selection, usually visual and of the kind associated with pedigree procedures, affecting maturity, agronomically desirable features, and presumably lodging, appeared in all of the methods, acting, so to speak, to reduce method variance. If F_2 seed amounts had been ample, a larger population base for B than that from two seeds from each F_2 plant might have been desirable. The yield differences among methods were small, and the small observed advantage to F would have to be weighed against the extra cost and labor of its F_4 yield and selection nursery. B would have shown labor and cost advantages over both P and F. The article is noteworthy for being one of the few research studies dealing with early (F_4) extraction of lines from a bulk population.

In the first of a series of papers celebrating Swedish plant breeding on its 75th anniversary, Mac Key (1962) discussed the Swedish experiences with the pedigree and bulk systems. The pedigree system, which in its pure form represents continuous selection, soon evolved into a discontinuous selection procedure wherein selection was restricted to the F_2, the lines were tested for some years, and further selection then made within the best lines. This modified pedigree system came into being as a way of reducing the costly and laborious pedigree system. Mac Key's views on this change are well-expressed in the following paragraph:

> As compared to the more elaborate method, this simplified new approach attached fundamental importance to the primary, early selection. The continued differentiation of near genotypes within the sometimes highly segregating lines was considered to be much more limited than the differences between the original lines. These may thus be characterized as populations with such a comparatively small range of variation that all the components of each F_2 or F_3 line may be judged by their average yield without the risk of losing anything of value.

If one looks carefully, one may catch glimpses of the ancestors of the single seed descent and the F_2-derived family methods in this description of an early modification of the pedigree method.

Heyne and Finney (1965) reported that F_2 progeny tests indicated that mixogram quality analyses were reliable as early as the F_3, and that there was a high correlation between mixing time and loaf volume.

The authors considered the F_2 progeny test to have advantages over the ordinary bulk method, which typically produced an inordinately high proportion of unwanted tall and late plants. Selection for agronomic type or disease resistance in the F_2 and F_3 can quickly reduce the content of undesirable lines. The use of "bulk" in these procedures refers only to the continuation of the bulk progeny of each F_2 plant saved. The authors suggested that a progeny line could become a cultivar without further selection.

Thorne, Weber, and Fehr (1970) examined six characters in soybeans with reference to their behaviour as they approached homozygosis, with special reference to the F_2-derived line method of breeding. The six characters, all in selections of midseason maturity, were yield, lodging, height, seed size, protein, and oil. The authors found that meaningful fixation for all characters occurred in the F_2, especially with height and lodging, less so with seed size and protein, followed by yield and oil content. Further fixation was observed in the F_3.

This method of breeding and selection is attractive because, logically, genetic variance within an F_2-derived line would be less than among F_2-derived lines. The greater similarity of lines eventually drawn from an F_2-derived family line has two implications: (1) The testing of the bulk progeny of the family is representative, thus predictive, of the lines which may be selected from it. (2) Any lines drawn from the family are closely enough related so that there is less danger of overlooking other superior lines from that F_2-derived family at the time of selection. The greater genetic variance among family progenies facilitates the discard of F_2-derived families, beginning with visual discrimination among the F_2 plants.

Luedders, Duclos, and Matson (1973) added maturity bulks of soybeans to a study expanding on the research of Voight and Weber in 1970, which evaluated efficiency of breeding methods. Luedders et al. found no statistically significant differences among four methods in ability to select high-yielding lines. Their methods were complete bulk, maturity bulks, pedigree, and early generation testing.

Ivers and Fehr (1978) credited H. H. Stine (via a personal communication in 1968) with proposing the *pure line family method* (PLF). Stine, working with soybeans, reasoned that because superior lines were used to identify superior crosses, they might also identify superior families in the cross. In practice, individual plants would be harvested from an F_2 population, presumably spaced. It would be reasonable to assume that selection, visual or otherwise, would be imposed on the collection of plants; however, in fairness to differing views, it is also quite possible that a random unbiased choice would be preferred. The F_2 bulked line was selfed until working homozygosis was attained (F_3–F_5). At this point a single plant (probably a head in view of the bulk growing conditions) was harvested and subsequently tested. If it met the standards of superior yield, retroactive access was made to its F_2 family (seed stored in reserve) and additional plantings for further selection were made. I like this approach but would suggest two modifications: (1) some selection of the original F_2 plants, and (2) taking more than one plant or head on the reasoning that broader representation of the F_2 family is required (otherwise the method has the same limited within-line flaw of the single seed descent method). This method contains the germ of an early screening device, to choose not only among crosses but among lines within crosses—with the added merit that even though reserves are held back, there is still useful production of lines out of the screening procedures.

Ivers and Fehr compared the PLF method with (1) early generation testing of F_2 plant-derived bulk progenies (BF), (2) pedigree selection (PS), and (3) single seed descent (SSD). An interesting modification was applied to SSD procedures: instead of advancing generations by one seed per plant from the F_2—which is required in the pure form of SSD to handle large numbers without changing population size—five seeds were selected from one randomly chosen plant in each generation. This would have the effect of producing highly homozygous individuals at F_5, but the important change is the expansion from one to five plants in the F_3, giving broader representation of the variance in the F_2-derived progenies. If planned selection rather than random selection were permitted later in the selection of the one among five choices, and this procedure were continued in each generation, the chances for genetic improvement out of that population would be enhanced. These points are not germane, however, to the research use of the SSD procedures in this study, where the emphasis was on the comparative worth of the pure line family method.

The studies allowed many insights and several deductive conclusions:

1. BF was superior to PLF in identifying the top (20) families, and showed a preponderance of F_5 lines outyielding PLF lines.
2. Even so, the two methods were similar in numbers of superior F_5 lines selected.
3. Neither method showed a decided advantage, although PLF with proper arrangement could produce cultivars one year sooner.
4. PLF produced more superior lines than SSD, but this finding was tempered by greater cost demands.
5. PLF would be superior to PS for yield improvement, particularly in situations where pedigree selection was on a one crop per year basis (in PLF F_2 families could be selfed overwinter).

6. Pedigree selection was particularly effective in eliminating low-yielding lines, primarily because of visual discrimination for yield.
7. Visual selection applied to PLF, BF, and SSD would improve efficiency.

Ivers and Fehr raised an important question that relates to all breeding and selection methods. They stated that the ability of the PLF method to identify superior F_2 families indicated a high level of genetic similarity for yield among progeny from an F_2 plant (this is particularly evident in the SSD method, which in its primary form does a good job of propagating high yield from an entirely random beginning choice of one seed per F_2 plant). Particularly, they cited the 1970 study of Thorne, Weber and Fehr, just reviewed, which showed important character fixation in the F_2, admittedly less dramatic for yield than for four other characters.

Valentine (1984) proposed a series of selection and handling procedures to tighten and speed up the early generation period. He called this *accelerated pedigree selection*, or *the APS system*. The method is based on his prior beliefs that (1) selection for yield on a single F_2 plant basis was inefficient, and (2) an emphasis on F_3 line selection was needed. The essence of the scheme lay in a rapid advance to the point of testing F_3 lines. Valentine compared procedures throughout to the chronology of other methods, particularly the pedigree and SSD systems. The plan worked most expeditiously with spring types but was readily adaptable to winter (cereals). The saving of time was one to two years for most system comparisons, except that the length of time required was the same as that of the doubled haploid method.

The time sequence for a spring cereal was seen in the following pattern:

Year 1: Crossing to produce F_0 seeds was done late in the winter in the greenhouse, or June in the field. The F_1 was grown the same season in field or greenhouse to produce F_2-embryo seeds. The F_2 population was grown in the greenhouse to produce F_3-embryo seeds over winter.

Year 2: F_3 lines were grown in rows in the field, yields taken, and selections made.

Year 3 and Beyond: Conventional yield and evaluation trials were conducted.

Valentine cited three advantages of APS over SSD: (1) it avoids the high mortality problem of SSD; (2) it has a lower resource requirement than SSD, that is, principally heated and lighted greenhouse space; and (3) APS begins early evaluation of lines whereas SSD does not.

If neither system used selection, the APS would be superior to the SSD because it maintains higher levels of both within- and between-line variability. The F_3 is an F_2-derived line emanating from a generous sample of the F_2, rather than the SSD's single seed. Dewey, Qualset, and Jensen, and perhaps others, have skirted this area by using single heads rather than single seeds. Valentine referred to "clumps" from as few as five seeds. These modifications would seem to fit nicely with Frey's hill plot procedures.

I would wish to add certain selection and discard provisions to the F_3 (and Valentine does not preclude these): truncation could proceed from visual selection for qualitative controlled traits and from the measurement traits of yield, components, and so forth. Valentine is to be commended for packaging efficient procedures into a system, and for his balance in recognizing that the APS should be thought of as a supplementary program reserved for elite crosses.

DISCUSSION OF PEDIGREE SELECTION

Because my breeding and selection techniques over the years evolved in directions away from pedigree selection, although never abandoning it, it seems worthwhile to assess the debits and credits of pedigree selection. It is easy in one breath, or one sentence, to describe one procedure as having a labor-cost or time advantage over another—mass selection was said to have such an advantage over pedigree selection—without really saying anything substantial about the meaning or the magnitude of the difference. In fact, the handling costs of pedigree selection over mass selection represent what today would be major economic costs, but in the 1940s with lower labor costs, they could be absorbed. Following is a detailed description of the wheat pedigree selection program as I observed it at Cornell:

The F_2-embryo seeds from each F_1 plant were hand-planted in rows previously furrowed and marked off with twine in 5-ft field sections. The seeds were spaced 2 in. apart by a worker bent over and backing down the section, row by row, until all the seeds from each F_1 plant were planted.

As many as 12–15 persons would be engaged in these operations, some laying out seed envelopes, keeping records, stringing lines, and marking out sections, but always trying to keep as many people as possible planting. As planting progressed the seeded rows—hundreds and thousands—would be covered by hand with a hoe.

At maturity, laborers with tined forks would loosen all the plants in each row. Next came the pullers, who pulled and piled all the plants in one F_2 family. Then the "first sorter" would pile all plants meeting a basic test, for example, having three or more culms. Next, an experienced breeder would make a final selection of plants and bundle them together. Incidentally, the photograph captioned "Single plant selection of oats in the field" in *Swedish Contributions to the Development of Plant Breeding* (see Akerman et al., 1938) could have been taken in Ithaca, New York—I have strikingly similar photographs taken about the same time.

The F_2 plants would be individually threshed and the F_3-embryo seeds placed in envelopes. At next planting, each envelope of seeds would be space-planted 2 in. apart in a rod row and covered by hoe. Selection at maturity would be made among F_3 rows. Each row saved again would be fork-loosened, then the plants pulled, selected, bundled, and later threshed. These F_4-embryo seeds would be hand sown, not spaced, in 5-ft plant rows. All F_4 rows selected at maturity would enter rod row trials at the next planting season.

Today, of course, some of these operations would be mechanized, but budgets that employed 12 laborers might now permit only 2 or 3. It is quite understandable why mass selection and bulk hybrid methods with their mechanized large plot procedures have an appeal to cereal breeders, especially to those who have had experience with large pedigree selection programs. It is important to note, however, that almost all breeders use pedigree procedures at some point in their operations.

REFERENCES

Akerman, A., I. Granhall, G. Nilsson-Leissner, A. Muntzing, and O. Tedin. 1938. Swedish contributions to the development of plant breeding. New Sweden Tercentenary Public., Stockholm. 111 pp.

Frey, K. J. 1954. The use of F_2 lines in predicting the performance of F3 selections in two barley crosses. Agron. J. 46:541–544.

Heyne, E. G. and K. F. Finney. 1965. F_2 progeny test for studying agronomic and quality characteristics in hard red winter wheats. Crop Sci. 5:129–132.

Ivers, D. R. and W. R. Fehr. 1978. Evaluation of the pure-line family method for cultivar development. Crop Sci. 18:541–544.

Love, H. H. 1927. 2. A program for selecting and testing small grains in successive generations following hybridization. J. Am. Soc. Agron. 19:705–712.

Luedders, V. D., L. A. Duclos, and A. L. Matson. 1973. Bulk, pedigree, and early generation testing breeding methods compared in soybeans. Crop Sci. 13:363–364.

Lupton, F. G. G. and R. N. H. Whitehouse. 1957. Studies on the breeding of self-pollinating cereals. I. Selection methods in breeding for yield. Euphytica 6:169–185.

Mac Key, J. 1962. The 75 years development of Swedish plant breeding. Hodowla Roslin, Aklimatyzacja I Nasiennictwo, Tom 6:437–467.

Mahmud, I. and H. H. Kramer. 1951. Segregation for yield, height, and maturity following a soybean cross. Agron. J. 43:605–609.

Thorne, J. C., C. R. Weber, and W. R. Fehr. 1970. Character fixation and its relationship to the F2-derived line method of soybean breeding. Iowa State J. Sci. 44:271–278.

Valentine, J. 1984. Accelerated pedigree selection: an alternative to individual plant selection in the normal pedigree breeding method in the self-pollinated cereals. Euphytica 33:943–951.

Voight, R. L. and C. R. Weber. 1960. Effectiveness of selection methods for yield in soybean crosses. Agron. J. 52:527–530.

6

BACKCROSS METHOD

Although backcrossing long had been known as a breeding procedure and used in some plant research, it apparently had not been used in a directed way, that is, directed at producing plant types, before the article by Harlan and Pope (1922). The authors sensed the potentials for progress in plant improvement in the backcross method, and chose the problem of creating a smooth-awned 'Manchuria' type barley as a test case. They were led in this direction by a consideration of the probabilities, considered rare, of finding a smooth-awned segregate, like Manchuria in all other respects, in the segregating generations of a simple cross. Harlan and Pope were concerned also about the effect of linkage in a backcross program, and concluded that it posed no greater a problem, and perhaps less because of the repeated crossing and opportunities for crossovers, than in any other breeding method.

EARLY USE OF THE BACKCROSS

Richey (1927) used the term *back pollinating* for backcrossing. Florell (1929) recommended the use of the backcross for one or two generations to precondition wide and complex crosses for bulked population procedures.

Mac Key's (1963) paper on procedures used by the Swedish Seed Association, discussed in Chapter 1, placed particular emphasis on the backcross method, not its use in the pure form as might be expected, but rather as a steady adjunct and participant in the mixture of procedures used over the years. For example, it was frequently used after wide crosses in order to channel the working population in the direction of the desired genotype. Often the backcross would be a compromise—an outcross to a genotype perhaps similar to the original desired parent. Another use of the backcross was the discontinuous backcross, wherein foreign germ plasm would be introduced into an ongoing series of planned adapted crosses by intercrossing the unadapted exotic genotype to selected lines, continuing the intercrossing among the progeny, and at some point backcrossing selected lines to an earlier adapted ancestor.

The versatility of the backcross method when combined with other methods is best illustrated by its use in convergent crossing, reminiscent of Harlan, Martini, and Stevens (1940) convergent cross procedures. These early "bridging" crosses rapidly introduced new parent varieties but in consequence suffered unknown germ plasm attrition from the oldest crosses in the series; that is, one only could hazard a guess as to the genetic contribution remaining from the parents in the first crosses made. Four of the many possible variations of the convergent cross, using the backcross, were diagrammed by Mac Key.

BRIGGS IN FOREFRONT OF BACKCROSS DEVELOPMENT

In an early report on the practical use of the backcross method in California, Briggs (1930) described its use in the ongoing development of a bunt-resistant 'Baart' wheat. The resistance donor parent was 'Martin', the cross was made to Baart, and the F_1 backcrossed to the recurrent parent, Baart. Selection of resistant plants was made during the following two generations (BC_1 and BC_2), and in the BC_3 several Baart-like resistant plants were crossed with Baart. Through selection and observation of Baart-like plants, any that deviated in appearance could be eliminated. This completed one cycle, and Briggs expected the last cross perhaps to produce a bunt-resistant Baart (if it did not, the cycle could be repeated).

Briggs expected that the resistant Baart-like strains might require less testing before release, and this has subsequently been shown to be the case. In the breeding program, plans were to put, early on, all new candivars (prospective releases) into a backcross program so that developmental lag-time in variety release would be minimized. He also emphasized that, after the first backcross completion, it would not be necessary to use germ plasm as exotic as the original source in order to capture the desired gene. This gene could be extracted from the acceptable background of the backcross genotype.

Briggs considered the problem of incorporating several desirable characters, and suggested separate backcross programs, followed by intermating and selection for desired combinations.

Briggs' (1935) basic paper on the backcross method was devoted to explaining its fundamental principles. For example, he pointed out that self-fertilization and backcrossing achieve parallel equivalent levels of homozygosity, but entirely different products are produced. In the first instance, every progeny individual is different, whereas under backcrossing all the individuals are of the same background genotype. If selection is practiced to retain a donor's traits, the backcross individuals differ in that respect from the recurrent parent. Briggs pointed out the advantages to stable crop production through a strategy of retaining basic successful commercial types with successively added improvements.

The reliability of backcross breeding to duplicate the recurrent parent's performance, with the added advantage of the new desirable genes, was shown by Suneson, Riddle, and Briggs (1941). Extensive field trials demonstrated that "White Federation 38" and "Baart 38" were similar to their recurrent parents when bunt and stem rust were absent, but were markedly superior in yield in epiphytotic situations. These new varieties demonstrated the correctness of the theory when they were released for commercial use without the benefit of the routine performance tests.

An extensive testing by Suneson (1947) of nine backcross-derived cultivars with their ancestor recurrent parent provided convincing proof of the similarity of the two. Exceptions were few and minor. The general procedure typically involved from four to six backcrosses, followed by testing of F_3 lines for homozygosity of the desired trait, after which the (resistant) lines were bulked to form the release stock.

The merging of separate backcross programs to produce populations having multiple gene transfers for desirable characters is especially interesting. A good example was the creation of 'Big Club 43' (C.I. 12244), a wheat variety carrying three resistant characters (for bunt, Hessian fly, and stem rust), each added separately from these sources, 'Martin', 'Dawson', and 'Hope', respectively, in concurrent backcross programs. In Suneson's words:

> It resulted from three independent breeding programs:
> (1) production of 'Big Club 37' as noted above [*ed.*: Martin × 'Big Club'^{7}].
> (2) Dawson-Big Club^4 × Big Club 37^2.
> (3) Hope-'Baart'^{4} × Big Club^2 × Big club 37^2.

The final merger involved random crossing of (2) on (3), and then of (2) on the F_1 of these.

Big Club 43 was assembled from 144 triply resistant lines, presumably F_3 lines, after multiple tests.

An excellent review of the place of the backcross in plant breeding was given by Briggs and Allard (1953). They listed the following points for consideration by breeders who wished to try the method:

1. The availability of a suitable recurrent parent.
2. Number of backcrosses: six if selection toward the recurrent parent is practiced to the fourth backcross, two more if no selection.
3. To recover the variability characteristics of recurrent parent: use many recurrent parent plants during the last backcrosses.
4. To incorporate additional genes: use separate backcross programs and merge the end products through hybridization.
5. For automatic continuous improvement: use the derived BC type as the new recurrent parent.
6. Evaluation trials before release: less stringent requirements because of knowledge of the recurrent parent's performance characteristics.

Additionally, the authors pointed out that any amount of heritability of a character makes it a candidate for backcross incorporation. The method is not limited to autogamous crops, but also works well with cross-pollinated crops where it becomes a case of general recurrent selection. It is important to have sufficient hybridizations in order to reproduce approximately the gene frequencies of the recurrent population. Readers with a special interest in this method should read Briggs and Allard's article in full.

ONGOING RESEARCH SOLIDIFIES BACKCROSS STATUS

Emsweller and Jones (1934) documented practical backcrossing procedures in programs to transfer simply inherited resistance (dominant) to rust in snapdragons. It was estimated that three to five backcrosses would be sufficient to recapture thoroughly the desirable commercial type.

Leininger and Frey (1962) showed that plant breeders can expect regression toward the recurrent parent to proceed in a satisfactory manner; nevertheless, there may be significant variance from theoretical expectations and different characters may regress at different speeds. In practice the deliberate choice of the individuals chosen for the next round of crosses to the recurrent parent exerts a bias. There seems little question that breeder choice can affect both the speed and direction of regression.

Wehrhahn and Allard's (1965) paper is important for presenting a technique that makes it possible to partition and identify portions of genes that collectively control a quantitative character. The procedure is called an inbred backcross line technique, able to detect the influence of loci providing only a small contributory amount to a trait, sometimes as low as a few percent. The article expanded the general idea of polygenes having only small similar effects to the possibility of their having both large and small effects.

The technique involved obtaining an F_1 from a cross of two pure line parents, followed by successive backcrosses to one parent of a large number of plants, which were then selfed until high homozygosity was reached. The presumption was that the partitioning of genes for the (quantitative) trait, for example, yield, may have taken place, and these genes could be identified in their isolation by suitable yield tests.

Meredith (1977) detailed two genetic relationships in cotton that seem to have their exact counterparts in the small grain cereals. These were (1) negative correlations between two desirable attributes, lint yield and fiber strength, and (2) quantitative genetic situations involving these complex traits and their components (fiber strength is considered to be a quantitative trait). In this study, he assessed the use of the backcross method as a procedure to advance gains in both lint yield and strength. Three cycles of backcrossing with selection for fiber strength produced several selections for use in a test against the recurrent parent and the parent source for greater fiber strength; no selection for yield was made. The donor parent for fiber strength had 19% greater strength but 32% less yield than 'Deltapine 16', the recurrent parent.

Selecting for fiber strength and ignoring lint yield, or rather, assigning the responsibility for advance in lint yield to the recurrent backcrossing technique, resulted in average fiber strength figures of 193, 230, 220–211–216, for Deltapine 16, the donor parent, and the three backcross-derived progenies, respectively. This is considered a satisfactory response. The lint yield was lower than expected, but the best family was only about 5% below Deltapine 16 and had above-average fiber strength. All in all, Meredith felt that backcrossing in this larger sense could be used in cotton breeding.

Bassett and Woods (1978) discussed procedures to select the best parents in an ongoing backcross program.

RECENT ADVANCES IN BACKCROSS USE

The backcross method normally would not be considered a promising approach to obtaining transgressive segregation in a cross; nevertheless, Smith (1966) has shown a successful use of the backcross to this end in durum and hard red spring wheats. Smith illustrated the procedures with five durum and two bread wheats released by North Dakota State University and the U.S. Department of Agriculture. In each of these varieties the breeding goal in its simplest form was the addition of at least two major desirable characters of the parents; use of the backcross would preserve the accumulated desirable characters present in the recurrent parent.

Smith's important contribution to methodology was a unique use of the backcross. After a first cross to a donor parent to acquire a particular gene, the segregating F_2 and F_3 populations were searched for individual plants having this desired character, plus any other "bonus" characters, which in the varieties described could be stronger straw, yield, early maturity, or shorter straw. The first backcrosses customarily were made in the F_3 or F_4, following which the backcross populations were searched and the desired combinations again identified for the next backcross. The method is effective in bringing into a cultivar additional transgressive characters and adds a new dimension to use of the backcross.

Flor (1971) used the backcross method as he moved to incorporate multiple resgenes (two or three) for flax stem rust resistance, as opposed to the previous use of a single resgene. It was expected that multiple resgenes would stop or delay rust epiphytotics.

USE OF BACKCROSS IN INTROGRESSION OF EXOTIC GERM PLASM

The value of the backcross method to transfer genes for simply inherited characters is universally known. Less widely known is its ability to do more, to carry over or introgress gene complexes for complexly inherited characters, although from its inception as a breeding method suggestions have been made for separate concurrent programs with bridging intermating and recombination (e.g., Briggs, 1930). Lawrence and Frey (1975) reported success in transferring significant *A. sterilis* germ plasm via backcross into recombinant recurrent *A. sativa* stocks. Two highly regarded *A. sativa* genotypes (one the 'Clintford' cultivar), the recurrent parents, were each crossed to four genotypes from the *A. sterilis* collection, the donor parents. The eight F_1s were backcrossed to the recurrent parents, followed by successive backcrosses of the BCF_1s without pause or selection through the BC_5. For each backcross about 10 backcross plants were used as females with pollen furnished by the recurrent parents. Their recommendation for future use was 10–15 plants of each parent for each backcross.

The evaluation was made on the produce of 50 populations (eight crosses six generations plus the two recurrent parents). Each of the populations represented about 80 lines, giving 3900 F_2-derived lines for test. Hill tests were used as the testing vehicle in a three-replication randomized complete block (RCB) design from which yields were recorded.

The results showed that the species crosses created substantial new genetic variation for yield of grain. A surprising conclusion comes from viewing the frequency histograms of grain yield in the different crosses which had the same recurrent parent, and in the different BCF_2 generations: there seemed to be little discernible loss of variance with generation until the BC_4 and BC_5, and little consistency in changes from generation to generation; that is, BC_1F_2 was not much different from BC_4F_2 in the Clintford recurrent parent series.

Lines that exceeded parental mean ranges were classed as transgressive. There was a noticeable difference in these numbers, traceable to the two recurrent parents. For example, crosses with the recurrent parent Clintford produced, in the five backcross generations, a much higher percentage of transgressive lines than was produced by the crosses to the other recurrent parent. All crosses were to the same donor parents. The magnitude of the potency difference can be seen from a summing of the percentage of transgressive lines in each generation of each cross for the two groups (drawn from Table 4): 378 and 107, respectively (Clinford group first).

The breeding of superior adapted spring wheats is a complex matter involving not only agronomic considerations but also detailed attention to breadmaking quality traits. There is an alternative to seeking always the comfortable company of potential adapted parents for crossing, and that is to reach out to foreign or exotic germ plasm and incorporate it into the grand crossing scheme. Plant breeders recognize the necessity of such moves but do not know thoroughly the advantages and the penalties. In this sense the research reported by Eaton, Busch, and Youngs (1986) is as much a study of breeding methodology as it is of spring wheat improvement.

The authors wanted to learn the relative efficiencies and values of three breeding methods applied to the task of introgressing unadapted germ plasm into their crossing and selection program. The methods were (1) single cross, (2) three-way cross, and (3) backcross either (a) one backcross to recurrent parent or (b) two backcrosses. Using three adapted domestic and two unadapted foreign germ plasms, three populations were formed: populations 1 and 2 explored, via the five different breeding methods (there were also two formulations of the three-way cross), the introgression of one of the exotics into the high-yielding 'MN7125', and the high-quality 'Len' genotypes, respectively—'MN7186' provided the swing genotype for the additional three-way cross. Population 3 involved the other exotic, notable for high protein content, in crosses to MN7186, with Len and MN7125 involved in the three-way crosses.

The mean results overall for yield, days to head, milling score, and water absorption did not differ significantly over the five breeding methods. This was believed due to type of cross × environment interaction. However, when other quality traits were examined, the observations pertaining to methodology were such that the following ranking could be argued:

1. *Three-way and First Backcrosses:* these were similar and were the preferred methods, that is, the best producers of desirable lines.
2. *Second Backcross:* these showed much reduced genetic variance and no advance in high yield line production, and they required more time (an extra generation).
3. *Single Cross:* these showed the lowest mean yield and fewest lines for high yield and breadmaking quality.

In summary, the conclusions relative to methodology were in favor of three-way crosses (which included a first backcross) over single crosses and continued backcrossing. Readers will recognize that the limited use of the backcross is in essence a separate category of backcross strategy that is designed to create a population for selection that fosters wider genetic variance and modest introgression as opposed to the other category of creating a backcross-derived cultivar.

The difficulties inherent in introgressing genes from wild populations into cultivated species were documented in a study with soybeans by Carpenter and Fehr (1986). They crossed two *G. max* cultivars with two wild *G. soja* accessions, then backcrossed to the respective cultivated parent to produce, without selection, backcross populations BC_0 to BC_5. In each population they scored the F_2 and F_3 for a series of traits, using the recurrent parent as the standard. This average score characterized the recovery of recurrent parent germ plasm by showing the percentage of segregates that returned to the agronomic rating level of the cultivated parents. The results by generation were BC_0 and $BC_1 = 0$, $BC_2 = 2\%$, $BC_3 = 22\%$, $BC_4 = 51\%$, and $BC_5 = 65\%$. In their general conclusions, Carpenter and Fehr recommended two to three backcrosses to provide a high enough working frequency in the population to be of real use in the breeding program. Of course, introgression of specifically desired traits/genes could be accomplished in shorter time by using selection with the backcross procedures.

STABILITY AND THE BACKCROSS METHOD

An important attribute of the backcross is its conservatism in preserving plant breeding gains. On various occasions I have commented on the risk plant breeders take in improvement through hybridization: the making of a hybrid destroys and scrambles the smoothly functioning genetic system of the recurrent genotype (Humpty-Dumpty!) and we, the king's men, proceed on the hope that we can put it together again—better. I was pleased to see this view expressed by Grafius, Thomas, and Barnard (1976) as follows: "...the backcross tends to preserve the integrity of an already proven system while adding elements of another system."

Their study supported another belief of mine that where feasible, minimum backcrossing should be practiced in a cross in order to maintain the greatest grem plasm breadth and population variance on which selection can be imposed. They found in their research that a single backcross was effective in cases where the parents differ distinctly in certain traits, in this case, components of yield: the highest-yielding segregates were found in the backcross populations.

A little heralded feature of the backcross method is its role in fostering stability. According to Qualset and Vogt (1980), stability is enhanced by removing defects of an adapted variety. Several decades of backcrossing initiated by Fred Briggs resulted in varieties that "provided improved stabilization against yield losses due to bunt, stem rust, and Hessian fly." It is accepted that inherent yield ability is not raised; nevertheless, the steadier yield line, owing to stabilization, represents higher over-years yield.

Qualset and Vogt also addressed a major disadvantage of the backcross program: the appearance of better varieties, or an unforeseen new hazard, can bring obsolescence to a long-time recurrent parent. Their solution is a two-pronged program that adds a productivity improvement goal to the ongoing backcross program, and merges the two whenever a superior recurrent parent appears. Admittedly, this eliminates the extreme program simplicity of the backcross method, but the necessity of this approach was brought home by the great germ plasm changes that occurred at the time of the stripe rust of wheat epiphytotic in the 1960s in California.

"NEAR-ISOGENIC" LINES ARE REAL

The appropriateness of the term *near-isogenic* line was neatly shown by Everson and Schaller (1955). A small segment of chromosome V in barley, containing the *Rr* genes conditioning awn barbing, had been identified earlier by Suneson, Schaller, and Everson (1952) as also bearing an association between semi-smooth awns and high yield, a marker arrangement of great interest to barley breeders. Everson and Schaller found in a study of segregates from the hybrid, 'Lion' × 'Atlas^{10}' 2× Atlas, that isolines identical in appearance to Atlas and each other, except for the small segment affecting awn barbing and rachilla hair length, still contained chromatin affecting and segregating for yield differences. Thus in some cases isolines may not be considered identical to the recurrent parent plus the added trait(s), and the breeder must consider the possibility that unknown associations, particularly of a negative nature, might be included in the block of chromatin. The authors suggested that with this in mind backcross-derived lines might receive more testing before release into commerce.

Frey and Browning (1971) produced two maturity series of isolines for the Iowa oat multiline program that were presumed isogenetic except for specified crown rust resgenes. Surprisingly, when they were tested under rust-free conditions, when yield performance would be expected to mirror closely that of the recurrent parent, they showed large deviations in grain yield associated with three of the crown rust resgenes. The yield responses were not the same: one resgene increased yield in both maturity groups; a second increased yield in early genotypes but not in midseason ones; and a third resgene, tested only in a midseason genotype, was associated with a 10% yield reduction (I believe an error in the penultimate paragraph of the article refers to this as a "9.6% increase"). The authors discussed the linkage possibilities.

The suggestion of linkage blocks accompanying the desired transfer of a single gene after five or six backcrosses and three selfing generations reinforces the meaning of "near isogenic," and the necessity of some performance testing of backcross-derived lines—generally not considered necessary. That the testing may be worthwhile is indicated by the 5% and 7% yield increases shown in two of the cases. Choosing the higher-yielding and rejecting the lower-yielding isolines could materially affect the performance of a multiline.

When is an isoline not an isoline? The answer: when it is a near-isoline. The term *near-isoline* implies either knowledge or uncertainty as to the content of the chromatin segment transferred by backcrossing from the donor to the recurrent parent. The doubt rests not with the target allele being transferred but with what I shall call *passenger alleles*. Likely candidates are closely linked alleles (so described because if they are not closely linked, one would expect separation in the backcross generations) or pleiotropism. The problem is an important one in plant breeding and there are known cases of both favorable and unfavorable associations.

Fujimaki and Comstock (1977) examined the undesirable situation of passenger alleles having negative effects in the new genotype. The theoretical aspects of their identity, magnitude, and nature were discussed. It was concluded that a search for unfavorable effects should be extended to an evaluation of all important traits; that is, it must be assumed that a donor gene for disease resistance, for example, might have negative effects on a quantitative trait. A general conclusion with plant breeding implications from the authors comments can be deduced from their statement, "Instead the more practical step will be to make the number of lines of descent in the backcross program large enough to make the probability of one or more recombinations between the target allele and alleles tightly linked with it (c as small as 0.02) very high."

Even at advanced backcross generations there is no assurance of complete return to the genetic structure of the recurrent parent plus the added gene(s), owing to linkage or pleiotropic associations. For this reason the term *near-isogenic* often is used. Dr. Glenn Burton (1968) originated this term, using it first in talks, and later in a paper (personal correspondence, 1987). More recently, several studies have shown that isolines often carry important "loads," some of which are very positive. To cite one case, Frey (1972) reported on comparisons of 12 early and 12 midseason oat isolines. Within each group the isolines were visually homogeneous in appearance (several traits). However, when tested for grain yield over a range of environments, 7 of the 24 isolines produced yields significantly different from their recurrent parents. Five were better and two poorer. This has breeding implications for the backcross method concerning the selection of the backcross individual to be used as parent for the next cross to the recurrent parent.

Confirmation of the suspicion that some near-isogenic lines may indeed be different from isogenic lines was obtained in wheat by Zeven, Waninge, and Colon (1986). They found differences between 'Triple Dirk' and several near-isogenic lines (developed by Pugsley from three to four backcrosses) in several characteristics. One explanation was that Triple Dirk was not truly homogeneous, so that the recurrent parents used may have been different. The study is useful in defining the nature of near-isogenic lines and in establishing research guidelines.

DISCUSSION OF THE BACKCROSS METHOD

The backcross method is an elegant breeding system. paradoxically, I have always had ambiguous feelings about it, believing it to be great in its occasional use as one of the tools in a breeding program but reluctant to accept it as the sole tool. The backcross must be associated with other breeding methods for the good of the whole program of crop improvement. This is because the backcross establishes a prototype, usually a successful cultivar. Unless new prototypes are developed independently, a cap is placed on the long-range potential of a crop improvement program. The incremental improvements possible over time with the backcross will inevitably be surpassed by the improvements possible through other more exploratory methods. This seems the case at least for the near future. An interesting aspect of the backcross method is that it is also the ultimate climax breeding method—the logical choice when the search for new superior prototypes no longer is successful.

Thirty-five years ago, in connection with the multiline variety, I advocated short—one and two—backcrosses in contrast to the longer backcrossing series with a high homozygosity end point. Today, there is a greater appreciation of the limited backcross with its greater heterogeneity.

My career experience with the backcross method was both limited and expansive, limited in the sense that only one variety was produced by the backcross method at the Cornell station, and expansive in that I used it extensively to introduce germ plasm into crosses via three-way crosses. The backcross has two functional aspects: (1) that of a prime breeding method for variety production, and (2) that of a recurrent procedure for occasional use as needed in a broad mix of methods. For example, in the latter instance, an initial cross might be made to a parent for a specific contribution, followed by a second cross to either the recurrent parent or, alternatively, an unrelated but desirable parent for an outcross. It was used extensively, for example, in the diallel selective mating procedures. Otherwise, I would characterize my use of the backcross as limited because of a reluctance to engage in a narrow channeling of germ plasm. I believed the backcross most useful when used sparingly—one or two backcrosses followed by selection and outcrossing.

Nevertheless, I used the backcross successfully in the creation of the 'Yorkstar' wheat cultivar. At that time, we needed an eastern soft white wheat semidwarf variety. I took a white-seeded selection from a dwarf red-seeded line obtained from Orville Vogel, and crossed it to 'Yorkwin'. The F_1 was immediately hybridized with 'Genesee'. Two additional crosses were made to Genesee on field-selected short plants in the BCF_2s. These F_2 plantings were drill-sown at very low seed densities in the field, so that short plants readily could be identified for crossing. As a consequence, the half-acre sites were overgrown with weeds, and I remember my frustration as I pushed through the knee-high weeds one summer day, attempting to show Richard Bradfield that this represented worthwhile research! In the end it was worthwhile, and Yorkstar became one of the historically important soft white wheat varieties in the eastern United States and Canada.

There is little need to extol further the merits of the backcross method. It is a solid, successful breeding and selection method, goal oriented and useful as a supplement to other breeding methods.

REFERENCES

Bassett, M. J. and F. E. Woods. 1978. A procedure for combining a quantitative inheritance study with the first cycle of a breeding program. Euphytica 27:295–303.

Briggs, F. N. 1930. Breeding wheats resistant to bunt by the backcrossing method. J. Am. Soc. Agron. 22:239–244.

Briggs, F. N. 1935. The backcross method in plant breeding. J. Am. Soc. Agron. 27:971–973.

Briggs, F. N. and R. W. Allard. 1953. The current status of the backcross method of plant breeding. Agron. J. 45:131–138.

Burton, G. W., J. B. Gunnells, and R. S. Lowrey. 1968. Yield and quality of early and late-maturing, near-isogenic populations of pearl millet. Crop Sci. 8:431–434.

Carpenter, J. A. and W. R. Fehr, 1986. Genetic variability for desirable agronomic traits in populations containing *Glycine soja* germplasm. Crop Sci. 26:681–686.

Eaton, D. L., R. H. Busch, and V. L. Youngs. 1986. Introgression of unadapted germplasm into adapted spring wheat. Crop Sci. 26:473–478.

Emsweller, S. L. and H. A. Jones. 1934. The inheritance of resistance to rust in the snapdragon. Hilgardia 8:197–211.

Everson, E. H. and C. W. Schaller. 1955. The genetics of yield differences associated with awn barbing in the barley hybrid (Lion × Atlas 10) × Atlas. Agron. J. 47:276–280.

Flor, H. H. 1971. Flax cultivars with multiple rust-conditioning genes. Crop Sci. 11:64–66.

Florell, V. H. 1929. Bulked-population method of handling cereal hybrids. J. Am. Soc. Agron. 21:718–724.

Frey, K. J. 1972. Stability indexes for isolines of oats (*Avena sativa* L.). Crop Sci. 12:809–812.

Frey, K. J. and J. A. Browning. 1971. Association between genetic factors for crown rust resistance and yield in oats (*Avena sativa* L.). Crop Sci. 11:757–760.

Fujimaki, H. and R. E. Comstock. 1977. A study of genetic linkage relative to success in backcross breeding programs. Jap. J. Breeding 27:105–115.

Grafius, J. E., R. L. Thomas, and J. Barnard. 1976. Effect of parental component complementation on yield and components of yield in barley. Crop Sci. 16:673–677.

Harlan, H. V. and M. N. Pope. 1922. The use and value of backcrosses in small grain breeding. J. Hered. 13:319–322.

Harlan, H. V., M. L. Martini, and H. Stevens. 1940. A study of methods in barley breeding. USDA Tech. Bull. 720, 26 pp.

Lawrence, P. K. and K. J. Frey. 1975. Backcross variability for grain yield in oat species crosses (*Avena sativa* L. × *A. sterilis* L.). Euphytica 24:77–86.

Leininger, L. N. and K. J. Frey. 1962. Backcross variability I. In oats. Crop Sci. 2:15–20.

Mac Key, J. 1963. Autogamous plant breeding based on already highbred material. *In* E. Akerberg and A. Hagberg (Eds.), Recent plant breeding research. Svalof 1946–1961, Stockholm.

Meredith, W. R. Jr. 1977. Backcross breeding to increase fiber strength of cotton. Crop Sci. 17:172–175.

Qualset, C. O. and H. E. Vogt. 1980. Efficient methods of population management and utilization in breeding wheat for Mediterranean-type climates. Proc. Third Intern. Wheat Conf., Madrid, pp. 166–188.

Richey, F. D. 1927. Convergent improvement in selfed lines. Am. Nat. 61:430–449.

Smith, G. S. 1966. Transgressive segregation of spring wheats. Crop Sci. 6:310–312.

Suneson, C. A. 1947. An evaluation of nine backcross derived wheats. Hilgardia 17:501–510.

Suneson, C. A., O. C. Riddle, and F. N. Briggs. 1941. Yields of varieties of wheat derived by backcrossing. J. Am. Soc. Agron. 33:835–840.

Suneson, C. A., C. W. Schaller, and E. H. Everson. 1952. An association affecting yield in barley. Agron. J. 44:584–586.

Wehrhahn, C. and R. W. Allard. 1965. The detection and measurement of the effects of individual genes involved in the inheritance of a quantitative character in wheat. Genetics 51:109–119.

Zeven, A. C., J. Waninge, and L. T. Colon. 1986. The extent of similarity between near-isogenic lines of the Australian spring wheat variety 'Triple Dirk' with their recurrent parent. Euphytica 35:381–393.

7

METHODS SHAPED BY COMPETITIVE FORCES

Four breeding and selection methods in particular have been influenced profoundly by the forces of competition acting within their populations. These are the bulk hybrid population, the evolutionary, the multiblend variety, and the multiline variety methods. The following section is intended as background to all of them.

The reader may well be forewarned about this chapter. It has been labeled "frustrating" and lacking in focus. As the author I accept this as "truth in labeling"—I had the same feelings while writing it. However, it is the goal of this chapter to bring together experiences and elements of Chapters 8–13 that were common to all, so that these chapters could be developed in a less cluttered manner. Chapter 7 does not lend itself to concise summations and I have attempted none. Nevertheless, I feel strongly that an awareness of the material in Chapter 7 is a necessary prerequisite to understanding the breeding methods that follow.

THE ROLE OF COMPETITION

Competition is interpreted here in the broad sense, and includes competing ability, aggressiveness, density, spacing, tillering, and other traits. Terms such as genotypic weaknesses and strengths are relative, meaningful only in the context of their placement in a proscribed environment, for example, one of several density arrangements possible within a population. We shall see in the following chapters on the bulk hybrid and evolutionary breeding methods and on the multiblend and multiline methods how the varying ways in which competition is perceived have been responsible for the continuing sense of ambiguity that surrounds these pairs of breeding methods.

THE NATURE OF COMPETITION

The subject of Sakai's (1957) paper was the relation between competition and relative frequency of different genotypes in a population. He stipulated that individuals of the same genotype are to be considered noncompeting and excluded from consideration. Others have made the same point that self-competition, although real, cannot be distinguished from environmental effects.

Sakai used two varieties of barley and two of rice, in each species one being a strong competitor and the other a weak one. Two types of experiments were organized, systematic and random. In the systematic arrangement with barley, for example, a weak competitor center plant was always surrounded by six plants, separately placed on the points of a hexagon. These surrounding plants were either of the weak competitor or the strong, and one replication included the seven arrangements from surrounded-by-all-weak to surrounded-by-all-strong. The systematic part was a restriction that all the changes kept the weak and strong competitors in the ring divided in their two groups. In the random arrangement with rice and barley the progressive plot arrangements from a ring of six all-weak to a ring of six all-strong took place with the added strong competitor falling at random into any of the remaining places—it did not have to fall next to its own kind. In both experiments six replications were grown.

The results are summed up simply: regardless of crop or arrangement, the effect of competition increased linearly in accordance with increases in competing individuals, for example, 6 weak: 0 strong, 5W : 1S, 4W : 2S, . . . , OW : 6S.

The results of experiments of Roy (1960) with rice in water, in which he found cooperative interactions of two varieties, are intriguing but raise questions that can be answered only with more research. He found yield gains from two varieties in alternate rows that amounted to 126% of their grown-alone means. he hypothesized that substances dissolved in the common water supply played a part. One comment of his suggests an interesting experiment in corn: "It would be of great interest to discover whether homozygous strains of *Zea mays* which produce very vigorous hybrids, influence one another favourably when planted alternately." In other words, what is the relation between heterosis and parental competitive interactions?

Lee (1960) suggested possible reasons that 'Atlas 46' held a competitive advantage in barley mixtures over 'Vaughn'. He observed phenomena at the beginning of the jointing stage that affected both varieties in the winter crop but were not particularly noticeable in the summer crop when Atlas 46 reached this stage. During this period Vaughn, but not Atlas 46, responded by losing tillers and adopting a slower growth rate. At the same time Atlas 46 responded by positive compensation moves, resulting in relatively greater productivity.

These observations led Lee to look at the root systems of the two cultivars. He found that at the stage of internode lengthening above the crown, Atlas 46 simultaneously began to produce dense masses of roots from the crown. This same phenomenon was not observed in Vaughn, and Lee speculated that this placed Vaughn under insupportable stress, a form of competition for soil nutrients.

Important evidence that heterozygote advantage may be responsible for the persistent survival of characters beyond expectation over time was provided by Jain and Allard (1960). They sampled the F_3, F_5, F_{13}, F_{15}, and F_{18} of Harlan's Composite Cross V of barley, and conducted census counts of certain visible marker characters to evaluate frequency changes. They first estimated the probable level of outcrossing in the self-fertilized species and found that it varied about a mean of 1–2%. This would allow the persistence of a few heterozygotes in the population over time, and could account for a small proportion above zero after prolonged generation cycles (F_{18}); however, the level found with certain traits was so high that a mechanism had to be hypothesized to account for their number. They concluded that only heterozygote advantage could explain the observed results. Furthermore, because the phenomenon was made evident from the census of marker loci, they concluded that the heterozygote advantage was due to fitness overdominance associated with the markers. In some cases, notably locus *B*, black–white pericarp color, no advantage was seen for the heterozygote; and in fact, only the homozygous recessive white pericarp color survived by the F_{18}.

The *shade tent* devised by Moss and Stinson (1961) to provide a screening technique for separating tolerant and intolerant genotypes might be adapted and be useful in multiline research to identify and measure differential responses of genotypes in competitive situations, for example, genotypes of different heights or canopy type.

Pendleton and Seif (1962) placed two corn varieties, 'U.S. 13' and its dwarf derivative, in competitive situations. They found that the increase in yield enjoyed by the tall (normal) corn when grown in various hill combinations was not great enough to compensate for the yield loss sustained by the dwarf. With tall and short rows placed in neighbor arrangements, the same results held; that is, normal bordered by dwarf yielded 6% more than when bordered by self, but dwarf bordered by tall yielded 30% less than when bordered by self. Light was considered the dominant competitive influence.

After describing the typical experience of researchers in studies of population changes in mixtures of pure lines, which was that the sequential procedures of mixing, growing, aliquot sampling, and an annual census led to dominance of the population by one pure line, Baker and Christy (1964) pointed out that the cause was usually attributed to competitive forces.

Acknowledging that a number of factors might be involved, they examined two, sampling errors and differential viabilities. Beginning with the statement that "it is well known that any finite multitypic population comprised of unchanging homozygosis lines will, due to sampling alone, eventually reach a state in which all individuals are of the same homozygosis type," Baker and Christy nevertheless concluded that sampling errors in reasonably sized samples, that is, hundreds of individuals per pure line, do not lead to large changes in relative genotypic frequencies. This assumes an idealized consistent sample-drawing technique, understandable to anyone who has tried to shake from a box a representative breakfast bowl of raisins and flakes!

The authors next turned to a consideration of random viabilities. Using appropriate formulas, they concluded that random viabilities do indeed increase the fluctuation in relative genotypic frequencies and especially in larger populations. Affected are both seed selection and growth of the subsequent population that harbors and reflects the random viability.

Baker and Christy applied their model to analyze the competition between 'Atlas' and 'Club Mariout' barleys in a mixture of four varieties (cf. Suneson 1949). They made an observation, which seemed to me to describe the situation that must obtain at times in the dynamics of endangered species, that "it frequently happens in experiments of this kind (though not for Atlas v. Club Mariout) that as one or more varieties near extinction, even with large populations and censuses, large uncertainties in their observed probabilities can occur."

They surmised that randomness of the relative viabilities, shown by wide swings, has greater influence on the fluctuations of relative genotypic frequencies than do sampling errors.

Jensen and Federer (1964) reported results of adjacent row competition in wheat nurseries that have important implications for testing procedures. The check variety, 'Genesee', in use during the four-year, 12-nursery experiment, is a tall wheat, taller than the averages of all flanking rows at the 12 sites. It was found that the yields of Genesee flanked by shorter wheats were enhanced by an average 5.0 bu/acre, whereas the yields of the shorter wheats flanked by Genesee were depressed an average 2.3 bu/acre. The fact that the competitive effects were not mutually compensatory meant that a field of alternate short and tall genotypes could produce a yield bonus—more than the average of either sown alone.

Doney, Plaisted, and Peterson (1965) conducted two experiments on competition in potatoes. If these experiments had been conducted singly and separated by time, the results of each would have been considered strong and definitive; taken together, however, they stood in opposition to each other. From the resolution of this conflict, the authors gained insight into the nature of competition.

The authors wished to find out whether ". . .progeny testing in potatoes with mixed plots duplicates the results from testing with pure plots." Because potato clones breed true, large numbers of "pure lines" were available from the breeding program, and pure plots were established merely by replicating any clone. The experiment was organized within the working framework of the potato improvement project at Cornell University, and selections were made from the overall populations.

In the competition experiment, 10-hill plots were used. The pure plots contained, of course, 10 hills of the same clone; the mixed plots contained one hill each of 10 different clones, randomly positioned. The results showed small and insignificant yield differences between pure and mixed plots, no matter how compared.

The second experiment, the "red and white," involved two white and two red potatoes tested in all two-pair combinations. Here the results were dramatically different: highly significant differences were found in mean yields and competitive influence, which is the ability of a genotype to influence its neighbors via competition. The overall mean yields of mixed plots were significantly higher than those of pure plots.

These results forced an intense review of possible differences in the two competition experiments. From that, these suggestions have arisen:

1. There may be major differences between pair measurements and multiple genotype arrangements. Internal compensations in multiple genotype plots may permit only the mean effects of (the 10) conflicts to surface. This masking phenomenon is in contrast to the head-to-head competition in pair plots, which is simpler, more direct, and more easily interpreted.
2. The genotypes used in the mixed plots were full sibs and the possible genetic similarity might have lowered the level of competition compared to what it might have been with 10 unrelated genotypes.
3. The unit of measurement was economic yield, tubers, which is but one manifestation of competition.

The subtle opportunities for competitive advantage may be illustrated by the common observations in transplanting cereal plants: a transplanted seedling, as from greenhouse to field, invariably is "set back," often with a traumatic delay consequence so great as to nullify attempts, for example, to replace a missing plot.

In a third paper of their 1968 series, Jennings and Aquino (1968) investigated the mechanism of competition that had been observed in the previous studies: this had convinced them that competitive ability and agronomic worth were negatively correlated. The question posed related in part to whether the observations were of a unique or general nature. In this study they attempted to identify basic sources and causes of competition through observations and measurements of plant development. They compared two phenotypic pairs with contrasting competitive relations, namely, a tall weak competitor with a shorter aggressive type, and a tall aggressive population with a dwarf weak competitor, in a series of trials in pure stand and mixture, in which measurements were taken at intervals from 32 to 88 days after seeding. Results were as follows:

1. Regardless of height, weak competitors mixed with strong had fewer tillers than in pure stand, leaf area index (LAI) declined, and dry weight was markedly lower, as were grain yields, grain/straw ratio, and spikelets per panicle. There were insignificant changes in leaf area or length, and height. These opposite effects were not compensatory before flowering. Although competition was severe early on, the advantages to the aggressive components did not accrue until they were about two months old, or one month after transplanting. The authors described strong competitors as follows: "Strong competitors have more tillers, longer and more leaves, and greater LAI, height and dry weight than weak competitors. The cumulative effect of these characters is, without exception, one of increased plant size. The strong competitor is a larger, more spreading plant than the weak competitor."

2. The critical importance of light in competition was shown with light measurements at two height locations in pure and mixed stands. In pure stands the weak competitors showed more light penetration than did the aggressive ones; in mixed stands the weak competitors received less light than in pure stands, whereas the aggressive ones received relatively more. Jennings and Aquino concluded that "in mixtures, the strong competitors intercepted more light at the expense of the weak competitors."

The overall deduction is that competition in rice was undesirable. Competition seemed to be for limiting factors, predominantly for light in this case. The seemingly obvious involved character, height, is to some extent misleading. It is not height per se but rather the intercepting leaves that interact with light penetration. Plants in mixed populations that are too "fast off the mark," although impressive, may not be the best in pure stands. One can see a relevancy here to the involvement in segregating populations of heterozygotes that seem to use their large size and vigor to advantage.

Mather's (1969) discussion of "selection through competition" is good background for an understanding of the effects of competition on individuals in segregating and homogeneous populations. Mather recognized three types of competition, one cooperative and two antagonistic. He explained that the fitness of an individual, or genotype, is relative both to its population and its environment. Thus the same individual may express different fitnesses, for example, as a homozygote in a segregating or heterogeneous population, or as a homozygote in its own replicated-self population such as a pure line cultivar.

Studying competitive responses in genetically heterogeneous soybean populations, Byth and Caldwell (1970) found that performances for populations could encompass a range traceable to the mixture and influence of heterogeneity. It seemed possible that populations could be average, lower, or higher in yield, for example, than expected due to over- and under-compensation. This has implications for both bulk and multiline breeding. In bulk population breeding, it is possible for the average yield of selected lines to be lower than that of the population itself. It is also possible that a fortuitous mixture of genetic heterogeneity might produce a lower-yielding population whose homozygous selections were better. The phenomenon observed may be seen as an encouragement to multiline formation, where the right combination of diversity could also provide a synergistic reaction.

John Hamblin and others have explored aspects of competition in a series of papers. Using a segregating population of F_3 barley plants, and a sampled selection of their derived F_5 lines, both grown in high and low nitrogen environments and at competitive spacings, Hamblin and Donald (1974) found no correlation between F_3 plant and F_5 line yields. In the F_3, taller plants with longer leaves were the highest yielders but in the F_5 monocultures the greatest yields were from the shorter plants with shorter leaves. They hypothesized that in the competitive F_3, tall plants and long leaves had an advantage over shorter plants with shorter leaves whose yields were suppressed, but in the F_5 monoculture it was the shorter, erect-leaved plants that excelled.

A graphic in their article of the relationship of F_3 plant height to grain yield in the F_5 showed that "if the shortest 25% of the F_3 population had been selected, then ten of the eleven highest yielding F_5 lines would have been retained."

Hall (1974) presented an interesting discussion of the complexity of the term *competition*, pointing out that both competitive and noncompetitive processes are involved, usually confounded. Harper (1964) applied the term *plant interference* to noncompetitive influences.

Phung and Rathjen (1977), in their study of frequency-dependent advantage in wheat, believed that competition took place principally between the root systems and was influenced by the varying distances between plants in a field situation. They concluded; "Thus the available resources of a certain environment seem to be exploited more efficiently by a mixture of several genotypes than when there is only one."

Jensen (1978) studied the competitive responses of a winter wheat, 'Genesee', and a spring wheat, 'Justin', when grown together in 26 seed planting ratios in the natural growing seasons of each. The effect of fall competition of spring Justin on winter Genesee, followed by the death of Justin over winter, had little influence upon the yield of Genesee. The matching fall competition used the same mixtures with the exception that the Justin seed was killed by autoclaving before mixing with Genesee; therefore Genesee had no fall competition at all. Again, Genesee yielded the same as when it had initial fall season competition with Justin. Genesee's average yields were 3079.5 g with live Justin, and 3095.2 g with dead Justin.

In the inverse mixtures of Justin and Genesee planted in the spring where in one case, both cultivar seeds were live, and in the other, Genesee was killed by autoclaving, a different result was obtained. The presence of live Genesee as nonheading plants in the mixed Justin plots represented a drain upon the resources available to Justin, and lowered Justin's average yield by 29%. Justin's average yields were 663.5 g with live Genesee and 853.5 g with dead Genesee.

The study showed that the advantage of winter wheat in the Northeast was due to its ability to establish a strong fall base for the earlier growth in the spring. Mean yields of each alone were Genesee 3287 and Justin 1360 g/plot.

When the genotypes in mixed stands of wheat represented three different heights and different yielding abilities, as determined from their pure stands, it was plant height and not high-yielding ability that influenced competitive ability in studies by Rajeswara Rao and Prasad (1984). The three wheats were a dwarf (60 cm), semidwarf (100 cm), and tall (120 cm), and their yields in pure stand were in the same order, thus another instance of a high-yielding dwarf in pure stand. The competitive ability of the genotypes (the percent of mixed stand versus pure stand yields) was in the reverse order, respectively. This was attributed to competition for nutrients and especially light.

Mixed stands did relatively better in yield than pure stands: from the two-year trials of 18 binary mixtures, there were four cases of yields that were superior to the better component's yield, by 4.4%, 2.7%, 3.3%, and 0.8%; and 11 out of 18 exceeded their components' means by 1–7.6%.

The authors identified mutual support in preventing lodging as a reason for the greater stability of mixed stands. This seems logical because the taller weaker genotype lodged frequently in pure stand, thus showing obvious instability in yield.

Dewey and Albrechtsen (1985) explored the relationship between tillering of winter and spring wheats when grown under space-planted and densely sown conditions. The relationship is important to breeders because candivars selected under space-planted conditions become cultivars grown under densely sown conditions. They found that the four winter wheat and two spring wheat cultivars with known tillering differences maintained their relative rankings when subjected to spaced planting, a conclusion of comfort to breeders seeking the perfect selection environment. In the segregating generations phase of their studies, they also found the following: "When F_5 plants selected for high or low tillering in spaced populations were subsequently grown in densely-sown F_6 plant rows, little relationship was observed between tillering characteristics in the two planting densities."

The latter finding may be linked to an adaptive heterotic response of segregating plants, related to the question of the validity of early generation selection for quantitative traits. The importance of these findings, that tillering differences observed in early generations are not a reliable selection criterion, is tempered by the observation that there was little association between degree of tillering and the subsequent grain yield of the plant. In other words, there was significant compensation among yield components.

Evidence from studies by Shanahan, Donnelly, Smith, and Smika (1985) showed that shoot or tiller survival was more important to winter wheat yields than any other tiller aspect, such as emergence. Grain yield had a positive correlation with shoot survival in all three years of the study. The authors' reasoning linked the factors in a chain of events in this manner: (1) a known positive relationship between kernel number and grain yield suggests that a limiting influence on grain yield is the lack of sink capacity during grain filling; (2) kernel number and spike number are positively correlated with shoot survival; and (3) higher shoot survival leads to increased spike numbers, kernel numbers, sink capacity, and finally, yield.

MIXTURE EFFECTS OF COMPETITION

Kiesselbach and Helm (1917) found a case of unusual competition efffects in studies with two cultivars of winter wheat, 'Big Frame' and 'Turkey Red'. Grown alone, Big Frame yielded 90% of Turkey Red's yield, but in competition (alternate plants from seeds of same size in rows) its yield was 55% as much. The next year under different climatic conditions, the situation was reversed, with Big Frame alone yielding 82% as much as Turkey Red but rising to 120% when under competition. Regarding seed size Kiesselbach and Helm believed that "competition between plants from large and small seeds sown in a mixture acts to increase the relative yield from the large seeds."

Half a century or so ago, Engledow and Ramiah (1930) contributed an interesting finding derived from intervarietal competition of the wheat varieties 'Squarehead's Master' and 'Yeoman'. The laid out two nursery beds, each with 100 rows 6 in. apart with 24 seed holes in each row. In one bed the two varieties were planted in alternating rows, two seeds per hole, one variety per row. In the other bed one seed of each variety was dropped in every hole of every row, thus also resulting in two seeds per hole. Data on tillering, ears per plant, yield per plant, and yield per ear were taken only from holes producing two plants.

At maturity, analysis showed that for every character studied, different varieties together in the same hole gave higher values than varieties in alternate rows. The calculated yield per unit area for the varieties together was 22% greater than for that of the means of the varieties in alternate rows. Thus in this case it was concluded that intense competition favored productivity.

Laude and Swanson (1942) provided a clear-cut case of one genotype eliminating another in a wheat mixture grown for nine years. The study, in

Kansas, involved equal-seeded mixtures of 'Kanred' with 'Harvest Queen' and Kanred with 'Currell'. The tests were grown at two sites: at Hays, both Harvest Queen and Currell had been eliminated by the eighth year; at Manhattan 6% and 2%, respectively, of these cultivars remained in the ninth year.

It is not clear what was the commercial relationship—the relative success—of the three cultivars grown in pure stand; however, it was known that Kanred was more winter hardy than either of the other two varieties. The authors discount this as a factor on the basis of benign weather conditions; however, it does not seem reasonable to separate this advantage from other unknown adaptive responses of Kanred. I am intrigued by the implications of these results: on the one hand it appears that the best genotype in pure stand best survived the competition. On the other hand, assuming this happened in a hypothetical heterogeneous hybrid population, I am unable to assess the loss if Harvest Queen and Currell had not been selected; and loss it would have been, because obviously they were good enough, for we know they had been selected and released as cultivars on their own merits. Therefore, their elimination by a dominating genotype must stand as an indictment of natural selection when the confrontations are extended beyond what would be considered practical. Still, it should be pointed out that if this were analogous to a practical breeding program where selections were removed from the F_5, the percentage of plants in the populations at that stage were Harvest Queen 31 and 26%, and Currell 31 and 8%, respectively, at Manhattan and Hays—adequate numbers for selection purposes. From a practical standpoint, the results could be seen as a method stressed beyond its limits.

An interesting observation from measurement by Laude and Swanson was that competition influenced genotype reductions in different ways: Kanred placed more pressure on Harvest Queen in the vegetative stage (number of plants), and more pressure on Currell in the fruiting period (number of grains). These physiological aspects need more research.

Suneson (1949) reported on a 16-year study of the survival of genotypes in a four-variety mixture. Beginning in 1933 with an approximate 25% input of each variety, the final 1948 census showed 'Atlas' 88.0%, 'Club Mariout' 10.5%, 'Hero' 0.7%, and 'Vaughn' 0.4%. The single-component yields were almost the reverse, namely, Vaughn, Hero, Atlas, and Club Mariout.

These dramatic results have been used extensively as evidence that survival in mixture cannot be used as an indication of a component's performance in a pure stand. The fact that the dominant component of the mixture had the worst leaf disease record and a below-median component yield had important implications limiting the power of the bulk hybrid population method.

Later, Suneson and Ramage (1962) reported that more extensive testing merited a second look at the conclusions drawn from the 1949 study. Vaughn was found to be very responsive to management for yield, and its relationship to Atlas moved from a 7% advantage to a 2% disadvantage with the addition of data from approximately 200 tests. Also the popularity of the varieties with growers, as measured by acreages grown, had shifted to 10 : 1 in favor of Atlas.

Hartmann and Allard (1964), in a continuing study of the competitive relationship between Atlas and Vaughn barleys, found that competition was the keenest under moderate moisture stress at above-average fertility levels:

> This differential in competitive ability disappeared at the highest moisture level but not at the highest nutrient level. Apparently, therefore, the competition between Atlas and Vaughn is primarily for moisture but the intensity of the competition at any moisture level depends on the nutrient level.

A likely morphological basis for Atlas's advantage lies in its larger root system.

Eberhart, Penny, and Sprague (1964) studied intraplot competition among corn single crosses to assess their relative performance in pure and mixed stands. Two sets of single crosses, one set having four and the other five single crosses, were used; a series of paired mixtures, both in hills and alternating in rows, were compared for performance in pure stands. One experiment used eight characters as measuring criteria. Certain competition effects were noticed but they were of a compensating nature so that no differences were found between performance in mixed or pure stands. Of interest was the finding that single crosses in mixed stands created smaller within-plot environmental variances than when in pure stands. It would be interesting to know whether this damping effect was an inherent feature of mixtures or an adaptive response deriving from the hybrid nature of the uniform hybrid single crosses.

Jensen and Federer (1965) used five-row plots with a center target genotype flanked by four rows of neighbor genotype to assess competitive effects among four genotypes of wheat. The four genotypes used showed distinctive height differences, namely, Genesee 49.45 in. '5207–34' 45.30 in. '5207–85' 42.13 in., and '53150–6' 28.55 in. In a previous experiment Genesee had experienced yield advantages related to height in neighbor associations. The 5207–85 selection was a heavy-tillering wheat with a history of aggressiveness. The authors proposed that Griffing's (1956) concept and formulas for general and specific combining ability might be used to measure competing ability simply by substituting "competing" for "combining."

The data from the 24 treatments (each variety and each type of neighbor) and an alternative 16 treatment group that was analogous to Griffing's diallel crossing array were analyzed with these general deductions:

1. General competitive effects were large and specific competitive effects small.

2. Extreme aggressiveness was shown by 5207–85: the lowest yields of each wheat occurred when Sel.–85 was the neighbor (including against itself).
3. Height could not be considered the sole factor influencing competition as measured by yield: the decreasing rank order by height was Genesee, Sel.−34, Sel.−85, and Sel.−6; the dominance rank order as measured by neighbor mean yield was Sel.−85, Sel.−34, Genesee, and Sel.−6. In other words, Genesee, the tallest wheat, ranked only third in its depressing influence on neighbors mean yield.
4. It appeared that a good measure of competing ability could be deduced from a strain's average yield under varied competitive target–neighbor environments: "The strain with the highest mean yield imposes the greatest mean yield depression when it is a neighbor."
5. The data showed that competition between genotypes resulted in an average 4.6% bonus over the mean yields of the genotypes grown alone. The authors stated: "For example, 1 acre each of Genesee/self flanking (3204) and 5207–85/self flanking (3486) would produce a total of 6690 pounds; the reciprocal arrangements, Genesee/5207–85 neighbor (3006) and 5207–85/Genesee neighbor (4014), would produce 7020 pounds, a 4.9% bonus."

The authors extrapolated these findings to suggest that field alternate row plantings could yield bonuses of 4–5%.

The 5207–85 line was of interest from a breeding viewpoint. Its known heavy tillering and aggressiveness resulted in its use as a parent in the Cornell wheat program for many years, yet its "get" never were outstanding. In the end, I used it infrequently. But even today I feel there should be some breeding approach that would ensure perpetuation of its good qualities. Its yielding abilities did not match its aggressiveness.

McGilchrist (1965) defined the competitive advantage of species i over species j as the average of two phenomena: (1) when grown together, the yield increase of i over its yield in monoculture, and (2) the yield depression in j over its yield in monoculture. He provided formulas to use in determining the competitive depression of i and j, which, of course, might be zero. The use of these concepts and analyses permit more detailed evaluation of the competitive suitability of components in mixture populations.

Important contributions to understanding the nature of competition in plants were made by Jennings and de Jesus (1968) in the first of a series of three papers in 1968. In essence, they compared two type groups of rice plants for their pure line behavior and behavior in mixtures. The least competitive in mixtures were the highest yielding in pure stand. These types were characterized as vegetatively small, erect, and sturdy. The highly competitive group had the common characteristics of tallness, leafiness, extreme vigor, and high tillering. These characterizations were drawn from equal-seeded mixture experiments involving five genotypes grown for four years under four environments each year. The results in the four environments were remarkably similar: one genotype, 'BJ', dominated, comprising 91–97% of the final population. This resulted in the almost total elimination of the other four genotypes.

In the second paper of the series, Jennings and Herrera (1968) concentrated on competition in segregating populations. The plant material for the study came from a cross of tall and short rice genotypes in which the height extremes were conditioned by monogenic action, with tallness being dominant and dwarfness coming from the double recessive. As segregating populations approached homozygosity a 50 : 50 ratio would be expected, building up from the 25% dwarf content of the F_2.

The ratios were monitored from F_2 through F_6 by classifying segregation each generation and allowing the unselected bulks to proceed. The procedures were tracked on separate plots involving two kinds of fertilizer treatment: 0 N for one, and 50 lb N in F_2 followed by 100 lb N thereafter for the other. In addition, a spacing experiment at one-half density was added in the F_5 and F_6, and homozygous tall and dwarf F_3 lines were selected for later yield trials.

The results of these separate competition experiments were these:

1. Survival from the orginal 25% F_2 level for dwarfs *declined* with bulk generation, and declined relatively more at higher density and especially more at the higher nitrogen level.
2. When grown alone, the dwarf parent exceeded the tall parent in grain yield with the difference being greater at the high nitrogen level (slight advantage only at 0 N).
3. The tall and dwarf bulks of homozygous segregates were not different in yield at 0 N but at high nitrogen both were intermediate in yield, the dwarf bulk approaching the superior-yielding dwarf parent.
4. The 36 tall and 36 dwarf F_3-derived lines grown with their parents at high density and high nitrogen produced mean grain yields about the same as their prototype parents. However, the five best of each gave the following comparisons of grain yield:

Tall parent, tall lines:	2.33 : 3.72 ton/ha
Dwarf parent, dwarf lines:	4.67 : 5.65 ton/ha

The authors presented a clear case of the effect of height on yielding ability and competitive ability. It was a negative relationship, with dwarfness showing higher yields alone but lower competitive ability in mixture, and with the competitive effect magnified under close spacing and increased nitrogen.

The slight yield advantage of a wheat variety surrounded by the other of a pair in hill plantings that was observed by Allard and Adams (1969) was observed again by Chapman, Allard, and Adams (1969) in blends of the same two wheats, sown with a range of proportionate mixes, and then planted at four densities in rows rather than hills. This study confirmed that the significant positive yield increase occurred at high densities, increasing from the lowest density figure.

The importance of seed size differences to the outcomes of competition in populations and the necessity to use uniform seeds in such trials were illustrated by Sandfaer (1970) in research that pitted two size classes, large 47.4 g and small 32.0 g/1000 kernels, of seeds from the same variety, 'Carlsberg II', against 'Freja' in 1 : 1 mixtures. In all categories examined, the larger seed was superior to the smaller, producing 16% more kernels. In the competition aspects, Sandfaer found that "Freja plants grown in mixture with Carlsberg II plants from large seeds had about 18 per cent fewer kernels than Freja plants grown in mixture with Carlsberg II plants from small seeds."

Early and Qualset (1971) uncovered an unusual form of complementary competition in winter barley. They studied three barley varieties, 'Dayton' (D), 'Hudson' (H), and 'Tenn. 60–34' (T), evaluating their performance in pure stand and in two-way and three-way mixtures, at three sites for two years. Early and Qualset were guided by the four categories of kinds of competition recognized by Schutz, Brim, and Usanis (1968). These groups were: overcompensatory, undercompensatory, complementary, and neutral. In addition to the three pure stands, there were three two-way equal-proportion mixtures, one three-way equal-proportion mixture, and one three-way mixture made up from 80% T and 10% each of D and H. In addition to yield, data were taken on number of spikes per square meter, numbers of seeds per spike, and weight per seed in milligrams.

To appreciate the surprising results found, it is necessary first to savor the quite ordinary ingredients, namely:

1. The three barley varieties were equal in yield in pure stand.
2. All mixtures of these three varieties in two- and three-way equal-seeded combinations, and in one case unequal-seeded combinations, showed no significant differences among themselves or with yields in pure stands.

Nevertheless, when the contribution of the components to the mixtures was examined after two years, it was found that the productivity of yield and spike number of one variety, Tenn. 60–34, fell more than 20% below expectations. There was no such suppression for Dayton and Hudson, which obviously had made up the deficit left by the competition effect on Tenn. 60–34. The actual reduction in yield of Tenn. 60–34 was found to be due to lower spike numbers, which further could be said to trace to tillering ability. The great shift in internal structure without a change in population productivity is unusual. There was no outward morphological indication that Tenn. 60–34 would react in this manner.

Relative to the multiline method, the results indicate caution in predicting, without testing, how different genotypes may react in mixtures. Even though Tenn. 60–34 was not close to elimination, there remains a concern for the possibility of loss of desirable genotypes in hybrid bulk populations through the action of natural selection. This loss can be avoided if genotype line withdrawal occurs at approximately the F_5 stage.

Khalifa and Qualset (1974) reported on a study of the interaction of two genotypes when grown in an equal-seeded bulk mixture for four generations, and when grown in a range of mixture frequencies. The two genotypes were 'Ramona 50' (R50), 100% and 80% relative to height and yield of the other dwarf wheat 'D6301', which were 75% and 100%; thus a taller lower yielder mixed with a shorter higher yielder, respectively.

The mixture bulk experiment showed strong competitive effects between the two wheats which, however, did not affect significantly the overall performance of the mixture at the end of the four generations. The effects were complementary. However, the lower total yield trend was evident by a 300 kg/ha drop over the period. Internally, the by-itself higher-yielding dwarf declined from a first-generation contribution of 3043 kg/ha to a fourth-generation 1247 kg/ha, whereas R50 increased from an initial 3043 kg/ha to an ending 4547 kg/ha. The advantage of R50 (lower yielding) tended to lower the overall performance as shown. It was clear that R50 was the better competitor and the explanation probably was height related; for example, competition for light forced physiological changes in both plants.

In the frequency study, the same genotypes were grown as pure stands and in seven two-way mixtures with genotypes appearing in frequencies of 87, 75, 62, 50, 37, 25, and 12%. The observed and predicted performances were remarkably similar—there were no significant differences. On the contrary, when the components within the mixtures were separated it was found that the complementary effects were the result of steady significant losses in yield of D6301, compensated by steady and significant greater-than-expected yields of R50. The impact of this was heightened by the performance of the two wheats in pure stands: D6301 was 24.5% higher in yield than R50. The effect on yield components was also studied.

The authors pointed to four implications for breeding and selection methodology:

1. Unless original genotype frequencies were reconstituted annually, there was a trend toward decreased productivity.
2. Genetic variability decreased rapidly because of the between-line competition.
3. Reproductive values were highly correlated with frequency.
4. There was progressive elimination of the most agronomically desirable genotype (based on its pure stand value).

The plant breeding implications are such as to raise a flag of caution, which may be expressed in different ways; for example, this seems to be a clear case of a weakly aggressive genotype (D6301), which consequently does well competing against itself in pure stand. Also, in hybrid bulk population procedures, it would be best not to defer selection too long because of the risk of losing such weak competitors, which in pure stand often are excellent cultivars. For multiline variety breeding the finding is that there is no substitute for actual screening field tests. Although in the case just discussed the obvious cause seemed to be height differences, the fuller explanation may be more complicated.

Hamblin and Rowell (1975) constructed a model to relate the competitive ability ("the difference between seed yield per plant in mixed culture and in pure culture") of genotypes in a segregating population to their pure culture yield. They used the data from the earlier Hamblin and Donald 1974 study. The inverse relationship between competitive ability (yield) in F_3 plants and yield in F_5 pure culture suggested an unacceptable plant breeding situation, namely, that the best competitors (yielders) in mixed culture were the lowest yielding in pure culture; or to express it differently, competition could be used to identify the poorest competitors (yielders), which would then be found to include the highest yielders in pure culture. The model yielded information that might be used to choose between the pedigree or bulk system of breeding and selection.

Hamblin and Rowell had an interesting application to developing multiline varieties: "If the high yielding genotypes are good competitors, then they are higher yielding in mixed culture and multiline varieties will yield more than pure line varieties." Thus a competition test could be held to choose lines for a multiline.

Hamblin and Rowell's paper, which suggested guidelines for selection in segregating generations and implications for breeding methods, merits further examination. The competitive effects noted were traced to height and leaf differences and thus might not represent a generalized situation for all cross populations. It would seem also that competitive effect could be influenced by nitrogen level, decreasing in low-nitrogen environments. One would like to know more, too, about the true genotypic nature of the F_5 populations, each descended from an F_3 plant, grown in pure culture. These populations, approaching homozygosis, nevertheless must be heterogeneous for all the genetic variance descending from the F_3 and might present internal and external competition and stability situations different from pure lines.

Perhaps greatest in interest to plant breeders is the question of whether the model should be considered decisive in choosing between the pedigree or bulk breeding systems. Equal consideration must be given to the suitability of a breeding system to the harmony of the entire breeding program, and to a consideration of whether adjustments to the system might be made. For example, implied flaws in the bulk system could be circumvented by (1) thinner seedings in the early generations, lessening competition and elimination, (2) choice of crosses for suitability for the system, (3) early removal of disadvantaged individuals, and (4) benign neglect followed by a search for disadvantaged individuals in later generations, knowing that the survivors then are the best of that group (see Jensen, 1978).

Roy (1976) studied intergenotypic plant competition in wheat under single seed descent breeding. He tested four genotypes grown in pure stand and in diallelic combinations of two in alternate rows under density pressure. Certain genotypes in mixtures showed a marked (8–14%) reduction in percent survival compared to survival in pure stand. One weak competitor nevertheless increased its grain weight under competition as compared to pure stand, suggesting an ability to compensate during the later stages of growth. Competitive ability was not related to genotypic differences in seed size in this study.

Gedge, Fehr, and Cox (1978) followed up the 1970 suggestions of Byth and Caldwell that there was a high coincidence of overcompensation with high-yielding, and undercompensation with low-yielding, heterogeneous soybean lines. Random distribution would be expected to distribute equally the three possible intergenotypic competition effects, namely, neutral, increase, or decrease in yields. To test this, they ran two experiments: Experiment 1 was essentially a repeat of the Byth–Caldwell experiment, slightly modified, in two crosses that included 40 random F_2-derived lines, four F_5-derived lines from each F_2 line, and an F_5 blend of the four lines. The yield of the F_2 lines and F_5 blends were compared, using linear regression, with the mean F_5 lines yield. In Experiment 2, six high- and six low-yielding F_2-derived lines were identified in each of the two crosses, and four F_5 lines selected from each category. These sets of F_5 lines were compared for yield in pure stands, all possible two-component mixtures, and in single four-component blends. Yield comparisons were between blends and pure stands.

The overall and pervasive findings were that yield compensation in either direction was neutral, that is, not related to the high- or low-yield status of heterogeneous lines. The inference to be drawn is that intergenotypic competition in heterogeneous soybean lines is not a bar to early generation testing. The findings of the authors, different from the Byth–Caldwell experiment, were linked to different statistical procedures and interpretations.

A clear case of unequal compensation by components in a mixed stand of wheat was found by Sharma and Prasad (1978). The mixed stand referred to two treatments of adjacent row arrangements of the three genotypes, tall, semidwarf, and dwarf, and the authors characterized the arrangements in terms of canopy structure of a whole plot; for example, a pure stand plot had a flat canopy.

No significant yield differences were found between the two forms of genotype arrangements of the mixed stands. The grain yields of mixed versus pure stands, however, were significantly higher in the mixed stands, approximately equivalent to the highest-yielding pure stand component. Unequal compensation was seen in the data, showing that the tall genotype yielded significantly more in mixed compared to pure stands, whereas the dwarf variety yielded significantly less in mixed than in pure stands (see also Jensen and Federer 1964, 1965).

A number of observations of a useful but inconclusive nature were made in soybeans by Sumarno and Fehr (1980). They were interested in the interaction of cultivars showing height differences, made possible through using the taller indeterminate and short determinate genotypes. They tested these pairwise in blends (3 : 1, 1 : 1, and 1 : 3 ratios) and in alternate row pure line combinations.

They found significant year, location, and entry interactions throughout; however, no combination tested had a significant yield advantage over the highest-yielding component, and only one combination significantly exceeded the predicted component mean yield. As has been found in many competition experiments, there was a trend in favor of higher yield—12 of the 18 combinations in blends and alternate rows yielded more than predicted.

Differences in competitive prowess were noted among cultivars, some of which were strong in blends and weak in alternate rows, and vice versa. In general, intergenotypic competition was less in alternate rows than in their counterpart 1 : 1 blends.

Plant height definitely was a factor in competition but was not consistent. Some mixtures of tall and short gave higher than predicted yields, whereas others showed no advantage. This confirms the hypothesis that components of mixtures cannot be chosen solely by general predictions based on limited trials. For example, in the final choice of the several components for a multiline variety, there is no escape from the necessity of doing extensive group testing, as contrasted to pairwise testing, in order to verify that there is internal harmony conducive to the highest productivity of the multiline. Even though the lines were developed by the backcross method, several studies have shown that there still can be distinct differences in the putative isogenic lines.

Row spacing was noted to influence competition but was not explored sufficiently owing to mechanical limitations of planting equipment. Because there was no clear-cut advantage of either alternate rows or blends, the choice would be decided on other grounds; for example, alternate rows require no mixing of seed.

DENSITY, COMPENSATION, AND SEED SIZE RELATIONSHIPS

Montgomery (1912) referred to the competitive aspects of the common farming practice of sowing larger quantities of seed than required for an optimal crop density. This led to heavy mortality of seedlings and plants, particularly among the smaller seeds. The competitive situation favored the larger seeds, possibly because of greater and longer-sustained vigor.

A plant breeder spends years working with selection under "controlled" density conditions, knowing only that the cultivar when released will be grown under "crop" density. This density is itself variable, being influenced in the case of cereals by seed size and variability, random placement of seed, viability, equipment, and a host of similar factors. In general, breeders consider this aspect of production outside their control, and rely on compensation within the crop to set the final yield of grain.

Nevertheless, even today it is instructive to know the range and magnitude of row gaps and thin spots that are found in the field after planting. The three ecosystems of Donald and Hamblin (1976) for spaced, mixed, and crop-density plants, produce matching ideotypes that will be successful only in their particular ecosystem. These ideotypes are, respectively, (A) the isolation ideotype, (B) the competition ideotype, and (C) the crop ideotype. For pure line varieties only the crop ideotype is desired. The breeding problem has always been to bring segregating material through appropriate ecosystems to produce the crop ideotype.

In general, the crop ideotype does match its dense monoculture ecosystem. This ecosystem, however, is not always consistent. Early research by Doughty and Engledow (1928), based on 100 1-ft row samples per acre, has shown a range of 5–21 cereal plants/ft, and from this it may be deduced that there are many wide gaps in the field rows—gaps that would be more appropriately filled by an isolation or competition ideotype. It is of interest also that per-acre yields based on these densities ranged from 28 to 96 bu/acre, with the average yield of any acre being the representation of all plants over all densities.

Smith (1937) observed that the effect of stand differences on yield in fields of pure line cultivars was minor, apparently compensated for by adjustments in tillering, faster growth, and individual plant yields. He conceded that there is an optimum density, "but the optimum is difficult to forecast before sowing." Therefore, if a reasonable seeding rate is chosen, variable density in the field can be seen as a form of assurance that varying densities would accommodate an unknown future soil and climate environment. These views do not, however, invalidate the use of spaced seeding in plot experiments, nor the precision spacing possible with modern field drills.

Christian and Gray (1941) drew important conclusions affecting the methodology of handling early generations of wheat from a spaced-plant and solid-stand study of interplant competition in two cultivars differing in seed size and maturity. Selected portions of what they found may be summarized as follows:

1. Initial seed size differences, large and small, of the same variety sown alone had little effect on characters measured with one exception: large seeds in the late variety had an advantage in producing plants with more early-season tillers.
2. When large and small seeds were placed as alternating spaced plants in

any competitive situation, there were profound effects of competition affecting all characters in both early and late varieties, but principally affecting tiller number, ear number, and grain yield. The competition could be traced to seed weight and rate of growth differences. The situation, even in case 1 above (same variety), changed when the large and small seeds were alternated. For example, large and small seeds when grown alone, yielded 74.61 and 83.86 grams per plot, respectively; in competition the yields were 90.58 and 62.60, respectively. Strong competition was associated with large seeds, late maturity, and large tiller numbers.

The effect of spaced-plant competition shown over so many combinations of seed size, variety, and maturity prompted Christian and Gray to state: "It is clear that in many wheat crosses, sown at close spacing, particularly in the early segregating generations, the actual yield of an individual plant may bear very little resemblance to the potential yield of its genotype."

Recognizing the unreliability of single plant measurement and selection, the authors made these suggestions to ameliorate interplant competition:

1. Grade seeds by weight or size and sow seed lots of one size.
2. Use wider spacings.
3. Take other traits into consideration; that is, either correct for or isolate individuals into groups that differ, for example, for maturity or height.

In studies with four soybean varieties, Probst (1945) found spacing to be less influential relative to a range of characters, including yield, than it is in the cereal grains. The closest spacings, that is, 1 in., had to be avoided to ease lodging and delayed maturity problems, and highest yields were obtained in the 2- and 3- in. spacings. Breeders should be aware, however, that a variety × spacings interaction existed that, although significant in only one year out of four, surprisingly was highly significant over the mean of the four years; nevertheless, yield rankings remained quite stable and consistent.

Sakai and Suzuki (1954) examined the association between population density and competition, and concluded that the relationship was simply a function of distance between individuals.

Kaukis and Reitz (1955) reported a two-year Nebraska study of spacing and tillering as it affected yield in five different oat cultivars. The spacings were 2.5 and 5.0 in. within rows 7 in. apart. Measurements were taken on total and grain-bearing tillers, height, and yield per plant. The average total number of tillers was apparently similar for cultivars, but there were significant differences in productive tillers formed by the cultivars. Three early varieties were 88% efficient, relative to total tillers, in producing grain, and the other two later-maturing varieties were at about a 60% level (maturity is relative; all the varieties would have been considered early at more northern latitudes). Varietal differences in tiller number lessened as density and competition decreased.

Plants produced an average 75% more grain at the 5-in. spacing than at the 2.5-in. spacing. However, it must be mentioned that there were only half as many plants at the 5-in. spacing, and that yield per area favored the 2.5-in. spacing. The increase in yield from the 5-in.-spaced plants over the more dense spacing was partitioned in this manner: 77% came from increased tillering, 16% from increased yield per tiller, and 7% from their interaction.

Grafius (1956) addressed the phenomenon of cultivar compensation when faced with changes in plant density. Using 40 varieties of oats and monitoring changes in panicle number per plant and unit area, he found that most of the 40 varieties readily adjusted for differences in stand by changes in panicle number (which traces to tillering). The situation was analogous to the relationship between rate of seeding and yield. The few exceptions noted suggested that some varieties respond differently at the optimum density.

An arbitrary population plant density must be specified for a corn yield trial. For any given nursery, where density is not varied, the response of each entry represents a G × E interaction, which may not be its best response in terms of highest yield. Duncan (1958) explored this situation and concluded, "The logarithm of the average yield of individual corn plants making up a population bears a linear relationship to the population. Therefore, only two yield-population values are needed to estimate yields at any other population within the linear range."

Kaufmann and McFadden (1960), in spaced tests of large and small barley seeds involving interplant and interrow competitive situations, found that the advantage of large seeds over small, shown throughout, including the control series, increased under competition. In general, a greater number of heads were produced under competition.

Distinct differences in the reaction of corn hybrids to population density have been recognized for some time. Stinson and Moss (1960) brought together 11 hybrids with known reactions to thick plantings, six tolerant (T) and five intolerant (I). These hybrids were grown under two conditions, shade and sun, in adjacent field areas. The density was equivalent to 13,500 plants/acre.

A typical shade reaction, increased height, was observed. The effect on yield was substantial: the yields in sun were 104 and 102 bu/acre, and in shade 84 and 60 bu/acre, for T and I, respectively. Thus the authors concluded that differences in reaction to high population density were related to differences in the capacity of varieties to use available light and translate this into grain production. In other words, there was a direct relationship between tolerance to thick plantings and tolerance to shade.

The research study with soybeans by Hinson and Hanson (1962) may not have direct application to cereals because of differences in plant structure and physiological response to variations in spacing. Response to photoperiod appeared to be the dominating factor in a soybean genotype's ability to make use of space and to compete. Perhaps there is an analogous situation in the cereals with respect to tillering and height. The authors attempted to differentiate between "spacing effects" in pure stands and "competition effects" in mixtures.

In the cultivar mixtures, significant yield differences (both directions) were found in six of eight mixtures; however, the effect was of a compensating nature in that the mean performance of mixtures was about the same as the mean of their components in pure stand. Overall, the dilemma of the findings was expressed by the authors: "At wide spacings the errors from competition effects become smaller, but the errors from a differential response to spacings are increased. It is apparent that yield data on individual plants selected from mixed populations at any spacing would have a very limited application to row plantings."

Siemens (1963) presented a review of the effect of varying row spacings, where within-row spacing was held constant, on tillering, yield, and other plant characteristics. Tillering increased with wider row spacing, as did the grain return per seed planted, whereas yield per unit area decreased.

Stringfield (1964) developed arguments for the position that a major objective of corn breeders should be breeding for tolerance to crowding. His major requirements for likely success were that genotypes show a range of variability for the trait, and that high density be a requirement for extracting highest yields from productive soils.

Studies in plant competition usually begin with designated and proportional seed numbers. Occasionally, after-planting happenings, such as differences in viability or emergence, can alter drastically the planned confrontations of individuals. Sometimes adjustments are possible, for example, replanting or replacing missing plots, and sometimes the researcher is interested principally in the main population effect.

Mead (1966) presented a neat concept to deal with after-the-fact situations, where the variations in stands can be measured, allowing the competitive effects study to continue. He did this by making each plant a focal space point and surrounding it with space determined by lines equidistant to all neighbors, or "the polygon of a plant, P, is the *smallest* possible polygon containing P, whose sides are perpendicular bisectors of the lines joining P to other plants."

The result looks like a botany textbook view of plant cells. Mead discussed a simple interpretation of a quantitatively determined area surrounding each plant, its relations with multiple neighbors, and possible interplant effects. I find the idea intriguing because it lends itself to a simple solution to the problems of variability in stands and of missing plots. Such cases need not invalidate a study; instead the existing points can be used to arrive at a solution. The application would work best in studies involving large populations.

An analysis of Kirby's (1967) study upon the effect of plant density upon growth and yield of four barley cultivars must be preceded by recognition of two influential factors that acted to shape general conclusions. These factors were the differential responses of varieties, and the differential responses of two-row and six-row types ('Moore', a six-row type, fell under the influence of both factors). Varietal differences were found in seedling establishment, maximum tiller number, percentage of tillers that formed ears, and other traits confounded with density.

The density treatments were named for the distance between seeds, 3.5, 5, 7, and 10 cm, which corresponded to sowing rates of 280, 140, 70, and 35 lb/acre, respectively (commercial sowing rates range from 100 to 150 lb/acre).

Total dry matter increased with increasing density, being highest, though only slightly, at the 3.5-cm distance. This was not the case with total grain yield, which was highest at the 7-cm treatment, thereafter declining and being lowest at the 3.5-cm distance (the lodging of 'Plumage-Archer' at 3.5 cm may have been a factor). From this it would seem that farmer practices in seeding are correct—perhaps a bit on the generous side. Number of ears per unit area increased with density at the same time that weight per 1000 kernels was decreasing.

Most impressive was the flow diagram (his Figure 6), shaped like an arrowhead, which showed the compensation occurring through plant development stages, resulting in a meeting at the point of all components at harvest time. The author discussed possible ways, such as early sowing, to disrupt the pattern so that the final product would be increased.

The density × variety interactions found emphasized the importance of testing cultivars in ways consonant with farming practices.

Puckridge and Donald (1967) examined the behavior of the 'Insignia' wheat cultivar over six growth harvest periods, in plots of five densities ranging from 1.4 to 1078 plants/m^2. The authors reported on a wide range of phenomena including leaf area index and photosynthesis. One phenomenon, closely related to density, was the height changes due to etiolation that were evident at the tenth week; and at the twentieth week the heights for the five increasing-density plots were 78, 89, 109, 116 cm, and lodged, respectively. High grain yields per plant at low density could not compensate for low stand. Interplant competition was evident at high densities Nos. 4 and 5 by the tenth week, density No. 2 showed interplant competition after the seventeenth week, and density No. 1 suffered no interplant competition at any time. The yield of grain per unit area was highest at intermediate densities.

I was intrigued by several of the observations and data accruing from the study which seemed to have unique relationships to the competitive situation. Table 1 of Puckridge and Donald was of special interest because the density No. 5 column (1078 plants/m^2) provided a measure of interplant competition that may be unique and specific to the pure line tested. In a 1967 paper (Jensen, 1967), I proposed the need for examining a problem by "exploding" it, and I have used this concept for understanding a cultivar, measuring its characteristics under spaced-plant and crop-density conditions. Puckridge's and Donald's paper provides some of these answers: the self-aggressiveness of Insignia in crowded conditions resulted in a loss of plants after 26 weeks, from 1078 to 447. One would like to know comparable performance figures of other cultivars. Perhaps we may anticipate the development of a standard measured competition test that incorporates other parameters such as those Puckridge and Donald studied.

Despite a wide range of spacings (18–72 cm) and population densities (75,000–346,000 plants/ha) of grain sorghum, Hegde, Major, Wilson, and Krogman (1976) were unable to detect any consistent effect on grain yield in a study reported in that year. Matching or canceling compensation among yield components was such that the authors concluded, "The choice of row spacing and population density in field production will, therefore, depend on convenience factors related to crop management."

Baker (1977) compared 'Neepawa' and 'Pitic 62' spring wheats; their 1:1 mixtures were sown at five rates of seeding, escalating from 12 kernels per meter row length to a density approximagely nine times greater. Tests were conducted in two years. The yield results from the two years were markedly different. In 1974 the mixtures were equal or exceeded the midparent average at all densities, averaging 9.9% above, and exceeded the higher-yielding parent at the two densest sowings. In 1976, however, the mixtures were equal or lower than their component means in all cases, showing an average loss of 5.0%, and they did especially poorly at the highest two densities. A striking difference in survival in pure and mixed stands was shown by the two cultivars in 1976: Pitic 62 had the same average survival, 41%, in both pure and mixed stands; Neepawa showed 56% survival in pure stand but 71% in the mixture. Pitic 62 consistently outyielded Neepawa. The yield of the mixture reversed itself at the two highest densities in the two years, in 1974 being higher than either cultivar and in 1976 lower than either.

These results are confusing and difficult to interpret. Obviously, there was a G × E interaction involving years. Neepawa showed high competitive ability in the mixtures without an expected rise in yields. Part of the answer may lie in tillering but more likely it lies in seed size differences. Neepawa had a much heavier kernel than Pitic 62, an 18.3% heavier average two-year weight per kernel. Although precise seeding rates were specified, it is not clear whether the 1:1 mixture was balanced by weight or by numbers, how the sowing was done, or how the actual placement numbers were realized. With different size seeds, hand planting or metered drilling would be necessary to produce equal densities. Equal weight of seeds would result in disproportionate numbers, as would the use of volume feed equipment typical of early farm drills. If seed size differences were not adjusted at the beginning, an unknown confounding might be introduced.

Briggs and Faris (1978) compared space-planted and solid stands of barley cultivars. They found that from 10 to 20 spaced plants were needed to gain consistency in yield response. If this is a general truism, the possibility of using single plants to measure yields would be discredited substantially. This is directly related to the question of whether the yields of single F_2 plants have real meaning. Regarding the relationship between spaced plants and solid stands, no conclusion could be drawn inasmuch as they found high correlation at one location and none at another site.

Martin, Wilcox, and Laviolette (1978) increased planting densities to encourage higher seedling losses in soybeans. Over four generations, plant losses from the original populations of three crosses were: low density, 26%, 18%, and 15%, and high density, 62%, 58%, and 47%, respectively. Observation confirmed that viability changes affected gene frequencies for different traits. From a methodology standpoint the large losses at high density are intriguing, for a breeder would want to know what *kind* of genotypes—desirable or undesirable—were being lost. With this knowledge the breeder could choose a favorable density level for early generation nurseries.

Nass (1978) concluded, after studies of selection efficiency in spring wheat crosses grown under two density regimes, that if lines after selection are to be tested under conditions comparable to commercial production, then selection under high population densities, rather than space-planted conditions, is the procedure to use.

Wilcox and Schapaugh (1978) used soybean isolines differing in maturity by 14 days to assess intergenotypic competition as affected by density in hill plots. The spacing ranged from 1 to 25 ft^2/hill and 15 spacings were tested. The four treatments were 'Clark' and 'Clark-e2', each alone, isolines in alternate hills, and two sets differing according to which isoline was in the innermost circle. The findings were that (1) high density resulted in one day earlier maturity; (2) at all densities, when the late isoline was bordered by the early isolines, its yield was statistically significantly higher than when self-bordered; and (3) the converse was true, early bordered by late yielded less than when self-bordered.

Spitters (1979), according to Langton (1985), found that although genetic variance in barley mixtures increased, environmental variance did not. It was also noted that there was a poor correlation between performances in spaced and in crop-density trials.

During my plant breeding career I had a continuing belief, perhaps irrational, that selection under stressed environments was worthwhile. This may have arisen from the fact that in New York we so frequently endured such environments, given the soil variability in the glacier-scraped fields. More than that, there was a genuine hopeful feeling that survivors among plants beset by adversity were special. In some cases significant gain was obvious; for example, in seasons when there was differential winter kill natural selection for cold hardiness was a positive and easily observed phenomenon. In fact, the greatest progress in genetic advance I ever witnessed came one winter as a result of a "disaster," when more than 98% of the winter barley nurseries suffered winter kill. The very few survivors constituted an elite continuing breeding population.

Another standard procedure in my breeding projects concerned density: all bulk hybrid population plots were sown at twice, and occasionally four times, the commercial sowing rate ("...we plant our composites at double the normal rate, thus creating a highly competitive situation"—Jensen 1978). This procedure violated the conventional wisdom on population density for segregating generations of cereals.

Thus it was that I viewed with special interest Troyer and Rosenbrook's (1983) article on corn performance testing at high plant densities, which described a testing and selection program deliberately designed to operate in a stressed environment. They compared the results of 84 corn performance tests grown over a nine-year period which featured significant density differences of 51,600 and 64,500 plants/ha. The higher figure is significantly above optimal density as defined during the period of this research. The high-density testing was chosen in the hope that it would aid in the selection of genotypes tolerant to the imposed stress for traits such as improved yield, lodging, and ear drop resistance.

The results at first glance appeared to be negative: yield test means and hybrid F values were reduced, whereas yield ranges among hybrids were increased, and there were greater incidences of barrenness, stalk breakage, and ear droppage. In reality, the increased range among entries, and the more evident deficiencies, resulted in more efficient selection. The authors found that a level of 56,000 plants/ha was excellent to differentiate these morphological traits, and fewer comparisons were found necessary to reveal the differences.

Critical tests of procedures must rest on their contribution to the ultimate objective of performance in the commercial arena. The authors presented the evidence of successful stress-resistant hybrids recently developed under high-density growing conditions.

Roos's (1984) actual and simulated study of the effects of storage and regeneration of seed stocks has important meaning to competition studies of genotypes in mixtures. To use an obviously extreme illustration to make a point, long-term storage of seed mixtures could produce crop census results in the first growing year that might show rapid elimination of one or more of the components. In the absence of knowledge about the seed storage, this loss might be attributed to intrapopulational competition in the field. Roos found that artificial aging tests identified certain bean genotypes with significantly lower seed-keeping qualities than other genotypes. When the germination level of a population fell to 50% or less, a lowered potential seed vigor of such genotypes resulted in their gradual elimination in storage without the influence of crop population competition. This suggests that competition studies involving regeneration of storage seeds ought to include a consideration of the storage factor. In the simulated studies, Roos developed equations that took into account the changing viability differences over time in mixed genotype stored seed. These formulas make it possible to adjust the planting sample to the original equal-seed composition.

Kempton, Gregory, Hughes, and Stoehr (1986) showed that interplot competition of different-height triticale and wheat genotypes was significant in *large* field plots consisting of seven rows 5.5 m long, spaced 17 cm apart, and separated from neighboring plots by a 50-cm alley. The competition effect amounted to average yield decreases of 1–2 g/m^2 for "every centimetre by which the plot was exceeded by the mean height of its two neighbors." Plots taller than neighbors showed yield increases. The authors cautioned that ordinary randomization or increased replication would not eliminate this bias. They made two recommendations: (1) use of guard plots, and harvesting only the center of plots, and (2) pretrial grouping of genotypes by height. *Apropos* methodology, the second recommendation to cluster genotypes is one that could be implemented from notes taken in nurseries just prior to lines entering yield nurseries.

RELATION OF SELECTION METHOD TO GENERATION AND DENSITY

Fuizat and Atkins (1953) analyzed six characters for genetic and environmental variation in barley crosses using the pedigree method in a study aimed at providing early generation guidance to the breeder. F_2 tests were made concurrently on the P_1, P_2, F_1, and F_2. Based on their data and observations, individual plant selection for earliness and height could be done in the F_2 with confidence; however, selection for number of heads, kernel weight, and the critical trait, grain yield, should be based on progeny tests. That is, spaced plant yields were not reliable.

Doggett (1972) used dense stands in sorghum selection experiments and found that the conditions forced up the height. He concluded that continual selection at high density was a mistake and suggested a better procedure would be a two-density–time scheme where selection first would be at low density and in later generations at high density. This seems a reasonable suggestion, but I would like to assure breeders that high density does not preclude selecting for characters that are at a disadvantage under those conditions. I found that there was no difficulty obtaining all the dwarf and semidwarf wheat plants wanted, simply by being aware of the high-density situations and "going in after them" (literally, spotting a desired plant and bending, reaching the plant at ground level to cut off a short culm). This can be done in F_2, but I found no necessity to pull short individuals from the populations before F_4 or F_5, except in the case where semidwarf types were the only type desired and they might as well be selected and rebulked at the F_2 level, and the remainder of the bulk plot discarded. This has not been the universal experience of breeders, however, and one must adapt procedures to specific crops and situations.

Doggett remarked also on the effect of high-density competition on male sterile plants. They formed a smaller proportion of the population than in less dense populations, indicating lower fitness or competitive ability.

An indication of the compensating influences inherent in G × E situations involving plant spacing was elucidated by Chebib, Helgason, and Kaltsikes (1973). They found that the higher-density conditions that increased competition between genotypes also increased genetic variance while at the same time reducing the nongenetic variance observed in spaced plant situations. Using seeds of uniform size increased selection efficiency at comparable densities.

Hamblin, Knight, and Atkinson (1978) used a moving average model to assess variation of individual plant yields within selection plots of barley. The study included Gardner-grid stratification, which governed the selection procedures. They found plant-to-plant variation to be less at low density than at higher densities. The higher variation at high densities was not expected, and the authors suggested it may have been influenced by seed size variation and greater interplant competition. Overall, they considered competition to be more important than systematic environmental variation within plots in determining individual plant yield.

I would add a note of hope to the comments of Hamblin et al., who stated:

> It would appear that single plant selection for yield is always liable to be confounded by genotype–seasons interactions, and if carried out at crop densities competition may also be important. If selection is carried out at non-competitive densities then the selected genotypes need not necessarily be high yielding at crop densities.

The bright spot is that the latter selected genotypes, if taken in adequate numbers, should include many that would be high yielding at commercial crop densities.

Baker and Briggs (1982) conducted a three-year test of the response of 10 cultivars of barley grown at five different spacings corresponding to 1.6, 6.2, 25, 100, and 400 plants/m^2. Overall interest was on the point of optimal density for selection; however, they did not restrict the research to only grain yields but evaluated six other plant characteristics, including harvest index. The lowest within-plot variance, an indication of stability, was found at two density levels for different characters: 6.2 plants/m^2 for height, kernels per spike, and kernel weight, and 400 plants/m^2 for spikes per plant, shoot weight, grain yield, and harvest index. The density difference is surprising for these two determinations. Harvest index and height were not significantly related to density; this, too, for height is surprising, because 400/m^2 should have resulted in etiolation. The authors concluded that optimal density for single plant selection would be 6.2 plants/m^2 (40 × 40 cm) because (1) reduced within-plot variance for all characters studied was found, and (2) cultivar differences were greatest at low densities.

There was significant cultivar density interaction for certain characters, but when the cultivars were clustered into two homogeneous groups of seven and three cultivars, this interaction within groups disappeared. Because of the absence of interaction within groups, no significant changes in ranks, and the initial absence of interaction for the other four characters, the authors concluded that there were no density-related changes in relative cultivar performance. I would have liked to see more information on this point because the authors have come close to producing a "fingerprint" profile for describing a cultivar's performance in different environments, something each research station might do for each cultivar it originates. Baker and Briggs's study is significant because it measures performance across five density environments (however, only the cultivar totals are given). It is clear that cultivars did respond differently to changes in density, as was shown by the separation into two homogeneous groups.

The paper by Kyriakou and Fasoulas (1985) is complex and important. It assessed the relation between yielding and competing ability in cross-pollinated rye by applying the same series of three selection intensities to the same populations grown under two densities. The densities were: "competition," plants 15 cm apart, and "no competition," plants 90 cm apart, implemented in Fasoulas' familiar honeycomb design. The three selection intensities were 14.3%, 5.3%, and 1.6%. The areas were delineated by moving hexagonal grids scaled for each selection intensity.

Of surpassing importance in their results were the yields obtained from growing the progenies of plants selected under the three intensities in "competition" and "no competition." Taking the source population yield of 1057 g/plot as a base, the yields went in opposite directions, as shown by the following graphic rendering of Table 1 of Kyriakou and Fasoulas:

Intensity	**Yield (g/plot)**		
1.6	1152	↑	90 cm
5.3	1114		
14.3	1100		no competition
Source:	1057		
14.3	1060		15 cm
5.3	1018		competition
1.6	1002	↓	

The implications for methodology if these findings are correct are sobering. The negative correlation between competing and yielding ability as expressed above means that selection for high yield should be done under noncompetitive situations where selection is based on yield data. Most efficient results were obtained at high selection intensity (low percent taken); however, a relaxation of intensity would be desirable because there were quite good yielding plants at 5.3% intensity. Adding visual judgment would also be desirable.

This apparent reversal of accepted selection strategy is not new. Kyriakou and Fasoulas listed three studies where breeders found that the strongest competitors were the poorest for yield when grown alone [Wiebe, Petr, and Stevens (1963) with barley; Kawano and Thung (1982) and Kawano, Tiraporn, Tongsri, and Kano (1982) with cassava]. I can add others: Jennings and de Jesus (1968), Hamblin and Rowell (1975), Rajeswara Rao and Prasad (1984), Langton (1985), and Sandfaer (1970). Sandfaer wrote, "In the mixtures of T. Prentice and Freja the variety with the lowest reproductive capacity in pure stand dominated the mixture when the mixture was repeatedly sown."

A second illuminating result was the different patterns of yield distribution emanating from the same populations under competition and no competition. Without competition (90 cm) the average plant yields were 7.5 times larger (range 0–210 g/plant) than yields under competition (15 cm, range 0–45 g/plant). It can be seen that the narrow range for selection under competition would cause problems for the breeder. Incidentally, the positive skewness evident in the competition graphic requires no sophisticated explanation; the narrow range and the abrupt left-side zero-yield barrier probably account for it.

A third element of this paper that intrigued me relative to methodology is the continuing use of Fasoulas' honeycomb arrangement. The use of moving hexagonal grids is sound, and reminiscent of Gardner's use of field gridding. Breeders like to say, "If only the environmental variances could be muted this would be a great system." However, we must not lose sight of the fact that the honeycomb system is a serious effort to inactivate or neutralize parts of the environment so that true genotypic responses might be seen. Recent papers have, in my opinion, enhanced the status of the honeycomb arrangement.

The experiences of Langton (1985) with intergenotypic competition should erect warning signals to plant breeders dealing with selection from heterozygous and heterogeneous populations. Langton took two commercial chrysanthemum cultivars, "Pollyanne" and "Hurricane", which in pure stands were judged to be of equal commercial value. However, when grown in 50 : 50 mixtures, aggressive Pollyanne suppressed the flower yield of Hurricane. An alternative explanation is that Hurricane enhanced its partner.

The denouement in this scenario rests on the point that Langton had prior knowledge that the two cultivars had similar economic value and opposing competitive abilities. The analogous situation, in early generations of a segregating population, involves intergenotypic competition, which at times pits an aggressive genotype against a nonaggressive one. The difference here is that the breeder knows neither the competitive abilities nor the potential economic values of the genotypes, and will, unless operating on a contrarian theory, select the competitive genotype and discard the weaker one, losing forever a possibly valuable genotype. Langton stated:

> Individual plant selection in Pollyanne type–Hurricane type mixtures would, inevitably, yield a heavy preponderance of Pollyanne types. Subsequently, the breeder would be disappointed in the performance of these growing in pure stands and would have little opportunity of seeing the enhanced performance of Hurricane types in monoculture.

Although Langton viewed this as an artificial situation, it cannot be denied that the bulk hybrid population might be a favorable environment for such no-win confrontations.

Incidentally, I referred to a contrarian approach—although not named as such, it is a valid approach to some selection situations. The paper by Kyriakou and Fasoulas, just reviewed, contains several references to cases where the indicated and seemingly obvious best plant selections were bypassed or overlooked and instead the *poorest* solution was advocated as the best strategic choice.

MEASURING THE EFFECTS OF COMPETITION

Schutz and Brim (1967) tested a nine-hill field plot that removed about 70% of the competition effect found in an unbordered hill test. The design is a three-plot × three-plot square: the center plot is competition bias-free, and the four other harvested plots are in the clock positions 3, 6, 9, and 12 o'clock and are 62.5% bias free.

Large competition effects among soybean cultivars were observed in seed yield, number, and efficiency (similar to harvest index). Certain genotypic combinations showed "specific combining ability" with overcompensatory effects. Schutz and Brim defined four types of compensatory response to competition, namely, neutral and complementary, where yield is equal to expected; undercompensatory where yield is less than expected; and overcompensatory, where yield is greater than expected.

The nine-hill plot proposed by Schutz and Brim is intriguing, harboring suggestions for other uses. It suggests ancestry tracing back to Hay's Centgener method and Fasoulas' honeycomb designs, while adding logical credibility. It might be used with the center plot surrounded by self for precise measurements of a genotype's traits. It could provide a one-on-one match for line evaluations for a multiline variety. It may have uses in early generation situations where seed amounts are limited. It might also serve as a standard vehicle for harvest index determination, based on the target genotype in the center plot. The authors suggested other uses.

Boffey and Veevers (1977) provided instructions for generating balanced designs for two-variety competition tests where hill plots (and presumably single plants) are used. Priorities of a balanced design are the full representation of all competitors, including the pure stand, and ease of repeatability.

The ideas of Fasoulas (he has written several articles but his 1973 and 1977 papers are representative) have fascinated plant breeders, who have been shown innovative field designs that appear to offer precision results unattainable in ordinary field nurseries. The designs, called "honeycombs," are based on hexagons and are recommended for a myriad of purposes ranging from competition testing to early generation yield testing. I have enjoyed reading his articles but recognize that there is lack of agreement as to the close relationship between theory and practice.

Mitchell, Baker, and Knott (1982) chose to make an independent evaluation of the honeycomb system by using it for selecting for single plant yield in durum wheat. The F_2-embryo seeds of three crosses were planted in the hexagonal honeycomb pattern under two interplant spacings, 30 and 60 cm, and single plant yield was determined at harvest. Bidirectional selection for high and low yield was practiced in two crosses, A and B. To be selected for high yield a plant had to equal or surpass "the maximum for its six neighbors." I interpret this to mean the highest yield among the six neighbors. Presumably the selection for low yield had to yield below the lowest-yielding of its six neighbors.

Tests for yield were made from five composites formed from the high- and low-yielding selections at 30 and 60 cm, respectively, plus one random group chosen equally from the two spacings and exclusive of the above selections.

In the third cross, C, all plants of the two spacings were harvested and weighed, and the lines increased for an F_4 yield test for correlations of F_2 plant yield with F_4 progeny yield. In addition, simulated selections for high and low yield were conducted on the F_2 results, 33/245 each for the 30-cm spacing and 36/267 each for the 60-cm spacing. The difference of the means between selected high and selected lows would be considered the selection response for each density. The results were as follows:

1. Selection for single plant yield was more effective at the wider, 60-cm spacing.
2. The correlations between F_2 single plant and subsequent F_4 plot yields were nonsignificant at the 30-cm spacing but significant at the 60 cm-spacing (r = .13–.17).
3. The heritability of single plant yield was 13–17%.
4. Selection of one out of seven F_2 plants (14.3% intensity) resulted in an eventual 4% gain in yield as measured in the F_4 plot yields.
5. Selections for high yield were more effective than for low.

The authors' conclusions gave some validity to Fasoulas's contentions, but because the gains were about the magnitude expected from mass selection, Fasoulas' methods would have to be discounted because of the enormous work load required. This, I think, is an acceptable conclusion to breeders who sincerely cling to the hope that precision single-plant testing has merit. The question arises as to whether the honeycomb system could be modified to lessen its load requirements. Suppose F_2 single plants were visually selected from a nonhoneycomb nursery (unidirectionally for high only, using all means to effect discards on any basis). The following season, the selections would be arranged in an F_3 hill plot honeycomb. The effort required would be only a fraction of the quintessential Fasoulas' honeycomb system. Or, a honeycomb hill nursery could be used as the first yield test of lines coming from a pedigree or bulk population system (F_4–F_6 lines), and could either replace or supplement head rows.

I am reminded of research on competition underway at Ithaca in the late 1960s, research never completed or published, but which was "state of the art" for that era. I invented a spiral planting system based on seed positions that expanded each next position by 1 mm, forward and outward, thus creating what I called an "incremental spiral." The seemingly miniscule movement of seed positions nevertheless resulted in plant spacings close to 1 ft at the outer edges of the largest template obtainable, a piece of vinyl floor covering. At the center, of course, the positions were closer than grain size, resulting in a piling up of seeds at the center, which gave a remarkable demonstration of etiolation as the height of the circle of plants was "pulled" upward to a point at the pivot center. Seed position holes were bored in the template, which was placed on the carefully prepared soil area, and over this was built a simple portable wooden superstructure for support as seeds were planted. The research was aimed at studying the behavior of putative multiline entries which, as randomized groups, could be repeated endlessly, that is, to the limits of the template. The spacing ranged from absolute crowding in the center to distant neighbors at the periphery. An interesting feature was that the space expansion was so slight, with only 1 mm being added to the distance from one position to the next, that a group of randomized entries could be repeated several times and still be considered as replications. Thus the entire template could be gridded into meaningful sections for certain experiments. At harvest time the circles of plants were simply "unraveled" by pulling plants from the outside and moving into the spiral.

These studies were not completed because I found that (1) despite extreme care the environmental variance in the field was too large, and (2) I did not have the staff to handle the magnitude of effort required over and beyond conducting the breeding program. I considered modifying a greenhouse room and installing a large turntable to control nutrient feed, but this proved not possible.

COMPETITIVE ABILITY AS AN INHERITED TRAIT

Sakai and Gotoh (1955) brought forth a distinction important to plant breeders: heterotic vigor of characters may not be the same thing as the competitive ability of a genotype; in other words, competitive ability may be a separately inherited plant trait. This was deduced from research between pure line parents and F_1s of barley, where certain characters showed heterosis, but competitive ability of the F_1s did not respond in the same fashion; in fact, it was generally inferior.

Tucker and Webster (1970) referred to earlier research (1965) which had shown that dominant white-seeded lima bean plants (*WW* and *Ww*) showed fitness levels only about one-half that of green-seeded plants (*ww*). They investigated this curious situation by removing lines from the population used in the earlier study after 10 and 11 years of bulk population growth. These lines were increased as a consequence of being grown for identification of the seed color genetic status, following which 14 lines were drawn at random from each category (*WW*, *ww*, and segregating), and yield trials conducted for three years. The results showed no significant differences in mean yields, leaving the authors with a paradox of extremely low fitness yet equivalent yielding ability. The authors concluded, "Results indicated no relationship between yielding in a pure stand and survival ability in a bulk population."

They stated these further conclusions: (1) bulk populations could produce highly competitive but not necessarily productive lines; (2) desirable traits, such as white seed color, should receive a different kind of competition; and (3) large populations should be grown to foster linkage breaks.

EFFECTS ON DISEASE RESISTANCE

A curious aspect of competition was revealed by Suneson and Stevens (1953): disease resistance does not necessarily enhance survival resistance in a mixture. On the basis of known parental input of resistance to barley scald into a population, about 11% resistance would have been expected in the progeny without any selective advantage. Yet in the F_{12} only about 5% resistant plants were found in random picks. One explanation might be a linkage of susceptibility genes with unknown genes having a survival advantage.

The same phenomenon was observed by Hanson, Jenkins, and Westcott (1979) in studies with barley, where "mildew was at the same level in high and low yielding groups." This is an unusual situation, where an obviously undesirable and discardable character is shown to have no correlation with yield. Of course, one could take a purist view that it would be better to have resistant selections and, because there was no yield penalty, why not?

Tiyawalee and Frey (1971) had somewhat the same experience with another pathogen, that causing crown rust of oats. Using mass selection for heavy seeds they were able to raise the gene frequency for resistance from 20–21% to about 35% over seven generations. What is intriguing and puzzling here is that the 35% level was attained after only three generations, and no further progress could be made in the following four generations.

Ross (1983) studied the dynamics of soybean biblend populations in which the two components represented resistance and susceptibility to soybean mosaic virus. Mixture proportions were 0, 10, 20, 60, 80, and 100% resistant, and 0, 20, 40, 60, 80, and 100% resistant. He found the following results:

1. Blend yields were related positively to percentages of resistant plants.
2. Yields per resistant plant increased as resistant plant percentage in blend decreased.
3. In contrast, susceptible plant mean yields showed little response to changes in blend proportions.

The differences in reaction by resistant and susceptible plants permitted resistant plants to show overcompensation.

THE EFFECT OF MORPHOLOGY (LEAF ANGLE EFFECT)

The relation of ideotype structure (leaf angle) to density and productivity was shown in wheat by Stoskopf (1967), using short-strawed wheats: two narrow upright-leaved Cornell selections and 'Wisc. 256', and the tall, droopy-leaved Canadian cultivar, 'Talbot'. Talbot was known to yield more than the semidwarf types but the research interest was in the response of the different types to density arrangements. Wide row spacings were 7 and 9 in. and narrow row spacings half those. The three seeding rates were 1, 2, and 3 bu/acre.

The results showed (1) narrow rows of all entries outyielded wide rows by 10%, (2) the upright-leaved wheats showed more yield increase in narrow rows than did the tall droopy-leaved cultivar, by 13–7%, respectively, and (3) the increase was most marked at the 2 and 3 bu/acre rates.

Upright-leaved cereals tolerate higher-density plantings because they have better light interception and less shading than droopy-leaved plants. The authors stressed that upright-leaved genotypes must first show competitive yielding ability, perhaps in routine standard-row-width nurseries, after which they might be tested in narrow-row and rate-of-seeding tests for evidences of still greater productivity.

HOW TO AVOID COMPETITION

As long ago as 1923 the simpler forms of competition in comparative nursery trials were recognized. Kiesselbach (1923) summarized his findings from a series of experiments with these advisory comments on how to avoid the consequences of competition:

1. Group similar sorts.
2. Use and discard border rows.
3. Overplant, then thin to uniform stand.
4. Use only hills with full stand and neighbors.
5. Sort entries according to optimal planting dates and rates.

Schwendiman and Shands (1943) reported on a unique characteristic of 'Vicland' oats, a type of dormancy manifested by delayed germination and emergence. This trait could be useful in competition studies. The Cornell cultivar 'Craig', derived from a cross with Vicland, also exhibited delayed emergence.

The following papers have not been seen, but I include them with short synopses of their contents:

Cardon (1921) found that cereal grain mixtures yielded less than the mean of their components.

Sakai (1961) discussed differences between intergenotypic and interplant (density) competition.

Reddy and Prasad (1977) brought out the fact that wheat genotypes that varied in height and lodging reaction in pure stands showed greater stability in mixed stands.

Reddy and Prasad (1980) showed that wheats differing in height exhibited yield advantages in mixtures.

The material in this chapter is intended to be absorbed as background to the next four chapters, and to slowly release its contents to the reader as the chapters are read.

REFERENCES

Allard, R. W. and J. Adams. 1969. Population studies in predominantly self-pollinating species. XIII. Intergenotypic competition and population structure in barley and wheat. Am. Nat. 103:621–645.

Baker, R. J. 1977. Yield of pure and mixed stands of two spring wheat cultivars sown at five rates of seeding. Can. J. Plant Sci. 57:1005–1007.

Baker, G. A., Jr., and J. Christy. 1964. Analysis of genetic changes in finite populations composed of mixtures of pure lines. J. Theor. Biol. 7:68–85.

Baker, R. J. and K. G. Briggs. 1982. Effects of plant density on the performance of 10 barley cultivars. Crop Sci. 22:1164–1167.

Boffey, T. B. and A. Veevers. 1977. Balanced designs for two-component competition experiments. Euphytica 26:481–484.

Briggs, K. G. and D. G. Faris. 1978. Influence of sample size on the relationship between the yielding ability of space-planted and solid-seeded barley cultivars. Can. J. Plant Sci. 58:263–266.

Byth, D. E. and B. E. Caldwell. 1970. Effects of genetic heterogeneity within two soybean populations. II. Competitive responses and variance due to genetic heterogeneity for nine agronomic and chemical characters. Crop Sci. 10:216–220.

Cardon, P. V. 1921. Grain mixtures and root crops under irrigation. Mont. Agric. Exp. Stn. Bul. 143.

Chapman, S. R., R. W. Allard, and J. Adams. 1969. Effect of planting rate and genotypic frequency on yield and seed size in mixtures of two wheat varieties. Crop Sci. 9:575–576.

Chebib, F. S., S. B. Helgason, and P. J. Kaltsikes. 1973. Effect of variation in plant spacing, seed size, and genotype on plant to plant variability in wheat. Z. Pflanzenzuecht. 69:301–332.

Christian, C. S. and S. G. Gray. 1941. Interplant competition in mixed wheat populations and its relation to single plant selection. Aust. Counc. Sci. Ind. Res. J. 14:59–68.

Dewey, W. G. and R. S. Albrechtsen. 1985. Tillering relationships between spaced and densely sown populations of spring and winter wheat. Crop Sci. 25:245–249.

Doggett, H. 1972. Recurrent selection in sorghum populations. Heredity 28:9–29.

Donald, C. M. and J. Hamblin. 1976. The biological yield and harvest index of cereals as agronomic and plant breeding criteria. Adv. Agron. 28:361–405.

Doney, D. L., R. L. Plaisted, and L. C. Peterson. 1965. Genotypic competition in progeny performance evaluation of potatoes. Crop Sci. 5:433–435.

Doughty, L. R. and F. L. Engledow. 1928. Investigations on yield in the cereals. V. A study of four wheat fields: the limiting effect of population-density on yield and an analytical comparison of yields. J. Agric. Sci. 18:317–345.

Duncan. W. G. 1958. The relationship between corn population and yield. Agron. J. 50:82–84.

Early, H. L. and C. O. Qualset. 1971. Complementary competition in cultivated barley (*Hordeum vulgare* L.). Euphytica 20:400–409.

Eberhart, S. A., L. H. Penny, and G. F. Sprague. 1964. Intraplot competition among maize single crosses. Crop Sci. 4:467–471.

Engledow, F. L. and K. Ramiah. 1930. Investigations on yield in the cereals. VII. A study of development and yield of wheat based upon varietal comparison. J. Agric. Sci. 20:265–344.

Fasoulas, A. 1973. A new approach to breeding superior yielding varieties. Dept. Genet. & Plant Breeding, Aristotelian Univ. of Thessalonika, Greece. Publ. No. 3:114 pp.

Fasoulas, A. 1977. Field designs for genotypic evaluation and selection. Dept. Genet. & Plant Breeding, Aristotelian Univ. of Thessalonika, Greece. Publ. No. 7:61 pp.

Fuizat, Y. and R. E. Atkins. 1953. Genetic and environmental variability in segregating barley populations. Agron. J. 45:414–416.

Gedge, D. L., W. R. Fehr, and D. F. Cox. 1978. Influence of intergenotypic competition on seed yield of heterogeneous soybean lines. Crop Sci. 18:233–236.

Grafius, J. E. 1956. The relationship of stand to panicles per plant and per unit area in oats. Agron. J. 48:460–462.

Griffing, B. 1956. Concept of general and specific combining ability in relation to diallel crossing systems. Aust. J. Biol. Sci. 9:463–493.

Hall, R. L. 1974. Analysis of the nature of interference between plants of different species. I. Concepts and extension of the de Wit analysis to examine effects. Aust. J. Agric. Res. 25:739–747.

Hamblin, J. and C. M. Donald. 1974. The relationships between plant form, competitive ability and grain yield in a barley cross. Euphytica 23:535–542.

Hamblin, J. and J. G. Rowell. 1975. Breeding implications of the relationship between competitive ability and pure culture yield in self-pollinated grain crops. Euphytica 24:221–228.

Hamblin, J., R. Knight, and M. J. Atkinson. 1978. The influence of systematic micro-environmental variation on individual plant yield within selection plots. Euphytica 27:497–503.

Hanson, P. R., G. Jenkins, and B. Westcott. 1979. Early generation selection in a cross of spring barley. Z. Pflanzenzuecht. 83:64–80.

Harper, J. L. 1964. The nature and consequence of interference among plants. Proc. 11th Int. Conf. Genet., The Hague, pp. 465–481.

Hartmann, R. W. and R. W. Allard. 1964. Effect of nutrient and moisture levels on competitive ability in barley (*Hordeum vulgare* L.). Crop Sci. 4:424–426.

Hegde, B. R., D. J. Major, D. B. Wilson, and K. K. Krogman. 1976. Effects of row spacing and population density on grain sorghum production in southern Alberta. Can. J. Plant Sci. 56:31–37.

Hinson, K. and D. Hanson. 1962. Competition studies in soya beans. Crop Sci. 2:117–123.

Jain, S. K. and R. W. Allard. 1960. Population studies in predominantly self-pollinated species. I. Evidence for heterozygote advantage in a closed population of barley. Proc. Natl. Acad. Sci. 46:1371–1377.

Jennings, P. R. and R. C. Aquino. 1968. Studies on competition in rice. III. Mechanisms of competition among phenotypes. Evolution 22:529–542.

Jennings, P. R. and R. M. Herrera. 1968. Studies on competition in rice. II. Competition in segregating populations. Evolution 22:332–336.

Jennings, P. R. and J. de Jesus. 1968. Studies on competition in rice. I. Competition in mixtures of varieties. Evolution 22:119–124.

Jensen, N. F. 1967. Agrobiology: specialization or systems analysis? Science 157:1405–1409.

Jensen, N. F. 1978. Composite breeding methods and the DSM system in cereals. Crop Sci. 18:622–626.

Jensen, N. F. 1978. Seasonal competition in spring and winter wheat mixtures. Crop Sci. 18:1055–1057.

Jensen, N. F. and W. T. Federer. 1964. Adjacent row competition in wheat. Crop Sci. 4:641–645.

Jensen, N. F. and W. T. Federer. 1965. Competing ability in wheat. Crop Sci. 5:449–452.

Kaufmann, M. L. and A. D. McFadden. 1960. The competitive interaction between barley plants grown from large and small seeds. Can. J. Plant Sci. 40:623–629.

Kaukis, K. and L. P. Reitz. 1955. Tillering and yield of oat plants grown at different spacings. Agron. J. 47 : 147.

Kawano, K. and M. D. Thung. 1982. Intergenotypic competition and competition with associated crops in Cassava. Crop Sci. 22:59–63.

Kawano, K., C. Tiraporn, S. Tongsri, and Y. Kano. 1982. Efficiency of yield selection in Cassava populations under different plant spacings. Crop Sci. 22:560–564.

Kempton, R. A., R. S. Gregory, W. G. Hughes, and P. J. Stoehr. 1986. The effect of interplot competition on yield assessment in triticale trials. Euphytica 35:257–265.

Khalifa, M. A. and C. O. Qualset. 1974. Intergenotypic competition between tall and dwarf wheats. I. In mechanical mixture. Crop Sci. 14:795–799.

Kiesselbach, T. A. 1923. Competition as a source of error in comparative corn yields. J. Am. Soc. Agron. 15:199–215.

Kiesselbach, T. A. and C. A. Helm. 1917. Relation of size of seed and sprout value to the yield of small grain crops. Nebr. Agric. Expt. Stn. Res. Bull. 11, 73 pp.

Kirby, E. J. M. 1967. The effect of plant density upon the growth and yield of barley. J. Agric. Sci. 68:317–324.

Kyriakou, D. T. and A. C. Fasoulas. 1985. Effects of competition and selection pressure on yield response in winter rye (*Secale cereale* L.). Euphytica 34:883–895.

Langton, F. A. 1985. Effects of intergenotypic competition on selection within chrysanthemum progenies. Euphytica 34:489–497.

Laude, H. H. and R. F. Swanson. 1942. Natural selection in varietal mixtures of winter wheat. J. Am. Soc. Agron. 34:270–274.

Lee, J. A. 1960. A study of plant competition in relation to development. Evolution 14:18–28.

McGilchrist, C. A. 1965. Analysis of competition experiments. Biometrics 21:975–985.

Martin, R. J., J. R. Wilcox, and F. A. Laviolette. 1978. Variability in soybean progenies developed by single seed descent at two plant populations. Crop Sci. 18:359–362.

Mather, K. 1969. Selection through competition. Heredity 24:529–540.

Mead, R. 1966. A relationship between individual plant-spacing and yield. Ann. Bot. N. S. 30:301–309.

Mitchell, J. W., R. J. Baker, and D. R. Knott. 1982. Evaluation of honeycomb selection for single plant yield in durum wheat. Crop Sci. 22:840–843.

Montgomery, E. G. 1912. Competition in cereals. Nebraska Agric. Exp. Stn. Bull. 127.

Moss, D. N. and H. T. Stinson, Jr. 1961. Differential response of corn hybrids to shade. Crop Sci. 1:416–418.

Nass, H. G. 1978. Comparison of selection efficiency for grain yield in two population densities of four spring wheat crosses. Crop Sci. 18:10–12.

Pendleton, J. W. and R. D. Seif. 1962. Role of height in corn competition. Crop Sci. 2:154–156.

Phung, T. K. and A. J. Rathjen. 1977. Mechanisms of frequency-dependent advantage in wheat. Aust. J. Agric. Res. 28:187–202.

Probst, A. H. 1945. Influence of spacing on yield and other characters in soybeans. J. Am. Soc. Agron. 37:549–554.

Puckridge, D. W. and C. M. Donald. 1967. Competition among wheat plants sown at a wide range of densities. Aust. J. Agric. Res. 18:193–211.

Rajeswara Rao, B. R. and R. Prasad. 1984. Intergenotypic competition in mixed stands of spring wheat genotypes. Euphytica 33:241–247.

Reddy, M. R. and R. Prasad. 1977. Studies on lodging in wheat in relation to tallness of wheat genotypes, their mixed culture and nitrgoen fertilization. Indian J. Agron. 22:99–103.

Reddy, M. R. and R. Prasad. 1980. Effect of nitrogen and row direction on yield and yield components in pure and systematic mixed stand of wheat varieties differing in plant height. Indian J. Agron. 25:332–341.

Roos, E. E. 1984. Genetic shifts in mixed bean populations. I. Storage effects. Crop Sci. 24:240–244.

Ross, J. P. 1983. Effect of soybean mosaic on component yields from blends of mosaic resistant and susceptible soybeans. Crop Sci. 23:343–346.

Roy, N. N. 1976. Intergenotypic plant competition in wheat under single seed descent breeding. Euphytica 25:219–223.

Roy, S. K. 1960. Interaction between rice varieties. J. Genet. 57:137–152.

Sakai, K. J. 1957. Study on competition in plants. VII. Effect of competing and non-competing individuals. Genetics 55:227–234.

Sakai, K. J. 1961. Competitive ability in plants: its inheritance and some related problems. Symp. Soc. Exp. Biol. 15:245–263.

Sakai, K. J. and K. Gotoh. 1955. Studies on competition in plants. IV. Competitive ability of F_1 hybrids in barley. J. Heredity 46:139–143.

Sakai, K. J. and Y. Suzuki. 1954. Studies on competition in plants. III. Competition and spacing in one dimension. Jap. J. Genet. 29:197–201.

Sandfaer, J. 1970. An analysis of the competition between some barley varieties. Danish Atomic Energy Comm., Riso Rep. No. 230:114 pp.

Schutz, W. M. and C. A. Brim. 1967. Inter-genotypic competition in soybeans. I. Evaluation of effects and proposed field plot design. Crop Sci. 7:371–376.

Schutz, W. M., C. A. Brim, and S. A. Usanis. 1968. Intergenotypic competition in plant populations. II. Feedback systems with stable equilibria in populations of autogamous homozygous lines. Crop Sci. 8:61–66.

Feedback systems with stable equilibria in populations of autogamous homozygous lines. Crop Sci. 8:61–66.

Schwendiman, A. and H. L. Shands. 1943. Delayed germination or seed dormancy in Vicland oats. J. Am. Soc. Agron. 35:681–687.

Shanahan, J. F., K. J. Donnelly, D. H. Smith, and D. E. Smika. 1985. Shoot developmental properties associated with grain yield in winter wheat. Crop Sci. 25:770–775.

Sharma, S. N. and R. Prasad. 1978. Systematic mixed versus pure stands of wheat genotypes. J. Agric. Sci. (Cambridge) 90:441–444.

Siemens, L. B. 1963. The effect of various row spacings on the agronomic and quality characteristics of cereals and flax. Can. J. Plant Sci. 43:119–130.

Smith, H. F. 1937. The variability of plant density in fields of wheat and its effect on yield. Council Sci. Indust. Res., Australia, Bull. 109, 28 pp. Melbourne.

Spitters, C. J. T. 1979. Competition and its consequences for selection in barley breeding. Agric. Res. Rep. 893. Wageningen, 268 pp.

Stinson, H. T. and D. N. Moss. 1960. Some effects of shade upon corn hybrids tolerant and intolerant of dense planting. Agron. J. 52:482–484.

Stoskopf, N. C. 1967. Yield performance of upright-leaved selections of winter wheat in narrow row spacings. Can. J. Plant Sci. 47:597–601.

Stringfield, G. H. 1964. Objectives in corn improvement. Adv. Agron. 16:101–137.

Sumarno and W. R. Fehr. 1980. Intergenotypic competition between determinate and indeterminate soybean cultivars in blends and alternate rows. Crop Sci. 20:251–254.

Suneson, C. A. 1949. Survival of 4 barley varieties in mixture. Agron. J. 41:459–461.

Suneson, C. A. and R. T. Ramage. 1962. Competition between near isogenic genotypes. Crop Sci. 2:249–250.

Suneson, C. A. and H. Stevens. 1953. Studies with bulk hybrid populations of barley. USDA Tech. Bul. 1067. 14 pp.

Tiyawalee, D. and K. J. Frey. 1971. Mass selection for crown rust resistance in an oat-M population. Iowa State J. Sci. 45(2):217–231.

Troyer, A. F. and R. W. Rosenbrook. 1983. Utility of higher plant densities for corn performance testing. Crop Sci. 23:863–867.

Tucker, C. L. and J. Harding. 1965. Quantitative studies on mating systems. II. Estimation of fitness parameters in a population of *Phaseolus lunatus*. Heredity 20:393–402.

Tucker, C. L. and B. D. Webster. 1970. Relation of seed yield and fitness in *Phaseolus lunatus* L. Crop Sci. 10:314–315.

Wiebe, G. A., F. C. Petr, and H. Stevens. 1963. Interplant competition between barley genotypes. *In* Statistical genetics and plant breeding. Natl. Acad. Sci. Natl. Res. Council, Wash., D. C., Publ. 982:546–557.

Wilcox, J. R. and W. T. Schapaugh, Jr. 1978. Competition between two soybean isolines in hill plots. Crop Sci. 18:346–348.

8

COMPOSITE METHODS—A PREAMBLE

As we enter the vast body of literature arising from considerations of competition, breeding, and selection methods, I hope a consensus will develop that we need to examine nomenclature. The amorphous nature of terminology dealing with populations specifically is a problem. Those methods that can stand alone should be granted independence and pride of place, rather than submerged in a larger ambiguous category. In my view, the major areas of concern are the two breeding methods, the bulk hybrid population and the evolutionary breeding system. These represent two methods whose first proposals were separated by several decades. During that period, and even today, they occupy the two extremes of an F time span. The confusion arises because the two methods overlap in the middle part of this time range (the overall time range might, for reasons of clarity, be described as extending from F_1 to F_{30} or infinity). The solution, paradoxically, may be relatively easy: a return to the ideas of the experts whose precedent by virtue of publication provides an answer. These authorities are Florell and Suneson, whose contributions are discussed below.

Whether terminology can ever be rendered truly unambiguous is doubtful. Many terms have the weight of synonyms. Even after the bulk hybrid population and evolutionary breeding methods are set aside, there remains the term *composite*, which embodies and contributes to parts of both methods, represents vast amounts of research done for research's sake, and embraces different kinds of populations covering varying ranges of generations. *Composite* might well be a generic term applied to the creations of Harlan and Martini, as well as to the numerous hybrid populations created by various breeders. A characteristic of these populations is their *timeless-*

ness, in the sense that they may be of any generation, or of any age, and still be of use. It is well known that several varieties of barley were obtained from Harlan and Martini's composite. These were selected at different times, from different generations, and especially from different environments. So these kinds of populations are different: they are usually acquired by a breeder, they are available upon request, and their status often dictates what a breeder does with them; that is, an F_2 would be handled differently from an F_{18}. A Harlan and Martini composite, in the forms still viable, would be just as valuable today as it was half a century or so ago.

I do not have the answer to the best way of handling all population terminology, nor is it necessary that everything be settled. My primary goal is to separate and establish the bulk hybrid population as a distinct breeding method, free of the restraints imposed by its association with populations of indefinite filial generations. To illustrate, the positive weight of evidence favoring yield increases in hybrid populations with advancing generations leaves the negative implication that such yields are unobtainable in early generations. Also, the evidence that competition can cause dramatic shifts in gene frequencies, even eliminating certain traits, leaves the implication that these phenomena can not be avoided in the low filial generations within which a practical breeder must work with the bulk hybrid population method. Arguments such as these have resulted in occasional recommendations against the bulk hybrid population method.

A basic difference between the bulk hybrid population method and the evolutionary breeding method is that the goal of the first is a concern for *individual* genotype selection, whereas that of the second is a concern for the *group* or population worth. The plant breeder is interested in both, but recognizes that they occupy different time frames. In this preamble, under the umbrella of composites, I have chosen to bring together much of the literature that has relevance to all population synonyms.

HARLAN AND MARTINI'S EARLY PIONEERING RESEARCH

In one of the earlier papers detailing changes in mixture populations through natural selection over time, Harlan and Martini (1938) grew mixture populations of 11 barley varieties in cooperative tests at 10 experiment stations in the northern and western United States. The original mixture had equal numbers of seeds of the 11 entries. The periods of annual tests ranged from 2 to 12 years. All entries were distinguishable for census purposes with the exception of two, 'Coast' and 'Trebi', which were treated as one entry.

Genotype changes were shown by the census for the last year grown at each station. The pattern of survival seemed to be similar at selected sites. From a clustering standpoint, the Nebraska, Montana, Idaho, and Washington sites showed similarities. Dramatic instances of domination of one genotype at a single site were found in Virginia. New York, Minnesota, and Oregon.

Although Trebi and Coast were difficult to separate for purposes of identification, they were known to be quite different in national adaptation, Coast to the west, and Trebi to the central and east. When the two were paired as one entry it seemed to me that the compensation occurring as a result of combining the two tended to give a stabler census line across sites.

Harlan and Martini developed curves to show the expected path of survival of 10 varieties over time. The census results showed good fits to these curves. Under actual conditions the oscillating census figures caused by a changing yearly environment tended to damp or slow the progress to the climax of domination, elimination, or some intermediate state of survival.

Harlan and Martini's paper is remarkable for its breadth, a scope that permits one to see ranges of interactions not ordinarily visible in a single breeding program or a single geographic region. Usually, competition is observed and evaluated under similar environments, often with genotypes selected for similar adaptation. Seldom are diametrically opposite results seen, as for example, the dominant genotype in one region being eliminated in another. This phenomenon alone is good background knowledge for researchers to have.

Some of the entries chosen for the mixture, for example, 'Deficiens', were known to be unpromising candidates. The research goal was not to produce a commercial blend. In this sense the population might be likened to a hybrid population where segregation patterns might produce unpromising individuals, which like 'Deficiens' were not likely to survive. The population, however, was homozygous and heterogeneous. As much as anything, the geographical site diversity has pointed out the uniqueness, and sometimes fragility, of the interaction of genotype and environment with population structure.

Earlier, Harlan and Martini (1929) had projected their plans and expected progress for the composite hybrid mixture that is examined in more detail below. There is little doubt that they were led to the formation of hybrid composites by the desire better to use and exploit the vast volume of barley germ plasm that resided in the USDA.'s world collections. The dilemma of choosing parents and using them in more restrictive breeding and selection procedures is mirrored in this statement:

> The originator of "Marquis" wheat could hardly have known that the two parents chosen held the possibility of producing that variety. If one were to take all the possible parents, the volume of work would become prohibitive.

RESEARCH PAPERS ON COMPOSITES

Suneson and Wiebe's (1942) paper was one of the early articles relating competition in homozygous but heterogeneous mixtures to possible breeding limitations of the bulk hybrid population method. Their research with two barley mixtures (four- and five-variety, equal-seeded) and one wheat five-component mixture, conducted for periods up to nine years, showed the same trend—component proportions in both directions were changed. All the varieties used were popular and productive, and considered to be essentially equal in pure stands. Yet the component proportions in percent at the end of the experiments were as follows:

Four-component barley mixture (ninth year): 65.5, 18.8, 7.7, and 7.5%.

Five-component barley mixture (fifth year): 31.6, 31.7, 24.1, 9.1, and 2.8%.

Five-component wheat mixture (fifth year): 40.0, 28.1, 13.5, 13.4, and 5.0%.

The final, worst figures of 7.5, 2.8, and 5.0%, represented declines from approximately 25, 20, and 20%, respectively.

Adair and Jones (1946) grew bulk populations formed by compositing F_2 seeds of several rice hybrids for eight years, following which random seeds were picked from each of the nine bulk populations and a census taken of seven characters. The bulks had been grown concurrently in three states.

Although differences and shifts of characters were noted in various populations, all characters survived at all locations, and furthermore, "Among the surviving plants in each lot at each station, the characters studied occurred in all combinations and in sufficient numbers so that strains possessing the combination of characters desired could be selected." The authors concluded that the bulk hybrid method was a useful one for rice breeding.

Two items of interest about this study were that the character census was made in F_{11}, a sufficiently long time for natural selection to have had an impact, and the formation of the initial F_3 bulks was not entirely random; in some cases, where all F_2 plants were not taken, only the seeds from the selected plants were included.

Suneson and Stevens (1953) studied changes occurring over time (6–24 generations) in six composite crosses of barley grown without selection. Their findings may be summarized as follows:

1. Survival was nonrandom: low survival or elimination was found for the traits, two-row, hooded, smooth awns, and black seeds. A significant elimination of an important character, earliness, was also reported; the earliest recombinations were eliminated by the F_8. Another finding was that scald resistance, which might be expected to enhance survival of progeny, actually decreased in the population. By the twelfth generation, less than half as many expected resistant plants remained in the population.

2. Continuous improvement in yield with time was found.

3. Linkage appeared to impose powerful limitations on genetic recombination.

4. Survival under natural conditions is not monolithic: the direction taken with regard to character survival may vary with environments.

Finkner (1964) examined the effect of natural selection on the winter hardiness of 20 hybrid winter oat bulks grown for six years in Kentucky and Indiana. At the start of the testing the bulks varied in generation age from F_2 to F_8.

The results showed alternate-directional (year-to-year) shifts in survival, and the survival in the final year was lower at all locations than in any previous year. This tells something about the hardiness level, which is often marginal, of winter oats. The authors concluded that natural selection was not effective in eliminating nonhardy types from bulk hybrid populations. This does seem surprising because one might assume that in a marginally hardy crop such elimination would occur. It should be noted also that a variable climate is a factor; seasonal changes often allow less hardy types to rebound after a bad year. The author suggested that a plant not killed actually may have other competitive advantages that restore its fitness by time of maturity.

In contrast to Finkner's results, reported above, Marshall (1966) found that natural selection among 23 winter oat bulk hybrid populations was significantly effective in eliminating less hardy types and in raising the general level of cold resistance when measured over a three-year period.

Reyes and Frey (1967) found that different seeding rates (1, 3, and 5 bu/acre), maintained for five generations, failed to modify significantly the genotypic balance in a heterogeneous hybrid oat population formed from a composite of 250 crosses. Five characters were measured, and in only one, panicles per plot, was there any evidence of shift. The implications seem favorable to the use of the bulk hybrid population method of breeding and selection, because essentially pure line status of individuals is achieved in the population without significant loss of variability or genetic traits.

Singh and Johnson (1969) studied natural selection in an F_4-embryo equal-seed number composite of 105 crosses from a diallel cross of 15 barley parents. The unselected composite was grown for five years at nine stations in Canada and Norway. Yearly seed samples were furnished to the central Edmonton station and compared with the original seed and population for changes in character frequency. Remembering that five years is a short time in the life of an unselected population, the authors noted the following changes in samples from different stations:

1. Significant differences from the original affecting heading date, maturity period, yield, spike and awn type were found.
2. Little or no change in plant height, seed size, or in several plant morphological characters could be detected.
3. There was a drastic shift against black kernel.

Two general conclusions were that natural selection produced locally adapted types at the different locations, and "the shift in the gene pool is predominantly in the direction of plant breeding objectives." An exception to the latter was noted in connection with the association of two characters, high yield and late maturity; at northern latitude locations, the short season mitigated against high yield because it restricted late types from reaching full maturity.

Doggett (1972) found that two sorghum populations were required to work with the developing potential of germ plasm, namely, a maintenance population with a selection intensity of about 50%, and a development population under maximum selection intensity. Doggett commented on the adverse effect on population breeding methods of certain unsuccessful panmixis proposals ("stir the pot and hope"). He believed, "Intense, deliberate human selection for the important cultivated characters is essential if crop plants are to be maintained and improved . . ." He feared that the powerful forces of natural selection might result in the dominance of wild-type characters. To counter these dangers, he suggested the alternatives of more selection pressure and direction, and the option of building separate composites with narrower objectives.

In a study patterned after that of Voight and Weber (1960), Luedders, Duclos, and Matson (1973) widened the parameters to include three maturity groups in an attempt to measure the efficiency of four breeding methods in soybeans. The methods were complete bulk (CB), maturity bulk (MB), pedigree (P), and early generation testing (EGT).

Differences in yield of F_6 and F_7 lines extracted from these methods were not statistically significant. This agrees with earlier studies in soybeans. The authors intimated that visual selection, which is not considered efficient in discriminating yield but was used throughout the experiment as a means of advancing the research, may have had an unknown but confounding effect, blurring distinctions that may have existed among methods. In their discussion the authors made a recommendation of significance to all breeders of the autogamous crop species: that attention be focused on a high percentage discard rate (they suggested 75% each year for soybeans). They had found in this study that selecting the top 25% each year retained the best lines, based on their two-year means. Although breeders may wish to adjust the percent cutoff level, the recommendation, nevertheless, addresses a key breeding concern, which is not to let numbers retained become the bottleneck in a program. With heavy discard, the authors pointed out, larger samples of populations could be evaluated. Equally important, additional crosses could be evaluated. It seems appropriate here to append Finney's (1958) comment in a discussion of selection: "Indeed, one index of the success of a plant breeder might be said to be the extent to which his products are discarded."

In a study on the influence of growing environments on natural selection and adaptation of lima bean populations for yield, Tucker and Harding

(1974), referring to yields of two unselected F_{11} bulk populations, one of which had 'Concentrated Fordhook' as a parent, stated:

> Neither bulk population tested in this study approached the yield of 'Concentrated Fordhook'. This indicates that the bulk population method may not be successful in the development of high-yielding lima bean cultivars. However, bulk populations are known to retain genetic variability (Jain and Suneson, 1966) and it is possible that genotypes are present in the bulk population that have higher yields than the bulk population averages.

These statements seem an unwarranted denigration of the bulk population method. In the reported research the populations had no artificial selection imposed, no lines were removed and tested, and the fact that the mean yields were intermediate to the parents is a normal expectation. Furthermore, it is more of a certainty than a possibility that the populations described would contain genotypes that would yield more than the population mean. Whether any would surpass the yield of the higher-yielding parent is another matter and is precisely the question that is common to all hybrid populations.

By dividing an F_1 bulked hybrid barley population and growing equal halves at two locations, followed by continued annual division and use of alternating sites, Choo, Klinck, and St-Pierre (1975) obtained a large number of populations, F_2–F_{15}, which had individual histories of continuous growing at either one site or a number of years of alternating site growth. On the basis of trait counts in F_{15} populations, significant differences among populations were observed. Yield tests were conducted at both locations using F_6, F_{10}, and F_{15} populations. Only in the F_{15} population were there found significant differences in two traits, head and awn length, and "there were no differences in yield, 1000-kernel weight, number of heads per plot, number of kernels per head, flag leaf area, flag leaf width, flag leaf length or lodging resistance in the three generations."

The authors' conclusions dealt with the unlikelihood that alternating generations and locations would improve yielding ability. However, there is another important observation to be made, namely, that the bulk population format over long periods had little effect either way on character (gene) frequencies or shifts. It is conceivable that the use of alternating locations may have influenced this apparent increase in stability through the dynamics of two different alternating environments, which could affect the direction of the annual genetic shifts.

The study of Quisenberry, Roark, Bilbro, and Ray (1978) on natural selection in a bulk hybrid population of cotton is different, in that it dealt with fiber rather than seed. Seeds readily are associated with survival, reproduction, and fitness, but fiber introduces an unknown element whose presence on seed is not known either to facilitate or to hamper fitness in a heterogeneous population. The authors examined ten generations (F_4–F_{13})

grown without selection, for seven traits, three of which involved seed numbers and size, and four of which were expressions of fiber traits. The results under natural selection showed that five of the seven traits changed significantly. In the seed category, seeds per meter increased by an average 3.8% per generation, and seed size decreased at an average 1.1% per generation. In the fiber category, lint yield increased at an average 3.3% per generation, fiber fineness increased at 0.5% per generation, and fiber strength decreased. No real changes were noted in number of seeds per boll, or fiber length.

THE PROBLEM AND SOLUTION IN PERSPECTIVE

The statement by Haldane (1957), "Natural selection is a statement of the fact that the fictitious parental population differs significantly from the population from which it was drawn," is relevant both to the bulk hybrid population and the evolutionary breeding methods, but especially to the latter. "Fictitious" parents are those whose progeny actually become the next generation; their seeds incorporate the preceding generation's genetic shifts. The element of sampling is crucial; only the sample of seeds chosen can perpetuate their progenies.

The crucial point is the probable magnitude of change that can occur in a few generations, as in a bulk hybrid population, versus what can happen over many generations, as in the evolutionary scheme. If the selections are made and bulked very early, as in the formation of a selected bulk in F_2 or F_3, or early as in F_4 or F_5 line withdrawal from a hybrid bulk population, the predicted genetic shifts rarely affect the practical outcome. It is only in the long evolutionary breeding methods that the succession of sampling of planting seeds for each next generation, the fictitious parents, plays a significant role in the census of traits in the climax population.

Simmonds (1962) in his excellent review of variability in crop plants stated: "The bulk method of handling single crosses does not seem to have been conspicuously successful. One would expect it to be about as successful as a pedigree treatment of the same cross."

I am indebted to Simmonds for his perceptive recognition not only of the similarities but of the profound differences inherent in population breeding methods, as illustrated in this statement.

> I conclude that the bulk method of handling single crosses and the composite-cross type of population, though superficially similar and based on the same assumption as to the nature of the survivors, are in effect quite different: The former is an operational device to ease the labor of pedigree breeding; the latter is a means of exploring, recombining and maintaining variability on a scale inaccessible to conventional methods.

I would add two things to this characterization of the differences: (1) the bulk method works equally well with multiple crosses; hence the distinctive difference cannot rest on the number of crosses. More important is (2) the time element: the bulk method is a short-range practical breeding procedure; the composite cross type of population is a long-range supplementary program (Simmonds recognized this in his statement that mass reservoirs are adjuncts to, and not in any sense a substitute for, plant breeding).

Unless this distinction is recognized, the bulk hybrid population method will continue to be judged to its disparagement by the weight of interpretations coming from the studies of high filial generations. The validity of these results is not in question when they are applied appropriately to high-F situations. In my view, these interpretations cannot be applied over the full filial generation time span, because they cast doubt upon a very successful bulk hybrid population breeding system that occupies only the first four or five F positions.

There is no reason for these two procedures to be in conflict; they are joined, unfortunately, by acquiescence to limited nomenclature. What is needed is new nomenclature. If Florell (1929) and Suneson (1956) represent the two distinctions discussed, as I think they do, we may have nomenclature already in place: Florell's *bulked population method* and Suneson's *evolutionary breeding method*. I have taken these terms as chapter headings.

REFERENCES

Adair, C. R. and J. W. Jones. 1946. Effect of environment on the characteristics of plants surviving in bulk hybrid populations of rice. J. Am. Soc. Agron. 38:708–716.

Choo, T. M., H. R. Klinck, and C. A. St-Pierre. 1975. The effect of location on natural selection in bulk populations of barley. Can. J. Plant Sci. 55:345 (abstr.)

Doggett, H. 1972. Recurrent selection in sorghum populations. Heredity 28:9–29.

Finkner, V. C. 1964. Effect of natural selection on survival of winter oat bulk hybrid populations. Crop Sci. 4:465–466.

Finney, D. J. 1958. Plant selection for yield improvement. Euphytica 7:83–106.

Haldane, J. B. S. 1957. The cost of natural selection. J. Genet. 55:511–524.

Harlan, H. V. and M. L. Martini. 1929. A composite hybrid mixture. J. Am. Soc. Agron. 21:487–490.

Harlan, H. V. and M. L. Martini. 1938. The effect of natural selection in a mixture of barley varieties. J. Agric. Res. 57:189–200.

Luedders, V. D., L. A. Duclos, and A. L. Matson. 1973. Bulk, pedigree and early generation testing breeding methods compared in soybeans. Crop Sci. 13:363–364.

Marshall, H. G. 1966. Natural selection for cold resistance in winter oat populations. Crop Sci. 6:173–176.

Quisenberry, J. E., B. Roark, J. D. Bilbro, and L. L. Ray. 1978. Natural selection in a bulked hybrid population of Upland cotton. Crop Sci. 18:799–801.

Reyes, R. and K. J. Frey. 1967. Effect of seeding rates upon survival of genotypes in oat populations. Iowa State J. Sci. 41:433–445.

Simmonds, N. W. 1962. Variability in crop plants, its use and conservation. Biol. Rev. 37:422–465.

Singh, L. N. and L. P. V. Johnson. 1969. Natural selection in a composite cross of barley. Can. J. Gen. Cytol. 11:34–42.

Suneson, C. A. and H. Stevens. 1953. Studies with bulk hybrid populations of barley. USDA Tech. Bul. 1067. 14 pp.

Suneson, C. A. and G. A. Wiebe. 1942. Survival of barley and wheat varieties in mixtures. J. Am. Soc. Agron. 34:1052–1056.

Tucker, C. L. and J. Harding. 1974. Effect of the environment on seed yield in bulk populations of lima beans. Euphytica 23:135–139.

Voight, R. L. and C. R. Weber. 1960. Effectiveness of selection methods for yield in soybean crosses. Agron. J. 52:527–530.

9

BULK POPULATION BREEDING METHOD

Florell's (1929) article, "Bulked-population method of handling cereal hybrids," may be considered the definitive statement of this breeding and selection system, and is the source of its nomenclature. In order to establish this point, and in view of the ambiguities surrounding the bulk population method, a discussion of Florell's paper inaugurates this chapter.

FLORELL'S LANDMARK PAPER

For the bulked population method to have impact as a practical plant breeding program it must be shown that it can compete efficiently in results with other widely used methods, such as the pedigree system. Florell thoroughly documented the practical short-time efficiency of the bulked population method of management. Throughout, Florell characterized the method as a six- or eight-generation procedure, with guided or artificial selection permitted in any generation: "Artifical selection toward a definite type usually is practiced, thus eliminating weak and undesirable plants."

Florell was first attracted to the bulk method by one of its marked advantages, efficiency in handling large numbers of hybrids. At Davis, California, the size of the wheat project was stretching the capacity of the pedigree method and an experiment was devised to evaluate the bulked population method. Nineteen wheat crosses were grown from 1923 to 1926. The Mediterranean type of climate permitted two crops per calendar year, a winter crop at Davis and a summer crop at Palo Alto or Davis. Selection for earliness was done in the summer crops, and for type in the winter crops where growth was more normal.

The first selections were made in 1926 at which time 9 populations were F_5, and 10 were F_6. Selection generations as early as these clearly are competitive with other practical breeding methods. In combination with a possible two crops per year, the elapsed actual time in the early generations is remarkably short for a breeding program.

Florell selected a total of 476 heads with a range per cross of 9–49 heads. These were planted in head rows and the best 45 rows saved. The number of heads saved from the crosses seems conservative, that is, low; however, this kind of procedure always reflects the breeder's evaluation of agronomic merit.

The 45 selections were advanced the following year to triple-row plots replicated three times with a check variety, 'Hard Federation', placed after every five selections. The results were as follows:

1. The mean of 33 check rows was 36.7 bu/acre.
2. Thirty-three of the 45 selections yielded more than the check.
3. The highest yielding selection produced 55.1, the second highest 52.9 bu/acre. Both of these had Federation as one parent; Federation's yield was 46.5 bu/acre.

Florell discussed the theoretical basis for determining the number of generations necessary before selection might be made and concluded that it largely depended on the number of genetic factors involved, tempered by the complexity of the cross. He made an interesting suggestion on preparation of wide or complex crosses: use the backcross for one or two generations before bulking.

Florell also recognized the sampling problem that involves the elements of plot area, planting seed sample size, and rate of seeding. His solution was a standard commercial rate of planting on an area of 1/50 acre (871 ft^2), later changed to 1/100 acre (436 ft^2).

In my use of the bulked hybrid population method in New York, I eased the sampling problem by doubling the rate of seeding, to 4 bu/acre and using standard drill-sown 6 × 16 ft plots. For simple hybrids one unit (one unit equaled one plot) was sufficient. For more complex formulations, for example, the diallel selective mating procedures that put many hybrids directed toward a common objective into one bulk population, the number of units was increased on the basis of one unit for each five hybrids in the population.

Florell's article is an admirable exposition of the bulked population method. I have only one suggestion for improvement and it addresses not a trivial point but one that in my opinion is crucial to the success of the bulked population method. Writing of head selection, Florell stated: "Head selections of desirable types were then made. Plant selections would have provided a larger supply of seed, but with drilled grain it is difficult to separate individual plants and from the standpoint of purity this practice is questionable."

This is an accurate statement, but I would strengthen it by stating that selection from a bulked, drilled population is in effect blind, and little better than random. Individual traits of a plant are seldom visible enough to make an impact on a person walking around or through a densely sown area. The solution to this problem, critical in the line withdrawal year, is documented in detail at the end of this chapter and has also been described by me (Jensen, 1978).

EARLY HISTORY AND LATER DEVELOPMENT

Mac Key (1962) credited the bulk population method to Nilsson-Ehle in 1908. Initially, individual selections were removed at about the F_7.

Harlan and Martini (1937), in their discussion of barley hybridization methods in the 1936 USDA Yearbook, mentioned only three methods: pedigree, bulk, and backcrossing. The bulk method, they indicated, uses natural selection to eliminate weaker lines, and moves individuals toward homozygosis. Two versions of the bulk hybrid method were given, namely, single-hybrid and multiple-hybrid bulks. The last mentioned is the least expensive but has the fault of loss of pedigree information. (I agree with their implied view, "The record of the parentage... is lost, if that is important.")

Harlan, Martini, and Steven's (1940) massive study of 378 barley crosses handled by the pedigree and composite method found the bulk method to have a slight advantage with an average yield of all selections of 480.4 g versus 463.4 g for the pedigree system. Furthermore, from the 2921 selections made from each method, a larger number deemed suitable for further testing were saved from the bulk (1269) than from the pedigree (965).

A further comment concerning their research study is that this bulk hybrid barley population became Composite II and was distributed widely to breeders around the world. In the United States three commercial barley cultivars were released from it, and by 1958, 17, or 10% of the 165 U.S.-listed varieties had originated as selections from this and other composites or bulks (Wesenberg, 1974). Even if some selections are similar or closely related to others, that still remains a remarkable figure.

Mac Key (1964) provided a literate analysis of differences between pedigree and bulk selection procedures. He pictured the pedigree system as a series of best plant selections, each selection based on that of the preceding generation, until satisfactory homozygosis was attained, usually between F_4 and F_7. The labor-intensive nature of the procedure was noted. Pedigree procedures sometimes are wasteful because they are based on phenotypic evaluation. He offered a discontinuous selection variation, substituting one generation of progeny testing followed by renewed selection in the retained lines.

On the other hand, the bulk method proceeds without selection in the F_2 or later generations, until lines are withdrawn in the period from about F_6 to F_8. Mac Key discussed the sampling problem in bulks. The increasing volume of seed from the harvest of bulk populations quickly requires a sampling procedure to obtain the small amount of seeds needed to plant the next generation. Sampling acts to reduce variation; it would be illogical to expect the same range or frequency of variation of genes or characters in a population arising from an aliquot sample as in a population arising from planting the entire harvested seed amount, which in any case becomes impracticable. Also, sampling may eliminate some of the very recombination segregants that the bulk method is expected to produce. Mac Key's solution was to substitute purposeful for random sampling, concentrating on qualitative traits such as disease resistance or earliness.

The pedigree method, with its controlled, every-generation selection, works well on observable and simply inherited traits, where homozygosity may be achieved early; it does less well in polygenic, quantitatively governed situations. Mac Key stated that "under conditions where heterozygosity may deceive the breeders," the bulk method could be more satisfactory because no selection is made until the population has passed through this stage. Mac Key bridged the gulf between the two methods by suggesting a compromise: a first selection in F_4 with reselection in F_8 or F_9.

Qualset and Vogt's (1980) description of the breeding methods in use in California covered the historical and philosophic background of wheat improvement, and included a detailed account of the practical application of the bulk method which, with the pedigree method, was the basic method in use. As is true of most breeding programs however, other procedures such as backcross and recurrent selection were used according to specific goals and needs.

Qualset and Vogt made a formal distinction between selected and unselected bulks (selected and random), but allowed cross populations to move from one category to the other depending on whether and when a population would benefit from selection. Although the authors admired the efficiency of the bulk method, they also pointed out that there are fewer opportunities to observe families and progenies than in the pedigree system, and there is a retention of more of the material that would be discarded earlier in the pedigree system. They have made an important contribution to the uderstanding of the bulk hybrid breeding and selection system by advocating that "the number of generations for selecting and bulking be small, perhaps no more than three."

Depending on whether bulking began in F_2 or F_3 this would place line removal at F_4 or F_5, which would place the lines in position for evaluation in head rows the following year.

Those of us who use the bulk population method as a practical system are grateful to Florell, who in 1929 gave a clear presentation of its efficiency, and whose procedures could be adapted today almost without change. I am sorry to say that only in recent years did I read Florell's paper. Had I done so when I started cereal breeding, I could have saved a decade or so of "reinventing the wheel." It is fitting that Qualset and Vogt, in California, now use the same method that Florell pioneered in California.

BREEDERS' EXPERIENCES WITH THE BULK POPULATION METHOD

Palmer (1952) clearly favored the bulk over the pedigree method, while admitting his study provided no clear-cut evidence. He gave these advantages to the bulk method:

1. There is an automatic increase in the proportion of homozygous plants with generation advance.
2. There is an increase in population mean yield from natural selection.
3. Any changes are to the breeder's advantage.
4. There is only one generation of single plant selection.
5. Final purification of lines is simplified by the absence of unwanted segregates.
6. Selection between crosses is possible before selection within crosses begins; therefore, the discard of poorer crosses allows more elite crosses to remain in the program.

In addition Palmer allayed the fear that unless desirable genotypes are preserved by selection in the F_2 they might be lost. He cited Akerman and Mac Key (1948), who stated that sufficient size of the F_2 bulk is adequate protection.

Palmer's study involved the parents, F_1, F_2, and F_8 of a wheat cross in which yield and its components were tracked. The only parent–offspring correlation of significance involved grain weight (single grain and per ear); this did not include yield. Otherwise F_2 selection was no better than random—a factor favoring the bulk method.

Atkins (1953) compared unselected with visually selected bulks from 11 bulk hybrid barley populations after three cycles of selection. Evaluation of eight of the populations showed yearly yield advantages of selected over unselected bulks as follows:

1948 F_4 : +1.9 bu/acre
1949 F_3–F_5 : +3.1* bu/acre
1950 F_4–F_6 : +1.0 bu/acre

The visual selection appeared not to be rigorous, and no attempt was made except by eye to coordinate the selection of many characters as would occur, for example, in index selection. A shift toward taller, later plants was observed in the selected bulk. A yield test of 20 F_7 lines, each of which was drawn from each climax bulk type, showed a statistically significant yield advantage to lines from the selected bulks. These results may be interpreted as suggesting the possibilities of applying selection to promising bulk hybrid populations during the early generations.

Taylor and Atkins (1954) made an extensive evaluation of the bulk hybrid population method with barley in Iowa. The plant materials under test were: (1) 20 bulk hybrid (single-cross) populations grown through F_2 to F_4 and F_5, and (2) the best 10 F_3-derived lines, visually selected from each hybrid. The bulks were grown at four locations with seed harvested from each site being used to plant again at that site. All generations and lines were brought together for final tests. The observations and results were these:

1. Natural selection resulted in a shift toward types adapted to that location.
2. Overall, population changes in the 20 crosses were essentially alike, suggesting that the widely different genotypic contents of the bulk populations responded similarly in the different environments.
3. There was a consistent linear relationship between the yields of bulk crosses and the yields of selections taken from them. For example, when the 20-cross bulk yields were placed in four groups, five highest to five lowest, the mean yield of the approximately 50 selections of each group corresponded with the group's position; for example, when groups were arranged according to order of bulk F_2 yields the mean yields were 225.8, 198.2 197.1, and 176.9 g/plot, respectively, and when arranged according to F_2–F_5 yield order the mean yields were 218.9, 212.5, 190.2, and 176.4 g/plot, respectively.

The authors concluded that "bulk yields of the crosses in this investigation gave a satisfactory evaluation of crosses that would be most likely to produce high-yielding selections."

Not mentioned but of plant breeding importance was their withdrawal of lines in the F_3 for experimental purposes. When it is realized that the F_4 and F_5 are also available and generally preferable because of higher homozygosity, it makes clear that the bulk hybrid method is a first-rate efficient system.

Torrie (1958) made a direct comparison of the bulk and pedigree systems in soybeans, using six hybrid populations, each divided into two parts with one going into an unselected bulk and the other used for pedigree selection. In the pedigree portion, the best rows were selected in F_3 and F_4, and two to three phenotypically outstanding plants taken from each, thus attempting to keep both between-line and within-line variance reasonably high. In the bulk part of the experiment, 50 visual selections were made in F_4 and grown in F_5 rows. An effort was made to select equal numbers in three maturity groups from both methods. Data were recorded on seed yield, maturity, plant height, lodging response—the main traits used in the phenotypic selection—and other characters such as protein and oil. The results were these:

1. No significant differences were found between the methods in height, lodging, bacterial blight reaction, oil and protein content, and oil iodine number.
2. Mean seed yields were similar, with exceptions found in two crosses: in these the bulk line yields were superior.
3. Later maturity was found in the early maturity group lines from the bulk method.

Torrie's paper was of special interest to me because it is one of the few records where the bulk method has been given a direct practical comparison with a recognized successful short-term system; in fact, the taking of F_4 selections for F_5 head rows is identical with the system I used at Cornell University. The positive results that Torrie obtained support the idea of the bulk population method as an efficient breeding and selection method.

Coyne (1968) investigated the relationship between yield components and selection in field beans. The findings were unusual because among 40 correlation coefficients only three showed significant negative correlations among yield components—a favorable selection situation. The prospect of rapid advancement was damped, however, by low heritability estimates for total seed yield and yield components, confirmed by finding no yield advance in selection of the top 5% in the F_2. Coyne suggested that bulk population breeding would be most efficient for situations of this kind. It would be easier and more accurate to withdraw and test homozygous lines.

Hamblin (1977) examined a key problem in plant breeding: the relationship between yield in pure culture and reproductive fitness in mixtures. The influence of density is such that an observed response cannot be given a general interpretation. These experiments led Hamblin to note:

> Bulk breeding does not lead to the elimination of high-yielding genotypes. Therefore, it would appear to be a better use of resources, when selecting for yield, to bulk breed the population until homozygosis is approached and then to yield trial as many lines as possible to identify the high-yielding genotypes.

MODIFICATIONS TO THE BULK POPULATION METHOD

For greatest selection efficiency, Sakai (1951) suggested a delay in beginning individual plant selection to later generations, and added "therefore the bulk-method will be rather preferable to pedigree-method so far as characters difficult to identify are concerned." Sakai is referring here to quantitative traits, which are difficult to measure, interpret, and reconcile with selection in early generations.

Raeber and Weber (1953) tested 25 cross hybrid bulks of soybeans and chose four to study further. Two were high and two were low, based on their F_2 and F_3 yields. Their objective was to compare the pedigree and bulk methods, and the selection procedures within each, by using final yield data in the F_6.

In the bulk method two kinds of selection were used, phenotypic superior and random. Phenotypic selection (visual) took place in the F_5 where plants were spaced 8 in. apart in rows 40 in. apart: 10 superior plants were taken from each cross. Random selection involved the taking of 40 "equally competitive" plants per cross.

The pedigree half of the study began with F_2 plants selected at random from each cross. These F_2-derived families, 75 per cross, were evaluated in an F_4 progeny yield trial. From the outcome, five high-yielding and five low-yielding lines were selected from each cross. These were space-planted in F_5 and two sister plants per line were selected for F_6 tests. The final comparison trials over two and three years also included parents of the crosses and the unselected F_6 bulk population. The results were as follows.

In the bulk experiments, the yields were phenotypic superior 38.3 bu/acre random 37.6, and the unselected F_6 bulk 36.0 bu/acre. For the pedigree series, the yields were high 38.7 and low 36.1 bu/acre. Thus the bulk phenotypic superior and pedigree high selections were equally good, both were better than random, and the pedigree low yielded the same as the unselected bulk.

Upon evaluation of the two sister lines extracted from the F_5 lines, it was found, first and as expected, that between-line variance was greater than within-line variance. Second, the distribution of superior lines followed the pattern of yields of the population from which the lines came. Their statement that "some lines from plants chosen at random were as high yielding as the best lines from plants selected as phenotypically superior" is an expected result inasmuch as a random draw from an unselected population ought to pick representatives from all along the yield range, including some superior lines.

The authors found that both the pedigree and bulk population methods shifted gene frequencies for yield in a positive direction. They also noted one instance when unselected bulk populations were penalized by certain components (late maturity leading to delayed harvest and frost damage). Consequently they proposed a combination methodology wherein "replicated drill rows and unreplicated spaced-plantings be grown and evaluated for agronomic characters simultaneously in F_3 and subsequent generations." I have quoted directly here because it is not entirely clear to me as to what the drill rows contained. Did they contain unselected bulk, or F_2-plant-derived F_3 lines? However, this seems less important than the illustration of the kinds of modifications that are optional for the breeder in any crop. Further illustrations of these options might include F_2 bulk yield trials to screen for desirable and undesirable crosses, followed by next-generation space plantings of retained crosses. Or, a breeder might reverse this and space-plant the F_2, sorting at harvest into meaningful groups, for example, maturity and height.

Sakai (1954) expressed a preference for the bulk population over the pedigree method: "Bulk method with the first individual selection in about F_6 immediately followed by progeny selection among a large number of plant lines regarding yield seems preferable to the pedigree method."

Sakai arrived at this conclusion from a realization that the heritability of selection for quantitative characters, for example, yield, as in the F_2 of a pedigree system, is so low as to be ineffective, but that it increases with advancing generations and homozygosity owing to the elimination of nonadditive gene action and the ameliorating influence of progeny mean values. I interpret the latter to mean that average values would damp the effect of environmental variation at any stage; for example, the mean yield value of an F_3 progeny would be more accurate and meaningful and have higher heritability than the single F_2 plant from which it derived. Practically, one must grant the likelihood that selection of reasonably large numbers from the F_2 of a pedigree system would garner a majority of the high-yielding plants that were in the F_2 population.

Mumaw and Weber (1957) in a study of natural selection and competition in soybeans observed internal population changes. Their observations were that the high-yielding ability of a genotype does not guarantee survival, and that certain characteristics influence population changes; for example, soybean branching types tended to dominate.

The authors believed these and other phenomena "impose a limitation on the bulk hybrid method of plant breeding. Thus, in order to prevent the early elimination of desirable types, it would appear advantageous to separate a bulk hybrid population in an early generation into 'like' groups of growth habit, maturity, height, lodging, etc."

This practical suggestion can be adopted informally whenever a breeder feels a particular cross population needs or would benefit from extra attention, as suggested by Florell, or formally as in Qualset and Vogt's (1980) "selected" bulks. The logical generation for clustering for a qualitative character would be the F_2.

Mac Key (1962) showed how mass selection procedures could be inserted, singly as in the F_2, or continuously as in the F_2 and F_3. The advantages and disadvantages of the bulk population system compared to the pedigree method were given as follows:

Advantages:

1. There are significant savings in handling cost and labor.
2. Deferring first selection to a generation later than F_2 eliminates or

minimizes the confounding effect of heterosis on phenotypic expression.

3. "Each available genotype is represented by several individuals with less chance to escape the breeder's eye."

Disadvantages:

1. There is a loss in time before the final type is fixed.
2. There is a risk of losing valuable types because of interplant competition.
3. Because all harvested material cannot be sown (the sampling problem), there is the likelihood of diminished variation from one progeny generation to the next.

Mac Key offered corrective measures to counteract these disadvantages. Lower seeding rates could lessen interplant competition. Growth in population size could be restrained by positive or negative mass selection, concentrating on qualitative characters.

Mac Key thought the worth of a cross would be revealed more quickly to the breeder in the pedigree than in the bulk system. He balanced this, however, by recommending the testing of segregating bulk populations and extrapolating from the results an estimate of their value as a source of superior lines. In both cases, the intent was to identify the best crosses with which to continue work and to discard the others.

The experiences of Jennings and Aquino (1968) led them to conclude that "in tropical rice programs, where wide crosses of distinct plant type are essential for breeding advance, the unrestricted bulk population method is futile."

Jennings and Aquino did not discard the bulk population method but instead modified it, thus giving an illustration of one of the many ways breeders can respond to methodological problems. Because one of the major difficulties was associated with competitiveness and early growth rate, they inserted a procedure in the F_2 and subsequent bulks of hand-cutting all tall, leafy, and spreading plants at water level before and after flowering. This, in effect, placed all individuals on a competitively equal status and might be though of as a form of biological handicapping. The analogy to horse racing may be apt because in both cases knowledge of strengths and weaknesses is used to attempt to give equal opportunities throughout the race, in order to provide a competitively equal finish (I am told that a handicapper's ideal, seldom realized, is a dead heat with all noses on the wire).

Tee and Qualset (1975) outlined an interesting accelerated generation scheme, utilizing greenhouse and field, which brought hybrid wheat material from the F_2 to field testing of F_4, F_5 and F_6 in about 1½ years. Two selection methods operating on the same hybrid material were used: single seed descent (SSD) and random bulk populations (BP). In the SSD, one seed per plant, all bulked, formed the planting seeds for each new generation; in the BP, the population was mass harvested and a sample drawn to sow the next generation.

The results showed the two to be essentially equal as selection methods. The similarities in the two methods working on the same biological hybrid material is not surprising. It is possible that unique characteristics of each tended to cancel the other; that is, in the SSD genetic variance is limited by the one seed per plant, which was also vulnerable to further attrition from loss of viability, whereas in the BP it is influenced by the chance draw of the sample for sowing because only a small aliquot of the available seed is needed. The authors chose the BP over SSD, primarily based on economics (lower cost); however, this was not an unqualified recommendation but was based on the absence of severe competition effects. If these effects were present, their choice favored the SSD. I thought an important contribution of the paper was its concern with efficiency, emphasized in the accelerated generation advance scheme.

Khalifa and Qualset (1975) suggested several modifications to the bulk population to lessen the problem of loss of desirable genotypes through competition. Increase in plant spacing should lessen competition. This was supported by the experiments of Sakai and Suzuki (1954) and, I feel, by logical expectation. Another variation suggested by Khalifa and Qualset was to use the single seed descent method, which ensures the continuance of all line progenies. A third modification suggested was stratifying the population according to individual or clusters of characters; that is, in this case removing dwarf plants from the F_2 and bulking them as a subpopulation of the cross.

Their study was more complex and wider-ranging than the above discussion indicates. It dealt with changes in a bulk population of a cross between a short and a tall wheat. The results were that population mean height increased with generations, and the frequency of short-statured plants decreased, reaching zero by F_7. Mean yields of grain increased steadily with the passing of generations; by F_7 some lines exceeded both parents. The authors detected both directional and stabilizing selection occurring naturally in the population. Directional selection was illustrated by the increase in height mean, and stabilizing selection by the decrease in variance for height as generations passed.

The results of measuring and comparing the correlations for grain yield of four early generations, F_2, F_3, F_4, and F_5, in *one season* represented a unique contribution by Whan, Rathjen, and Knight (1981). All generations were plots of derived lines, thus bulks. The great merit of doing this in one season, rather than in successive years, is that the effects of years, sites, and seasonal differences are eliminated.

The results showed that correlations between generations increased as generations advanced. Their illustration shows the following general picture:

F_2–F_3 $r = .51$, F_3–F_4 $r = .68$, and F_4–F_5 $r = .78$, all highly significant. The correlation between the F_2 and later generations, for example, F_5, was variable, and in general was not a reliable predictor. Whan et al., acknowledging the effect heterozygosity and variability of the F_2 play on the low correlation between its yield and that of its progeny, stated,

> It might be more desirable to test progeny from later generations such as the F_3 and F_4. We would expect to get closer associations between performances in two late generations than between two early generations as the greater homozygosity achieved would result in greater genetic similarity between lines in the successive generations.

The authors' recommendations included (1) isolating F_2 or F_3 lines, (2) increasing seed of lines, (3) conducting replicated tests for yield at multiple sites, and (4) repeating over years with best lines.

Another alternative might be drawn from the research, namely, using the bulk population method itself. The F-derived lines are continuing bulks, and the recognition of the desirability of selecting later-generation lines is a bulk population feature, as is the availability of large amounts of seed for multiple testing. The size of bulk populations can be trimmed by a judicious visual selection for discard in the F_2.

The experiences of Ntare, Aken'ova, Redden, and Singh (1984) with breeding methods for cowpeas were favorable both to the SSD and early-generation (F_3) yield testing, the latter using elements of bulk and pedigree methods. In this research visual ratings in the field for yield were almost as effective as the actual yield data. From their experiences I can visualize a modified bulk population system suitable for the small grain cereals, namely, visual discard in F_2, optional F_3 (F_2-derived line bulk) visual rating and/or yield test on selections, one bulk of all selected lines remaining for F_4 and F_5, and removal of lines in F_5 or F_6. There is room here for the exercise of different breeders' judgments as to procedures, but meaningful and positive population control would be accomplished.

THOUGHTS ON THE TIME LINE

I have never comprehended the basis for the statement, accompanying many discussions of the bulk method, that the bulk hybrid method results in a loss of time and that selection should be delayed two or more generations beyond that in the pedigree system. In my paper (Jensen, 1978) I questioned this: "It is impossible, however, to escape the fact that the same population carried by the mass method would contain precisely the same individuals of the same homozygosity at any given familial generation."

Of course, this statement could not be accurate if competition in bulk altered gene frequencies and population composition. Because it is known that competition does act in this manner, the breeder is left with the question whether the competition in segregating populations—the breeder's crosses in that particular environment—is of a kind and magnitude to invalidate F_4–F_6 withdrawal of lines designed for use in a homozygous commercial crop.

Although I considered the idea of lost time indefensible on theoretical grounds, on practical grounds I simply ignored it, removing lines in F_4 or F_5. Allard (1966), however, has offered the explanation of heterozygote advantage to account for the recommended delay:

> It will be recalled that the average percentage of homozygosity under selfing is very high by the F_6 generation and approaches 100 per cent by the F_{10} generation, even when a very large number of genes are segregating. Should heterozygotes be more productive than homozygotes, the rate of return to homozygosity is not accurately represented by a descending Mendelian series, but will be delayed.

Allard considered early-generation competition in bulk to be primarily among heterozygotes, with the result that the desirable forces of natural selection would have insufficient time to be fully effective by F_6 line withdrawal. Consequently, he recommended increasing the number of selections. These practical adjustments seem eminently reasonable.

In a later paper Simmonds (1968) gave a most favorable review of barley composite crosses, beginning with those of Harlan and colleagues in California. He concluded with these remarks:

> The main implications are therefore conceptual rather than immediately practical. A composite cross is a long-term complement to cereal breeding, not a substitute for it; even plant breeders want results in less than twenty years, however willing they are to plan that far, or further ahead.

Indeed, Simmonds' article cuts across so many boundaries (adaptation, competition, performance of mixtures, and others) that it warrants close reading. His list of references is comprehensive.

Sandfaer (1970*) accepted the premise that natural selection acts in bulk populations to increase the frequency of agronomically desirable types, but stressed the long period of bulk handling necessary when

> ... only in the last part of the bulk period does the natural selection act among homozygous or nearly homozygous plants. It therefore seems doubtful that natural selection in a population of this structure and in such a limited period is a really effective force able to bring about an important shift in the genotypic constitution of the population in the desired direction.

*This and other quotes from Sandfaer (1970) throughout the book are reprinted by permission of Jens Sandfaer.

Sandfaer's remarks were directed at "material in bulk for 5–8 generations after which desirable individual plants are selected from the population and continued as lines."

Sandfaer himself made a distinction between short- and long-period manipulations of bulk populations, continuing the above discussion as follows:

> On the basis of the gradual yield increase found in some composite crosses Suneson (1956) proposed an evolutionary plant breeding method. The method involves the growing of populations of the composite type, i.e., populations containing a very large genetic variation, under competitive natural selection for many generations. The final "variety" might be either the population itself or a pure line selected from it. In this case the probability for natural selection to be effective is greatly increased, mainly owing to the prolonged bulk period. The experimental results from varietal mixtures as well as from composite crosses show that it is a slow and rather inefficient breeding method to leave it to natural selection alone to sort out of a composite bulk a final, usable "variety." The use of long-continued bulk handling as a method of handling hybrid populations up to the state where the populations are exploited by selection of individual plants seems more promising, but the long-time requirement will undoubtedly greatly reduce most plant breeders' interest in this method.

It seems to me that the important and crucial point is the division of the time period into the short-term bulk hybrid population portion, and the long-term evolutionary breeding portion. When this division is accepted the bulk hybrid population will stand on its own as a breeding method, and the evolutionary breeding method will be seen as a desirable germ plasm reservoir and breeding supplement to other breeding systems.

The terminology that surrounds the bulk population method includes terms often used as synonyms, such as *composites* and *mass reservoirs*. These populations are sometimes formed as adjuncts to theoretical research. They are not, however, synonyms for the bulk hybrid population breeding system. One characteristic of these other kinds of populations is the range or latitude accepted in their formation, which extends from hybrids to accumulations of germ plasm made for divers reasons. Another characteristic is their timelessness. It seems to me that the key to their use by breeders hinges on their F status when obtained by a breeder. If obtained early enough, that is, by F_2–F_6, they can be adapted to the bulk hybrid population method; if obtained at a higher F status, the breeder receives the earlier research work as a bonus, and selections may be withdrawn at any time; however, these populations gradually have moved beyond the time efficiencies of a practical breeding and selection system.

In a study on selection for yield and yield components in wheat, McNeal, Qualset, Baldridge, and Stewart (1978) noted the rapid reduction of genetic variance as generations advanced. This phenomenon provides an incentive to remove selections at relatively early generations. Extrapolated to a bulk hybrid procedure, line removal in F_4 or F_5 is practical.

Busch and Luizzi (1979) pointed out that a deterrent to early-generation bulk testing has been the fear of losing wanted genotypes through competitive elimination. Their study of height, heading date, and yield changes in bulk populations of six hard red spring wheat crosses showed that the changes followed expected genetic segregation and would have negligible effect on selection. Only when data from all crosses were pooled could a statistically significant shift be detected (toward taller plants and later maturity), but this was deemed to be of minor importance; that is, selection for semidwarfism or early heading was not impeded.

Bulk grain yields, data pooled across generations, showed a positive and linear correlation from F_2 to F_6. Crosses differed for yield, as might be expected, but the generations did not (the high correlation of F_2 yields with subsequent generations lacks individual cross predictive value because of the pooling).

Apropos the concern over early-generation loss of desirable genotypes because of natural selection pressures, Busch and Luizzi ended on a reassuring note in their conclusion that several years in bulk populations are not inimical to selection:

> Since the plants are more homozygous as generations are advanced, more semidwarf plants are present in F_4 or F_5 than in F_2 from tall by semidwarf spring wheat crosses, thus allowing options in the breeding program as to the generation for plant selection.

COMMENTS ON THE BULK METHOD

The bulk population has been extolled in literature and practice both as the bulk hybrid population breeding and selection method, and as the long-range evolutionary breeding method. The problem that has arisen stems from the fact that the bulk similarities of the two otherwise different methods intrude on each method by virtue of the fact that the methods represent two extremes on the same time line without a sharp central boundary. The result is that research extrapolations from one method, whether explicit or implicit, are often applied to the other. It is my belief that this situation, which has existed for a long time, has hurt the acceptance of the bulk hybrid population breeding method rather more than it has affected the evolutionary breeding method.

Jensen (1978) discussed the merits of the bulk hybrid population breeding method in a paper showing its use in the diallel selective mating (DSM) system in cereals. He pointed out that plant selection under bulk population breeding is carried out on exactly the same time schedule as in other recognized efficient methods, that is, in F_4–F_6, depending on the breeder's preference. In practice, the bulking began with the F_2-embryo seeds, regardless of whether they were from a simple cross or a multiple hybrid DSM cross. Rates of sowing were doubled in the Cornell programs to increase the numbers of individuals in plot areas of fixed size; this also minimizes the sampling problem effect. No attention was required for these field nurseries until the selection generation, except that it was difficult not to do some mass selection where it seemed desirable.

The choice of selection generation is optional: I preferred F_4 or F_5. In general, the more advanced the filial generation when selection is made, the higher the percentage of true-breeding lines. Planting the selection generation involved two steps: (1) mixing dead (autoclaved) seeds with the live seeds to provide a desired field spacing for best observation of the individual plants, and (2) planting twin rows of a bulk lot over a desired field distance, for example, the length of a field (see diagram at end of chapter). Space for walking was left along the twin rows. My average time for harvesting selections was 20–30 minutes per population. Surprisingly, even at this stage whole populations were occasionally discarded without selecting any plants; for methodology this suggests that the visual appearance of solid stands can be misleading, and that the appearance of the individual plants is important for significant selection. Promising bulk populations, however, usually produced hundreds of head selections.

My experience with the bulk population method was that it was the most efficient and flexible of breeding and selection systems tried. Fifteen of the Cornell project's grain varieties were produced under the bulk hybrid method. Aside from routine large-plot planting and harvesting, and seed processing (we screened for a medium-large seed, which involved a buffing treatment of oats in a paint shaker, deawning of barley, and color sorting of red–white wheats), there were only the optional procedures to concern the breeder until the year of selection. An optional feature sometimes used was the selection and rebulking of desired segregants in the F_2 or F_3—an example would be saving only short or disease-resistant plants—then discarding the rest of the original bulk.

The bulk hybrid breeding method is a "fail–safe" system, providing the breeder with little likelihood of missing useful segregants. Selections from my bulks always included multiple versions of "look alikes," which became the superior lines in performance tests. These near-clonelike individuals were the result of the expansion of F_2-derived progenies and were the products of within-line genetic variance. "Look alikes" could also be interpreted to represent the breeder's biases, which is a positive way of saying that the breeder is a project's most important screen.

Muehlbauer, Burnell, Bogyo, and Bogyo (1981) used computer simulation to estimate whether bulk population (BP) or single seed descent (SSD) populations retained more additive genetic variance after four generations from the F_2, that is, by F_6. Different assumptions were made for the different methods; for example, the breadth of line continuation over generations was based on line survival or elimination (no offspring) in the single seed descent method, and variation in fecundity (fitness in seed reproduction terms) of individuals in the bulk population method.

By the F_6, the authors found the SSD populations to have greater additive variance than the BP ones. The greater loss of variance in the bulk population was attributed to loss during generation advance occasioned by competition and fecundity.

Some comments are in order relative to this excellent paper. It is unfortunate that there is often a general mutual communication barrier, which operates in both directions, between theoretical and practical research. Muehlbauer et al. are absolutely correct that the single seed descent method exploits or enhances between-line variances whereas the opposite, within-line variances, are favored in the bulk population method. They did not make a choice between the methods except to conclude that the SSD method would be appropriate for the practical use envisaged by them for breeding dry-edible legumes. I want to discuss here some of the issues that lie beyond these theoretical findings.

The first point concerns the meaning of genetic variation. Retention of between-line variance is foreordained in the SSD method: assuming no lethality the population at homozygosis contains the same number (not kind) of variant individuals as the F_2. From a practical standpoint, we know that the SSD method compares favorably with other breeding methods, yet I would have to say that it does so with a certain amount of good fortune. The genetic variance of the SSD exhibits two flaws: (1) There is no within-line variance; thus all the promise inherent in an F_2 line has been lost by the time homozygosis is reached save for the one surviving individual (hence the element of luck). (2) The between-line variance, depending on variations in the overall excellences of different crosses being observed, represents a preponderance of individuals that might better have been eliminated in F_2 or F_3 (or carried with little cost in bulk). A third weakness should be mentioned: the one-seed-per-line scheme is vulnerable to a rather high rate of mortality.

In the case of the bulk population method, the opposite dynamics are presumed to operate: differential fecundity and competition act to decrease between-line variance and concentrate within-line variance. In both these situations practical accommodations are available: simple modifications involving, for example, seed numbers or heads, selection, sampling, recombining after selection, density changes, and more.

Muehlbauer et al. stated: "The substantial within-line component in the BP method indicates considerable duplication of genotypic combinations; such duplication reduces potential for selection and increases probability of selecting similar genotypes."

In a 1978 paper I documented the existence of this phenomenon from practical experience with bulk populations, but with opposite conclusions: "This redundancy of basically similar material with small differences is a boon to the breeder in the final stages of the search for a new cultivar, because it protects him against inadvertently overlooking good material represented by a single genotype sample. . . ."

I would like to call attention to differences in definition and perception of the term *bulk population breeding method*. It is interpreted by some as a method that uses natural selection to sort out superior lines, a definition that relegates the method to a long-range program of 10–20 or more years, or whenever it is felt that the population has matured; indeed, there were many uses of this nature in an earlier period. Tucker and Webster (1970) stated, "Bulk population breeding is a method which depends upon natural selection to sort out superior breeding lines."

They removed lines after 10 and 11 years. This is actually a better description for the evolutionary breeding method. There is evidence also that natural selection does not always sort out the superior lines; for example, Suneson and Wiebe (1942) conducted an experiment, using mixtures of barley varieties and mixtures of wheat varieties, each of which was a successful cultivar in its own right, where the "primary purpose of the experiment was to permit natural selection to operate." The result after five and nine years was differential survival: dominance of some varieties and the practical elimination of others. Frey (1954) viewed the bulk hybrid method not so much as a breeding-selection vehicle but as a sorter of random crosses, that is, a system for choosing among crosses. Indeed, it can play this role.

I have recorded many descriptions in the literature of the bulk hybrid method. It is unfair to single out one statement but I do so because it has the dubious merit of listing more deficiencies in a short space than others:

> Another method which is now little favoured involves growing in bulk up to about the F_7 in order to test almost pure lines. The procedure is time-consuming, favourable genotypes are easily lost and even many years of multiplication provide no information on genotype × year interaction; the advantages of a possible natural selection pressure, even in cereals, do not outweigh all this" (Sneep, J. and A. J. T. Hendriksen, Eds. 1979).

On the other hand, bulk breeding can be seen as a method on a par and competitive with the pedigree, backcross, and single seed descent schemes. In my breeding programs with several grain crops at Cornell University, the bulk population method served as the vehicle for efficiently carrying hundreds, even thousands, of segregating populations through the early generations. Lines were extracted for head rows in the same elapsed-time period as in the pedigree system, that is, F_4 or F_5.

Khalifa and Qualset (1975) refer to the conflicting opinions of ability to survive in bulks and the agricultural merits of cultivated plants, and I interpret this to mean a question exists as to whether genotypes developed under bulk (competitive) conditions are the best, considering other breeding path options, when grown as cultivars under cultivation. The final answer to this question, which has been much studied, is still in the future.

The bulk hybrid population method lends itself to extended and cooperative research efforts because of its great seed production, which offers opportunities for early-generation selection and testing in diverse environments, or for sharing with colleagues.

The bulk hybrid population method is unique among breeding systems because it is the only one in which all members are subjected to the full competitive pressures of the environmnet on a population that is at maximum heterozygosity and heterogeneity. All other systems, through selection, tend to isolate individuals or families through the formation of population subsets in which heterogeneity and heterozygosity forces are weakened and muted.

The full force of environmental pressures on a population is a phrase synonymous with natural selection, a process that eliminates the nonfit and produces the fit individual or population. Fitness in this evolutionary sense means ability to survive to reproduction time, to compete successfuly in leaving survivors. The plant breeding relevance is, in what way is fitness, naturally derived, related to agronomic productivity, artificially defined by humans?

Competition has different consequences, some of which are favorable. Although there also may be heterozygote advantage and natural selection forces acting against the desired direction of selection, I subscribe to the practical view that these forces, operating within the bulk population's short-range F_4–F_6 time requirement, do not have either the persistence or magnitude to eliminate, or reduce below recoverable levels, the population's content of wanted genotypes. I write this mindful of the special cases to the contrary that have been recorded in the literature on intergenotypic competition. The plant breeding relevance of these documented cases is that each breeder must reflect and decide whether his environment, or crop, falls within the parameters of these special cases.

The implication in existing literature is that the extraction of superior lines from a bulk population poses an unduly difficult problem if attempted at a time point that would allow the method to be considered a competitive breeding method. Here lies the nub of the recognition problem: (1) it is not believed that lines can be removed equally as easily from a bulk population as from a pedigree or other breeding system, and (2) it is not believed that these F_4–F_6 lines can be as good, especially for yield, as those selected generations later from the same hybrid populations.

That hybrid populations increase in yield with time, and that this is related to natural selection and fitness, is a reasonable position, soundly documented, and scarcely to be doubted. Neither is it to be questioned that superior lines may be withdrawn from advanced-generation populations, bearing in mind that the yield of a heterogeneous population is the sum of the interactions of near-homozygous individuals that may perform differently when grown in pure stands after selection. What is to be questioned is whether it is necessary to wait up to two decades to obtain these superior lines. For 30 or more years I withdrew lines from F_3–F_6 bulk populations, following which the bulks were discarded. Were there potentially better "superior" lines to be found in these discarded bulks had they been grown for more generations before selection? The answer is not in the confirmed fact that hybrid populations increase in yield with generations. The increase in yield with time is an expected phenomenon, and is due to many of the aspects of a dynamic population; for example, low-fitness types disappear and high-fitness types increase. One reason why so much time is required if left to nature is that these and other interactions with the environment generate a continual back-and-forth surge. My belief is that the answer lies in a direct comparison of a breeder's F_5 selected lines with, for example, the F_{20} selected lines from the same population. Despite a wealth of research data, I think this comparison has not been adequately tested.

One recognizes that with the passing of generations there is a steady progression toward homozygosis of the individuals in a bulk population. We are dealing with F_4, F_5, and F_6 genotypes, well within the range of a plant breeder's "working homozygosity." Each individual comes from a family, an F_2-derived family, and if it is a reasonably fit line it probably has numerous sibs scattered throughout the population. This explains why released varieties from the Cornell wheat, oats, and barley programs had to be chosen from scores of "look-alike" candivars from the same cross.

I referred to the need for further testing. If this were done I would expect the relationship between the yields of lines withdrawn at F_5, the yields of the continuing bulk population, and the yields of lines withdrawn after F_{15} to be somewhat like this:

1. The mean of F_6 selections grown in head rows after F_5 line withdrawal would be greater than the F_6 bulk population. This is logical because breeder experience and visual judgment should exert a positive spin to the yield of selections.
2. The mean of retained F_8 selections after two years of plot tests for yield and discard should be substantially greater than the F_8 bulk population. This is logical because retention of lines is now based on yield data.
3. The mean of the few elite F_{10} candivar lines should be substantially greater than the F_{10} bulk population. Although the bulk population may be increasing in yield, it still may be 5–10 years from achieving its maximum yield. Furthermore, these lines will have been the result of natural selection, and their suitability for commercial growing must still be determined.

My favorable bias toward the bulk population breeding method is evident in my attempts to separate it from the evolutionary breeding method; however, this does not imply bias against the evolutionary breeding segment. In fact, as is evident in the next chapter, the attributes of the long-range system are quite impressive. I have felt, however, that much of the theoretical research on population dynamics is applicable more to the long time range (evolutionary) than to the short time range (bulk population) aspects.

Some personal comments on my use of the bulk population system may be helpful. From the outset of my career, I opted to conduct large crossing programs. This was a direct result of a fascination with the possibilities of germ plasm. There was an obvious conflict between large numbers of crosses and the program requirements of the pure pedigree system that I inherited. I "evolved" slowly toward the bulk population system, searching for expeditious ways to handle larger population numbers. It was easy to handle populations in bulk in the field, and soon we had acres of F_3 and F_4 plots, many of them of great length, inasmuch as we tended at that time to plant all of the seeds from the F_2; our breeding nurseries occupied 33 acres one year. A great amount of mass selection effort was expended on these plots, for example, for height, disease reaction, and so forth, using scissors, rechargeable hedge cutters, pulling of plants, inoculating susceptible plants, and similar procedures. This extended to mass grooming of the seed supplies, a very effective way of reducing stocks both for sowing and for inventory. In wheat this involved rigid use of screens and, because we were breeding for a soft white winter wheat region, hand-picking of white (recessive) kernels. In barley and oats, respectively, we used deawning and friction (paint shaker) procedures.

All of this was a learning experience, which could have been shortened had I critically read Florell's benchmark paper at that stage. I was "reinventing the wheel"—not entirely a wasted effort—and finally I was using a bulk system that was similar to Florell's basic recipe. One modification worthy of reiteration is related to the planting arrangements of a population in the year of line selection, F_4 or F_5. We referred to this in the Cornell project as the "23–78" arrangement, which designated twin rows formed from seeds planted by a nine-hole farm drill using, however, only the second and third, and seventh and eighth drill feeds. A mechanical but variable space planting was achieved by mixing the seeds with appropriate amounts of the same crop seeds that had been killed by autoclaving. This procedure is critical because a selection made from a bulk crop-density population is, in my estimation, little better than random. A typical cross-field planting had this appearance:

drill wheel

1
2 Pop. 1 .. ..
3 Pop. 1.. ... .
4
5 sowing direction →
6
7 Pop. 2 .... Pop. 3 .. ..
8 Pop. 2. .... Pop. 3 ....
9

drill wheel

The breeder uses the open pathways to walk around each cross population, collecting a head from each desired plant. The head may be either dropped in a bag or left on a length of straw to make a bundle.

Ambiguous as is the role of competition in bulk population and evolutionary breeding, I think we can make a further distinction between these two methods, and it is this: the bulk population method *avoids* natural selection, whereas the evolutionary breeding method does not. Put another way, the bulk population breeder has and uses options that allow him to avoid the *penalties* of natural selection. I had a practical reason, the sampling problem, for doubling and in some cases quadrupling the planting rate, thus creating a dense, competitive environment. It was necessary to get as large a number of seeds representative of each hybrid population in each planting area unit as was possible. Even when bulking began with F_2-embryo seeds, the maximum generations in bulk were only three, otherwise two, before planting for line selection. A surviving F_2 or F_3 plant is going to advance some of its progeny into the next bulk population, and natural selection (competition) is unlikely to eliminate all the within-line representatives.

A procedural point that had a favorable bearing upon the Cornell use of the bulk hybrid system is this: the stringent processing of seed lots for uniform seed size acted to remove one of the sources of unequal competition in the populations.

In summary, the great merit of the bulk population system is that it bypasses all of the questions and problems related to spacing, density, tillering, heterosis, heterogeneity, correlation between generations, and early generation testing—except one, the effect of negative competition effects on genotype survival, and this is an element subject to monitoring and preventive action if necessary.

It is my belief, after surveying the literature on plant breeding and selection of the past century, that the bulk hybrid population method has suffered from a crisis of identity. Specifically, the practical breeding method has not been separated from research more applicable to the long-range evolutionary breeding method. This has resulted in a confrontational relationship, wherein research results, particularly in the areas of natural selection and extended time situations, are seen as mutually exclusive in this circular reasoning. The two parts must be seen as separate entities. I have discussed the need for this division, suggesting that it be made at the F_6 (Jensen 1978, p. 623):

> The ambivalence, in my opinion, is due to a failure over the years to make clear distinctions between short range practical breeding programs and long range supplementary programs involving composites. I loosely define these categories as follows: short range programs employ procedures that permit extraction of "pure" lines no later than F_6; long range programs delay extraction of lines beyond F_6

We must make this separation, and I have taken a first step by treating these two methods in separate chapters, and suggesting that Florell's *bulked population method* and Suneson's *evolutionary breeding method* be adopted, respectively, as official designations.

REFERENCES

Akerman, A. and J. Mac Key. 1948. The breeding of self-fertilized crops by crossing. Svalof 1886–1946, 46–71 (Lund).

Allard, R. W. 1966. Principles of plant breeding. John Wiley and Sons.

Atkins, R. E. 1953. Effect of selection upon bulk hybrid barley populations. Agron. J. 45:311–314.

Busch, R. H. and D. Luizzi. 1979. Effects of intergenotypic competition on plant height, days to heading, and grain yield of F_2 through F_5 bulks of spring wheat. Crop Sci. 19:815–819.

Coyne, D. P. 1968. Correlation, heritability and selection of yield components in field beans, *Phaseolus vulgaris* L. Proc. Am. Soc. Hort. Sci. 93:388–396.

Florell, V. H. 1929. Bulked-population method of handling cereal hybrids. J. Am. Soc. Agron. 21:718–724.

Frey, K. J. 1954. The use of F_2 lines in predicting the performance of F_3 selections in two barley crosses. Agron. J. 46:541–544.

Hamblin, J. 1977. Plant breeding interpretations of the effects of bulk breeding on four populations of beans *(Phaseolus vulgaris* L.). Euphytica 26:157–168.

Harlan, H. V. and M. L. Martini. 1937. Problems and results in barley breeding. USDA 1936 Yearbook Separate 1571:303–346.

Harlan, H. V., M. L. Martini, and H. Stevens. 1940. A study of methods in barley breeding. USDA Tech. Bul. 720, 26 pp.

Jennings, P. R. and R. C. Aquino. 1968. Studies on competition in rice. III. Mechanisms of competition among phenotypes. Evolution 22:529–542.

Jensen, N. F. 1978. Composite breeding methods and the DSM system in cereals. Crop Sci. 18:622–626.

Khalifa, M. A. and C. O. Qualset. 1975. Intergenotypic competition between tall and dwarf wheats. II. In hybrid bulks. Crop Sci. 15:640–644.

Mac Key, J. 1962. The 75 years development of Swedish plant breeding. Hodowla Roslin, Aklimatyzacja I Nasiennictwo, Tom 6:437–467.

Mac Key, J. 1964. Selection procedures in wheat and barley breeding. Inform. Bull. Near East Wheat Barley Improv. Prod. Project. Vol. II, No. 1:1–5.

McNeal, F. H., C. O. Qualset, D. E. Baldridge, and V. R. Stewart. 1978. Selection for yield and yield components in wheat. Crop Sci. 18:795–799.

Muehlbauer, F. J., D. G. Burnell, T. P. Bogyo, and M. T. Bogyo. 1981. Simulated comparisons of single seed descent and bulk population breeding methods. Crop Sci. 21:572–577.

Mumaw, C. R. and C. R. Weber. 1957. Competition and natural selection in soybean varietal composites. Agron. J. 49:154–160.

Ntare, B. R., M. E. Aken'ova, R. J. Redden, and B. B. Singh. 1984. The effectiveness of early generation (F_3) yield testing and the single seed descent procedures in two cowpea *(Vigna unuiculata* (L.) Walp.) crosses. Euphytica 33:539–547.

Palmer, T. P. 1952. Populations and selection studies in a *Triticum* cross. Heredity 6:171–185.

Qualset, C. O. and H. E. Vogt. 1980. Efficient methods of population management and utilization in breeding wheat for Mediterranean-type climates. Proc. Third Intern. Wheat Conf, Madrid. pp. 166–188.

Raeber, J. G. and C. R. Weber. 1953. Effectiveness of selection for yield in soybean crosses by bulk and pedigree systems of breeding. Agron. J. 45:362–366.

Sakai, K. 1951. Studies on individual selection and selective efficiency in plant-breeding. Jap. J. Breeding 1:1–9.

Sakai, K. 1954. Theoretical studies on the technique of plant breeding. I. Value of heritability in bulks and plant progenies of successive hybrid generations in autogamous plants. Jap. J. Breeding 4:145–148 (English abstr.)

Sakai, K. J. and Y. Suzuki. 1954. Studies on competition in plants. III. Competition and spacing in one dimension. Jap. J. Genet. 29:197–201.

Sandfaer, J. 1970. An analysis of the competition between some barley varieties. Danish Atomic Energy Commission, Riso Rep. No. 230:114 pp.

Simmonds, N. W. 1968. Report by the Director, Scottish Plant Breeding Station, Pentlandfield, 47th Ann. Rep., 52 pp.

Sneep, J. and A. J. T. Hendriksen (Eds.). 1979. Plant breeding perspectives. Centennial publication of Koninklijk Kweekbedrijf en Zaadhandel, D. J. Van der Have Seeds, 1879–1979, Centre for Agricultural Publications and Documentation, Wageningen.

Suneson, C. A. and G. A. Wiebe. 1942. Survival of barley and wheat varieties in mixtures. J. Am. Soc. Agron. 34:1052–1056.

Taylor, L. H. and R. E. Atkins. 1954. Effects on natural selection in segregating generations of bulk populations of barley. Iowa State J. Sci. 39:147–162.

Tee, T. S. and C. O. Qualset. 1975. Bulk populations in wheat breeding: comparison of single-seed descent and random bulk methods. Euphytica 24:393–405.

Torrie, J. H. 1958. A comparison of the pedigree and bulk methods of breeding soybeans. Agron. J. 50:198–200.

Tucker, C. L. and B. D. Webster, 1970. Relation of seed yield and fitness in *Phaseolus lunatus* L. Crop Sci. 10:314–315.

Wesenberg, D. M. 1974. Composite cross breeding. Ninth Am. Barley Res. Workers' Conf., 17 pp. mimeo.

Whan, B. R., A. J. Rathjen, and R. Knight. 1981. The relation between wheat lines derived from the F_2, F_3, F_4 and F_5 generations for grain yield and harvest index. Euphytica 30:419–429.

10

EVOLUTIONARY PLANT BREEDING

Suneson (1956), long an admirer of Harlan and Martini's pioneering work with bulk hybrid populations, proposed the evolutionary breeding method in an attempt to incorporate increased genetic diversity into the bulk system. He noted:

> It requires assembly and study of seed stocks with diverse evolutionary origins, recombination by hybridization, the bulking of the F_1 progeny, and subsequent prolonged natural selection for mass sorting of the progeny in successive natural cropping environments....
>
> ...The core features of the suggested breeding method are a broadly diversified germplasm, and a prolonged subjection of the mass of the progeny to competitive natural selection in the area of contemplated use.

Suneson suggested 15 generations of natural selection:

> Thereafter there can be repeated recourse to three methods of breeding, (1) continued natural selection with prospects for significant gains in yields to accrue throughout a working lifetime; (2) cyclic hybrid recombinations with intervening natural selection to give a kind of recurrent selection; or (3) resort to conventional selection and testing.

I often had wondered why Suneson, who first reported the existence and potential use of the barley male sterile genes, did not incorporate the male sterile technique into his evolutionary breeding proposal, turning it in effect into a male sterile facilitated recurrent selection (MSFRS) method. In the 1956 study, however, one of the four bulk hybrid populations did contain the *ms* gene, a fact that was recognized but not considered relevant to the study, and no extraction of *ms*-derived seeds was considered. Suneson merely stated, "Its more persistent heterozygosity was not considered in this paper."

Bal, Suneson, and Ramage (1959) examined changes in 21 characters in three F_{30} versions of barley Composite Cross II, namely, no selection, selection for large seeds, and selection for small seeds. The shift in characters noted over generations was seen as affirming the concepts of evolutionary plant breeding.

Jain (1961) reviewed the bulk population breeding method using certain generations from Harlan's Composite Cross V of barley which evolved from a bridging series of crosses among 30 varieties. His objective was to assess the effect of natural selection on genetic variability, as well as the method's usefulness. Examining four aspects of the method he ascertained the following:

1. *Yielding Ability and Fitness.* Using the following year's mean yields of randomly chosen individuals, he found a linear relationship between yield and fitness (seed numbers produced) over a range of advancing generations; for example, yields were F_4 307.2 g, F_7 304.7 g, F_{14} 353.2 g, and F_{19} 394.5 g. The correlations with fitness were highly significant.

2. *Components of Fitness.* Jain described fitness under natural conditions as composed of three attributes, namely, germinability, survival to reproduce, and number of seeds produced. He charted the changes observed in eight marker characters in five generations between F_3 and F_{18}, and noted different patterns of frequency change.

3. *Genetic Variability.* Although some of the traits observed conformed to the expected per-generation halving of heterozygosity, two characters did not, showing 10 and 13% heterzygotes as late as 18 generations after the cross.

4. *Response to Selection.* A form of stabilizing selection pressure was noted where, under natural selection, extremes were avoided and a more moderate expression became the fitness model. Jain pointed out that breeders tend to select for maximum expression of desirable traits (the use of an index selection seems in this sense to be more nearly like the natural selection response).

Jain referred to a phenomenon, *environmental slippage*, which would influence bulk hybrid populations more than artificial selection programs. Slippage might be said to be the forward–backward oscillation that is likely as a population experiences a series of varying environments. This would seem to relate more to a bulk hybrid population in the evolutionary, many-generation sense where there would be ample time involved.

Jain acknowledged that progress under natural selection might be too slow. His suggestion of isolating superior family lines by F_{10} or F_{15} still leaves the method in the category of a long-range, evolutionary breeding system.

Allard and Jain (1962) used existing F_3, F_5, F_6, F_{13}, F_{15} and F_{18} populations of Harlan's well-known barley Composite Cross V in a statistical study of changes in quantitative characters. With one exception (spike density) generation means shifted slowly in the direction of reduced height, earlier heading, larger seeds, and shorter spikes. Variance decreased with time, and this was considered a stabilizing movement inasmuch as it was accomplished by dropping extreme individuals from the tails of the curve. Even so the expressed variability seen in individuals by F_{19} was very high—few alike when carefully scrutinized—and Allard and Jain attributed this to residual heterozygosity owing to heterozygote advantage.

It seems reasonable that fit survivors will pour more progeny into the next generation, which would lead to a heterogeneous population with many homozygous individuals showing small differences. This was exactly what was observed in the F_4 and F_5 plant (head) selections from my composite breeding programs (Jensen, 1978).

Suneson, Ramage, and Hoyle (1963) extended Suneson's 1956 concept of an evolutionary plant breeding method to include mutant populations. They analyzed four generations of an equal mixture of 16 X-ray-derived mutants and 'Hannchen' barley, and concluded that the better mutants survived. There were two characteristics of the mutant populations that favorably predisposed them to evolutionary breeding: (1) sterilities associated with irradiation favored intrapopulational crossing, and (2) the increased heterozygosity promoted a further advantage.

Allard and Hansche (1964), in their wide-ranging discussion of population variability, stated, "The problem is to find the most efficient method of managing the hybrid gene pool so as to maximize the probability of obtaining optimal gene combinations."

This is particularly relevant to evolutionary breeding methods in view of the elements of time and natural selection.

Wesenberg (1974) presented a detailed review of the history and use of composite crosses in barley. In the beginning they were formed from single crosses, sometimes derived from diallel or partial-diallel parental arrangements. Then came a move to a bridging cross system involving double crosses (F_1s as parents). Spring barleys were involved almost exclusively to this stage. The third stage arrived with the discovery and use of genetic male sterility. Today the use of composites has cut across spring and winter types and shows no geographic limitations. The importance of composite cross breeding is evident clearly from Wesenberg's statement that a 1958 survey of 165 U.S. and Canadian barley cultivars included 17 (10%) that were selected from composite crosses.

Evolutionary plant breeding may be thought of as a "stretched" bulk, resting on the dynamics of natural selection, and unique in the length of its development period. Theoretically and practically sound, its greatest value would seem to be as an adjunct to a conventional program; that is, any breeding program would be enriched by such long-range resource-forming composites. On the other hand, because of the long wait before selection could begin, it is not feasible as a single-method breeding program.

A compromise approach would be to treat all crosses by the bulk hybrid population method, withdrawing selected lines in F_4–F_6 for the breeding project. If the population is judged useful for the future, it can be continued as an evolutionary breeding entity.

REFERENCES

Allard, R. W. and P. E. Hansche. 1964. Some parameters of population variability and their implications in plant breeding. Adv. Agron. 16:281–325.

Allard, R. W. and S. K. Jain. 1962. Population studies in predominantly self-pollinated species. II. Analysis of quantitative changes in a bulk hybrid population of barley. Evolution 16:90–101.

Bal, B. S., C. A. Suneson, and R. T. Ramage. 1959. Genetic shift during 30 generations of natural selection in barley. Agron. J. 51:555–557.

Jain, S. K. 1961. Studies on the breeding of self-pollinated cereals. The composite cross bulk population method. Euphytica 10:315–324.

Jensen, N. F. 1978. Composite breeding methods and the DSM in cereals. Crop Sci. 18:622–626.

Suneson, C. A. 1956. An evolutionary plant breeding method. Agron. J. 48:188–191.

Suneson, C. A., R. T. Ramage, and B. J. Hoyle. 1963. Compatibility of evolutionary and mutation breeding methods. Euphytica 12:90–92.

Wesenberg, D. M. 1974. Composite cross breeding. Ninth American Barley Research Workers' Conference, 17 pp.

11

SYNTHETIC LINE POPULATIONS—A PREAMBLE

There is a danger of elevating the term *multiline* to generic status covering any mixture, as witness the following statement: "The use of multiline varieties in self-pollinating crops has been advocated repeatedly in the literature since the late nineteenth century..." (Marshall and Pryor, 1978). What has been advocated repeatedly is crop and species mixtures.

MULTILINE AND MULTIBLEND VARIETIES ARE NOT THE SAME

Increasingly today, the differences between crop mixtures or blends and multiline varieties are being blurred inaccurately. It is my intention in this book to reverse this tendency, and I have chosen to do that by coining *multiblend* as a term on an equal footing with *multiline* [we already have *uniblend* and *biblend*; see Federer, Connigale, Rutger, and Wijesinha (1982)], and devoting a separate chapter to each. A first task, surely, is to show that a distinction does exist.

I am reminded of an old baseball story involving, perhaps apocryphally, the famous umpire Bill Klem. A crony was describing a controversial pitch, giving an exact description of how the ball passed over the plate, and ending with this question, "Now, Bill, what was that?" Klem, glowering, took his time, then replied, "It ain't NOTHING until I call it." As the author of the first paper on the multiline (Jensen 1952), and coiner of the very word, I have a right to "call" this pitch. The multiline variety is clearly a breeding proposal for developing a variety in a new way, dealing with breeding lines out of a breeding program, clearly differentiated from variety and crop mixtures.

When my multiline paper was written data from a pilot program did not exist, but there were numerous results from crop mixture studies that could, and still do, provide analogous information. Even my presentation of these data made clear the distinction between the new proposal and the crop mixture studies (*underlines* added):

> One of the basic pieces of information plant *breeders* should have before *designing a multiline variety* is data on the relative performance of *lines* when grown singly and in a mixture. To be successful a mixture should yield at least as well as the average of its components grown alone. Any consistent depression of yield of a mixture below this level would indicate a lack of efficiency for the mixture. Some information which sheds light on this subject is *available from studies with mixed grain crops*.

It is unnecessary, perhaps, to note that even today, a third of a century after this proposal, the principal vehicle for research on component lines remains the cultivars usually associated with variety mixtures and blends. Research on component lines uniquely identified with multiline breeding is naturally restricted to those very few places that have multiline variety breeding programs. The fact that cultivars and independent genotypes are used as a source of information on how multiline varieties might perform is not a valid excuse for equating variety mixtures or blends with multilines. It is a canard that my 1952 multiline paper was "merely a paper on variety blending," as has been said.

The contention that the multiline rests on a breeding concept has an interesting corollary in practice; namely, it is almost impossible to conceive of a multiline variety developing outside of a research and breeding program. On the other hand, crop mixtures or multiblends can be and have been put together, marketed, and used by farmers, seedsmen, or anyone with a reason for doing so (an undesirable reason for a commercial mixture might be the disposal of surplus seed inventories). The technical requirements alone for a multiline, for example, the detailed and specific development of lines for balanced protection against pathogens, make it distinctly different from a multiblend.

Murphy, Helsel, Elliott, Thro, and Frey (1982) called attention to another matter that confuses the two terms: "The Crop Science Society of America has a policy of registering multilines as 'cultivars', but certain seed regulatory agencies insist that multilines that enter into commerce be labeled 'blends'."

This policy apparently is concerned with the possibility of internal population, that is, component, shifts within the multiline as compared to a pure line variety. Here is an ironic paradox indeed: a great merit of the multiline is its stability and promise of longer useful life, yet it is downgraded to the status of a blend on the basis of suspected internal instability!

There is an interesting distinction to be made between populations, based on their origin and internal structure. In populations such as multiblends and multilines, which are homozygous and heterogeneous, genetic shifts can be controlled through annual composition of new planting seed. Populations that are hybrid in nature, however, are variably heterozygous and heterogeneous and, except for the F_1 hybrid, internal genetic shifts cannot be controlled or moderated through the medium of new seed composition annually. In hybrid populations the competitive environment is continuous, and each new generation faces the risk of elimination of noncompetitive yet desirable individuals.

There are also enough differences between multiblends and multilines to suggest other distinctions between the two. For example, component lines for a multiline are chosen on the basis of several factors other than performance in a monoculture. Also, they need not be totally homozygous. A critical difference is that multiline breeding requires a research and development breeding program. Mixtures or multiblends do not, although it would be desirable. Multiline breeding requires the close cooperation of plant specialists, for example, plant pathologists and plant breeders; this should also be the case in multiblends, but unfortunately, I know of no case where multiblends receive this research attention.

DISEASE PROTECTION—A POWERFUL STIMULUS

The capacity of one component of a mixture to take advantage of an "abnormal" environment was shown in experiments with durum–bread wheat mixtures conducted by Klages (1936) in South Dakota. The abnormality of season was due to a destructive epiphytotic of wheat stem rust; all wheats were susceptible to some degree but the rust developed more slowly and was less destructive on 'Mindum' durum.

Stevens' (1942) prescient paper called attention to the hazards to crop production from disease- and insect-caused losses and linked these hazards to inherent consequences of plant improvement programs. He pointed out that historically a relatively stable agriculture was based on established and adapted varieties, and when losses occurred they were usually due to one of two factors, changes in climate or changes in parasites. With no derogatory intent he added a third factor, the work of plant breeders.

Today we are all too aware of the correctness of his views. One can even say that plant breeders are the victims of their very successes because the popularity of new varieties often blankets an agricultural area with a single genotype. On the other hand, considerable progress has been made in the strategies of creating and deploying varieties.

In a reply to Stevens in the same year, Briggs (1942) was quick to extol the backcross method for breeding disease-resistant crops. He cited the security of adding genes for corrective purposes to a known genetic background.

GENERAL INFORMATION ON MIXTURES

Glenn Burton (1948) aptly referred to the subject of his studies in pearl millet as "controlled heterosis." He pointed out that the difficulties of sufficient F_1-embryo seed production for crops requiring high seeding rates are a deterrent to commercial hybrid production.

Burton was intrigued by an earlier report (Tysdal and Kiesselbach, 1944) involving the inclusion by mixing of various proportions of seed of a high-yielding alfalfa variety, 'Ladak', and its first-generation selfed inbred. The respective proportions of the two were 100 : 0, 75 : 25, 50 : 50, and 0 : 100. The forage yields, in the same order, were 4.24, 4.09, 3.80, and 2.46 tons/acre. There were no significant differences among the mixtures and cultivar, despite the low yield of the inbred alone. The significance of Tysdal and Kiesselbach's findings are that a population can tolerate a high amount of selfs without reduced performance. Extrapolating these findings to a hybrid population, which would be expected to contain a proportion of selfs, there would seem to be a further opportunity to ease seed production by including seed of varieties without detracting from the overall performance of the hybrid.

Burton verifed that cross pollination was the predominant natural reproduction pattern in millet, that selfing resulted in reduced vigor, and that F_1 yields were the maximum obtainable, combining ability considered. He calculated that a mixture of a selected four lines would produce 75% hybrid and 25% selfed or sibbed seeds, and that the lower vigor of the latter would result in their natural elimination because of the competition imposed by the high seeding rates. His study, then, became research on the yield response of seed mixtures of hybrid and parent (inbred) seeds.

The proportions of hybrid and parents mechanically mixed and tested for 6 years are given below along with actual and expected yields in percent (parent yield was base of 100%), taken from Burton's Table 1:

Hybrid : Parent	Actual %	Theoretical %
100 : 0	119.1	119.1
90 : 10	120.7	117.2
80 : 20	121.2	115.3
50 : 50	118.1	109.6
20 : 80	108.5	103.8
0 : 100	100.0	100.0

These dramatic results showed that none of the mixtures, 50 : 50 and higher on the hybrid side, differed significantly in forage production; furthermore, all mixtures performed above expectations.

In a rate of seeding or density experiment, it was found that hybrid plants were more vigorous, survived better, and a 10 lb/acre seeding rate (3–3½ seedlings per inch of row) would provide the best mix for population performance. Burton recommended at least three or four parent lines to the seed-producing mixture, anticipating that, for example, four lines would produce a 75% hybrid : 25% selfed or sibbed result, whereas a mixture of only two lines would be expected to yield only at a 50 : 50 ratio.

These findings in a cross-pollinated crop are relevant, understandable, and readily transferable to multiline and heterogeneous or heterozygous variety formation.

Research on variety mixtures or multilines most often has been on yield and disease aspects. Grafius (1966) measured the effects of mixtures of three oat cultivars on lodging resistance, yield, and test weight. He concluded that some mixtures improved the "worth" of the population; that is, weak and strong points can be merged into a mutually advantageous cultivar. The mixtures were made by 10% increments and Grafius stressed that "blind mixing in equal proportions" be avoided because there was no reason to believe, without testing, that the advantage of a mixture would follow mixing proportions.

Frey and Maldonado (1967) reported on a study of mixtures that was invaluable to their developing research and use of multiline varieties. They took six oat cultivars and constructed 57 mixtures, testing them for yield at two sowing dates approximately a month apart for three years. Oats is always planted as early as possible in the spring, and many studies have shown measurable and consistent daily losses in yield if the sowing date is delayed; thus Frey and Maldonado created two rather distinct environments for their tests. Their findings were as follows:

Mixture Advantage: There was none at the first (earliest) sowing date; mixture yields approximated component means. At the second (late) sowing all yields were reduced but the mixture yields showed an average 4% advantage over their component means. These findings may be interpreted to show that heterogeneous populations thrive better in stress environments than do homogeneous ones.

Yield Stability: Mixtures were shown to be more stable across sowing dates than were cultivars, as was shown by substantially higher dates × cultivar mean square versus dates × mixtures mean square.

Number of Cultivars in Mixture: At both sowing dates no association was noted between the yield of a mixture and the number of its components; for example, mixtures of two were as effective as mixtures of more.

Specific Component Effects: No component was found to contribute a specific productivity influence; that is, mixture yields were a function of combination rather than component.

Sage (1971) gave the results of spring wheat (four varieties) pure stands compared with their equal numbers of plants mixed by pairs, all in spaced plantings. Overall, the average yields of the mixtures was 10.5% more than the pure stands; however, in a separate test over a range of sowing rates the mixtures exceeded midparental values only at the lowest rate, and yielded less at all other increased density levels.

According to Sage, 'Rothwell Perdix' was a highly successful commercial winter wheat multiline variety in England. It was short-lived, however, falling prey to a new race of leaf rust. Described as "a mixture of a number of pure lines," its demise suggests either the catastrophic, unavoidable impact of a new race, or a flaw in the backup research program to prepare reserve lines in anticipation of need.

Sage's findings that mixtures exceeded midparental values only at low-density plantings is analogous to most findings on the behavior of early-generation spaced plants relative to their later behavior in pure culture at high density.

Sage's general conclusions suggest that mixtures per se will not lead to higher yields in benign environments, but may have an advantage in unfavorable environments. This fits the hazard and insurance features that are a basic part of the multiline concept.

Sage's experience with mixtures led him to suggest a multiline approach to hybrid wheat: blending male fertile (selfed) and hybrid seeds in an approximate 50:50 mixture, either through harvesting the separate seed strips together or by actually incorporating the pollinator plants in the male sterile population. The result would be less expensive production, larger seed supplies, and perhaps a "kicker" in yield increase, particularly if, as Sage suggested, the seed mixture were sown at lower density, where heterotic and heterogenetic forces would be favored. A similar approach was suggested by Pfeifer (1966) and Jensen (1967).

In a study of natural selection, Blijenburg and Sneep (1975) compared eight barley varieties in monoculture with an eight-line initial equal-seed-number mixture over six years. The monocultures exhibited a reasonable steady-state reaction, but the variety components in the mixture changed dramatically with the final pattern already visible after the first growing year. The correlation between yield in monoculture and in mixture (competing ability) was high, although there was one exception, the variety 'Alasjmoen'. In this case there was a ready explanation: Alasjmoen is a semi-winter type and showed slower early growth, to its disadvantage; consequently, it showed to better advantage in monoculture than in mixtures. In general, the authors' conclusions were that monoculture yields were in good agreement with competitive ability.

Extensive research studies comparing mixtures of pure lines with monocultures by Shorter and Frey (1979) resulted in the conclusion that mixtures did not provide an advantage over monocultures.

Using hill plots and measuring grain and straw yields, the authors conducted two experiments, both on a massive scale. Both involved the same group of eight cultivars and 20 random F_9-derived lines from a composite hybrid population. Experiment 1 dealt with two-component mixtures, in three diallel sets arranged as follows:

Set 1: 378 1:1 mixtures of the 28 base components.

Set 2: 105 1:3 component frequency mixtures among lines 1–15 (four cultivars and 11 random lines).

Set 3: 105 3:1 component frequency mixtures among lines 1–15 (same group).

Mean straw yields within sets were slightly more variable than grain yields, which overall were remarkably uniform, varying a maximum of 0.6 quintal/ha and not statistically significantly different. There were, however, several cases of statistically significant differences, reflecting lines and interaction effects. Mixture sets exhibited greater buffering as shown by fewer significant environment interactions. Unexplained was the observation, from calculations, that the genetic variance among 1:1 mixtures was greatly subdued below expectations—from one-fifth to one-half of those found among monoculture components—but it seems this may also be related to populational buffering.

General mixing ability predominated and was statistically significant, in contrast to specific mixing ability, which was not significant.

The authors did find several mixtures that showed substantial gains in both grain and straw yields over their high components. However, using a two-tailed *t*-test the authors attributed these few cases to chance deviations. Applied to a breeding program, however, it would seem that these extended-tail cases might merit further examination on the premise that they might not be chance deviations.

A very interesting finding was that mixtures of lines were more effective than mixtures of cultivars in generating high yields of both grain and straw. The authors suggested this was due to natural selection favoring survival in mixtures—an apt description of the F_9-derived lines' experience. This is in accord with my original concept of multine formation, where diversity would be favored by the selection of genotypes out of the breeding program.

The relation of high-yielding components to a mixture was shown by these comparisons:

1. *Straw Yields*. Twenty-three of the highest-yielding 1:1 mixtures always had at least one of the five best components.

2. *Grain Yield*. Twenty-three of the highest 27, of the three-line mixtures always had one of the three best components.

However, in almost all cases, mixtures were not superior either to their best component or to the highest-yielding line grown alone.

Overall, mixtures performed more impressively than monocultures relative to greater stability. On balance, mixtures showed no advantages for yield over monoculture components; however, it was not entirely clear that the few exceptions attributed to chance deviations were indeed that.

This study by Shorter and Frey indicated that random lines from bulk populations, and presumably advanced generations, are more effective in mixtures than are cultivars. The study also showed that superior mixtures must include high-yielding components. There is no conflict here: random lines must be tested fully as carefully as lines originating, say, by single seed descent or pedigree selection. The difference among lines may be the result of the type of competitive advancement and natural selection that they have experienced.

From my experience with multiline theory I would agree that the performance of a synthetic is less concerned with number of components than with their quality, potency, and combining ability. This concern with quality rather than quantity can be addressed by the breeder through a preliminary choice of potential parents based on the desired traits. The final desired number may be determined by a form of restricted diallel testing of the lines in mixture.

Rajeswara Rao and Prasad (1982a, 1982b), failed to find any yield advantages of spring wheat mixtures.

Baker and Briggs (1984) conducted a diallel competition test of 10 barley cultivars in their 45 possible 50 : 50 combinations, measuring yield, grain and shoot weight, and harvest index. The three-year average results showed no significant differences between mean performance in pure stands and in biblends. There were, of course, some year and trait differences but the authors opted for the mean effects and concluded that intergenotypic competition would likely not be a serious factor in a breeding program. There was one exception: where relatively similar genotypes are involved it could be a factor. There remains a difference, however, between the range of diversity of genotypes possible in a biblend as compared to the range in a population of segregating genotypes.

GENETIC DIVERSITY AND STABILITY PLAY A PART

A review of genetic diversity in crop plants as a protection against pathogens and insects was made by Suneson (1960). Long a champion of germ plasm diversity and its uses in cereal breeding, Suneson chronicled some of the aspects of types of diversity, for example, nonuniform commercial cultivars, multiline varieties, host–pathogen relations, and so forth. Practical examples from his own barley breeding projects were given.

An important study on the relationship between population genetic diversity, stability, and performance was presented by Allard (1961). In review he cited the case with corn, where the highest yields in particular environments were usually from single crosses, which are as genetically uniform as pure lines with the added kick of heterosis. Today, however, maize hybrids are predominantly single crosses, many as widely adapted as older double crosses. Allard wished to test the validity of these associations in the self-pollinated crops. To do this he constructed 10 lima bean populations organized from the following three groups: (1) three pure line cultivars, (2) four synthetic mixtures including the three possible cultivar two-line combinations and one three-line combination, and (3) three hybrid unselected bulk populations (F_7 and F_9) from the three possible crosses of the three cultivars. Testing of the 10 populations covered a four-year period.

Four comparisons were made to assess productivity (yield) and genetic diversity; these are listed with results:

1. *All Pure Lines versus All Mixtures and Bulks*: the yield order of differences, highly significant, was bulks, pure lines, mixtures.
2. *Each Mixture versus Its Pure Line Components*: mixtures were generally intermediate to pure line components.
3. *Each Bulk Hybrid versus Its Parents*: each bulk equaled its better parent in yield and yielded significantly more than its parents' mean.
4. *Mixtures versus Bulks Derived from the Same Pure Lines*: bulks showed a decided advantage over mixtures (39 out of 48 paired comparisons).

In sum, the overall yield advantage was to the hybrid bulks. Mixtures did not show well, either compared with hybrid bulks or pure lines. The study seemed to provide a fair comparison based on a common genetic base.

In an exhaustive study Clay and Allard (1969) examined the yields and stabilities of a number of mixtures of pure lines of barley, and in the second year, a group of nine F_2 populations from crosses between some of the pure lines used in the mixtures. Ten primary cultivars with a range of type and adaptation were chosen. Twenty-three mixtures were formed using the primary cultivars; there were thirteen 2-component, two 3-component, five 4-component, two 5-component, and one 10-component population. The mixtures were made up each year with equal (proportionate) component seed contribution. The results were as follows:

1. *Yields:* none of the mixtures yielded significantly more than the best component. Three were significantly higher yielding than their midcomponents. The overall performance of the 23 mixtures, however, averaged 1.5% above the midcomponents.

2. *Number of Components:* no relation was found between number and diversity of components and mixture performance.

3. *Stability:* Eberhart and Russell's 1966 modification of the Finlay and Wilkinson (1963) model was used. Finlay and Wilkinson's model basically

involved using the mean yield of all entries as an environment indicator and regressing each entry yield on this base, noting the mean squares of deviation from regression. Two stability parameters are derived from this modification. Interestingly, the two methods used alone led to diametrically opposite interpretations. Clay and Allard's resolution of these two views indicated that mixtures were less stable than their components.

4. *F_2 Population:* yields averaged 14% over midcomponents (5% over better parents), but were less stable than their parents.

The authors concluded that simple mixtures were not exciting prospects for commercial use; however, they recognized certain possible limitations in their study, namely, only two years of testing, and components mixed only in equal proportions. They suggested that the breeding and selection method used to create components for a mixture, as contrasted with using existing pure lines developed for single line commercial use, might be important. Specifically, they thought that hybrid bulk populations might produce individuals with interactive properties favorable for use in a contrived heterogeneous population, for example, a multiline.

In pair mixtures of two commercial oat cultivars in ratios of 0, 25%, 50%, 75%, 100%, and reciprocals, respectively, grown over 10 environments, Qualset and Granger (1970) found differences in yield and lodging from that predicted by component participation. All of the ratio mixture yields exceeded the component means—a common finding—however, none was higher than the higher-yielding component. The highest yield, 4.1% above midcomponent, came from a 75 : 25 mixture. Also, lodging in mixtures, a negative factor, was greater than predicted. Nevertheless, the lodging was so much less than in the weaker parent that it was felt that the better straw strength of the mixture over that of 'Blount' might balance the slight loss in yield from growing the stronger variety alone. The two cultivars fill different agricultural needs, for example, differences in forage quality for use in different seasons.

A finding important to multiline variety development was that stability in mixtures was greater than in the component pure line populations. Furthermore, this stability was frequency-dependent, increasing as the content of Blount increased in the mixture. In fact, the highest stability, found in the 75% Blount: 25% 'Forkedeer' mixture, was seven times that found in Blount, the higher stability component. The reason for this was not apparent.

Hamblin and Rosielle (1978) presented a model to show the potential importance of intergenotypic competition on genetic parameter estimation, and fitted it to published data on rice and barley. They concluded, "Estimates of genetic parameters obained in mixtures may have little relevance for pure culture."

The gray area between multilines and multiblends was entered in a suggestion (Sneep and Hendriksen, Eds. 1979) that mixtures of sister lines from multiple crosses might have advantages over variety mixtures because they offer the breeder a population for selection that has elements of both uniformity and diversity.

ALTERNATIVES TO PURE LINE VARIETIES PROPOSED

An excellent review of potentials for intracultivaral variation was presented by Heyne and Siddig (1965). Their list of breeding objectives for self-fertilizing species covers many of the alternatives to single pure line varieties as currently viewed. They suggested a breeder might develop varieties over this range:

1. Wide adaptability.
2. Different maturities.
3. Narrow adaptability, suited to specific environments.
4. Variety blends.
5. Multilines.
6. Direct release of early generation lines: F_2, F_3, or of whole cross populations. An example of the latter is the cultivar 'Scout', described as an F_3-derived line release.

The great range of research results contained in this chapter are applicable to synthetic cultivars in general, and are thus placed here as a prelude to a discussion of two important synthetic types, the multiline and the multiblend.

REFERENCES

Allard, R. W. 1961. Relationship between genetic diversity and consistency of performance in different environments. Crop Sci. 1:127–133.

Baker, R. J. and K. G. Briggs. 1984. Comparison of grain yield of uniblends and biblends of 10 spring barley cultivars. Crop Sci. 24:85–87.

Blijenburg, J. G. and J. Sneep. 1975. Natural selection in a mixture of eight barley varieties, grown in six successive years. I. Competition between the varieties. Euphytica 24:305–315.

Briggs, F. N. 1942. Breeding disease resistant crops. Science 96:60.

Burton, G. W. 1948. The performance of various mixtures of hybrid and parent inbred pearl millet, *Pennisetum glaucum* (L) R. BR. J. Am. Soc. Agron. 40:908–915.

Clay, R. E. and R. W. Allard. 1969. A comparison of the performance of homogeneous and heterogeneous barley populations. Crop Sci. 9:407–412.

Federer, W. T., J. C. Connigale, J. N. Rutger, and A. Wijesinha. 1982. Statistical analyses of yields from uniblends and biblends of eight dry bean cultivars. Crop Sci. 22:111–115.

Finlay, K. W. and G. N. Wilkinson. 1963. The analysis of adaptation in a plant breeding programme. Aust. J. Agric. Res. 14:742–754.

Frey, K. J. and V. Maldonado. 1967. Relative productivity of homogeneous and heterogeneous oat cultivars in optimum and sub-optimum environments. Crop Sci. 7:532–535.

Grafius, J. E. 1966. Rate of change of lodging resistance, yield, and test weight in varietal mixtures of oats, *Avena sativa* L. Crop Sci. 6:369–370.

Hamblin, J. and A. A. Rosielle. 1978. Effect of intergenotypic competition on genetic parameter estimation. Crop Sci. 18:51–54.

Heyne, E. G. and M. A. Siddig. 1965. Intra-cultivaral variation in hard red winter wheat. Agron. J. 57:621–624.

Jensen, N. F. 1952. Intra-varietal diversification in oat breeding. Agron. J. 44:30–44.

Jensen, N. F. 1967. Agrobiology: specialization or systems analysis? Science 157:1405–1409.

Klages, K. H. 1936. Changes in the proportions of the components of seeded and harvested cereal mixtures in abnormal seasons. J. Am. Soc. Agron. 28:935–940.

Marshall, D. R. and A. J. Pryor. 1978. Multiline varieties and disease control. I. The "dirty crop" approach with each component carrying a unique single resistance gene. Theor. Appl. Genet. 51:177–184.

Murphy, J. P., D. B. Helsel, A. Elliott, A. M. Thro, and K. J. Frey. 1982 Compositional stability of an oat multiline. Euphytica 30:33–40.

Pfeifer, R. P. 1966. Science for the Farmer 13:3.

Qualset, C. O. and R. M. Granger. 1970. Frequency dependent stability of performance in oats. Crop Sci. 10:386–389.

Rajeswara Rao, B. R. and R. Prasad. 1982a. Studies on productivity of seed blends of two spring wheat cultivars under rainfed conditions. Z. Acker. Pflanzenbau 151:17–23.

Rajeswara Rao, B. R. and R. Prasad. 1982b. Productivity and nutrient uptake by two spring wheat cultivars in pure and mixed stands. Z. Acker- Pflanzenbau 151: 235–244.

Sage, G. C. M. 1971. Inter-varietal competition and its possible consequences for the production of F_1 hybrid wheat. J. Agric. Sci., Cambridge 77:491–498.

Shorter, R. and K. J. Frey. 1979. Relative yields of mixtures and monocultures of oat genotypes. Crop Sci. 19:548–553.

Sneep, J. and A. J. T. Hendriksen (Eds.). 1979. Plant breeding perspectives. Centennial publication of Koninklijk Kweekbedrijf en Zaadhandel, D. J. Van der Have Seeds, 1879–1979, Centre for Agricultural Publications and Documentation, Wageningen.

Stevens, N. E. 1942. How plant breeding programs complicate plant disease problems. Science 95:313–316.

Suneson, C. A. 1960. Genetic diversity—a protection against plant diseases and insects. Agron. J. 52:319–321.

Tysdal, H. M. and T. A. Kiesselbach. 1944. Hybrid alfalfa. J. Am. Soc. Agron. 36:649–667.

12

THE MULTIBLEND VARIETY

Multiblend varieties are generally simple mixtures of existing lines or varieties. They may be blended for the simplest of reasons, for example, to attempt to capture the performance of two or more genotypes, or, in a commercial sense, to reduce inventories by providing an additional outlet for sales. On the other hand, serious research programs would attempt to fathom the performance dynamics of specific blends, hoping to discover higher performances. Blend varieties, successful in commerce, have been produced.

In the next chapter on multilines, I am going into considerable detail to show the ways in which multiblends are often confused with multilines. A distinct difference between the two is that multiline activities rest on a solid research base in the breeding program. This is mandated particularly by the breeder–pathologist involvement in the disease protection aspects of multiline variety creation. The component lines for a multiline variety are generated out of a breeding program, and are then prepared for the complex addition of resistance genes and genes for other traits in a separate and parallel backcross breeding program. An institution beginning a multiline breeding program does not do so lightly—such a program represents a long-term commitment. On the other hand, multiblend variety creation rests on a more casual foundation, vulnerable to the demands of expediency within a breeding program. Although there are breeders who have insisted on a firm research and development approach to multiblend variety breeding, in general this has not been the case.

Experimental blends serve a useful research purpose for multiline breeding. They are a major source of information on the internal dynamics and the possible performance results of populations. In fact, they furnished the first data that could be extrapolated to multiline performance, and they continue to do so.

BLENDS HAVE A LONG HISTORY

If we say that pure line varieties were preceded by landrace varieties, in an analogous way blends followed the old agricultural practice of mixed cropping. The older literature contains liberal references to crop mixtures such as oats and peas, different forages, and barley, oats, and wheat. It is a natural progression in sophistication to the next step, the blending of varieties within the same species. The following is a sampling of these experiences.

Nuding (1936) used a diallel mixing scheme of four varieties in paired 50:50 blends and compared these with the parent components. According to Frankel (1939) the tests were extensive: three years and seven sites. In general, the blend yields exceeded component means, especially when both components were adapted to the environment. Visible changes were observed in blends over time (succeeding crops were sown from harvested blend seeds), especially when the varieties used differed in their adaptation.

Similarly, Heuser, according to Frankel (1939), found that blends of five locally adapted wheats performed about the same as the pure line varieties. Also, Engelke grew blends mixed in different proportions of two wheat varieties. In all cases the blend yields were greater than the component means and in a few cases exceeded the higher-yielding variety.

Frankel (1939) reported on results with wheat blends. Overall objectives for the blends were maintenance of a high grain quality level and yields surpassing the components' mean and/or equaling the standard, 'Tuscan'. Frankel emphasized the similarity of the components used with the hope that a successful blend would fit smoothly into the farming and milling markets. He was also interested in the effect of blending on yield components.

Frankel had available the Tuscan cultivar and 11 F_6 or F_7 lines from three crosses. In a large-scale field trial he compared blends of two lines, each providing 25% of the blend with 75% Tuscan. The observed and calculated results showed the A-line blend exactly equal in yields to components, whereas the B-line blend was 3.7% higher than expected.

The small-scale trials were more extensive, being nine in number with each trial comprising five treatments: Tuscan, a breeding line, and blends in proportions of 90:10, 80:20, and 70:30, respectively. One trial with the L-line varied the proportions slightly, namely, 90:10, 75:25, and 50:50.

To paraphrase Frankel's results from the small-scale trials, (1) seven blends showed close agreement between observed and expected values on the basis of calculated component means, (2) no blend was significantly below expectation, and (3) two blends yielded above expectations but not significantly.

Frankel noted that one line, D, was greatly depressed in blends with Tuscan. Even so, Tuscan provided superior yield performance and the blend was one of the two that exceeded expectations. Thus the pattern seen by many later researchers is evident: the preponderance of evidence favors equivalency of performance as between blends and component means, but there is in addition a slight skew, usually not statistically significant, in favor of higher than expected blend yields. Blend performances exceeding the best component were rare.

Jensen (1952) used performance data from crop species and variety mixtures to support his proposal for a multiline breeding method. He presented his own research results with blends or crop mixtures, and also cited similar studies by Warburton (1915), Zavitz (1927), and Bussell (1937).

Gustafsson's (1953) research on varietal mixtures arose from his interest in genotypic competition, which he referred to as "the cooperation of genotypes," and represented a slightly different and optimistic viewpoint. He used three barley varieties, apparently in two-parent 50:50 combinations, with different fertilizer and spacing arrangements, and measured not only yield (total kernel weight) but some components of yield traits, namely, number of spikes and kernels, and 1000-kernel weight.

The results were that mixtures were superior to the variety component means in every instance. The yield advantage was about 4%. The variety 'Maja' was especially potent in mixtures: if present, the mixture exceeded the best parent. Substantial interactions were found in the density and fertilizer trials, so much so that environment exerted a major influence on the relative performance of mixtures and varieties.

Soybean variety blends gave a good account of themselves in experiments reported by Probst (1957). He used three commercial varieties: 'Blackhawk' (B), early and relatively low yielding; 'Hawkeye' (H), midseason and high yielding; and 'Lincoln' (L), later with equivalent yield to H, tested alone and in two- and three-way combinations for four years.

The yield data showed four-year component yields as B = 32.9, H = 40.2, and L = 39.7 bu/acre; these ranked sixteenth, first, and third, respectively, out of the total 16 combinations or pure stands. Thus it may be seen that many blends did well; for example, rank #2, a 1B–1H–1L blend, yielded 40.1 bu/acre. The general experience with blends is that they slightly exceed midcomponent yield—here the comparison is with highest-yielding component—so it is evident that they did well. Blends that contained the lower-yielding B generally reflected that in their low ranks.

Maturity of a blend was scored on the maturity of the latest component. A general shift toward earliness in the later components was noted, and the whole population was influenced according to the proportion of early components. At the same time the early components were slightly delayed in ripening.

Regarding lodging, the pattern followed that of the most susceptible component with the degree proportionate to that component's inclusion in the blend. This pattern has been seen in other research and other crops. The authors also concluded that blending tended to stabilize yields over years.

Mumaw and Weber (1957) used soybean variety blends to study natural selection and competition. The blends emerged from five years of testing with an average 1% increase (range +0.7 to +1.5) in yield over the mean of their components. Branching habit was identified as one of the characters showing change: it tended to dominate over the nonbranching character. It was believed that branching types received relatively more light in mixed populations.

Other traits that increased were height, maturity, and lodging. Large shifts occurred with time; in fact, one variety, 'Adams', was nearly eliminated. The authors suggested this had important negative implications for the use of the bulk hybrid population breeding method. The greatest yield increases over means of pure lines were found in bulks of "unlike" varieties, for example, those showig differences in maturity or height. The implication of this is not clear inasmuch as high yield and high diversity are acting synchronously in a positive manner at the same time as negative character shifts raise doubt as to the suitability of the bulk population format.

A general yield superiority of blends over component yields was observed (40 of 45 blends were superior) and there were five cases of mixtures that exceeded the highest-yielding component.

Stringfield (1959) conducted a mixture experiment involving 19 double crosses, four single crosses, and one open-pollinated variety of corn. These were planted both alone and mixed, to form 42 hybrid pairs. Some adjustment was required because tests were in groups of four hybrids and their six mixtures. The results showed no advantage, the mixtures very closely yielding the same as the mean of components. In addition, no matter how pairs were grouped, for example, high- and low-yield categories, the results were the same—no differences. Stringfield pointed out that a mixture of two hybrids of equal yielding capacity would be satisfactory, but otherwise ". . . a superior hybrid grown separately will outyield any mixture of it with a mediocre hybrid."

All the tests were grown in a single year and thus sampled only a portion of the environmental variance that would be expected over a longer period of time. Stringfield mentioned one situation that might favor mixtures: in hot weather, mixtures extend the period of pollen shedding.

Patterson, Schafer, Caldwell, and Compton (1963) reported on oat variety blends and their yields and lodging resistance. They used six cultivars, alone and in all 15 paired combinations, tested over four years. The blends showed no significant yield advantage over the mean of the components sown separately. Lodging, not a predictable event, occurred in only one year at a level showing differential responses. Generally the blends were superior in standability to the mean of their two components, though not always significantly.

Funk and Anderson (1964) reported on studies of mixtures of field corn. Five series were included in the research with these results:

Series I: Twelve Double Cross Hybird Blends Grown over Two Years. Yields of the mixtures showed no significant difference from average yields of the components grown separately. The blends showed increased stability as evidenced by a higher hybrid × location interaction than blend × location interaction.

Series II: Intrarow Competition between Two Single Cross Hybrids Differing in Height, Maturity, and Leaf Area. Differences in competitive ability were found but effects were compensatory, with no significant differences in yield.

Series III: Interrow Competition. It was found that the yields of alternate rows of hybrids equaled the average yields of the solid component blocks.

Series IV: Competition Involving Contemporary and Delayed Planting. Yields were taken on different combinations of 16,000 plants/acre involving early and late interplanting (three weeks later). The best yield was produced by 16,000 plants planted early. All other combinations of delayed interplanting resulted in lower yields regardless of whether the interplant was the same or created a blend. Competition seemed to be a function of the seasonal planting date rather than genotype.

Series V: Single Crosses from Six Inbred Lines Crossed in All Fifteen Combinations. These were grown at three levels of competition. Despite clear differences observed in competitive ability, no significant advantage or disadvantage was shown in grain yields for blends, which presumably were a series of paired single cross mixtures.

In sum, except for greater stability, blends showed no advantage over components grown separately.

In studies with mixtures of varieties of oats, Pfahler (1965a) concluded that high grain and forage production with increased stability was possible in bulks. However, he qualified this with the following statement, which has methodological significance to multiline breeding:

> The selection of varieties for inclusion in a composite would require actual testing of combinations since an interaction of varieties within the composite may occur and probably prediction of the composite response from the response of the variety when grown separately would not be reliable.

Pfahler (1965b) gave a generally favorable report of variety mixtures. Two commercial oat varieties were mixed 3 : 1, 1 : 1, and 1 : 3, and yields of grain and forage compared. The grain yields of all bulk populations were equal to ("not significantly lower than") the higher-yielding component and significantly higher than the lower-yielding component. The forage production relationships varied according to the proportions of the component varieties included. Pfahler felt that bulked varietal mixtures might provide greater flexibility for multiple use situations, for example, grain, forage, and grazing.

Ross (1965) tested mixtures of grain sorghum hybrids for yield performance. In general, single cross hybrids are like pure line varieties as regards uniformity and stability. Ross used five single cross hybrids grown alone and in equal blends of 10 two-hybrid mixtures over a five-year period in Kansas. Yearly environmental conditions showed a range of distinctly different seasons. Five-year mean yields of hybrids alone ranged from 3410 to 4280 lb/acre; blend yields ranged from 3630 to 4200 lb/acre. Overall, blends outyielded single crosses alone by an average 50 lb/acre (not significant). No blend outyielded the highest-yielding single cross. In general, the blends showed no particular advantage or disadvantage.

Caviness (1966) reported on experiments with two-line soybean mixtures. His results showed that, depending on the mixing ratio, yields ranged from below that of the component mean to as high as the better pure line.

Shalaan, Heyne, and Lofgren (1966) reported on a seven year study of three-component wheat mixtures of two types, (1) two mixtures of adapted cultivars and (2) two mixtures of similar-phenotype advanced lines. The results were inconclusive as to whether pure lines or mixtures were superior in yield.

Annual reconstitution of seed for planting was found desirable, enhancing stability and reducing year-to-year variation. The authors described the favorable performance, and commercial reception by growers, of 'Rodco', a blend of two different hard red winter pure lines.

Brim and Schutz (1968) reviewed the literature on competitive influences in soybeans, and examined whether hill and row plots might serve as a basis for predicting multiblend performance. They compared predicted and observed performances of the two-line and three-line mixtures of four soybean cultivars. Among 10 mixtures observed, seven were superior to the component mean and none was inferior. Four of the mixtures were higher yielding than the highest-yielding component (by 2–4%). However, because the gain in productivity of the best competitor seldom compensated for the loss occasioned by the poorest competitor, the authors concluded that the component mean level would not often be exceeded without prior knowledge of component interaction, an area that is not well understood. This does leave open, however, the possibility that certain genotypes might be found that have favorable interaction dynamics. Brim and Schutz found that the preponderance of observed competitive effects were traceable to within-plot phenomena, and that between-plot effects were minor. They found hill plots to be satisfactory measures of mixture effects.

Qualset (1968) examined population structure as related to productivity and other performance criteria in wheat. The population arrangements were seven parents (P), 21 F_2s (F_2), and 21 1 : 1 mixtures of parents (M).

Comparison of the 21 F_2s with the 21 mixtures showed that the F_2s outyielded the mixtures in 16 of the 21 cases, averaging 4.4% greater yield. Furthermore, the F_2 yields were higher than the midparent but those of the mixtures were not. An interesting case of specific combining ability was 'Tenn. 9', only modestly high in yield, whose F_2 population from 'Knox' × Tenn. 9 was significantly higher in yield than the highest parent in the test. The author pointed out that the F_1s of this experiment had been examined for heterosis and combining ability by Gyawali, Qualset, and Yamazaki (1968), and they had found that hybrids with Tenn. 9 were the highest yielders. Thus we have a coincidence of indicators showing in both the F_1 and the F_2.

Reich and Atkins (1970) evaluated stability and grain yield in four populations of grain sorghum, namely, parents, parental blends, hybrids, and hybrid blends, grown over nine environments in two years in Iowa. The mean yields in order were:

Hybrid blends	3041 kg/ha
Hybrids	2985 kg/ha
Parental blends	2429 kg/ha
Parents	2388 kg/ha

Thus hybrids exceeded parents and blends exceeded components by about 2%—a common observation and magnitude in many studies. The highest-yielding individual entry in eight environments was a hybrid; in the other environment a hybrid blend topped the list. Of the total of 32 blends, 22 yielded more than their component means. Twelve of the 16 hybrid blends exceeded their component means. An uncommon occurrence was that six blends outyielded their most productive component. The highest stability was found in the hybrid blends but otherwise the rankings for stability were confounded by inconsistencies found in the regression coefficient and deviations from regression values.

Kannenberg and Hunter (1972) mixed 10 commercial corn hybrids of the same maturity in all possible paired equal-seeds ways, and compared them with their components in pure stands for two years. Planting densities were varied to induce some stress conditions. The results showed no significant influence of densities, nor were there significant differences between yields of mixtures and average yields of their components. It was noted that the highest pure stand yielders contributed more, and lower pure stand yielders less, to mixtures than would be expected based on their component performances.

Abel (1974) evaluated variety blends of saffflower, using two biblends of four different parents plus the parents alone. The proportions of the two parents in a blend were varied by changes in seed allotment. Seed and oil yields and oil percentage were measured. The results in general showed three things:

1. Significant year-to-year changes
2. Blend performances intermediate to the parents
3. A specific case where a 50:50 blend tested for three years averaged 3.6% overcompensation for seed yield, and 3.9% for oil yield.

In one of the three years oil yield of the blend exceeded that of both parents.

Fehr and Rodriguez (1974) reported results of a study in soybeans of the effect of row spacing and genotypic frequency on yield and responses of soybean blends. Five soybean cultivars were used and evaluated in all possible 10-pair blends, each blend at five frequencies, 4:0, 3:1, 1:1, 1:3, and 0:4. All possible three-component blends were also tested. All were tested at two row spacings, 70 and 100 cm. the results were as follows:

1. *Row Spacings:* Narrower rows produced greater yields per area but the mean response was essentially identical for pure stands and two- and three-component blends. There were, however, significant deviations of certain blends at certain spacings, which led the authors to suggest final field testing of blends of interest using desired spacings.
2. *Compensatory Responses:* The averages were practically identical for both row spacings.
3. *Genotypic Frequency:* In a test where genotype frequencies of two-component blends were varied in 10% increments over a 100-point range, it was found that a yield-frequency dependency existed, or as the authoıs wrote, "All blends had their highest yields when the highest yielding cultivar made up at least 70% of the blend."

Trenbath's (1974) massive review of productivity of mixtures is concerned with biomass yield as contrasted to yield of seeds. He concluded that most binary mixtures yield between the yields of the two components. The data in his Table I, however, do show a skew toward the higher side; that is, 40% of the cases yielded below the overall mean and 60% yielded above. There were 45 experiments with mixture yields below the lower-yielding component and 83 with yields over the higher-yielding component. This approximately parallels the experience found in cereals with yields of grain.

Trenbath did not rule out the possibility of transgressive yields even though the evidence was sparse, because the mechanisms exist that could account for their occurrence. From a study of models he concluded that transgressive yields would occur "only if the monocultures are sufficiently similar." Trenbath noted the paucity of experimental evidence to test hypothetical mechanisms.

Glenn and Daynard (1974) reported on a research study in corn that included single crosses and their mixtures. In general, little difference was found between the mean yields of the mixtures and the mean yields of their mixture components. The one exception involved a case of yield depression involving a mixture containing a much shorter single cross as a component.

McKenzie and Grant (1975) mixed six spring wheat cultivars (equal-seed mixtures) and grew the mixtures and components for six and seven years under (1) irrigation, (2) no irrigation, and (3) wheat stem sawfly attack. The results for the three environment, respectively, were:

1. The highest yielder in pure stand declined slowly in mixture; the other five performed as in pure culture.
2. The cultivars performed in mixtures as in pure stands.
3. The two resistant cultivars almost eliminated three of the four susceptible ones. One susceptible survived by escaping infestation because of late maturity.

Gedge, Fehr, and Cox (1978) examined heterogeneous F_5 lines and soybean blends with the objective of measuring yield compensation (over-, under-, and neutral). In each of two crosses, six high- and six low-yielding F_5 lines of F_2 derivation were established and four lines selected from each. These four-line sets were tested for yield in (1) pure stands, (2) all possible two-line blends, and (3) four-line blends. The mean yield of components in pure stand was the comparison criterion. The results showed that the most common response to intergenotypic competition in the blends was neutral or complementary. There was little deviation from chance expectations relative to over- or undercompensation. There was no statistically significant difference in response to intergenotypic competition between the high- and low-yielding groups. Two-component lines performed similarly to four-component lines in their response. Actual yields were not presented so that further interpretation of blends is not attempted.

Walker and Fehr (1978) examined the question of what strategy a grower might employ to realize highest yield and stability in soybean culture. They studied the optimal number of cultivars to be used in a mixture. Their material consisted of 28 high-yielding cultivars. With these they prepared 80 entries including 14 pure lines, their up-to-14-component mixtures, and for diversity, additional mixtures drawn from the full 28-component pool. Twelve environments were created by using six sites over two years of testing: the location year mean squares were found to be highly significant, suggesting, in effect, that 12 different environments were sampled. Tests were made for stability.

The results were ambiguous concerning stability. There was a lack of consistency in response; that is, some pure lines performed in as stable a manner as did mixtures. Overall, however, there was a general trend favoring mixtures over pure lines, leading to the conclusion that probabilities for stable production would suggest growing a mixture of several cultivars rather than growing one pure line. However, the authors were unable to ascertain an optimum number of pure lines for a mixture.

Regarding yield, it was shown that (1) yield was not related to stability, and (2) yields reflected the mid component mean (mixtures had a nonsignificant 0.4% advantage over component means). Mixtures and pure lines performed relatively the same across environments differing in quality.

Fehr and de Cianzio (1980) reported on a comprehensive study of the relationship of component frequency (involving the 41 2.5% increments) to the yield responses in soybean blends. The two varieties were 'Corsoy' and 'Amsoy 71', which were evaluated in four environments over three years. Compensatory response is defined as the deviation between actual and predicted yields.

The difficulties of interpretation of single site-year results become obvious when the interactions are available; for example, overall, Amsoy 71 outyielded Corsoy 3409 to 3280 kg/ha, but in one experiment, in one year, Corsoy outyielded Amsoy 71. There were wide swings in yield ranks when component frequencies across environments were examined. It was helpful for interpretation that the 41 ratio entries were averages of all tests. The results showed that there were no significant deviations from predicted yields. The authors' graph of observed and predicted yields showed what could be interpreted as purely random deviations. Indeed, it would be difficult to conclude otherwise: with 2.5% increments, for example, a wide deviation swing, such as −111, logically could not be rationalized as having any explanation other than a random deviation.

I would like to call attention to a phenomenon that has appeared in many studies: mixtures often yield more than the average of their components. Although seldom statistically, economically, or practically significant, the trend is almost always in a positive direction. In Fehr and de Cianzio's study the average yield of blends was greater than the predicted yield by 20 kg/ha, or 0.6%.

The authors concluded that tests of soybean blends for the purpose of finding overcompensatory or undercompensatory combinations cannot be justified in a breeding program. Instead it seemed more practical to predict on the basis of pure stand performance.

Gorz and Haskins (1982) explored blend performances of tall, medium, and short forage sorghums, involving three sets of genotypes and seven height blend combinations in each set in replicated trials over two years. Blend performances between genotype groups were similar and the combined data were used for analysis. Twenty-three traits were measured.

When the data for all the possible seven height combinations were assembled it was found that tall alone produced the highest total dry matter in both years; thus there was no yield advantage to growing blends for forage. Although some of the blends exceeded the yield of their components, it was nevertheless obvious that the highest obtainable yield was with the tall variety alone.

Federer, Connigale, Rutger, and Wijesinha (1982) used the terms *uniblends* and *biblends* to refer to pure lines in self competition and 50:50 varietal mixtures, respectively. Presumably additional terms could cover other combinations, and *blends* might be an all-inclusive term for varietal mixtures. However, I have coined *multiblend* to title this chapter for two reasons: to fill the need for an equivalent-level term to be set against *multiline* and to make clearer the characteristics that differentiate these two breeding entities.

The interest of Schweitzer, Nyquist, Santini, and Kimes (1986) in soybean cultivar mixtures was triggered by the recent adoption of narrow-row, non-cultivable production practices. They made 16 two-variety mixtures, using four cultivars in determinate–indeterminate combinations. In each pair there were six levels of seed mixtures. Emphasis in the results centered on two combinations: in one, there was undercompensation resulting in lower yield, and in the other, overcompensation. It was concluded that the undercompensation was due to cultivar similarity, and the overcompensation was due to diversity in height and relative maturity. The authors cautioned that too-wide maturity differences could create harvesting problems.

In the overcompensation pair it was possible to show that the maximum yielding ratio came from seeding proportions of 54:46, essentially equal amounts. At this level the population yield was 11% higher than the higher-yielding cultivar. Although such favorable results are infrequent it is important to acknowledge their occurrence so that breeders of both multiblend and multiline varieties know that such phenomena are possible.

CONCLUDING REMARKS

In general, it is difficult to extract from the long research on multiblends any strong consensus of support for their development. The strongest pattern of yield is positioned near that of the components' mean. The slight positive yield skew is not exciting. There is merit, however, in exploring the occasional blends that exceed all components in yield, in the hope of finding the rare superior specific combiners. Yield aside, there appear to be benefits which derive from blends, affecting such attributes as lodging resistance, disease resistance, stability, and special crop uses.

REFERENCES

Abel, G. H. 1974. Cultivar blends in safflower. Crop Sci. 14:276–277.

Brim, C. A. and W. M. Schutz. 1968. Intergenotypic competition in soybeans. II. Predicted and observed performance of multiline mixtures. Crop Sci. 8:735–739.

Bussell, F. P. 1937. Oats and barley on New York farms. Cornell Ext. Bul. No. 376.

Caviness, C. E. 1966. Performance of soybean varietal mixtures. Ark. Farm Res. 15:2.

Federer, W. T., J. C. Connigale, J. N. Rutger, and A. Wijesinha. 1982. Statistical analyses of yields from uniblends and biblends of eight dry bean cultivars. Crop Sci. 22:111–115.

Fehr, W. R. and S. R. de Cianzio. 1980. Relationship of component frequency to compensatory response in soybean blends. Crop Sci. 20:392–393.

Fehr, W. R. and S. R. Rodriguez. 1974. Effect of row spacing and genotypic frequency on the yield of soybean blends. Crop Sci. 14:521–525.

Frankel, O. H. 1939. Analytical yield investigations on New Zealand wheat. IV. Blending varieties of wheat. J. Agric. Sci. 29:249–261.

Funk, C. R. and J. C. Anderson. 1964. Performance of mixtures of field corn (*Zea mays* L.) hybrids. Crop Sci. 4:353–356.

Gedge, D. L., W. R. Fehr, and D. F. Cox. 1978. Influence of intergenotypic competition on seed yield of heterogeneous soybean lines. Crop Sci. 18:233–236.

Glenn, F. B. and T. B. Daynard. 1974. Effects of genotype, planting pattern, and plant density on plant-to-plant variability and grain yield of corn. Can. J. Plant Sci. 54:323–330.

Gorz, H. J. and F. A. Haskins. 1982. Performance of blends of short, medium, and tall sorghum for forage. Crop Sci. 22:223–226.

Gustafsson, A. 1953. The cooperation of genotypes in barley. Hereditas 39:1–18.

Gyawali, K. K., C. O. Qualset, and W. T. Yamazaki. 1968. Estimates of heterosis and combining ability in winter wheat. Crop Sci. 8:322–324.

Jensen, N. F. 1952. Intravarietal diversification in oat breeding. Agron. J. 44:30–34.

Kannenberg, L. W. and R. B. Hunter. 1972. Yielding ability and competitive influence in hybrid mixtures of maize. Crop Sci. 12:274–277.

McKenzie, H. and M. N. Grant. 1975. Population shifts among spring wheat cultivars in artificial mixtures. Can. J. Plant Sci. 55:345 (abstr.).

Mumaw, C. R. and C. R. Weber. 1957. Competition and natural selection in soybean varietal composites. Agron. J. 49:154–160.

Nuding, J. 1936. Leistung und Ertragsstruktur von Winterweizensorten in Reinsaat und Mischung in verschiedenen deutschen Anbaugebieten. Pflanzenbau. 12:382–447.

Patterson, F. L., J. F. Schafer, R. M. Caldwell, and L. E. Compton. 1963. Comparative standing ability and yield of variety blends of oats. Crop Sci. 3:558–560.

Pfahler, P. L. 1965a Genetic diversity for environmental variability within the cultivated species of *Avena*. Crop Sci. 5:47–50.

Pfahler, P. L. 1965b. Environmental stability and genetic diversity within populations of oats (cultivated species of *Avena*) and rye (*Secale cereale* L.). Crop Sci. 5:271–275.

Probst, A. H. 1957. Performance of variety blends in soybeans. Agron. J. 49:148–150.

Qualset, C. O. 1968. Population structure and performance in wheat. Third Intern. Wheat Genetics Sympos., Aust. Acad. Sci., Canberra, pp. 397–402.

Reich, V. H. and R. E. Atkins. 1970. Yield stability of four population types of grain sorghum, *Sorghum bicolor* (L.) Moench, in different environments. Crop Sci. 10:511–517.

Ross, W. M. 1965. Yield of grain sorghum (*Sorghum vulgare* Pers.) hybrids alone and in blends. Crop Sci. 5:593–594.

Schweitzer, L. E., W. E. Nyquist, J. B. Santini, and T. M. Kimes. 1986. Soybean cultivar mixtures in a narrow-row, noncultivatable production system. Crop Sci. 26:1043–1046.

Shalaan, M. I., E. G. Heyne, and J. F. Lofgren. 1966. Mixtures of hard red winter wheat cultivars. Agron. J. 58:89–91.

Stringfield, G. H. 1959. Performance of corn hybrids in mixtures. Agron. J. 51:472–473.

Trenbath, B. R. 1974. Biomass productivity of mixtures. Adv. Agron. 26:177–210.

Walker, A. K. and W. R. Fehr. 1978. Yield stability of soybean mixtures and multiple pure stands. Crop Sci. 18:719–723.

Warburton, C. W. 1915. Grain crop mixtures. J. Am. Soc. Agron. 7:20–29.

Zavitz, C. A. 1927. Forty years' experiments with grain crops. Ontario Agric. College Bul. 332.

13

THE MULTILINE METHOD

The following excerpts from my paper on the multiline variety concept (Jensen 1952*) describe the basic thoughts inherent in the proposal.

> **The multiline variety**
>
> It may be possible to produce a satisfactory multiline variety through a blend of multiple pure lines, each of which is of a different genotype. Such a variety would be composed of pure lines chosen from the oat breeding project for uniformity of appearance—particularly height and maturity—resistance to diseases, and other characteristics essential for a basic desirable agronomic type. Each line used should contribute additional desirable genetic factors without detracting from the phenotypic uniformity of the composite. Through a knowledge of the characteristics, of the individual component lines, and with performance data gathered in the usual way, the plant breeder would be able to blend compatible lines in the proper proportion. He would retain the individual lines and could effect withdrawals or make additions of promising new selections in later years. The seed blends would proceed through established certified seed channels to the grower. A multiline variety thus would change little in outward appearance with the passing of time but might change in genotypic composition.

The expectations for the multiline variety expressed below make it clear that this was a complex breeding program, not a simple variety mixture procedure:

* Reprinted from N. F. Jensen (1952), Intravarietal diversification in oat breeding, *Agronomy Journal*, Volume 44, No. 1, January 1952, pages 30–34 by permission of the American Society of Agronomy, Inc.

> A multiline variety would be expected on theoretical grounds to possess the characteristics of longer varietal life, greater stability of production, broader adaptation to environment, and greater protection against disease. Small losses from diseases would be expected perhaps to occur oftener than in a single pure line variety, but the multiline variety would present a buffer system of checks and balances against the inroads of any one particular disease. Plant pathologists, through an annual program of identification of rust collections, have maintained for years a file of up-to-date information on the geographic distribution and trends in population of various races of cereal rusts. This information has prediction value and is used extensively by plant breeders in planning their hybridization and testing programs. Similarly, this information could serve as a guide to insure that the multiple line variety is at all times balanced with respect to resistance to the potentially-dangerous races. Changes in the composition of the multiline variety at the seed supply source could accomplish the objectives of variety rotation without its practical disadvantages and with no interruption in seed supplies. The concept of a multiline variety, therefore, carries the implication not only of the variety in being but also of a reserve of stock lines having known resistances to various races of plant pathogens.

Borlaug (1953) proposed a *composite variety* that would be "a mixture of a number of phenotypically similar lines, which are genotypically different for resistance." The lines would be developed by a series of crosses of a commercial wheat variety to different sources of (stem rust) resistance, followed by backcrossing to the commercial variety: "As many lines can be developed and held in reserve as there are types of resistance." Borlaug's proposal of the backcross method to develop lines has become the standard.

Gustafsson (1953) implied that competition of genotypes, like a coin, has two sides, the obverse being cooperation. His paper, "The cooperation of genotypes" appeared contemporaneously with my 1952 paper on the multiline variety, and with Borlaug's 1953 composite variety proposal using lines developed by the backcross method. When Gustafsson wrote, "There is successively spreading in agriculture praxis a belief that a superior cereal stock should not be built upon a single genotype, but rather on a couple of or on numerous genotypes, selected so as to cooperate," he was independently expressing a consensus, although we may not have realized it then, that Time and an idea were in a favorable conjunction.

BACKGROUND DEVELOPMENTS THAT LED TO THE MULTILINE

Stevens (1949), noting that crop rotation was an effective means of controlling insects and diseases, suggested that the same principle would hold if varieties of the same crop were rotated. Plant pathologists and plant breeders had long observed the seemingly inevitable buildup of virulent pathogens on new "resistant" host varieties. Stevens pointed out that the withdrawal of an "old" variety is accompanied by a decrease in at least some of its virulent pathogen races. He believed that after an interval of time certain good "old" varieties could be brought back for a few years of production, until the expected buildup of pathogens again would necessitate their recall. Stevens' views were sound, but the necessary planning and management has never been addressed on an organized basis. Specifically, there would have to be changes in seed merchandizing patterns. There is nothing to prevent an individual grower from implementing the principle involved by saving seed of a favorite cultivar and bringing it back into production a few years later. This would be especially effective (because no one else would be doing it) if the favored "old" variety were sufficiently genetically different from the current varieties in its pathogen-resistance genes.

T. E. Stoa once told me with some satisfaction that they were quite content in North Dakota to be so geographically situated that a variety barrier was created through their use of the midseason Canadian oat cultivar 'Garry'. The earlier varieties to the south were often markedly different in their stem and crown rust resgenes.

Went's (1950) article describing the Earhart Plant Research Laboratory at the California Institute of Technology showed the relationship between climate and worldwide plant distribution. This relationship was on the macro level; however, on the micro level Went showed photographs of homozygous pea plants whose uniformity was remarkable in the climate-controlled laboratory. At the time I thought that the observed plant-to-plant variability of a pure line that we see in nature was indicative of the wide variations in environment that existed, even in adjoining spaces. Different genotypes would be needed in order to utilize and exploit these different micro environments; thus this was an argument favoring the multiline. However, because the system operates in the field on the basis of a chance seed distribution pattern we obviously cannot match each genotype seed with its optimal space. Viewed this way, a multiline would always operate at less than top efficiency.

LATER DEVELOPMENTS AND MODIFICATIONS

Jensen and Kent (1963) projected a two-pronged proposal to change the New York state pattern of one-variety culture by "... substituting a series of varieties that act as genetic barriers to the spread of pathogens. The second and longer range objective is the development of what are known as *multiline* varieties (see Jensen 1952), which have genetic barriers built into them."

A projection to 1970 was given for oats; however, the multiline stage was not implemented because as time passed it became evident that the prototypes and effort required to add a multiline breeding project to the Cornell program were not available. Instead, it was more efficient to extract the steady productivity gains from breeding pure line varieties. In the end, it was not possible to expand and conduct two parallel breeding programs.

In this paper is an early reference to the protection against disease afforded by a multiline: "A population of plants (that is, a variety) need not be 100 percent resistant to a pathogen in order to resist damage; if as few as 40 percent of the plants in a population are resistant, the population may have full protection against damage."

This research was done by Professor Kent and graduate students in the Department of Plant Pathology at Cornell University.

I discussed the multiline with Dr. Dobzhansky, who frequently visited Cornell, and I believe he was in sympathy with its theoretical basis. In 1963 he wrote as follows (1963): "Genetic analysis shows that, at least as a rule, it is not a single genotype but an array of genotypes that fit a Mendelian population to secure and to maintain its hold on its environments." and, further noted:

> There are, however, at least two possible ways of adapting to environmental heterogeneity—genetic variation and developmental homeostasis (the ability to react adaptively to the whole range of environments which a population encounters). These two methods of adaptation are not mutually exclusive, and both are made use of in evolution.

The CIMMYT program involving wheat multiline varieties was described in detail in an article by Breth (1976). At that time the outstanding '8156' cross was available as a recurrent prototype. Its high yield, stability, and broad adaptation were known worldwide. An important change in breeding procedures to produce multiline components was detailed: instead of using rigid backcrosses where the donor rust resistance genes represent practically the sole variance in the final multiline, CIMMYT wheat breeders used double crosses (of single crosses, each of which had 'Siete Cerros' as one parent). The other parent in each single cross was different. The double cross was monitored, selected, rogued, and eventually was stockpiled as a line or component. This system has much to recommend it and opens the door to genetic gains in productivity on top of the protection against diseases. Breth said of the procedural changes: "This system is not the full backcross approach originally conceived for developing multiline varieties. Repeated backcrossing would make the component lines almost identical genetically except for their rust resistance and would be a slow process."

STRUCTURAL AND DYNAMIC ASPECTS OF THE MULTILINE

Sprague and Tatum's (1942) article on general combining ability (GCA) and specific combining ability (SCA) in hybrid combinations of corn inbreds is relevant to the multiline breeding method because (1) the GCA and SCA terms have been modified to fit competition situations, and (2) the paper provides an analogy for a discussion of certain aspects of multiline formation, principally the testing of lines.

Jensen and Federer (1965) used Griffing's (1956) concepts and formulas on combining ability to apply to research on *competing ability* in wheat. By dropping the word *hybrid* from definitions, GCA becomes the "average performance of a line in combinations," and SCA refers to instances where combinations deviate from expected average performance of *competitive* ability.

In a multiline variety created by the backcross method, which is the only method currently in use, only general competing ability is relevant inasmuch as all lines have a very-close-to-recurrent parent genotype. Any line interactions based on the added backcross genes would not be expected to be large. On the other hand, specific competing ability would be important in other methods of forming a multiline culitvar, as the following examples illustrate.

1. *Use of Single Backcrosses to Form Component Lines.* Selection for the desired resistance gene to be added would be more rigid, and the selected BC_1 lines would be verified for resistance in order to choose either one line or bulk-resistant lines to form one line. The amount of recurrent parent germ plasm would be less than in the full backcross, but variability and the chances for line interaction in the multiline variety would be increased.
2. *Use of Unrelated Lines.* Here there is no recurrent genotype buildup at all, except that due to coincidental relationships, and testing for specific competing ability would be necessary. Although this has not been used to develop a commercial multiline, it is a valid option and would offer the opportunity for yield increases from specific line interactions. This method would be suitable especially for a large breeding program where the frequency of finding suitable lines ready for use would be enhanced. Desired additions of genes as for disease resistance could be accomplished through backcross procedures.
3. *Use of Different Selected Recurrent Parents.* If we may extend the concepts of the backcross and recurrent parent, a three-way cross could be made to a genotype similar to the recurrent parent that has been idealized as the prototype for the multiline. In such procedures a full backcross could be used, selecting for the added gene throughout. If a number of genotypes similar phenotypically to the recurrent parent (but different genetically) were available, line diversity would be increased, and interactions between lines in the multiline could be substantial.
4. Use of the double-cross procedures practiced by CIMMYT.

If any method other than the repetitive full backcross to a single recurrent parent is used to develop lines, the breeder needs to be aware of possible specific competing ability interactions among lines. Certain combinations of lines could differ greatly from the expected general competing ability. Also, competition must be viewed as intrapopulational, not simply as the more limited pairwise competition. It is this possibility of increased yield from the interaction of unrelated genotypes that holds the hope of adding a yield bonus to the other expected advantages of a multiline. To date, only disease resistance has been a major reason for forming a multiline variety.

The practice of extrapolating results exclusively from paired line tests to predict the situation in a multiple line population is subject to criticism. Alternatives to pairwise component testing are not numerous. Sprague and Tatum reported that Beard (1940) proposed that a highly heterozygous single cross could be used as a tester parent (in hybrids). Perhaps an analogy can be sketched for self-pollinated crops whereby a standard population, composed of a group of highly competitive component lines, could serve as a tester population into which candidate lines might be inserted in order to measure their reaction and interaction under "crowd" conditions. It would seem the results would have more general relevance than a series of pair comparisons. In short, the component line interactions within a multiline of more than a few lines are likely to be so complex that they cannot be estimated or gauged accurately without actual testing.

A multiline variety has points of similarity to synthetic varieties, as in corn and forage crops, a principal difference being that in the latter crops inbreds and cross pollination are part of the variety structure. Some of the synthetics have been studied into advanced generations, and various studies made, for example, on performance as related to number of lines. Without attempting a detailed review, it is worthwhile nevertheless to call attention to this allied field.

Kinman and Sprague (1945) considered the question of the number and nature of corn parental lines (inbreds) deemed most efficient for the creation and performance of synthetic varieties. They compared the yields of the diallel cross of 10 inbreds (45 single crosses), F_1 and F_2, with their parents. Interpretation leaned toward a preponderance of additive (arithmetic) gene action. The authors used the yield data of the 10 inbreds (their average combining ability) and single crosses to compute the theoretical yields of synthetic varieties that included from 2 to 10 lines.

The results, based on F_2 means, both arithmetic and geometric, showed that five or six lines were most efficient; that is, they produced the highest mean yield of the hypothetical synthetic. There is an interesting paradox here: it might seem simple logic to involve the mean yields of F_1 and parents, in which case (see Kinman and Sprague's Table 4) the descending yields for two, three, four, five, and six lines were 97.9, 93.4, 93.8, 91.6, and 89.2 bu/acre, respectively. In the authors' words,

> If any series of inbred lines is arranged in order of mean combining ability, the mean yield of all possible single crosses will be greater for the best two, three, or four lines than when a larger number of lines are involved. . . . in general, the

most efficient number of lines to be included in a synthetic variety will vary with the range in combining ability among the inbreds available as parents.

Therefore, the most efficient number is obtained from the F_2 means, which would attach to the arithmetic series shown above as 65.3, 76.1, 79.3, 80.2, and 80.2 bu/acre, respectively, thus establishing five or six lines as most efficient.

Kinman and Sprague speculated upon the effect of using vigorous lines that have not been subjected to extensive inbreeding as compared with their more highly inbred progeny. The difference is in breadth of germ plasm and has a corollary in cereal multiline variety formation, where different levels of homozygosity/heterozygosity are possible, as for example, in the number of backcrosses imposed during line formation.

Frey and Browning's (1971) paper disclosed important information vital to the preparation of component lines for a multiline variety. Breeders of self-pollinated crops have been aware for some time that isogenic lines sometimes exhibit traits that could only have appeared because, as in backcrossing,the chromosome segment carrying the desired gene also carried additional genetic information. Cautious breeders use the term *near-isogenic*. Frey and Browning provided case histories of this phenomenon where three crown rust resgenes that had been backcrossed into lines were found to be associated with large deviations in grain yield when compared with the recurrent parent under rustfree conditions. The associations found included both yield increases and reductions. The specific importance to the multiline variety is that backcross lines must be observed and tested, perhaps also produced in larger numbers, in order to choose the best line to become a component line in a multiline. The source of the resgene is important too; that is, different genetic backgrounds of the resistant gene may harbor different associations with yield.

Boerma and Cooper's (1975) paper on early generation selection in soybeans has been reviewed in the chapter on single seed descent. However, the point about F_2-derived heterogeneous lines is worth noting for its relevance to multiline components or, indeed, separate heterogeneous cultivars, as suggested by Rosen (1949) as well as Boerma and Cooper. The final F_2-derived heterogeneous lines remaining after three years of yield truncation (which involved discard of lines but no selection within the retained lines) resulted in a group of lines containing individuals yielding up to 109% of the high-yielding parent in the cross. Boerma and Cooper discussed the merits of F_2 heterogeneous advanced lines as cultivars. On the positive side were (1) reduced development time, (2) capture of competitive yield responses, and (3) essentially the equivalent productivity of pure lines. On the negative side would be the problems associated with marketing, maintaining, and certifying such a cultivar, and certain nonuniformity problems, for example, maturity. However, in the latter cases there is no reason why a certain amount of mass selection and grooming might not be practiced.

A comment about crop uniformity may be appropriate here. It is easy to document my development in thinking on this subject. In my 1952 paper I considered phenotypic crop uniformity a necessary requirement for a multiline variety. In 1966, referring to the 1952 paper, I wrote, "I have changed my opinion" (Jensen, 1966). I have always treasured H. A. Wallace's (1938) comment that the goal of breeders should be "functional excellence" rather than "show-ring excellence."

For a multiline cultivar there is no reason why diversity need be restricted to between or among lines. Intraline diversity is also possible through hybrid heterogeneity, where planned segregation might be used as a vehicle to distribute resgenes through a population. The use of restricted-generation backcrosses is another way of introducing additional diversity into a multiline.

Hamblin, Rowell, and Redden (1976) have given the theoretical basis for a way to select lines to be used in mixed cropping. They argued that such lines should be specifically bred and selected for mixed culture rather than by the present practice of using lines selected for monoculture. The proposed scheme uses a checkerboard plot arrangement where one species is overplanted by another. In the simplest case this is done by planting the block in strips of one species and then planting strips of the second species at right angles. Rather than using pure lines as, for example, in the small grain cereals, a breeding situation is featured which begins with F_2 populations from numerous crosses. The best combinations of the two-component plots are chosen, and the tests are then continued to select the best lines and combinations within these crosses.

If we turn our attention to multiline varieties, Hamblin, Rowell, and Redden's proposals do address a critical problem in line development, namely, how to test all the possible combinations of available lines. There exists no program to screen for lines developed out of competition situations where the competition is tailored to produce "cooperative" components. The authors' proposal of using F_2 populations to provide competitive interaction responses in mixture may encounter the familiar low correlations found in many F_2-to-derived-line evaluations. It may be a questionable assumption to expect derived lines from two successfuly interactive populations to perform similarly when placed together later as population components. Nevertheless, the proposal is intriguing and deserves a test (the paper was concept only). It may be that early-generation hybrid progenies can project later pure line competitive propensities; if so, the breeder would have a powerful new tool. It should be pointed out that the checkerboard arrangement of oversown biblends lends itself equally well to tests of pure lines in mixtures, something that is needed on a larger scale for population studies.

A weakness of line-forming techniques for a multiline variety lies in the acceptance of biblend interactions to characterize the complex interactions that are likely to occur in a multiline of a dozen or more lines. Perhaps the checkerboard arrangement might be adapted to measure second-order interactions. For example, suppose the biblend tests identified a few promising combinations: perhaps these biblends could be tested against a series of third lines, or a (diallel) series of biblends. In each case the seed amounts for sowing each component would be adjusted so that the amount sown per plot equaled a designated commercial rate.

The use of multilines in peanuts has an aspect unique to the method, the use as components of siblings from the same cross. Schilling, Mozingo, Wynne, and Isleib (1983) stated that industry requirements dictated the use of sibling lines in order to assure uniformity for quality factors among components. This is interesting because the sacrifice of genetic heterogeneity appears to counter the multiline concept of genetic diversity of components. Although siblings need not be similar they would be expected to show a narrower range of genetic variance, which the peanut breeder desires, than genotypes from different crosses. In this study the authors examined the relation between closely related components, where each line traced to a separate F_2 plant, and the stability of the multiline.

The authors found that if the components were too similar and lacked variation, they might as well be pure line cultivars. In such cases equivalent yield or stability has little meaning. On the other hand, if the components show significant variability for yield, regression coefficients, and deviations from regression, it is possible to put these variable units together and achieve a stable multiline.

This use of the multiline is intriguing. It is analogous to a cross where the progeny individually show parts of the whole genotype desired in one individual but no one individual has enough of the package to be worthy of cultivar status. This use of the multiline permits such individual lines to be combined into a satisfactory population. Schilling, Mozingo, Wynne, and Islieb found siblings of low genetic variance to provide sufficient diversity for multilines. Examples of multiline peanut varieties are 'Florigiant' and 'Florunner'.

The research on multilines exposes a weakness in that many studies lead to generalized deductions based on paired genotypes. Although it is possible to have a one-pair multiline variety, I consider a likely number to be 9 to 12, perhaps a range of 5 to 20 lines. From a research-for-information standpoint it is manifestly a more daunting problem to factor the contributions of several individual genotypes to a multiline than it is to work with pairs. Nevertheless, in multiline populations, there is no alternative to the testing of all of the possible entry configurations. The serious breeder would look also for the special combination, perhaps rare, which has to be found only once.

STABILITY IN THE MULTILINE

Marshall and Brown's (1973) paper adds much to our understanding of populational stability in multilines and multiblends. They used statistical models to determine the "conditions under which a mixture will be superior to its best or better component in stability of yield." As part of the background they recorded Allard and Bradshaw's (1964) concept of stability mechanisms, and individual and populational buffering, roughly illustrated by homogeneous and heterogeneous populations, respectively. The interaction found in populational buffering may have different genetic origins, for example, segregation from hybridity or mixture of pure lines. Marshall and Brown derived the following expectations of yield and stability under specified conditions:

1. *Yield.* Their conclusion, given the assumptions of pure lines, equal-seeded mixtures, similar environments, and all lines ecologically independent, was that "in the absence of interactions among genotypes, the yield of the mixture is necessarily less than the yield of the most productive line in pure stand."

2. *Stability.* Their conclusions under the same assumptions were that with components performing similarly in each environment populational buffering is nonexistent, zero, and it is implied that stability would be seen as an average. However, if there is any deviation such as one line with a different response, the populational buffering component turns negative; the mixture must be more stable than the component average, and has the capability of greater stability than the most stable line. The mathematical proofs are given for these conclusions. I interpret this as importing a general positiveness about stability because there is greater likelihood of differential responses of some components to different environments than there is of all components having similar responses. Furthermore, for any level of mean variance and covariance in yield, the populational buffering component becomes increasingly negative with increase in number of components. Thus increases in diversity of a mixture via components favor the increase of stability.

So far, Marshall and Brown have been dealing with the absence of interactions. The expected, more typical case with multilines is that there would be interactions and thus populational buffering. This case holds open the possibility that mixture yields may exceed the mean or highest yields of components. This would occur if there were a net positive interaction effect among lines that offset the contributions of lower-yielding components.

Regarding stability when intergenotypic interactions are present, Marshall and Brown found the model explicit—stability of performance *will* be altered—but less explicit as to direction and magnitude. From experimental results of several researchers it was known, however, that intergenotypic interactions are relatively small when compared to the pure stand yields of components, and the low variance and covariance yield figures suggest a small influence on stability. Apropos of this, the authors draw a conclusion with strong breeding implications:

> If this assumption is valid, the plant breeder interested in achieving increased *stability* through the use of multiline varieties will be able to predict the potential stability of a line blend with reasonable accuracy from the mean variance and covariance in yield of the components grown in pure stands. As a result, he would not be faced with the task, as are breeders interested in exploiting the potential of mixed populations for increased *yields* of growing all possible pairwise combinations of the likely components of the mixture to estimate intergenotypic interaction effects.

The authors summarized the theoretical results in two predictions, namely:

1. Where component lines give dissimilar responses to different environments—not unusual—there exists the possibility that the mixture could exceed the stability of the most stable component line.
2. It is easier to obtain increased yield stability than increased yield itself ("improved mean yield requires net positive intergenotypic interactions, improved stability has no such requirement").

I have devoted considerable attention to Marshall and Brown's paper, first because of its admirable and logical balance between situations and assumptions on the one hand and results and interpretations on the other. Second, their interpretations and predictions show general agreement with what is known about mixture population responses, specifically, many cases of increased stability and few of increased yield, although there is a preponderance of cases showing statistically insignificant yield increases of the order of 1–3%. I mention this latter point not to weaken the general expectation that increased yield is difficult to achieve but to hold forth the hope that it may be possible occasionally with certain rare component combinations. This flies in the face of Marshall and Brown's conclusions that predictions can be made without testing all combinations pairwise, in which case the rare combination of components that defies the rule might not be found.

The emphasis in their paper has been on the effect of diversity on stability. The authors pointed out that adapted genotypes—the kind most likely to be used as multiline components—grown in stable environments show little promise of increasing stability. I think further that increased diversity among the components of a multiline does offer hope for yield increases. Any increment above the midcomponent mean yield floor would be desirable. Where many components are involved, however, it is not clear that prediction without testing can identify the rare synergistic meld of genotypes and proportions that would mark the rare interaction. Increased diversity could be obtained by selecting lines from a single backcross population, combining them to form one component if desired, or by choosing genetically different lines.

I am not sure there is a direct link in van Emden and Williams' (1974) paper to breeding for insect or disease resistance, or to the multiline variety breeding method, yet I found the philosophic review of insect stability and diversity relative to the environment interesting reading.

The question of whether multilines act to stabilize pathogen populations was addressed by Groth (1976), using additive and multiplicative versions of a mathematical model. The additive model assumed that virulence genes in the pathogen additively reduce within-component fitness, whereas the multiplicative model assumed that individual fitnesses combine through multiplication rather than addition. The effect of the additive model is to show that relative reproductivity of simple and complex races is a function of the within-component selection coefficient and is not related to the number of lines in a multiline. The difference between results of the two models is that for any given value of the selection coefficient, the multiplicative one allows more complex races to occur but each added virulence gene has diminishing effect.

In discussing the problem Groth acknowledged the complexity and the dependence on situation assumptions. For example, it might increase stability to add a susceptible component to a multiline, but this would have to be balanced against the practical hazard of economic loss in the host population. It would be difficult to disagree with his statement that "only experimental evidence will tell us the optimum composition of a multiline."

A multiline variety composed of near-isogenic lines could show populational shift in two ways, as a result of the action or interactions (1) of the added traits or (2) of unknown "passenger" genes linked or carried on the added chromatin. Murphy, Helsel, Elliott, Thro, and Frey (1982) set up an experiment to determine the existence and magnitude of such change as a way to gauge the stability of populations. They formed an experimental oat multiline from five near-isogenic lines that differed in genes for crown rust resistance and could be identified by their responses to tester races. The common seed lot was divided so that one population could be grown in a rustfree environment while the other was subjected to crown rust attack. The respective populations were grown through four generations of the same treatment, and the composition of each was calculated from the identification of lines. Any shift in component share would be measured from the original approximate 20%. The results after four generations were these:

Rustfree: Stability measured by regression analysis showed one line, C.I. 9192 (which declined from 22 to 10% in share), to have a persistent, significant *b* value. Another line, C.I. 9184, increased its share from 20 to 38%. These changes occurred in confirmed rustfree environments.

Rusted: Rust was present in epidemic quantities in all generations. Again, C.I. 9192 showed the greatest share decline, from 22 to 15%. This line, however, was not the most susceptible of the five. C.I. 9184,

> which was the least susceptible to crown rust, again made a highly significant (stability) advance, its share going from 20 to 28%.

In general the population shifts were not too different in the two groups, but they were overshadowed by the changes within groups. Obviously the five lines were showing different responses to the total environment, not just to the rusts. The implications to seed increases of a multiline variety were discussed.

Murphy et al. noted that any recognizably inferior component lines, such as C.I. 9192, would be excluded during development: "Ideally, only NIL's with equivalent competitive capabilities would be included in a multiline cultivar." I understand the authors' sense here, but we might have to know more about why lines such as C.I. 9184 perform as they do. Are they the cause of the decline of other lines, and is the gain–loss a "wash"? Such knowledge may make the difference between multiline yields at the component mean level or at a positive transgressive level. An interesting thought is that the most productive multilines might lose a measure of stability because they must include lines such as C.I. 9184, whose aggressiveness might be upsetting to internal order.

Gill, Nanda, and Singh (1984) designed three experimental wheat multilines for each of the two recurrent prototypes 'Kalyansona' and 'PV 18'. The component lines were developed by two different methods, both involving three to four backcrosses to the prototype parent. In the first method, brown rust-resistant plants were chosen for crossing in each BC generation, thus producing near-isogenic lines differing in resgenes only. Although in the second method "the components were selected from among the segregating materials of the crosses involving the recurrent parent," it was not clear to me what this meant in actual line preparation (I suspect it means non-backcross segregating generations). Similarly, the composition of the different multilines was not evident, nor was there any explanation of the ways in which the three multilines of each prototype cultivar differed. Consequently, we cannot examine the results in terms of the objective of their research, a stability-across-environment analysis, but instead will interpret them in terms of the comparisons of several multilines with their recurrent cultivar prototypes. Comparisons were made in tests at nine locations over four years with several criteria measured.

Singling out important results, it was seen that all six multilines showed increased seed size and four of the six showed higher number of tillers than the recurrent parents. The second-order interaction of G × Y × L was exceptionally high. The authors suggested dividing the region into subareas and producing varieties for each area, at the same time increasing the number of locations and reducing the testing years. Although this is a possibility, the second suggestion does not necessarily follow the first: a clustering analysis could lead to fewer locations and lower second-order interactions.

Regarding stability, Gill, Nanda, and Singh used Eberhart and Russell's 1966 three criteria of a stable variety, namely, maximum yield, unit regression, and small deviations from regression. The findings were:

1. *Yield:* all Kalyansona multilines were higher yielding than Kalyansona. All PV 18 multilines were lower yielding than PV 18, but not significantly lower.
2. *Regression:* all multilines were close to unity.
3. *Deviation from Regression:* the largest deviations from regression were in the two cultivars.

Gill et al. stated:

> Summarising on three stability parameters it is seen that the multilines of Kalyansona yield better and those of PV 18 less than their respective recurrent parents. From a regression coefficient point of view the multilines are as good as the recurrent parents except for KSML 3. The deviations from regression are smaller for the multilines than for their recurrent parents. It may, therefore, be inferred that the multilines are better than the recurrent parents.

The KSML 3 multiline had an above-unity regression coefficient, and showed the maximum deviation from unity, which the authors interpreted to mean that it would do better under high-yielding environments.

Relative to disease resistance, Gill et al. found all multilines "much more resistant than their recurrent parents." Both parents, however, were highly susceptible to the brown rust disease. Infection scores of the multilines ranged from 0 to 30. Apparently the multilines were designed according to the "dirty crop" approach; that is, components were not resistant to all races, and the population had a mixture of components–resgene combinations. The authors clearly favored this concept.

Several multiline cultivars have been released in *Arachis hypogaea* L., the domestic market peanut. Norden, Gorbet, Knauft, and Martin (1986) examined population stability in four multilines. The study was suggested by observations that certain stable Florida peanut multilines produced less stable pure lines when lines were removed and examined.

The plant materials used in the experiment consisted of four multilines (I–IV), specially formulated from 16 genotypes selected from four crosses: four genotypes from each cross constituted a multiline. The replicated comparison nursery included each multiline and its four components, thus there were 20 populations in total. Eight environments were sampled (two locations, four years).

To measure stability Norden et al. used the 1963 Finlay and Wilkinson formulas and the 1966 Eberhart and Russell model on the traits of pod yield and five quality factors. The results brought out significant G $\times$ E interactions for the six traits. When linear regression was used to analyze stability it was further found that in general there were not significant differences between the multiline and its components. The interpretation of stability was complex, influenced by the number of traits considered. In groups I and II the multilines showed stability for yield equal to or better than their component lines; however, when all six traits were included there was always a component line better than the multiline. The advantage of this line was not great and did not equal the advantage of the multiline over the less stable components. In Groups III and IV the multilines were stable for as many as or more traits than any component line. The identification of certain component genotypes with excellent stability, for example, in five of six traits, seems noteworthy.

Norden et al. concluded that the odds of improving yield stability and market acceptability of peanut cultivars are increased through the multiline approach. Their reasoning introduces a novel idea, namely, that a multiline cultivar can be fine-tuned whereas a pure line cultivar cannot. It is part of the concept of a multiline that components may be added or deleted; they pointed out that further research on identifying stable genotypes might warrant such a change.

RESULTS OF RESEARCH WITH MIXED POPULATIONS

In two years of tests of 50 : 50 mixtures of two oat varieties, 'Clintland' and 'Mo. 0–205', the first susceptible to stem rust race 7 and the second resistant, Browning (1957) reported that under induced epiphytotic conditions the average yields for the two varieties in the two years were 28 and 48 bu/acre (mixture yield 45 bu/acre), and 82 and 137 (mixture 124) bu/acre, respectively. In both years the mixtures exceeded the averages of the components by 7 and 14.5 bu/acre, respectively, and it was noted that much less rust developed on Clintland in the blend than in pure stand.

The implications to multiline breeding of Sakai's (1957) findings must be interpreted in the light of certain qualifications and restrictions. His research dealt with populations of two genotypes only. Also, the preparation of commercial multiline variety seed carries a range of options; for example, the component lines may be newly mixed annually, or not. Sakai envisaged the possibility that even with a very strong competitor an equilibrium state might ensue:

> However, the keener the competition between these two races, the faster the strongly competing race increases within the population. As the relative frequency of the strongly competing race becomes higher and higher, the benefits it receives from competition against the weakly competing race should become less and less. Thus it may be expected that when a few individuals with a strong competitive ability happen to grow in a plant population, members of which are all weak in competition, they will increase at a high rate in the beginning, but slow down after they have propagated to some degree.

This implies that elimination of the weaker individuals might not take place; however, if their frequency is very low at equilibrium, it would be almost certain that with time they would be eliminated because of variations in environment or in sampling for the planting seeds. I have thought that there might be a place in a multiline for a small percentage of a strong competitor that could add to productivity by exploiting a previously unused environmental niche, for example, 5–10% of a taller individual occupying a stratum of higher canopy space. If in fact there were a net productivity gain, it would be captured in the harvested crop, which would be the end of it, inasmuch as the planting seed for the next generation—Haldane's fictitious parents—would revert to the original proportions of components.

Sakai's findings have relevance to evolutionary breeding populations with their long string of generations, but are not of serious consequence to the bulk hybrid population method, where lineout occurs anywhere from F_3 to F_6, and where in any case early selection with rebulking can produce less competitive groups of fictitious parents.

The role of epistasis in breeding and selection is known sketchily if at all. Sprague, Russell, Penny, Horner, and Hanson (1962) defined epistasis as interaction between alleles at nonhomologous loci. The problem rests in the fact that epistasis can be estimated but principally is unknown or unpredictable until it appears. It is a bonus well worth searching for in lines for a multiline, and may well account for a part of yield gains without indicating the source. Thus epistasis can be said to be synergistic, the whole greater than its parts, and certainly at the present time it is serendipitous also.

Using six corn inbreds and all their single and three-way crosses, Sprague et al. were able to test and arrange all the components of three-way crosses in an orthogonal manner. The difference between observed and predicted results in sets of crosses can be interpreted as due to epistasis. Using their illustration, the predicted yield of the three-way cross, (A × B) × C, can be obtained from the mean of the two single crosses, (A × C) and (B × C). Using the condensed two-year results for comparison, 5 of 20 tests showed statistically significant gains of actual over predicted, three at the 1% and two at the 5% level. These top three increases were relatively large, about 6–8%.

Jensen (1965) drew upon routine nursery results of oat mixtures at the Cornell University Agricultural Experiment Station to project possible multiline performances. The first general case considered showed that 124 composites grown over an eight-year period throughout New York State exceeded their components' mean by 3.23% (2297 versus 2225 lb/acre). The second specific case involved a five-line composite formulated on the basis of expected performance. The test multiline was one of a 13-entry nursery grown at 10 locations in the state, and ranked first in yield, averaging 3472 lb/acre, exceeding significantly by 237 lb the component lines' mean, and by 147 lb the next-highest entry. Its yield was significantly better than three of its components, and approached statistical significance for the other two.

Except for the results of the specific case, which are not often encountered, the general results are in line with many other findings. Specifically, blends and multilines usually do better than component means but are rarely superior to the best component. This phenomenon is specifically related to their yield performance and does not consider perhaps more important features of the multiline, such as greater stability and disease resistance.

Davis and Rutger's (1976) heterosis and mixture studies illustrate a little-discussed aspect of multiline populations. They found none of the 12 two-parent mixtures to yield more than its high parent yield (no statistically significant difference). Nevertheless, three mixtures equaled or exceeded the high parent's yield, and this pattern has been found in several other research studies, sometimes with individual mixtures statistically significantly outyielding the high parent. It seems to me that the breeder, in the search for a successful multiline, must accept the general concept that many disappointments may occur in the choosing and mixing of lines, while holding to the possibility of occasional successes for transgressive yields. The mutual interaction of components may be unique and not understood, but they should be seen as the shining hope that excites further probing research.

DISEASE PROTECTION A MAJOR FEATURE OF MULTILINES

Van der Plank (1965), discussing the dynamics of epidemics, alluded to the protective features embodied in multilines. In conclusion he stated, "Two other factors become increasingly important as the epidemic proceeds: the proportion of healthy susceptible tissue remaining available for infection, and the degree of uniformity of the population of host plants and of their environment."

It does not seem feasible to review Van der Plank's book (1968); nevertheless it is imperative to call attention to his many important contributions, perhaps best exemplified by the terms horizontal and vertical types of resistance.

Leonard and Kent (1968) used two races, 6F and 7A, of stem rust in mixed plantings of susceptible and resistant oat cultivars. This was organized in such a way as to provide all combinations of hosts and pathogens over a range of host mixtures, which provided population susceptibilities of 100%, 75%, 50%, or 25% to each race. The results showed that the rate of rust increase provided a good match to van der Plank's formulas.

Two contrasting combinations threw light on host–pathogen relations (infection levels encountered ranged from 5 to 25%):

Cultivars	*6F*	*7A*
Clintland A	S	R
Clintland 60	R	S
Clintland D	S	S
Garry	R	R

In the first host pair the separate races increased without any mutual interference. In the second host-pair mixture there was mutual interference, explained simply on the basis of two predators competing for the same territory, that is, leaf area on susceptible Clintland D.

Insight into the tolerance some oat varieties, for example, 'Cherokee', display toward oat crown rust infection was gained from a study by Torres and Browning (1968). Cherokee and susceptible "Clinton' were inoculated with equivalent spore loads of three races of *Puccinia coronata* f. sp. *avenae*, and both sporulating area and urediospore yield were measured. Two of the three races showed no differences; however, after three weeks the uredia on Cherokee were significantly larger than on Clinton. Despite this, Clinton showed a much higher yield of spores (per primary leaf unit length), and the spore yield was sustained longer; thus Clinton received the greater damage. This was interpreted to mean that tolerant varieties delay rust development.

Cournoyer, Browning, and Jowett (1968) reported on a study of crown rust *(Puccinia coronata* var. *avenae)* spread in pure line and multiline plots. The multiline variety had seven components. They stated, "Each plot was inoculated with a 'super race,' race 264, virulent on all components of the multiline, or with a mixture of six rust races, including race 264."

The pathogenicity relations of the five races relative to the components or pure line were not stated. The results were as follows:

1. A sigmoid curve represented rust buildup in both rust treatments in the pure lines, but did not appear for the multiline.
2. The multiline showed less early season rust and showed a slower epiphytotic buildup. The epiphytotic did not reach climax and was terminated by crop maturity.
3. The six rust race mixture grew slightly more in the multiline than in the pure line.

Leonard (1969a), in the first of a series of three papers, studied the rates of increase of two races of stem rust in oat host mixtures of susceptible and resistant plants, and in their components grown singly. The results from the model equation used were found to agree well with observed results. Leonard found that when the two races were introduced into a mixture where each race was virulent on only one of the components, the increase was independent; however, in mixtures where both races were virulent on the same variety, "the amount of rust produced was less than the sum of the amounts produced by separate populations of 6F and 7A in separate host mixtures with corresponding proportions of susceptible plants."

In a second paper Leonard (1969b) showed that heterogeneous oat stem rust races cultured for eight uredial generations on two different host cultivars, namely, 'Craig' (susceptible) and 'Clintland A' (containing the A resistance gene), produced pathogen populations with distinctly different components. Two other populations were subsequently lost due to a decline in spore production. The tests for adaptation were made by growing the initially identical heterogeneous populations separately on each host for seven uredial generations. In generation 8 this was continued with the additional switch of culturing the rust also on the other host. The results (derived from Leonard's Table 5) were:

	Pustules/plant on	
	Craig	*Clintland A*
Craig		
Host gener. 1–7 & Craig 8	29.56	21.95
Host gener. 1–7 & Clintland A 8	36.23	25.50
Clintland A		
Host gener. 1–7 & Clintland A 8	35.99	34.26
Host gener. 1–7 & Craig 8	14.75	17.04

The interpretation of the results were (1) that the population cultured on Craig produced more pustules on Craig than on Clintland A, (2) that one generation on the other host did not reverse this pattern, and (3) that the spores cultured on Clintland A produced about the same number of pustules on both hosts. Leonard's hypothesis was that Craig was the more susceptible of the two hosts, and that the culturing on each resulted in adaptation of the races to each different host environment.

The research offers hope that multiline varieties can create equilibrium among rust races, thereby reducing the overall attack. Although there is only a suggestion here that this might be so, and it is unclear in what way to proceed to achieve the best results for a multiline, I interpret Leonard's discussion to rest on the interaction of simpler races with the excessively virulent races by denying both an open field for attack. The success of this approach would hinge on the judicious choice and distribution of resistant plants in the multiline.

In his third paper of the series, Leonard (1969c) expanded the discussion and constructed a model for how genetic equilibria could be reached in host–pathogen systems such as that represented by a multiline population. Leonard envisaged the presence of a certain proportion of simpler races in the environment, and expected that the components of multiline varieties would include a proportion of susceptible lines. The aim would be to provide competition and barriers to the otherwise unimpeded buildup of more complex races that might have combinations of different genes for virulence. Drawing on the results of his second paper, Leonard commented:

> In a multiline variety, the predominant biotypes of each simple race of the pathogen should be those biotypes which are best adapted to their respective host varieties. Complex races would attack several of the component varieties in the multiline variety. If these component varieties had different genetic backgrounds,

> the complex races should not be well adapted to any one of them. For maximum effectiveness in disease control, multiline varieties should be composed of a series of compatible varieties with different genetic backgrounds, rather than a series of backcross lines from the same recurrent parent.

Leonard's last statement deserves attention: it is one of the few reservations expressed about the use of the backcross method for line formation. It makes the implied point that the major part of the genetic background of plants in a multiline population, after the allocation of resgenes to the different components, would be identical. One concedes that other genetic material may accompany the chromatin segments containing the resgenes. The backcross method is a proven efficient means of developing lines; nevertheless I reiterate my long-held view that the possibility of different ways of choosing component lines should be kept open and tried. Leonard's objectives could be met by choosing lines from single backcross populations, or by using phenotypically similar recurrent parents, rather than a single recurrent parent.

Browning, Cournoyer, Jowett, and Mellon (1970) compared spore spread and grain yields in plots of two early and two midseason multiline cultivars, and in resistant and susceptible isogenic lines of equivalent maturity. The plots were inoculated with four selected races of crown rust. The results were as follows:

1. The time of maximum spore release was delayed three to four days in the multilines compared with the susceptible midseason line.
2. In the same comparison, the quantity of spores released was twice as great in the susceptible lines as in the multilines.
3. Seasonal totals, spores/100 liters of air, were of this order (figures approximate): susceptible midseason line 5600, susceptible early line 3734, multilines less than 1860.
4. Grain yields in percent of respective checks were: susceptible midseason line 53%, susceptible early line 73%, multilines 81 and 86%. A high negative correlation (−0.90) was found between grain and spore yields. The effective buffering shown by the multilines was attained against "a crown rust epiphytotic much more severe than one likely to occur in a farmer's field."

In a wide-ranging review of host–pathogen relationships, and variation in pathogens, Watson (1970) contributed to lay understanding by defining three terms: pathogenicity, virulence, and aggressiveness. Pathogenicity, under polygenic control, refers to the several observed evidences of a pathogen on a host (pustule size, spore volume, growth rate). Virulence, under specific gene control, is a separate phenomenon relating to a specific host; if virulence is present it means either the host lacks a particular resgene, or the resgene is nullified by the fungus's gene for virulence. Aggressiveness refers to a competition factor where, in mixed-strain populations, one strain can multiply at the expense of others. As might be expected there may be overlap in these three concepts but there is general agreement that they exist.

Watson made the point important to plant breeding that in an unchanging or slowly changing milieu of host and environment—such as must have existed over periods of evolutionary time—a form of dynamic and sustainable equilibrium exists between host and pathogen. Plant improvement projects continually disrupt whatever successful form of stability might be in place, and from those have evolved our pathogen, pest, and host problems. The placing of genetic barriers in the host has resulted in population shifts in the parasites.

Watson believed that single genes for specific resistance are generally unsatisfactory, and I presume this is founded on the known short lives of cultivars with such resistance. He favored using single genes in different combinations. This practice has given satisfactory control of the rust *(P. graminis f. sp. tritici)* in Australia and other parts of the world.

Watson discussed most of the strategies that have been considered to influence the pathogen–host confrontation in a direction favorable to the continued and uninterrupted ideal of a successful cultivar. His concluding remarks may encourage a reading of his paper in full:

> ...The author proposes the following as a working model to explain the host–pathogen relationship. Virulence and avirulence is controlled by specific genes whose products, if any, interact with those of corresponding resistance genes in the host, under a relatively simple genetic system. Growth rate, lesion size, spore production, aggressiveness and characters related to pathogenicity are controlled by a polygenic system. Survival ability is probably related to characters controlled by a genetically similar, but different system. It would be theoretically possible to select genotypes of the pathogen which combine genes for virulence with those for aggressiveness, and hence with those for fitness. If a gene for virulence is associated with genes for fitness it will be retained in the population, even though it may appear to be unnecessary.
>
> Population shifts of plant pathogens will not be eliminated in the future because both cultivar and environment will continue to influence strain frequency. In any one region, however, fluctuations will be minimized by encouraging cultivars that have a broad genetic base which comprises genes for specific resistance and, where possible, those for nonspecific resistance.

Flor (1955 and earlier) proposed the influential gene-for-gene hypothesis of host–pathogen relationships (for each resgene in the host there is a corresponding pathogenicity gene in the pathogen). In a later paper (1971) he chronicled and assessed the changes noted in flax in approximately a quarter of a century. The combination of successful flax varieties with a consequent concentration of genetic material exposed to the flax rust has resulted in resgenes succumbing one after another to the successive buildup and onslaught of new pathogenic races. Of the 25 identified resgenes considered to give satisfactory protection, only six remained effective. Flor said, "One by one, the reservoir of genes conditioning resistance to North American races is being depleted."

Flor's ideas are of interest to those with an interest in multiline breeding, suggesting an alternative use for the lines containing the resgenes. One also wonders whether Stevens' ideas of variety rotation might be used to bring back "old" resgenes if the genes for that specific pathogenicity have disappeared in the pathogen. I confess I have not encountered a discussion of what happens to a gene conditioning pathogenicity in the pathogen when the corresponding resgene in the host disappears. Is it lost along the way as the pathogen alters its breeding direction to counter a new resgene? If feasible, an updating feature could be incorporated by backcrossing the "old" resgene into the latest recurrent prototype, which might be a successful variety or new release.

The subject of Flor's 1971 paper is a different approach to the problem we have been discussing, which he expressed as follows: "The probability of accumulating two or three pairs of mutant recessive genes in the dicaryophase is much lower than the occurrence of a single pair. Therefore, the use of two or more genes for resistance should delay if not prevent the establishment of the new mutants to virulence.

Flor developed flax lines that incorporated two-gene and three-gene resistances, as contrasted to the typical monogenic backcross approach. The number of resgenes incorporated represented a recognition that (1) two or three resgenes might present the pathogen with an impenetrable or slowly yielding barrier, and (2) there were diffiulties with tester race combinations to identify plants carrying more than three to five different resgenes. Flor gave details of creating the multiple resistant lines, which proceeded by adding a third resgene to lines having two. Because the resgene races being added were not present in the region, great care had to be exercised under controlled greenhouse conditions, and all susceptible plants destroyed. Flor produced 20 lines that when tested were about equal in yield among themselves and equal or superior to the recurrent "Bison" parent. With their other satisfactory characteristics these lines are all potential cultivars.

Marshall (1977) covered the advantages and hazards of homogeneity as seen from a global point of view. He discussed the question of what to do about the present hazardous imbalance, and accepted the need for a reintroduction of diversity into world crops, focusing on the special urgency for designing breeding strategies to deal with disease and pest resistance. Rediversification deals with:

1. *Intravarietal Diversity:* specific (vertical) and nonspecific (horizontal) resistance, tolerance, multilines, bulk hybrids.
2. *Intervarietal Diversity:* gene deployment and mixed monocultures and cropping.

It is clear that so vast a subject cannot be summarized here in a few neat phrases; Marshall felt that breeders are at an early stage in reaching a unified field concept and that more knowledge and resources are needed.

Marshall used several terms that occur infrequently in the literature and thus may not be recognizable instantly to all readers. *Flypaper effect* (attributed to Trenbath 1975) refers to what happens to spores of a wind-borne pathogen produced on susceptible plants in a multiline that also contains resistant plants. The spores are trapped as though they fell on sticky flypaper (if one can remember that product!). Then there are also the "clean" and "dirty"crop approaches to multiline breeding methods. The clean crop idea is applied to procedures used by Borlaug (1958), in use at CIMMYT (but since changed, I believe), and pursues the ideal of complete resistance in the population to be achieved by multiple resgenes in each component line. The dirty crop view, according to Marshall, is espoused principally by the Iowa group (Frey, Browning, and Simons, 1973 and later). Each component is viewed as basically carrying a single vertical resgene; thus the oat population may be expected to show (crown) rust, but at benign levels owing to the protective spore trapping and impedence of spread because of resistant plants.

The perceived differences in the clean or dirty crop approach are illustrative of the general interest in research on multilines and mixed crops. Trenbath (1977) worked with two models, one for mixed crops with a pest parasite (nematode) and one for multilines with a wind-borne pathogen (rust races). I was particularly impressed with his handling of assumptions in developing the multiline model. For the model to work, stabilizing selection needed to be reduced to zero. Instead of using assumed data in order to get a parameter value that would achieve this, he turned to Flor's gene-for-gene hypothesis and the results from Watson and Singh's (1952) stabilizing selection study to get actual data. His simulated study was then calibrated to accommodate these data.

Of course the nematode pest introduced a new dimension, the capability of the parasite to move and seek out host plants. A wind-borne pathogen can move, but the movement is passive and not directional except as wind currents dictate. Results were far from conclusive, but Trenbath felt that the deduced predictions from the mixed-crop model hinged on influx of inoculum. With influx mixed cropping was superior; without influx the best strategy was rotation of pure (crop) cultures. For the multiline situation the model attached importance to stabilizing selection for increasing horizontal resistance. I interpret the findings about resgenes and manner of incorporation as a "draw" between the clean and dirty crop approaches.

Gallun (1977) was also concerned about single or multiple genes for Hessian fly resistance in wheat. Based on many years of observing and breeding fly-resistant wheats he stated,

> Because of the extreme variability that already exists in the Hessian fly and the potential for mutations for virulence, it should be recognized that incorporating

> too many genes for resistance into any one variety will set the stage for genetic vulnerability and possible epidemics. It would be best to use single genes for resistance in our wheats rather than combining genes, and only change sources of resistance when populations of a specific biotype begin to build up in the field.

Frey, Browning, and Simons (1977) presented an *integrated management system*, which analyzed the history of host–pathogen relationships in the United States over the past three-quarters of a century. The authors specifically discussed the oat crown rust problem. Chronologically, they covered the gradual loss of inter- and intraspecific diversity, the development and spread of pure line varieties, the use of vertical resistance and its failures, the discovery of pathogenic races, the search for tolerance, and gradually, the dawning realization that new approaches were needed still. The situation reminds me of a morality play—valiant struggles, victory within grasp, the end of the Age of Innocence, the dawning realization of the true wickedness of the adversary, and finally, the uphill winning struggle.

But this is no morality play. The plan the authors present is logically impressive. The management system is a continental program to manage genes conferring resistance to crown rust. The program has three phases, namely, (1) incorporating genes for tolerance in host cultivars, (2) provision for interregional diversity (protection) through selection of genes for vertical resistance, and (3) development of multiline cultivars as the program vehicle. These three ingredients are discussed in detail. The paper is "must" reading for those following this long-running play.

Marshall and Pryor (1978) discussed two philosophies of disease control that earlier had been designated as the "clean crop" and "dirty crop" approaches. The first envisaged all component lines resistant to all prevalent races of the pathogen; the second permitted diversity, each line carrying a different single gene resistance with none resistant to all races. They used theoretical models to ascertain the effect of the dirty crop approach on changes in the pathogen population.

The authors found the key points to be (1) the magnitude of selection against unnecessary virulence genes, and (2) the number of component lines making up the multiline. They found that different levels of the first point above led to different outcomes, with the general conclusion being that "dirty crop" multilines "will provide stable disease control in crop plants only in limited and relatively rare circumstances."

As time goes by, information will accumulate from practical experience with multiline varieties to help answer this question. Frey (1986) has written, "The family of crown rust resistance genes being used in the Multiline E and Multiline M series nearly 20 years ago seem to be as effective today against crown rust as they were at that time. This speaks well for the longevity of multiline resistance."

The results of Luthra and Rao's (1979) research on leaf rust spore spread in multiline and pure line varieties of wheat can only be construed as positive proof that the multiline variety configuration works. Using Van der Plank's terminology of vertical and horizontal resistance, they made the point that the first reduces the initial inoculum and slows the buildup of races to which the plant is resistant, but does not act against virulent races to reduce their infection rates. Just the opposite happens with horizontal resistance: the initial inoculum is not reduced but the rate is. Luthra and Rao's results showed that multiline varieties reduced both the initial inoculum and the infection rate. And if the rate of spread is reduced, an epiphytotic is delayed. The positive relationship between the delay or reduction of spore generations and reduced crop damage is well known.

Luthra and Rao prepared seven multilines from combinations of a successful commercial cultivar, 'Kalyansona', and 12 backcross lines that were derived from crosses of Kalyansona with other varieties chosen as donors of leaf rust resistant genes. Because the lines were chosen from the first or second backcrosses, significant nonrecurrent parent germ plasm would still be present. Not many combinations of 12 components or resistances are possible in only seven multilines. The combinations of the components they chose presented population resistance fronts to the 12 leaf rust races as follows (figures are number of components—percentage resistant): 5—30%, 5—40%, 12—42%, 4—50%, 7—60%, 7—72%, and 3—100%. Unaccountably, there was no combination or pure line representing 0% resistance (R). A mixture of the 12 leaf rust races was introduced on susceptible spreader rows throughout the plots. Severity of rust was recorded on five different dates, the first representing initial infection. In addition the average infection rate (*r*) was recorded. With this information and using an equation of van der Plank's, the amount of leaf rust at any time was determined from the initial infection and the rate of infection. The results were these:

1. The initial infection was significantly lower in multilines than in the average of their components. The greatest reduction was in the 4—50% multiline (ML) where initial infection was only one-fourth of that in the components' mean.
2. The average infection rate over the five recording periods was lower in the multilines than in component means in five of the six MLs having R:S variability.
3. In every multiline the severity of rust was lowered significantly over the components' mean. The results at the initial and final recording dates were as follows, where the ratio is that of final severity of ML and components' means:

No. lines—%R	*Initial Infection*	*Final Severity*	*Ratio*
5—30 ML	0.0039	35.0	
Components	0.0138	53.6	0.65

5—40 ML	0.0036	28.9	
Components	0.0113	42.6	0.68
12—42 ML	0.0019	10.0	
Components	0.0080	20.7	0.48
4—50 ML	0.0023	8.8	
Components	0.0089	31.5	0.28
7—60 ML	0.0035	7.8	
Components	0.0071	25.2	0.31
7—72 ML	0.0028	5.8	
Components	0.0051	18.0	0.32
3—100 ML	no infection		

There is much to think about in these results. For one thing, there was some similarity in multilines with the same number of components (5—30 and 5—40, 7—60 and 7—72%), yet the results were also in the direction of percent resistance (lower severity in 40% than in 30% R, and in 72% than in 60%). This observation is refuted by the 4—50% ML, which had few components yet low final severity. The authors' ratios are very convincing as to the population effect of combining sources or resistance to leaf rust, and they summarized their total impressions as follows: "The severity of leaf rust in a multiline is not only due to the proportion of susceptibility but it also depends upon the number of components and the percentage of resistance to each race."

The authors recommended the use of multilines in India, where quick replacement of varieties is beset with problems, and because of the generally large contemporary number of leaf rust races.

Marshall and Pryor (1979) used theoretical models to test the efficiency of resistance genes according to their number and placement in a multiline via component lines. The emphasis was on the stabilizing effect on the pathogen populations in contrast to the effect on the host. Three types of multilines were postulated: (1) all components carrying a single resistance gene; (2) all components carrying overlapping gene pairs, that is, AB, BC, AC; and (3) all components carrying disjoint gene pairs, that is, AB, CD, EF.

The results from the models showed that the poorest strategy was No. 2, overlapping, whereas No. 3, the disjoint set case, had a slight advantage over No. 1. Furthermore, this difference between the overlapping and disjoint sets widens as the number of available resistance and virulence genes are fed into the system on a gene-for-gene basis. The disjoint sets seem to pose an increasingly difficult problem to the pathogen's virulence genes, and pathogen stabilization occurs at a lower level than with the overlapping sets. The authors pointed out that the degree of disease control depends upon the number of resgenes, the level of stabilizing required to balance the virulence genes, and the allowable amount of susceptibility to be carried in the multiline. Examples are given of the number of genes required for levels of stabilization.

Marshall and Pryor hold the hope that if pathogen buildup can be slowed significantly, that is, on a time scale equivalent to an average useful variety life, then a form of rotating variety replacement involving different sets of resgenes might deal a further destabilizing blow to the pathogens. Properly administered, a more lasting form of disease control might be obtained. In this scenario the disjoint gene sets would be considerably more effective than single resgenes per component.

THE IOWA EXPERIENCE WITH OAT MULTILINES

The background and genesis of the Iowa multiline oat project was described by Browning, Frey, and Grindeland (1964). Historical milestones were cited in a plant pathology framework (crown rust was the greatest threat to oat production in Iowa). The role of the plant breeder exacerbated the risk situation when common resistance genes were used in cultivars which spread over wide usage areas. As the authors commented, "Each new resistant variety changed the rust races, and the newly prevalent rust races in turn caused the development of a new variety."

In these conditions the average useful life of an oat cultivar became approximately five years.

The authors showed that the multiplication of crown rust on susceptible plants was appreciably slowed when it took place in two-cultivar mixtures containing two resistance gene sources. The epidemic stage was delayed four days, which is a critical period in terms of grain filling and yield. Breeding earlier maturing varieties would also help avoid the later epidemic stages.

Browning and Frey's (1969) paper is an excellent review of the whole multiline breeding approach, with emphasis on the control of "crowd" diseases. Evidence of control was given. They provided a needed rationale for engaging in the multiline approach: for example, the research must be backed up by a sound "conventional" program able to produce high-performing prototypes. To meet the requirements of the multiline variety, these prototype lines would be altered through backcrossing. An important consideration is whether a successful conventional program would be willing to pause, organize, and add the additional procedures necessary. It is a mark of the Iowa group's professional competence that they have been able to do this. The authors noted that the multiline approach was a conservative one, embodying a number of fail–safe elements. The threat of crop hazards, especially that of disease epiphytotics, is a factor to be considered in evaluating the need for a multiline program. Another important factor in the success of the Iowa program is the authors' backgrounds: Browning is a plant pathologist and Frey a plant breeder—both disciplines are essential for multiline success.

Frey, Browning, and Grindeland (1970) announced the first oat multiline varieties to Iowa growers. a major thrust for their cooperative development by breeders and pathologists was the protection afforded against crown rust, the most destructive disease of oats in Iowa. Two series of releases were made: an early series (E), and a midseason series (M). Their article provides a good background for understanding the theoretical and practical problems of multiline variety development.

In two papers in 1971, Frey, Browning, and Grindeland (1971a, 1971b) registered three early and three midseason multiline oat cultivars for Iowa. Their basic composition formulas included these features: two maturities, 7–11 near-isogenic lines, successful prototype recurrent parents, two basic stem rust resistance genes (Pg_2 and Pg_4), and varying line resistances to three crown rust races that provide populational resistance.

Near-isogenic line seed was periodically grown to provide component seeds in an amount sufficient to form the multiline mixture planting needs for three to four years.

AUTHOR'S COMMENTS ON THE MULTILINE VARIETY

As a beginning young plant breeder in the 1940s, I was intrigued by the genetic and breeding problems inherent in the genotype–environment relationship. This was a vast complex of dynamic forces embracing competition, population structure, stress, crop hazards, genotypes, adaptation, niches—in fact, most aspects of plant breeding. The only ideal in variety formation at that time was the single-genotype pure line variety. The vulnerability of pure line varieties to environmental hazards in the broad sense was generally recognized and appreciated. Plant breeders and their agricultural constituency recognized that not every new variety represented a technological advance; sometimes an emergency replacement—a stopgap—was required. This general problem of the vulnerability of single genotype crop varieties was being considered by others; for example, in 1949 Stevens suggested rotating old varieties, bringing them back into production when the particular pest to which they had succumbed had disappeared, and in the same year Rosen suggested utilizing "mixed populations of any one cross and not breeding for uniformity." Stakman (1947) and other pathologists and breeders had chronicled the close relationship between host variety populations and their pathogens.

My ideas on the multiline concept were influenced by Beardsley Ruml (1950), who originated the "pay as you go" tax scheme (withholding) and also was a pioneer in the development of mutual funds, then in their infancy. Ruml's subject was investment and the mutual investment company—in 1950 mutual and stock funds were new enterprises on the American economic scene. The relationship between this business article and the development of a new plant breeding method is certainly not obvious at the moment, but will become apparent from these statements:

> The investment company provides three important advantages that the ordinary citizen cannot provide for himself, or at best can provide only imperfectly. These advantages are: (1) diversification of investments; (2) experienced and continuous management supervision; (3) liquidity.
>
> All [investment companies] . . . have the element of diversification, and all are therefore protected against overwhelming loss as a result of a single error of judgment.
>
> Next, there is the advantage of experienced and continuous mangement. The investment company is much more than a pool of diversified securities. These securities which the company owns are under constant watch by specialists who are always prepared to recommend purchase or sale of investments in accordance with their judgments as to how developments affect the relative values of the securities of different companies.

Here, by analogy, is the heart of the concept of the multiline variety. Make these substitutions:

"plant breeder" for "manager"
"variety" for "portfolio of stocks"
"genetic lines" for "individual securities"
"crop hazard" for "error of judgment"
"removal and addition of genetic lines" for "buying and selling,"

and we have the multiline concept. In fact, the above Wall Street analogy is from my first draft (not published) of the multiline paper.

I considered the multiline manuscript to be an "idea" paper; however, when I circulated it among my faculty colleagues at Cornell University for comments, it was suggested that its chances for publication would be enhanced if data could be included. Of course, no data on a multiline variety existed because this was the first presentation of the hypothesis. The inclusion in the published article of data on crop mixtures, not a part of my original paper, had a mixed impact on the way the paper was received. Some readers perceived the paper as a simple variety blending idea; however, it has seemed unfair to denigrate the multiline concept in this way. The article clearly presented the multiline variety as a new breeding concept.

The criticism that the article was vague about line-forming procedures is valid. I believed, then and now, that there were several ways and variations of ways by which component lines for a multiline could be formed. The reasons why I did not expand and develop such procedures in the multiline paper are cause for reflection even today. At the time I was influenced by two things in this respect:

1. Because Dr. Love was a staunch advocate and practitioner of the pedigree method of breeding and selection, little attention was given to the backcross in the Cornell project.

2. My views on germ plasm diversity led me to want to avoid the almost identical background germ plasm that would accompany line development by the backcross method. I would have accepted a limited backcross, that is, a single backcross to the recurrent prototype with selection for component lines within the segregating population.

For these reasons, and because I believed that there were other feasible procedures to obtain components, I left the design considerations for constructing multiline components open to all possibilities.

My article on the multiline variety appeared in the January, 1952, issue of the *Agronomy Journal*. Within the month I had a letter from Coit Suneson, written in his enthusiastic and generous style. His letter was undated but because my reply of February 6, 1952 referred to his letter of "a week ago," it can be dated approximately. In this letter he applauded the idea, gave his views on its use, and alluded to three ways in which multiline varieties might be built. Suneson discussed at some length his experiences with the backcross programs at California. In conclusion he stated, "Our experience suggests, however, that it is better to tolerate some degree of diversity in disease reaction and retain a broader genetic base in the composite of lines."

In my reply I said this:

> Thank you very much for your kind letter of a week ago. I was very glad to have your views on this subject of variety development, especially because of your long experience with the backcross method of breeding.
>
> While I did not go into detail in the article on how the different lines would be developed, I had in mind that there were several possible approaches. Certainly one of the important ways to get similar appearing lines would be through the use of the backcross method, *at least for one backcross progeny* [italics added]. Through the use of the same desirable parent in different crosses, a group of similar-appearing lines of different genotypes could be developed.

In summing up for the multiline breeding method, I think it is too early to estimate its future role and place in crop breeding. In my own conventional breeding programs at Cornell University, there were sound reasons not to develop a multiline project: (1) New York had no pressing "crowd" disease problems and (2) good genetic advances were being made in the existing programs. The use of male sterility for the barley and wheat programs offered more immediate gains.

In the much larger programs at CIMMYT, Borlaug and colleagues had the same situation of outstanding advances coming from the conventional programs. Furthermore, the worldwide scope of their activities was not conducive to the centralized development in Mexico of multilines adapted to and suitable for other geographic areas in the world. Therefore CIMMYT changed its strategies and concentrated its efforts on the development of large numbers of prospective component lines that could be requested and used for multiline development in the country of intended use. John Gibler produced an early multiline wheat variety in Colombia for that specific area. The availability of the CIMMYT stockpile of lines could spawn the formation of multiline varieties in different parts of the world. Again, the existence of a competent research staff including pathologists, breeders, and biometricians is necessary for successful development.

The long-running Iowa experience of producing multiline oat varieties provides a good illustration of how multiline breeding can be matched with and integrated into an existing other-methods program. Frey (1986) found that two parallel programs are required, that is, one for pure line cultivars and one for multiline cultivars. The two programs complement each other. Two pure line varieties, 'Grundy' and 'O'Brien', have been released in recent years, and a third, based on *A. sterilis* cytoplasm, is in prospect. On the other hand, this program provides the lines for multiline development: the original ML-M is no longer grown; however, the updated ML-E occupies about one-fourth of the Iowa oat acreage. Frey portrayed the interworking relationships between the two programs in this way:

1. Lines coming out of the conventional breeding program are selected for yield and agronomic characteristics. No attention is given to crown rust resistance at this time; however, selection for barley yellow dwarf virus is included.

2. When the lines have advanced to candivar status, they also are placed in the second, multiline program, where the process of moving 8 to 10 crown rust resistant (CRR) genes into the lines via the backcross is begun. The 'Webster' multiline, recently released, has nine CRR-gene lines; an interesting note is that all of its CRR genes are from A. *sterilis*. Further in development are three new pure line candivars, also undergoing evaluation for use as multiline prototypes.

Frey and his colleagues have found that the CRR gene complexes have given adequate protection against crown rust for two decades, certainly evidence and confirmation of the theoretical expectations. The conversion of lines, that is, getting them ready for a multiline, takes about two years and is seen as a purely technical task.

REFERENCES

Allard, R. W. and A. D. Bradshaw. 1964. Implications of genotype-environmental interactions in applied plant breeding. Crop Sci. 4:503–508.

Beard, D. F. 1940. Relative values of unrelated single crosses and an open-pollinated variety as testers of inbred lines of corn. Doctoral dissertation No. 3:9–16. Ohio State University Press (abstr.).

Boerma, H. R. and R. L. Cooper. 1975. Effectiveness of early-generation yield selection of heterogeneous lines in soybeans. Crop Sci. 15:313–315.

Borlaug, N. E. 1953. New approach to the breeding of wheat varieties resistant to *Puccinia graminis tritici*. Phytopathology 43:467 (abstr.).

Borlaug, N. E. 1958. The use of multilineal or composite varieties to control airborne epidemic diseases of self-pollinated crop plants. First Int. Wheat Genet. Symp. Proc., University of Manitoba, Winnipeg, pp. 12–27.

Breth, S. A. 1976. Multilines: safety in numbers. CIMMYT Today 4 : 1–11.

Browning, J. A. 1957. Studies on the effects of field blends of oat varieties on stem rust losses. Phytopathology 47:4–5 (abstr.).

Browning, J. A. and K. J. Frey. 1969. Multiline cultivars as a means of disease control. Ann. Rev. Phytopathol. 7:355–382.

Browning, J. A., B. M. Cournoyer, D. Jowett, and J. Mellon. 1970. Urediospores and grain yields from interacting crown rust races and commercial multiline cultivars. Phytopathology 60:1286 (abstr.).

Browning, J. A., K. J. Frey, and R. L. Grindeland. 1964. Breeding multiline oat varieties for Iowa. Iowa Farm Sci. 18:5–8.

Cournoyer, B. M., J. A. Browning, and D. Jowett. 1968. Crown rust intensification within and dissemination from pure-line and multiline varieties of oats. Phytopathology 58:1047 (abstr.).

Davis, M. D.. and J. N. Rutger. 1976. Yield of F_1, F_2 and F_3 hybrids of rice *(Oryza sativa* L.). Euphytica 25:587–595.

Dobzhansky, T. 1963. Evolutionary and population genetics. Science 142:1131–1135.

Flor, H. H. 1955. Host–parasite interaction in flax rust—its genetics and other implications. Phytopathology 45:680–685.

Flor, H. H. 1971. Flax cultivars with multiple rust-conditioning genes. Crop Sci. 11:64–66.

Frey, K. J. Letter. November 4, 1986.

Frey, K. J. and J. A. Browning. 1971. Association between genetic factors for crown rust resistance and yield in oats (*Avena sativa* L.). Crop Sci. 11:757–760.

Frey, K. J., J. A. Browning, and R. L. Grindeland. 1970. New multiline oats. Iowa Farm Sci. 24(8):571–574.

Frey, K. J., J. A. Browning, and R. L. Grindeland. 1971a. Registration of Multiline E68, Multiline E69, and Multiline E70 oat cultivars. Crop Sci. 11:939.

Frey, K. J., J. A. Browning, and R. L. Grindeland. 1971b. Registration of Multiline M68, Multiline M69, and Multiline M70 oat cultivars. Crop Sci. 11:940.

Frey, K. J., J. A. Browning, and M. D. Simons. 1977. Management systems for host genes to control disease loss. Ann. N.Y. Acad. Sci. 287:255–274.

Gallun, R. L. 1977. Genetic basis of Hessian fly epidemics. Ann. N.Y. Acad. Sci. 287:223–229.

Gill, K. S., G. S. Nanda, and G. Singh. 1984. Stability analysis over seasons and locations of multilines of wheat (*Triticum aestivum* L.). Euphytica 33:489–495.

Griffing, B. 1956. Concept of general and specific combining ability in relation to diallel crossing systems. Aust. J. Biol. Sci. 9:463–493.

Groth, J. V. 1976. Multilines and 'super races'—a simple model. Phytopathology 66:937–939.

Gustafsson, A. 1953. The cooperation of genotypes in barley. Hereditas 39:1–18.

Hamblin, J., J. G. Rowell, and R. Redden. 1976. Selection for mixed cropping. Euphytica 25:97–106.

Jensen, N. F. 1952. Intravarietal diversification in oat breeding. Agron. J. 44:30–34.

Jensen, N. F. 1965. Multiline superiority in cereals. Crop Sci. 5:566–568.

Jensen, N. F. 1966. Broadbase hybrid wheats. Crop Sci. 6:376–377.

Jensen, N. F. and W. T. Federer. 1965. Competing ability in wheat. Crop Sci. 5:449–452.

Jensen, N. F. and G. C. Kent. 1963. New approach to an old problem in oat production. Farm Rearch (N.Y.) 29(2):4–5

Kinman, M. L. and G. F. Sprague. 1945. Relation between number of parental lines and theoretical performance of synthetic varieties of corn. J. Am. Soc. Agron. 37:341–351.

Leonard, K. J. 1969a. Factors affecting races of stem rust increase in mixed plantings of susceptible and resistant oat varieties. Phytopathology 59:1845–1850.

Leonard, K. J. 1969b. Selection in heterogeneous populations of *Puccinia graminis* f. sp. *avenae*. Phytopathology 59:1851–1857.

Leonard, K. J. 1969c. Genetic equilibria in host–pathogen systems. Phytopathology 59:1858–1863.

Leonard, K. J. and G. C. Kent. 1968. Increase of stem rust in mixed plantings of susceptible and resistant oat varieties. Phytopathology 58:400–401 (abstr.).

Luthra, J. K. and M. V. Rao. 1979. Multiline cultivars—how their resistance influence leaf rust diseases in wheat. Euphytica 28:137–144.

Marshall, D. R. 1977. The advantages and hazards of genetic homogeneity. Ann. N. Y. Acad. Sci. 287:1–20.

Marshall, D. R. and A. H. D. Brown. 1973. Stability of performance of mixtures and multilines. Euphytica 22:405–412.

Marshall, D. R. and A. J. Pryor. 1978. Multiline varieties and disease control. I. The "dirty crop" approach with each component carrying a unique single resistance gene. Theor. Appl. Genet. 51:177–184.

Marshall, D. R. and A. J. Pryor. 1979. Multiline varieties and disease control. II. The "dirty crop" approach with components carrying two or more genes for resistance. Euphytica 28:145–159.

Murphy, J. P., D. B. Helsel, A. Elliott, A. M. Thro, and K. J. Frey. 1982. Compositional stability of an oat multiline. Euphytica 30:33–40.

Norden, A. J., D. W. Gorbet, D. A. Knauft, and F. G. Martin. 1986. Genotype × environment interactions in peanut multiline populations. Crop Sci. 26:46–48.

Rosen, H. R. 1949. Oat parentage and procedures for combining resistance to crown rust, including Race 45, and *Helminthosporium blight*. Phytopathology 39:20.

Ruml, Beardsley. 1950. Collier's Magazine: Jan. 21.

Sakai, K. J. 1957. Study on competition in plants. VII. Effect of competing and non-competing individuals. Genetics 55:227–234.

Schilling, T. T., R. W. Mozingo, J. C. Wynne, and T. G. Isleib. 1983. A comparison of peanut multilines and component lines across environments. Crop Sci. 23:101–105.

Sprague, G. F. and L. A. Tatum. 1942. General and specific combining ability in single crosses of corn. J. Am. Soc. Agron. 34:923–932.

Sprague, G. F., W. A. Russell, L. H. Penny, T. W. Horner, and W. D. Hanson. 1962. Effect of epistasis on grain yield in maize. Crop Sci. 2:205–208.

Stakman, E. C. 1947. Plant diseases are shifty enemies. Am. Sci. 35:321–350.

Stevens, R. B. 1949. Replanting "discarded" varieties as a means of disease control. Science 110:49.

Torres, E. and J. A. Browning. 1968. Yield of urediospores per unit of sporulating area as a possible measure of tolerance of oats of crown rust. Phytopathology 58:1070.

Trenbath, B. R. 1977. Interactions among diverse hosts and diverse parents. Ann. N. Y. Acad. Sci. 287:124–150.

Van der Plank, J. E. 1965. Dynamics of epidemics of plant disease. Science 147:120–124.

Van der Plank, J. E. 1968. Disease resistance in plants. Academic Press, New York, 206 pp.

Van Emden, H. F. and G. F. Williams. 1974. Insect stability and diversity in agro-ecosystems. Ann. Rev. Entomol. 19:455–475.

Wallace, H. A. 1938. Corn breeding experience and its probable eventual effect on the technique of livestock breeding. Spragg Memorial Lecture, Michigan State University, East Lansing, Mich., April 21.

Watson, I. A. 1970. Changes in virulence and population shifts in plant pathogens. Ann. Rev. Phytopathol. 8:209–230.

Watson, I. A. and D. Singh. 1952. The future for rust resistant wheat in Australia. J. Aust. Inst. Agric. Sci. 18:190–197.

Went, F. W. 1950. The response of plants to climate. Science 112:489–494.

14

METHODS SHAPED BY RECURRENT FORCES

Recurrent selection in its various forms has played a major part in the development and use of the following breeding and selection methods. Of course, it also plays a role in certain other methods; for example, the backcross is basically a recurrent selection procedure, but its distinct origin and use suggest inclusion under the historically primary methods. As a preamble to recurrent selection we open this section with a general consideration of intermating and linkage, and later the importance of genetic male sterility will be introduced.

THE ROLE OF INTERMATING AND LINKAGE

The question of intermating, as following a first hybridization, is of considerable theoretical and practical interest to plant breeders.

Sewall Wright (1931) obviously was not considering index selection in his discussion of "control of evolution" (pp. 151–153) but he certainly was describing a form of intermating and recurrent selection. After first stating about control that "little is possible either within a small stock or a freely interbreeding large one," he commented:

> The only practicable method of bringing about a rapid and non-self-terminating advance seems to be through subdivision of the population into isolated and hence differentiating small groups, among which selection may be practiced, but not to the extent of reduction to only one or two types (Wright 1922a). The crossing of the superior types followed by another period of isolation, then by

> further crossing and so on *ad infinitum* presents a system by means of which an evolutionary advance through the field of possible combinations of the genes present in the original stock, and arising by occasional mutation, should be relatively rapid and practically unlimited.

Anderson (1939) discussed the barriers imposed on free recombination by linkage in mathematical terms. He considered linkage interference to be a serious and general phenomenon. Although the prospects were discouraging, Anderson implied that the recourse for the breeder is to use intermating and many hybrid generations.

The need for intermating was documented in the conclusions of Suneson and Stevens (1953) who noted:

> There is a non-random survival of recombination characters in hybrid mixtures. The complete loss of some characters, the impotence of disease-resistance factors, and the general cohesiveness of the populations all suggest that the limitations on recombination into a favorable adaptation complex imposed by linkages are larger than most barley breeders have realized. This suggests that all breeders should employ more backcrossing or grow bulk populations for long terms to insure more complete recovery of proved gene associations.

One of the strategies proposed by Palmer (1953) in order to capture the highest expression of quantitative gene traits (e.g., yield), was to intermate the high-yielding selections from a cross so as to bring together different superior genes.

A case of beneficial linkage in barley was shown by Everson and Schaller (1955). The association of semi-smooth awns and increased yield was shown to be due to linkage rather than pleiotropic action of the semi-smooth awn gene. Recombination within the chromosome segments was indicated by substantial crossing over and segregation showing different factors for yield. From a breeding standpoint, the awn character can be used as a marker to identify the quantitative trait, yield, but testing must be done to identify the line having the most productive accumulation of yield genes.

Hanson's (1959) paper was the culmination of a series in which he studied changes in average segment length per chromosome. His conclusions are germane to plant breeding:

> It appears evident, based on the analysis of the breakup of linkage blocks, that a breeding program for a self-pollinated species should include at least one, and preferably three or four, generations of intermating if at all feasible. Genetic recombination within linkage groups is extremely limited, especially for short chromosomes unless intermating cycles are included in the breeding program. Furthermore, at least four parents should be used to synthesize the intermating population. The inclusion of four or more parents in the population increases the genetic potentials of the population and effects a greater reduction in average block lengths as compared to a population synthesized from two initial parents.

A few years later Hanson and Hayman (1963) found in the case of crosses of two homozygous parents that the additive genetic variance in their homozygous progeny lines was a function of their average recombination. They considered the effect of adding intermating after the F_2. When linkage equilibrium was unity, the variances did not change with intermating, and with intermating the variances tended to move towards the equilibrium value. Thus "unless one dealt with a cross between two extreme parents, it would be extremely difficult to establish changes in genetic variability with intermating."

The interpretation here, as the authors acknowledge, weakens the practical impact of Hanson's 1959 paper, which considered intermating before selfing, but it does not withdraw the thesis that intermating is necessary for real and long-term gains in the populations. It might be added also that in a practical breeding program with stated goals, intermating recombination is not a random process, nor is it concerned only with breaking linkages.

Andrus (1963) characterized all breeding methods as occurring in the range between maximum inbreeding (pure line) and maximum interbreeding (bulk with mandatory sib crossing). Although his article is primarily a review of breeding systems, the paper nevertheless distills Andrus's views and experiences in muskmelon breeding. Andrus would soften the rigid sequential breeding cycle of pure line to pure line by introducing sib crossing to bring together partly balanced genotypes in early generations. The hope here was that recombination would produce the desired individuals. The process was aided by the largely cross-pollinated nature of the muskmelon (pollinated by insect vectors), facilitating recurrent selection procedures. The sib-crossing was not random although there was an element of randomness owing to the vector and a certain amount of selfing that occurred. Selection pressure was guided through the use of a multiple index procedure (14 traits), with the twice-annual selections intermated to create a continuous sib-crossing pattern. Generally, Andrus believed that appropriately balanced polygenic individuals are most likely to evolve through delayed inbreeding, which can be fostered by intermating or sib-crossing in early generations of a cross.

An early-generation analysis by Allard and Harding (1963) to predict the probable gain in bidirectional selection for heading date indicated a close agreement between the observed and theoretical distributions. The authors concluded that a single gene pair was involved and little advance could be made through selection. However, using the same cross material propagated by pedigree selection, with examination of the F_4–F_7 generations of two groups, namely, random choice and selection of earliest and latest 10%, a strikingly different result was obtained. In the random group the earlier bimodal pattern all but disappeared and transgressive extremes appeared in both directions. In the selected families there was rapid early-generation advance, moving the populations well beyond the parent standards of heading. A between-parent range of 20 days was expanded to 40 days in the selection extremes. Further observed gains were 22% (early) and 55% (late) greater than predicted gains.

Allard and Harding came to the conclusion that genetic variability found in closely linked gene complexes made its appearance only after several generations of selfing and genetic recombination involving parts of or whole chromosomes. They concluded. "Intercrosses among early and among late families are being studied in an attempt to assay the extent of such variability."

In a theoretical and statistical analysis of the effect of linkage on selection, Felsenstein (1965) showed that selection creates linkage disequilibrium. If selection is positive, tight linkage increases the rate of gene frequency changes; if negative, the reverse occurs. The extrapolation of these reasonable findings to artificial selection of an additive phenotype leads to the result that artifical selection generates negative linkage disequilibrium, and therefore tight linkage reduces the response to artificial selection.

Hanson, Probst, and Caldwell (1967) created a random mating population from 56 selected F_3 lines taken from 28 crosses of an eight-parent diallel cross of soybean cultivars. Their principal interest was in forming and evaluating a more productive gene pool for selection. Careful and tedious procedures were used to assure random crossing (a "free" benefit in populations where genetic male steriles are available).

The authors referred to the stepwise truncating-of-germ plasm process familiar to breeders of self-pollinating crops. They pointed out that intermating broadens the coefficient of parentage. Using the limited cycling of their study, they postulated that if the top-yielding 20 selections were intermated and then selfed, a gain of 3.5 bu/acre over the base population mean would be obtained.

Faced with a wide cross of cotton which continued to show a preponderance of parental type reproduction, Miller and Rawlings (1967) reasoned that persistent linkage blocks might be the cause. Reproducing the cross, they maintained the population for six generations in an isolation block in an area where natural crossing was of the order of 50%. At the end of this time the intermated and selfed populations were sampled, and the means of seven traits were compared with those found in the original populations. Six of these traits were thought to be in coupling phase linkages and one in repulsion phase linkage. If their surmise were correct, genotypic variances would be expected to decrease for the former and increase for the latter under intermating or breaking up of linkage disequilibrium, and this shift was observed.

The case for benefits from intermating is persuasive but not conclusive. The researchers considered that the intermated population after six cycles gave them a better pool for selection than the original F_2. One would have to conclude, too, that the high level of sustained outcrossing itself would have a marked effect upon breaking linkages.

I had always thought of cotton as showing about 85% selfing; 50%

outcrossing does create a quite different breeding situation. Of course, 50% outcrossing also means 50% selfing, and this combination would slow genetic advance. Nevertheless, the two cultivars used as parents were considered highly homozygous, possibly a consequence of continual roguing and grooming for commercial use.

ALTERNATIVES TO AND MODIFICATIONS OF INTERMATING

For varying reasons it is not uncommon for breeders to encounter hybrid populations that show unexpected responses. There are alternatives to intermating a difficult cross population. The cross could be abandoned, and other crosses examined for the same objective. Rather than intermating, a program of backcrossing or recurrent selection might be instituted; in this case, knowledge about the linkage situation and the parental sites of the coupling and repulsion phase linkages would be helpful. Manual crossing could be used to hasten intermating. Advantage could be taken of the 50% outcrossing by surrounding rows of the space-planted F_2s of the difficult cross with plants of desired genotypes. These would serve as pollen donor parents and genetically "overwhelm" the more-distant population pollen donors, thereby increasing the chances of outcrossing over selfing. Genetic diversity could be increased and linkage equilibrium hastened by introducing parents different from those chosen for the first cross but that exhibited similar valuable traits.

In a computer simulation comparison of mass selection with single seed descent methods, Bliss and Gates (1968) found linkage to exert an important influence on genetic gain in both schemes. They approved some form of intermating:

> Since additive variance is independent of population size, increasing the population beyond the point where sampling errors cease to be a factor is unlikely to reduce linkage bias or greatly increase realized genetic gain. A system of intermating prior to selection, which allows the additive variance to approach values obtained for random gene assortment, would be advantageous in these families.

Matzinger and Wernsman (1968) practiced mass selection in a hybrid population of tobacco; however, the linear increase in yield over four cycles cannot be attributed solely to mass selection because the procedures also involved "controlled random mating" and recurrent selection. These procedures used together formed a powerful breeding and selection method. At the start, the eight-line synthetic had been random-mated for three generations. The base population used was formed from bulking equal amounts of seeds from about 100 crosses from the third intermating. Green weight of tobacco leaves, which is known to be highly correlated with cured weight, was used to measure yield. A form of gridding was imposed on the 36 blocks of rows, and the top-yielding three plants from each block were identified prior to anthesis. Immediately after identification, controlled random mating of the 54 pairs (among the 108 selected plants) was made. Equal numbers of seed from each of these crosses was bulked to form the next cycle, which was repeated through four cycles. The result was a linear increase of 44 g/plant per cycle with no apparent reduction in genetic variability. Population appearance and other characters changed during the four cycles. One would have to credit the intermating sequences with a share of the advances made in this study.

Baker (1968) conducted a computer simulation program to assess the effect of intermating on a breeding program, recognizing that intermating imposes a cost and a delay of one generation. With two linked loci he found that intermating of 10 or 30 pairs of random F_2 individuals would significantly increase the retrieval probability of the desired recombinants. With more loci (nine) an interesting situation develped: with tight linkage, recombinant gametes were so few that they had little chance of finding each other, and instead, united with parental gametes causing an increase in heterozygosity. With looser linkage, however, the recombinant gametes were more numerous; thus the chances of two unions were increased. Therefore, the success of intermating depends heavily on linkage intensity. However, in the case with linkage intensity, $p = .20$, the increase in gene frequency from intermating seemed small (.013). Nevertheless, intermating translated into an increase of effectiveness of selection of 18.8%, a considerable amount. Baker recognized the need for a recurrent selection program with intermating in each cycle. This suggestion would transfer simple intermating effects into powerful "pyramiding" effects in the successive populations thus generated.

The complexity of the linkage problem as it relates to the more general problem of breeding self-pollinated crops was cogently stated by Baker:

> In practical breeding programs it is common to cross two inbreds each of which has a different set of desirable traits. If there is linkage among the various genes which control these traits, we should expect that the majority of the linkages among genes for different traits will be in repulsion. If two genes governing a single trait are in coupling and by chance lie on either side of a third gene in repulsion to the two, it will be necessary to break the coupling phase chromosome twice in order to obtain the three genes on one chromosome. Thus there is a need for breaking linkages in order to get the new combinations that a breeder would like to see come out of a breeding program. Intermating enhances the breaking of linkages whether they be in coupling or repulsion phase.

In 1970 I proposed a modification of recurrent selection for use with inbreeding crops called a *diallel selective mating system*, or DSM. It is discussed in Chapter 16.

In cotton, a predominantly self-pollinating species, it has been difficult to increase yield along with fiber strength. Meredith and Bridge (1971) wondered whether intermating would influence, that is, reduce, this troublesome negative correlation, a likely explanation for which was linkage. Two generations of crossing were imposed in an appropriate cross beginning with F_3 plants, and from the final 250 crosses a bulk was formed using one seed from each cross. From this bulk, 96 plants were individually selfed and became the intermated group. The control group represented remnant seed of these 96 plants selfed once, thus F_3. The comparisons to be made from nursery boll samples, therefore, were between two F_3-progeny groups, one which was the "original" and the other the original plus two intermatings. Parental controls were included.

The results showed a high degree of similarity between the two populations; however, the economically important negative genetic correlation between lint yield and fiber strength was decreased by intermating. In fact, 8 of the 10 highest correlations in the original population showed reductions in the intermated population, implying a loosening of bonds.

Meredith and Bridge discussed ways of breaking linkage, and mentioned four modifications of the conventional breeding system: (1) random, (2) a diallel selective mating system (Jensen 1970), (3) selection index, and (4) backcrossing. Although not mentioned, it would seem that radiation and, where feasible, the introduction of male steriles might also be useful tools in breaking up linkage blocks. Finally, the authors suggested a modified bulk wherein one locule would be harvested from each plant, the seed bulked, and the procedure repeated, with selection delayed until F_4. This is similar to a procedure I used in the winter barley project, where the composites were built around the male sterile gene: I collected one head each from selected plants in the population, and bulked these for the next generation. Intermating would occur because of the segregating male sterile trait. I termed this *SHD* for *single head descent*, because it is reminiscent of single seed descent, but it has the added advantage of supplying the seeds from a group of plants, each with its own range of within-line variances, rather than from a single plant, to the next generation.

Doggett (1972), in an article on recurrent selection in sorghum, noted an interesting pattern that was found in every population and every breeding system tested; namely, recombination following inbreeding resulted in higher yields. This alternating pattern of heterosis was noted in cotton also by Walker (1963). Doggett thought this pattern might be found in other self-pollinated species.

The ease of supplying intermating to modern programs is illustrated by the release by Nordcross, Webster, Gardner, and Ross (1973) of three sorghum random-mated germ plasm stocks.

Two cycles of random intermating in wheat and barley mass selection programs to increase tillering were shown to be very effective by Redden and Jensen (1974). Three populations were generated from one F_2 population in wheat, and three more were generated from one F_2 population in barley. One population (hybrid series) involved two cycles of random mating among plants selected before anthesis for tiller number. A second population (selfed series) was mass-selected without intermating. A third population (control series) was selfed without being subjected to either mass selection or intermating.

When tested at putative F_5, and averaging the hybrid and selfed series, mean selection response over two sites was 10.3%/cycle for wheat and 6.3%/cycle for barley. The responses partitioned by site and type were for wheat (one site only), 22.6% hybrid and 18.5% selfed; for barley; Site 1, 17.1% hybrid and 8.7% selfed; and for barley, Site 2, 13.9% hybrid and 10.3% selfed.

These sizable advantages from both mass selection and intermating, but especially the latter, demonstrate the power of selection and intermating. Efficiency was increased, of course, by the fact that selection and intermating could be done sequentially in the same generation, respectively. One wonders if the use of directed mating, that is, application of DSM among individuals showing the highest expression of tillering rather than random, might have increased efficiency. Still, the selection intensity was high for the selected group used in the random matings. The authors discussed plant breeding possibilities using recurrent selection in crops where male sterility is available.

Pederson (1974) cautioned against the indiscriminate use of intermating before selection in self-fertilized species. A basic consideration was whether the alleles brought into a cross are in the coupling or repulsion phase. Only in the repulsion phase is there likelihood of benefit inasmuch as the coupling phase already has the desired arrangement (the familiar "if it ain't broke, don't fix it" syndrome). Pederson gave examples and case documentation for two or more linked loci, and showed that if the loci are dispersed over more chromosomes the odds are even more daunting against receiving benefits. He preferred ordinary truncation selection to intermating immediately in the F_2.

MALE STERILITY AND RANDOM VERSUS NONRANDOM MATING

An interesting situation is the use of the genetic male sterile trait in a self-fertilizing species, where the intermating, either random or directed, is on a large scale and continuous. But the question may well be asked, is this still a self-fertilizing species?

Bailey and Comstock (1976) used a simulation model to consider the theoretical and biometrical aspects of the breeding problems: (1) how to collect "into one line all the favorable alleles present in one or the other of any pair of pure lines," and (2) what are the effects of linkage. The study showed that the linkage effect depended on the linked relationship of superior and other genes. For example, superior genes in the coupling phase enhance recovery, whereas a superior and inferior gene in repulsion phase lowers recovery probabilities. In general, lines of apparently equal genetic value have little influence on recovery probabilities. These findings are not favorable to the objective of getting a transgressive line from a two-parent cross. The authors proposed these two methods to improve the probabilities:

1. The repetitive breeding cycle should be continued but be fueled by double crosses of four selections from the preceding cycle.
2. Cross pollination and population size should be increased.

Stam (1977) published a thoughtful article on selection response under random mating and selfing. It seemed to me that he posed the plant breeder's question, which is, paraphrased, "What benefit is to be expected from intermating in terms of both increased selection response and pay off?" Included in the presentation is a computer simulation study involving different two-locus and multilocus situations in progenies of crosses of homozygous parents. Stam's conclusions were:

1. Up to about F_4, and regardless of number of loci, random mating showed no advantage over selfing; indeed Stam postulated that selfing held a selection response edge because of environmental variance owing to deviation from Hardy–Weinberg frequencies. A surprising finding was that the results for linkage groups were quite similar to those for independently segregating loci.

2. "Random mating always is superior to selfing" over many generations, the advantage increasing with many loci.

3. The effect of a single round of intermating depends on the generation of application; the computer simulation showed the optimal time to be F_4 or F_5. Presumably this involved the factors of greater homozygosity and visual selection.

Response to selection involving random mating is influenced by population size, number of loci, environmental variance, and selection intensity.

Bos (1977) marshaled the statistical arguments against the random intermating of F_2 plants and the consequences relative to later optimal genotypes. His conclusions were that with random mating in the F_2, fewer desired genotypes were found in F_3 than in normal non-intermated F_3s; and in the ultimate advanced generation population with independent segregation, the numbers were comparable. In the case of linkage, intermating was superior, producing about 25% more desirable genotype plants in the ultimate population.

Bos's theoretical and statistical treatment of intermating situations emphasized random intermating of F_2 plants, a practice that is little used by breeders because a random approach is wasteful, and because the F_2 is only one of the possible stages for intermating and perhaps not the best. Also, in this study the extent of intermating was not specified; it would matter whether the level of intermating were low or high.

The availability of male sterility creates a crossing situation dramatically different from that of conventional hand crossing. Bos refers to the use of male sterility as an option to aid in the random crossing of F_2 plants ("by pollinating male sterile F_2 plants with male fertile F_2 plants. This results in an F_3' population, which after continued spontaneous self-fertilization transforms in an F' population."). I grant that Bos's terse description of the dynamics of change in such a population may serve the objectives of his paper; however, the details of what happens in such a population over several years after the male sterile gene is inserted via a cross are of considerable interest in themselves. For example, in barley an F_2 population of male sterile plants and male fertile plants is a population segregating one-fourth male sterile plants and three-fourths fertile plants, with the fertile group split, in whole population terms, into a one-fourth homozygous fertile part and one-half heterozygous fertile portion. This population will experience continuous intermating far into the future, with the sterile plants being pollinated by either the fertile heterozygotes or homozygotes. The proportion of male sterile plants gradually declines in such a population, and in time male sterile plants disappear. The situation cannot in any sense be considered a straightforward single intermating situation followed by generations of self-fertilization.

In a sense, the availability of male sterility in some of our self-fertilizing economic crops has reduced the pertinence and application of many studies to situations involving conventional hand crosses. The research questions must be rephrased to focus on repeated populational intermating, which if uncontrolled is random and akin to that in cross-pollinated crops.

Bassett and Woods (1978) discussed strategies that are fundamental to parent choice for intermating projects. For details see Chapter 30, "Predicting and Choosing Crosses and Lines".

Verma, Kochhar, and Kapoor (1979) recounted the familiar litany of the plant breeder's early generation problems with conventional selfing, ending with the common situation where the combination of desired genes are not found in any individual among the selected genotypes. The solution to this problem involved random crossing among F_2 lines. The intent was to generate greater genetic variance for certain traits, and compared with F_3 lines, this was accomplished for grain yield, rust reaction, and grain weight in wheat crosses. They interpreted this to mean that linkages were present and intermating provided opportunities for crossing over. The breaking of repulsion phase linkages was more to be desired than the breaking of coupling phase breakages.

Considered from the standpoint of methodology, I see two aspects worthy of comment. The first involves the nature of the applied crossing, whether it is random or nonrandom. Without some form of male sterility, random crossing makes little sense, especially in view of the large numbers that are necessarily involved, the tediousness of the preparations, and the incumbent delay. Indeed, Verma et al. sense this, for they stated, "Intermating of the selected plants possessing extreme phenotypic expression for various characters is likely to be more rewarding than intermating the plants at random. . . ."

This is tantamount to a recommendation for nonrandom visual selection for prospective parents among the plants of the F_2 population. The second point involves timing: I would favor the intermating at a later generation, for example, F_4 or beyond, when selections of the parents for intermating would be more meaningful, and when the potential of the cross population itself was more evident. Intermating is a procedure not suitable for all crosses in a breeding program.

Using a dominant male sterile gene to facilitate crossing, Altman and Busch (1984) compared lines taken from each of three cycles of random mating before selection in three single-cross spring wheat populations with lines directly selected from the populations (no intermating). The results showed similarities rather than differences in the comparisons, and bearing in mind the additional cost of intermating procedures, the authors concluded that "Random intermating within single cross populations resulted in insufficient useful recombination to justify its use as a primary breeding procedure prior to selection."

Altman and Busch's article includes a review of recent research on intermating, and offers an opportunity for a wider discussion of its role. I agree with the authors' conclusions and believe further that there is a general consensus among breeders that random intermating before selection in a single cross would not be an appropriate or efficient procedure. Does this breeding situation offer any appropriate place for the interposition of intermating? Perhaps *after* selection. There are cases of released cultivars arising from sib crossing of good-but-not-releasable lines out of the same cross.

The appropriateness of using intermating will be clearer if we separate the functional attributes, random and nonrandom. When this is done we have only two applications:

1. *Random Intermating.* This is designed to increase the likelihood of useful recombinants.Theoretically and practically, any increase in genetic variability is desirable, but random mating, excluding male sterile-based programs, is seldom used for this purpose in practical breeding programs. Other alternatives such as irradiation or use of chemical mutagenic agents might be considered to increase genetic variability.

2. *Non-random Intermating.* This is planned or selective intermating, that is, recurrent selection pure and simple. The acronym DSM in my 1970 paper stands for "diallel *selective* mating" and nowhere involves any random mating. The expanded genetic variance comes from an initial diallel of F_1s as parents followed by periodic selective matings.

The distinction between random and planned nonrandom intermatings is important. In a later paper on the DSM system merged with the bulk hybrid population system (Jensen, 1978), I tried to make this clear: "There were special reasons for adopting objective bases as an operational concept. First, random crossing or compositing has little to recommend it. A breeder needs directional goals."

Breeders will note that a large number of DSM objective base composites [15 were listed in my Table 1 (1978)], plus elite three-way and double crosses, would constitute a breeding program. Considering the demands of the conventional breeding program already in place, little time could be spared to practice selective intermating. The objective base composites were formed on DSM principles and handled in the field as bulk populations for F_5 line withdrawal. Due regard was given to the need for increased sampling as described in Chapter 16. Nevertheless, the opportunity remains for intermating in special cases, as illustrated by Redden and Jensen (1974), and for increasing wheat seed protein, as illustrated by my 1978 paper, both involving recurrent selection via selective, not random, intermating.

Frederickson and Kronstad (1985) practiced bidirectional selection for heading date over two cycles in two wheat crosses, comparing the efficiency of two procedures implemented after visual selection of F_2 plants for early and late heading. The two procedures were selfing following visual selection and intermating following visual selection. The authors listed studies where recurrent selection with intermating had been successful. However, the results of their studies showed that the two procedures were essentially equal in effectiveness, meaning that selfing would be the prudent choice if one abided by their guideline: "If recurrent selection is to be an effective tool in breeding of self-pollinated crops, it should offer advantages over conventional methods in order to offset the added time and expense of intermating selected plants or families."

On the other hand, the authors pointed out that traits such as heading date often have high heritabilities, which make them amenable to visual selection procedures. Visual selection could identify much of the available genetic variance before the first cycle had begun. It also should be noted that where male sterility is available, intermating and recurrent selection have an entirely different aspect. This and other papers emphasize the importance of the timing of intermating to its success.

REFERENCES

Allard, R. W. and J. Harding. 1963. Early generation analysis and prediction of gain under selection in derivatives of a wheat hybrid. Crop Sci. 3:454–456.

Altman, D. W. and R. H. Busch. 1984. Random intermating before selection in spring wheat. Crop Sci. 24:1085–1089.

Anderson, E. 1939. The hindrance to gene recombination imposed by linkage: an estimate of its total magnitude. Am. Nat. 73:185–188.

Andrus, C. F. 1963. Plant breeding systems. Euphytica 12:205–228.

Bailey, T. B., Jr., and R. E. Comstock. 1976. Linkage and the synthesis of better genotypes in self-fertilizing species. Crop Sci. 16:363–370.

Baker, R. J. 1968. Extent of intermating in self-pollinated species necessary to counteract the effects of genetic drift. Crop Sci. 8:547–550.

Bassett, M. J. and F. E. Woods. 1978. A procedure for combining a quantitative inheritance study with the first cycle of a breeding program. Euphytica 27:295–303.

Bliss, F. A. and C. E. Gates. 1968. Directional selection in simulated populations of self-pollinated plants. Aust. J. Biol. Sci. 21:705–719.

Bos, I. 1977. More arguments against intermating F_2 plants of a self-fertilizing crop. Euphytica 26:33–46.

Doggett, H. 1972. Recurrent selection in sorghum populations. Heredity 28:9–29.

Everson, E. H. and C. W. Schaller. 1955. The genetics of yield differences associated with awn barbing in the barley hybrid (Lion × $Atlas^{10}$) × Atlas. Agron. J. 47:276–280.

Felsenstein, J. 1965. The effect of linkage on directional selection. Genetics 52:349–363.

Frederickson, L. J. and W. E. Kronstad. 1985. A comparison of intermating and selfing following selection for heading date in two diverse winter wheat crosses. Crop Sci. 25:555–560.

Hanson, W. D. 1959. The breakup of initial linkage blocks under selected mating systems. Genetics 44:857–868.

Hanson, W. D. and B. I. Hayman. 1963. Linkage effects on additive genetic variances among homozygous parents. Genetics 48:755–766.

Hanson, W. D., A. H. Probst, and B. E. Caldwell. 1967. Evaluation of a population of soybean genotypes with implications for improving self-pollinated crops. Crop Sci. 7:99–103.

Jensen, N. F. 1970. A diallel selective mating system for cereal breeding. Crop Sci. 10:629–635.

Jensen, N. F. 1978. Composite breeding methods and the DSM system in cereals. Crop Sci. 18:622–626.

Matzinger, D. F. and E. A. Wernsman. 1968. Four cycles of mass selection in a synthetic variety of an autogamous species *Nicotiana tabacum* L. Crop Sci. 8:239–243.

Meredith, W. R., Jr., and R. R. Bridge. 1971. Breakup of linkage blocks in cotton, *Gossypium hirsutum* L. Crop Sci. 11:695–698.

Miller, P. A. and J. O. Rawlings. 1967. Breakup of initial linkage blocks through intermating in a cotton breeding program. Crop Sci. 7:199–204.

Nordcross, P. T., O. J. Webster, C. O. Gardner, and W. M. Ross. 1973. Registration of three sorghum germplasm random-mating populations. Crop Sci. 13:132.

Palmer, T. P. 1953. Progressive improvement in self-fertilised crops. Heredity 7:127–129.

Pederson, D. G. 1974. Arguments against intermating before selection in a self-fertilizing species. Theor. Appl. Genet. 45:157–162.

Redden, R. J. and N. F. Jensen. 1974. Mass selection and mating systems in cereals. Crop Sci. 14:345–350.

Stam, P. 1977. Selection response under random mating and under selfing in the progeny of a cross of homozygous parents. Euphytica 26:169–184.

Suneson, C. A. and H. Stevens. 1953. Studies with bulked hybrid populations of barley. USDA Tech. Bull. 1067. 14 pp.

Verma, M. M., S. Kochhar, and W. R. Kapoor. 1979. The assessment of the biparental approach in a wheat cross. Z. Pflanzenzuecht. 82:174–181.

Walker, J. T. 1963. Multiline concept and intra-varietal heterosis. Empire Cotton Growers Rev. 40:190.

Wright, S. 1931. Evolution in Mendelian populations. Genetics 16:97–159.

15

THE RECURRENT SELECTION METHOD

It seems appropriate that much of our early information on recurrent selection came from a cross-fertilized crop, maize. Today recurrent selection can be worked across type-of-pollination boundaries, and in autogamous species it is used equally well with or without the aid of male sterile genes. It is also true in the self-pollinators that the availability of the genetic male sterile character has advanced the use of recurrent selection because of the ease with which both random and selective matings can be made.

HISTORY AND EARLY DEVELOPMENT

Richey (1927) is generally credited as the originator of the recurrent selection concept, proposing convergent improvement as a means of improving the vigor of two parents without changing their combining ability.

Merle Jenkins' (1940) paper on yield prepotency in corn has been included for several reasons. His suggested alternatives to mass selection and methods of production of synthetic varieties represent early forms of recurrent selection. His views on parent selection, the advantages of selecting among rather than within lines, the tests for yield prepotency, the confounding effect of environmental variation on precision of measurement—all are relevant also to self-pollinated crop methodology.

Penny, Russell, Sprague, and Hallauer's (1963) "where we stand" paper is valuable for its detailed and authoritative chronological presentation of the development of recurrent selection. The paper takes the reader through general combining ability (Jenkins, 1935, 1940, specific combining ability

(Hull, 1945), and reciprocal recurrent selection (Comstock, Robinson, and Harvey, 1949). The common thread running through all procedures embraces successive cycles of selection with selected recombination. Because recurrent selection originated in the cross-pollinated crops (maize), it is natural that the inbred phases of corn breeding are the link with the self-pollinators.

Penny et al. list the accumulated results of research on recurrent selection with corn at the Iowa station. A clue to the source of greatest genetic variability was that the highest gains came from the application of recurrent selection to open-pollinated varieties. This was a bit puzzling because genetic variability was known to be high in hybrid populations (from crosses of inbred lines), but the explanation may lie in the fact that single crosses have fewer deleterious genes than open-pollinated varieties. Also linkage effects, which would be greater in the F_2, might have been responsible. The authors cited three problems in evaluating advances via recurrent selection:

1. The existence of genotype × environment interactions, which tend to confound results from year to year.
2. The possible absence of negative published results—a seldom-mentioned reason.
3. Yield differences must be large or number of replications many in order to reach significance. This point is interesting because it was about this time that Gardner proposed the gridding that facilitated the selection of best plants.

PROCEDURES AND USES OF RECURRENT SELECTION

A significant feature of Ferwarda's paper on rye breeding (1956) is its clear statement that methodology must be tailored to the crop and its idiosyncrasies. Although superficially rye breeding and corn breeding are similar, Ferwarda found that the practices so successful in corn were ineffective in rye, largely because rye was unsuitable for the heterotic approach. Programs based on mass selection, family selection, and the polycross test allowed only limited progress. Ferwarda concluded that recurrent selection, especially reciprocal recurrent selection, held the most promise. The use of two populations gradually approaching and complementing each other had the additional advantage of reducing inbreeding, which was undesirable because it led to selection for self-fertility. Seed for a final variety was obtained from a final mass intercrossing of clones from the two populations, forming in effect a synthetic "varietal hybrid" exhibiting heterosis.

Khadr and Frey's (1965) paper on recurrent selection in oat breeding has wider crop application because of a breeding procedure derived from the use of radiation. Recurrent selection was practiced in four populations for the trait 100-seed weight. The populations were the irradiated 'Clintland' and 'Beedee' cultivars, and the irradiated and nonirradiated hybrids of the cultivars. The initial parents for the crosses were selected high-seed-weight individuals from the populations. The results showed that the greatest advance in genetic variability by selection was made in the irradiated hybrid, whereas the irradiated cultivars and their nonirradiated hybrid gave about equal results. Projecting for two consecutive recurrent selection cycles, a six-year time requirement, the expected mean genetic advance for the four populations was 25%, with the irradiated hybrid estimated at 33%.

Significant observations from this study were that the radiation-induced variability in the pure lines was equivalent to that in the nonirradiated hybrid of the two, and even greater in the irradiated hybrid—a strong suggestion that irradiation must be considered along with hybridization as a source of variability. Recurrent selection resulted in good retention of variability in cycle progenies with little diminution after two cycles. The authors suggested, first, the use of radiation, and second, alternating cycles of pedigree and recurrent selection because of the cost and difficulty of crossing oats in a rapid cycling procedure.

Compton (1968) proposed a recurrent selection scheme based on an extension of the single seed descent method whereby the single seeds represented the F_1s of different crosses rather than the F_2s of one cross. Outstanding selections were periodically crossed to begin another cycle in the same manner.

Jensen's diallel selective mating system (DSM), as presented in two papers (1970 and 1978), reflected the melding of recurrent selection with a diallel cross composite base.

Doggett (1972) wrestled with the problem of recurrent selection applied to sorghum germ plasm breeding composites. His solution was to maintain two bulks sufficiently different so that suitable heterosis resulted when they were crossed, and to improve each by recurrent selection [cf. also Doggett and Eberhart (1968) and Eberhart, Harrison, and Ogada (1967)]. This implied the use of different germ plasm and logical objectives. A series of subpopulations was indicated to advance certain selection objectives. These procedures paralleled the continuing work in barley in the United States.

McNeal, McGuire, and Berg (1978) used recurrent selection to increase grain protein content in spring wheat. They randomly chose and grew 2000 entries from the World Wheat Collection, analyzed them for protein content, and subjected the highest in protein content to milling and baking tests. Nine genotypes were selected and nine crosses were made among these and three other western U.S. wheats.

In the first cycle, the F_2-derived F_3 progeny rows of the nine crosses were measured for protein content by the Udy method, and the two best lines from each cross became the 18 parents for the second cycle. The lowest protein lines were also selected. Diallel crossing produced 149 of the 153 possible combinations. The F_3 progenies were tested for protein in 1973, but because of unusually high readings the test was repeated the following year on F_4 progenies.

Grain protein percentages increased from first to second cycles, but this was accompanied by a decrease in grain yield *except* for three crosses where both protein percent and grain yield rose. Two other crosses showed promise with protein percentages high enough to compensate for slightly lower grain yields. From these crosses 27 lines were selected that met a tandem criteria index of 17.5% protein and protein yields of 40.0 g/2.4 m row; 26 of the lines were superior to the parents. Presumably these lines would form the nucleus of a third recurrent selection cycle.

Of added interest was the comparison of high and low protein lines—the bidirectional selection emphasized the gains from high protein selection.

Bassett and Woods (1978) discussed some of the problems inherent in the choice of sequential parents for use in backcrossing programs.

Brim and Burton's (1979) paper is especially important because it addressed key points of efficiency in recurrent selection programs, for example, population size, number of parents, and index selection. The study centered on percent protein, often inversely related to yield. Therefore, soybean breeders must keep in mind not only been yield but protein content, both percent and yield. The method explored was recurrent selection, which can be costly and laborious in self-pollinating crops without the aid of male sterility.

The authors used two basic populations, I and II, which differed principally in the nature of the germ plasm used. Population I began with a cross of two highly adapted soybean lines, whereas Population II involved crosses of nine high-protein unadapted plant introductions to a highly adapted line, followed by backcrosses of the F_1s to the recurrent good line. The two populations were each further subdivided into A and B populations, which differed from each other mainly in size, IA and IIA being about three times the size of IB and IIB. The procedures involved the sequential testing for protein, intermating the highest percent protein lines, and selfing. Four to six cycles were completed. Final selections were compared in performance nurseries for yield, protein, and oil.

In all populations percent seed protein showed linear and statistically significant increases as shown below:

Pop.	IA	(six cycles)	46.3–48.4%
Pop.	IB	(four cycles)	46.4–47.6%
Pop.	IIA	(five cycles)	42.8–46.1%
Pop.	IIB	(four cycles)	43.2–45.9%

Also, there was a general pattern agreement in all populations toward a decrease in yield and oil.

The rates of percent improvement of seed protein differed, attributable to the considerable germ plasm differences in the basic populations. Nevertheless, all procedures demonstrated the capability of recurrent selection to improve an important trait. Because cost and efficiency were important, the authors concluded that for short-term gains the number of parents intermated and lines tested should be relatively small, whereas for long-term programs, the numbers might be increased to counter genetic shift and loss of genetic variability.

The inverse relationships suggest yield should be incorporated, properly weighted, into an index selection where simultaneous selection would move to improve both, though at a somewhat lower rate. Selecting for total protein (yield × percent protein) was suggested as a possible strategy.

One exception in the results was found in Population IA, where both yield and total protein increased during the first two cycles. The authors suggested capitalizing on this (when it can be recognized) by shifting from recurrent to tandem selection, thereby attempting to hold both at the highest levels.

McConnell and Gardner (1979) showed that selecting for seedling cold tolerance in corn under laboratory conditions did not coordinate well with field emergence. Bacon, Cantrell, and Axtell (1986) decided to base selection on field conditions, and put together procedures that formed a neat recurrent selection scheme that incorporated several seemingly unrelated items. The basic biological material was a grain sorghum bulk population containing the ms_3 male sterile alleles. Pressure for cold tolerance was imposed by early planting, and the survivors were recombined via crossing on male sterile plants identified at anthesis. It is presumed that the male sterile plants were in the population and themselves survivors; if not, the scheme would still work but at a slower pace. Seed aliquots from each crossed plant were bulked to form the next cycle. Advance in cold tolerance was measured after three and four cycles, the combined results indicating a 2.8% gain per cycle. Measurement of other traits showed that the population had become slightly earlier in maturity, and that there was a large gain in population yield (197.7 kg/ha/cycle).

Guok, Wynne, and Stalker (1986) crossed a domestic tetraploid peanut to a diploid wild species, restored the sterile triploid to fertility at the hexaploid level with colchicine, and used the fifth generation following for a search for high-yielding derivative lines. After yield testing, the best 10 were diallel crossed to form the base population for a recurrent selection program to improve yield and other traits. Three cycles of selection were conducted, using similar procedures each time to select the highest-yielding 10 individuals as parents for the next cycle.

The results showed a significant yield increase from cycle 1 to cycle 2; however, cycle 3 was not significantly higher than the cycle 2 parents. No best family from either C_2 or C_3 was significantly higher in yield than the best of C_1, but the number of high-yielding families was greater in C_2 and C_3. Another measure of improvement was that the best family in each cycle was superior in yield to the check cultivar, 'NC7'.

The apparent limitation on variability after cycle 2 is somewhat different from the situation in maize, where variability often continues with little diminution. The yield testing and selection of the best 10 lines after five generations, which formed the base population, may have preselected a genetically similar group of lines. Aside from this aspect, however, the superior yielding ability of the interspecific-derived lines recommended the procedure as a means of improvement.

PROCEDURES ARE ADAPTABLE TO SELF-POLLINATED CROPS

Hull (1945) proposed by name the first recurrent selection scheme in corn. The procedures fitted a three-year cycle as follows:

First Year: Self-pollinate 100 random plants in the "crossbred lot" and take pollen from each to the silks of a tester line.

Second Year: Test 100 hybrids for yield.

Third Year: Grow ear-rows from selfed seed of 10 or more plants from the highest-yielding hybrids and make crosses *between* rows. The selection of plants for intercrossing may involve screening for "pest and weather damage," which I interpret as insect, disease, and lodging deficiencies.

If we were to adapt this first recurrent scheme to self-pollinators, it would appear somewhat as follows:

First Year: Tag 100 plants or heads in a space-planted F_2 nursery of wheat or barley, and use these identified pollen donors to pollinate a male sterile tester. Plants may be random or selected or both.

Second Year: Test the F_1 hybrids for yield and select the highest yielding.

Third Year: Plant the highest-yielding lines (now F_2) in rows. Random-intermate between rows, using male steriles as females.

The choice of the tester line would be a matter of judgment: it could be a backcross-derived cultivar male sterile. To broaden germ plasm participation, the scheme could be applied simultaneously to more than one base population. I would suggest also a multiparent hybrid F_2 as a base population.

McGill and Lonnquist (1955) described recurrent selection in corn as "a breeding method in which inbreeding is held to a minimum." It is easy to advance these ideas as procedural goals in self-pollinated crop breeding also, given the wealth of available procedures such as outcrossing, backcrossing, mutagens, and male sterility.

Good background on the practical use of recurrent selection in cross-pollinating crops was provided by Eberhart, Harrison, and Ogada (1967). Although principally concerned with maize, Eberhart (1969) extended the breeding procedures to self-pollinated crops for the development of either hybrid or pure line varieties. Two population procedures were recommended for cross-pollinated (hybrid) situations, and one for pure line development. The influence of choice of parents for populations and the role of heritability was stressed. Eberhart refers to the differences in making cross pollinations, which are easy in corn, but laborious in self-pollinated species. However, there have been great changes in some of the self-pollinated crops, for example, in barley with the use of genetic male sterility. In fact, cross pollination and automatic progression to homozygosity, which accompanies the use of male sterility, have made recurrent selection quite practical in barley and wheat.

Prohaska and Fehr (1981) found recurrent selection for resistance to iron chlorosis deficiency successful in soybeans. The breeding objective was to attain a higher level of resistance to the condition, complete resistance not being known to exist. A cycle 0 population was formed by a series of three intermatings, beginning with 10 domestic cultivars or lines that showed a high expression of resistance, each crossed to one of 10 selected-for-best-resistance genotypes found among plant introductions (i.e., foreign). In the F_2 each population was crossed with two of the 10 two-parent populations. The third bridging mating took place among the hybrids from the second mating. S_1 lines were obtained from the third mating, which became the base population designated cycle 0. Two additional cycles, 1 and 2, were successively formed by analysis of S_1 lines, choosing the 10 best, and intermating them as earlier described.

The final tests for genetic gain took place over two years when the cycle 0 (20 parents), cycle 1 (10 parents), and the cycle 2 (10 parents) were compared. Gain was expressed through changes in chlorosis score, a lower score indicating greater resistance. The mean gain per cycle amounted to 0.2 units per year and represented a 9% gain. The relation to selection for other characters was not explored, because this was seen as the development of a germ plasm bulk, and in fact, was later released as such.

Using kernel weight as the selection character, Busch and Kofoid (1982) made a diallel cross of 10 spring wheats already screened for high kernel weight. A selection intensity varying from 1.5 to 2.9%/cycle produced approximately 22 high-kernel-weight individuals per cycle, which served as parents for the next round of crossing. Toward the end of the study, random lines from the base population, maintained at the approximate original variance via a single seed descent procedure, and representative lines from the selection cycles were compared for their kernel weight and other character measurements. The genetic gain per cycle was estimated at 3% for lines and 7% for populations. There was little evident decrease in genetic variance for most characters after two cycles. A convincing facet of the efficiency of recurrent selection was that two cycles produced lines with kernel weights higher than *any* lines from the base population.

There were some negative indirect effects, suggesting that some form of index selection could be helpful. The fact that grain yield and test weight remained stable over the two cycles suggested also that compensation was taking place among yield components. The authors reiterated that significant genetic advancement took place at high selection pressure in populations derived from parents already upgraded for the trait.

Early heading in winter wheat was the trait chosen to determine the efficiency of recurrent selection by Avey, Ohm, Patterson, and Nyquist (1982). The cycles were completed under greenhouse conditions using 16 different parents for the initial crosses. Because early heading is a pre-anthesis character, it was possible in the same generation to select early-heading plants and use them as parents in crosses to begin the next cycle. Three cycles were completed and each cycle featured a large number of parents; for example, cycle 1 used the 135 earliest-heading plants (27% of population), cycle 2 used 195 (36%) and cycle 3 used 234 (35%). The tests for gains were made under field conditions. The results showed substantial gains for recurrent selection. At two locations the average number of days to heading, after an April 30th base, under 10% selection intensity were:

Location 1 (base measurement population 14.2 days): cycle 1 12.0 days, cycle 2 11.7 days, and cycle 3 11.2 days.

Location 2 (base measurement population 23.4 days): same order, 22.2, 21.6, and 21.6 days.

Selection was conducted also at two lower intensities with somewhat similar results, although better gains were achieved with 10% pressure.

Early heading is an easily observed trait responsive to selection. The authors assumed that it was controlled by both major and minor genes and interpreted the larger gains in the first cycle to be due to a quick garnering of the major genes.

The authors stressed the importance of achieving minimum environmental variance within each cycle. They selected greenhouse areas with uniform lighting and also used Gardner's gridding selection scheme to counter environmental area differences.

Recurrent selection for seed yield in soybeans has generally been evaluated in early generations, often F_1. Sumarno and Fehr (1982) reasoned that a delay to F_4 or F_5-derived lines would provide not only an evaluation but also an array of lines ready for individual testing; that is, for each cycle the evaluated best lines would represent the parents for the next cycle and also could be entered in performance nurseries. The four cycles evaluated followed this unique expansion–contraction arrangement whose sequences were as follows:

Cycle 0. Forty parents were chosen for partial diallel crosses. The partial diallel crosses were bulked. Plant-to-plant crosses were made for the third intermating, and 300 F_5 lines were tested.

Cycle 1. The 10 best each of early, midseason, and late soybeans were used as parents in diallel crosses. Procedures were as in cycle 0 except that the third intermating was excluded and the lines tested were F_4-derived.

Cycle 2. Procedures were similar except that crossing diallels were set up for the 10 parents in each maturity group rather than across the maturity range, and 100 F_2 families were established from each. In the F_4 of each group 200 lines were saved, representing equal representation of all lines. These were tested and reduced to 100 F_2-derived lines, one from each F_2 ancestor.

Cycle 3. Ten lines from each maturity were the parents. Procedures were as above.

For evaluation the authors had three maturity groups, namely, (1) the 40 original parents, now divided into 13 early, 13 midseason, and 14 late, (2) the parents of each cycle, and (3) composites of the parents of each cycle. The results were these:

1. *Mean Yield of Parents:* The early cycle 3 showed gains per cycle of 120 kg/ha, or 60 kg/ha or 2.1%/year of cycle 0 parents (a cycle was two years). The midseason cycle 3 fell below the cycle 0 level; this was unexplainable. The late cycle 3 gave a 12 kg/ha/year or 0.4% gain over cycle 0.
2. *Highest Yield of Parents:* None yielded significantly more than the highest yielder of cycle 0 parents, which was disappointing. After evaluation "none yielded as well as the best public cultivars available."
3. *Composite of Parents:* The estimates of selection response were similar to those obtained from the mean of individual parents, suggesting that intergenotypic competition was not an important factor.

I have given considerable detail concerning the procedures used by Sumarno and Fehr to evaluate a few cycles of recurrent selection, not only to indicate the complexity and effort required but also to present the plant breeding innovations used. The authors used combinations of single plant, hill, and plot tests, and combinations of choice and visual selection, all designed and successfully manipulated to maintain both between-line and within-line variance, somewhat akin to the single seed descent concept. As a result, a balance was maintained over generations giving proper recognition to the maturity groups, and to the original population genetic variance. Although none of the test lines appeared good enough to become a cultivar, the results were favorable to the use of recurrent selection. Any number of variations are left open for the breeder, for example, changes in number and characterstics of original parents, earlier stratification for maturity, greater reliance on visual discrimination in selection, bringing superior lines from a cycle back into crosses with original parents, looking closely for important character advances, use of selection indices, and so forth.

Two cycles of recurrent selection to increase wheat grain protein percentage were deemed qualifiedly successful by Löffler, Busch, and Wiersma (1983). High-protein F_3 lines, comprising 7%, 5%, and 4% selection intensity in each of the two populations, were selected as parents for cycle 1 and cycle 2. Gains per cycle were estimated at 0.5% for lines and 0.35% for bulks. The greatest gains came from cycle 2.

A second objective of the research was to observe indirect effects on other traits. These seemed negligible except for two important traits, height and yield. The authors concluded that recurrent selection for protein by itself would be insufficient to select appropriate trait-endowed individuals for cultivar release. They stressed the need for additional multiple trait selection to overcome the negative relationships between direct selection for high protein and grain yield. This seemed to be a situation where index selection would be appropriate.

Using only yield gain as a criterion, Payne, Stuthman, McGraw, and Bregitzer (1986) conducted a recurrent selection program in oats over three cycles. They reported detailed results of associated changes in 12 other characters. The study was a continuation of an earlier two cycles of recurrent selection. A recapitulation of procedures showed that there were an original 12 genotypes in a near-diallel series of crosses. Single seed descent was used to advance 10 lines from each of the 64 crosses to the F_4, followed by a one-year increase of seed. This led to a hill plot yield nursery, in which a line from each of the 21 highest-yielding crosses became the parents for the beginning of the second cycle. Procedures were similar for the second and third cycles.

The average yield gains achieved were 11.5%, or 3.8%/per cycle, when the third cycle selections were compared to the original 12 parents. The authors expressed this another way: each cycle time period covered four years; therefore, there was a steady 1% annual gain, deemed to be substantial.

The study was done with a minimum of selection. Changes noted were greater phytomass, higher grain-filling rates, later heading and maturity, increased height, more kernels per panicle, and heavier kernel weight. Harvest index, grain-fill period, and panicles per hill remained essentially unchanged. Especially noted in the authors' conclusions, "Our results suggest that increasing biomass may be necessary to achieve future yield increases, and that more attention should be given to grain-filling rates than to vegetative growth rates."

Even though not qualified to interpret Silvela and Diez-Barra's (1985) paper on recurrent selection under forced random mating in self-fertilizing species, I think it important to call attention to it because it clearly deals with processes and situations familiar to the plant breeder. To illustrate, the computer simulation model considered three systems of selection: (1) SS, selection under self-fertilization; (2) RS, recurrent selection under forced cross-fertilization; and (3) CS, conventional selection. SS reminded me of a single seed descent procedure with truncation selection added. RS reminded me of a genetic recessive male sterile system with beginning random mating, followed the next generation by selective mating (best by best phenotypes). The third system, CS, is frequently encountered; an illustration would be eight cycles of SS selection followed by one cycle of RS.

Silvela and Diez-Barra made these points:

1. RS was generally superior to SS.
2. Random mating in a population in linkage equilibrium is ineffectual, it "does not change anything."
3. Epistasis can be especially important in autogamous species because, over evolutionary time, natural or artificial selection has acted predominantly upon gene combinations that have become homozygous and presumably have been preserved because of improved fitness. Furthermore, epistatic effects can be fixed.

The following statement by the authors accurately describes many breeding programs:

> In actual practice, most breeders start with a large number of inbred lines of top performing ability. In genetic terms, this is equivalent to a large number of different genotypes with approximately the same genotypic value or to a population with genetic variability but without genotypic variance.

Also, their description of introgression rings true: "A ... different case ... occurs when one tries to introduce genes from inferior into superior strains. The base population, loaded with positive linkage disequilibrium will have a largely inflated genotypic variance."

RELATION TO MALE STERILITY AND MSFRS

It is not always possible to separate the by-products of male sterility, recurrent selection, and intermating, because the use of one often involves the other. Brim and Stuber (1973) illustrated this in their article on the use of genetic male sterility in recurrent selection programs in soybeans. Recessive male sterile genes were employed to produce a three-generation step program of population synthesis, which then became the base for the next cycle.

McProud (1979) analyzed three major world barley improvement programs, namely, from The Netherlands, Japan, and North Dakota. He researched the origins, assigned base populations to each, noted germ plasm additions, and constructed a chronological framework of selection cycles. From the turn of this century, the evolution of all programs could be described as phenotypic recurrent selection. The North Dakota program showed a base population of 16 genetic sources (12 introduced in the beginning years), which generated 11 cycles of recurrent selection averaging 6.5 years each. The Netherlands program involved a base population of 13 genetic sources, only one of which was introduced after 1921, which spawned seven cycles of recurrent selection by 1968 and had an average cycle length of 9.75 years. The Japan program contained a base population of 14 genetic sources (nine introduced prior to 1921), led to seven cycles of recurrent selection, and averaged 10.5 years per cycle. McProud recommended expansion of the genetic bases and reduction in cycle time.

McProud's analyses and conclusions are well taken. The programs mentioned nevertheless have been successful and have served agriculture well. There is more awareness today of the need for broadening germ plasm use, and for better tools for accomplishing it. One only need mention the availability of genetic male sterility. The increasing use of several base populations in a program makes the reduction in cycle time less important today.

Recurrent selection is a proven performer. It has been shown to be effective across species and type-of-pollination boundaries. The degree of effectiveness is influenced by the choice and timing of available options.

REFERENCES

Avey, D. P., H. W. Ohm. F. L. Patterson, and W. E. Nyquist. 1982. Three cycles of simple recurrent selection for early heading in winter wheat. Crop Sci. 22:908–912.

Bacon, R. K., R. P. Cantrell, and J. D. Axtell. 1986. Selection for seedling cold tolerance in grain sorghum. Crop Sci. 26:900–903.

Bassett, M. J. and F. E. Woods. 1978. A procedure for combining a quantitative inheritance study with the first cycle of a breeding program. Euphytica 27:295–303.

Brim, C. A. and J. W. Burton. 1979. Recurrent selection in soybeans. II. Selection for increased percent protein in seeds. Crop Sci. 19:494–498.

Brim, C. A. and C. W. Stuber. 1973. Application of genetic male sterility to recurrent selection schemes in soybeans. Crop Sci. 13:528–530.

Busch, R. H. and K. Kofoid. 1982. Recurrent selection for kernel weight in spring wheat. Crop Sci. 22:568–572.

Compton, W. A. 1968. Recurrent selection in self-pollinated crops without extensive crossing. Crop Sci. 8:773.

Comstock, R. E., H. F. Robinson, and P. H. Harvey. 1949. A breeding procedure designed to make maximum use of both general and specific combining ability. Agron. J. 41:360–367.

Doggett, H. 1972. Recurrent selection in sorghum populations. Heredity 28:9–29.

Doggett, H. and S. A. Eberhart. 1968. Recurrent selection in sorghum. Crop Sci. 8:119–120.

Eberhart, S. A. 1969. Efficient plant breeding with recurrent selection. Invitational paper, Genet. Plant Improve. Symp., XI Int. Bot. Congr., Seattle, Wash., Aug. 28, 1968.

Eberhart, S. A., M. N. Harrison, and F. Ogada. 1967. A comprehensive breeding system. Der Zuchter 37:169–174..

Ferwarda, F. P. 1956. Recurrent selection as a breeding procedure for rye and other cross-fertilized plants. Euphytica 5:175–184.

Guok, H. P., J. C. Wynne, and H. T. Stalker. 1986. Recurrent selection within a population from an interspecific peanut cross. Crop Sci. 26:249–253.

Hull, F. H. 1945. Recurrent selection and specific combining ability in corn. J. Am. Soc. Agron. 37:134–145.

Jenkins, M. T. 1935. The effect of inbreeding and of selection within inbred lines of maize upon hybrids made after successive generations of selfing. Iowa State Coll. J. Sci. 9:429–450.

Jenkins, M. T. 1940. The segregation of genes affecting yield of grain in maize. J. Am. Soc. Agron. 32:55–63.

Jensen, N. F. 1970. A diallel selective mating system for cereal breeding. Crop Sci. 10:629–635.

Jensen, N. F. 1978. Composite breeding methods and the DSM system in cereals. Crop Sci. 18:622–626.

Khadr, F. H. and K. J. Frey. 1965. Effectiveness of recurrent selection in oat breeding (*Avena sativa* L.). Crop Sci. 5:349–354.

Löffler, C. M., R. H. Busch, and J. V. Wiersma. 1983. Recurrent selection for grain protein percentage in hard red spring wheat. Crop Sci. 23:1097–1101.

McConnell, R. L. and C. O. Gardner. 1979. Selection for cold germination in two corn populations. Crop Sci. 19:765–768.

McGill, D. P. and J. H. Lonnquist. 1955. Effects of two cycles of recurrent selection for combining ability in an open-pollinated variety of corn. Agron. J. 47:319–323.

McNeal, F. H., C. F. McGuire, and M. A. Berg. 1978. Recurrent selection for grain protein content in spring wheat. Crop Sci. 18:779–782.

McProud, W. L. 1979. Repetitive cycling and simple recurrent selection in traditional barley breeding programs. Euphytica 28:473–480.

Payne, T. S., D. D. Stuthman, R. L. McGraw, and P. P. Bregitzer. 1986. Physiological changes associated with three cycles of recurrent selection for grain yield improvement in oats. Crop Sci. 26:734–736.

Penny, L. H., W. A. Russell, G. F. Sprague, and A. R. Hallauer. 1963. Recurrent selection. Stat. Genet. Plant Breeding, NAS–NRC 982:352–367.

Prohaska, K. R. and W. R. Fehr. 1981. Recurrent selection for resistance to iron deficiency chlorosis in soybeans. Crop Sci. 21:524–526.

Richey, F. D. 1927. Convergent improvement in selfed lines. Am. Nat. 61:430–449.

Silvela, L. and R. Diez-Barra. 1985. Recurrent selection in autogamous species under forced random mating. Euphytica 34:817–832.

Sumarno and W. R. Fehr. 1982. Response to recurrent selection for yield in soybeans. Crop Sci. 22:295–299.

16

DIALLEL SELECTIVE MATING METHOD

Jensen (1970) proposed a modification of recurrent selection for use with inbreeding crops. Called a *diallel selective mating system* (DSM), it essentially dealt with the F_1s as parents and contrived to get increased parental input into populations by diallel crossing of the F_1s. Thereafter, random crossing was avoided through a series of selective matings of F_1s and mass-selected F_2s. Recurrent selection occurred through the repeated (over generations) use of reserve F_1s, and desired cultivars were taken from the original homozygous parent list. The system was designed for both male sterile and non-male sterile situations.

The DSM system was built around each stated objective of the breeder. Parents were chosen with this objective in mind. A limit of 21 single-cross combinations was suggested (a diallel of seven parents would produce 21 combinations); however, more parents could be included by using selected three-way or four-way crosses and by using variations of the complete diallel scheme. In this crossing stage the breeder might wish also to skew the contribution of parents by imposing his preferences and feelings about the objective.

Once the basic crosses were in hand, the important F_1 diallel crossing stage could begin. A complete diallel was a goal, but only an ideal one. Often it would be more efficient for the breeder to settle for an incomplete diallel than to delay the program. After completing the diallel crosses all hybrid seeds were composited, and the system took the form of the bulk hybrid population system. Mass selection was practiced toward the planned objective. On the basis of perceived progress, selective mating could be inserted into the procedures to form a new recurrent selection population.

In most cases this would take place among individuals within the population; however, it could also be interpopulational, because the system envisaged the formation of a few or several populations (P_1, P_2, etc). Lines were withdrawn from the F_5 generation using the special field precautions I have described elsewhere for the bulk population method. Of course, each time a new recurrent population is begun it must be taken to about the F_5 for line withdrawal, or later if built around the male sterile character.

I described the aims of the DSM system as follows (Jensen, 1970):

> The diallel selective mating system should ease some of these problems currently found in conventional systems for self-pollinated crops. Specifically, it provides for broad use of germplasm, simultaneous input of parents, creation of persistent gene pools, breaking of linkage blocks, freeing of genetic variability, and general fostering of genetic recombination.

The DSM breeding system concentrated crossing efforts on the F_1, first as a diallel, and then through selective mating of F_1 and F_2 plants. Random mating was excluded. The populations proceeded through generations with mass selection and compositing until normal extraction of lines at F_4 or F_5. The system was designed to utilize and modify the elements of conventional crossing by minimizing the limiting influence of the two-parent cross with its accompanying (often low-range) genetic variance, by excluding automatic inbreeding, and by increasing recombination potential through the breaking of linkage blocks.

The DSM, although described in conventional terminology, was proposed for both male sterile and non-male sterile handling, and illustrations were given for both situations. The Cornell barley breeding program was exclusively a male sterile-facilitated recurrent selection (MSFRS) program. Today I would stress the male sterile situation in barley and wheat, because it has the advantages of scale, labor and cost efficiency, and greater flexibility, in preference to the non-male sterile case for use with the DSM. Compare, for example, the ease of diallel crossing using the genetic male sterile gene over obligatory hand crossing. It may be argued that there is more "slippage" in the male sterile system because it is started with heterozygous carriers. This slippage can be minimized through the use of backcross-derived lines isogenic for the male sterile gene, or by careful selection of *ms* plants from populations of known origin. The use of the male sterile in the DSM softens and blurs the precise procedural steps described in my 1970 article for the non-male sterile situation, which is a great advantage for barley and wheat breeding. I have described the use of the genetic male sterile gene in barley breeding in association with the section on MSFRS. For those who wish more practical applications and information on bulk population breeding, the DSM, or recurrent selection, the crop newsletters are often a good source (e.g., see under New York in the 1971 Oat, Wheat, and Barley Newsletters).

In a later paper (Jensen, 1978) I showed how the DSM fit into the bulk population breeding system as used in the Cornell cereal improvement program. Integration principally involved the question of scaling. The simplest formulation is the one-hybrid bulk population. The larger DSM formulations, one of which involved a bulk of 82 hybrids, had to be scaled upward in order to get equivalent field representation of the included hybrids. This was done by increasing the number of F_2 field units from a multiple-hybrid DSM population according to the arbitrary rule of one field plot unit for every five crosses in the base bulk population. Because these samples for planting are putatively similar, their use also may be thought of as a form of replication or a use of the Gardner grid principle. Extended over years, the yield results could be used to measure precision and calculate variances within populations. The 1978 article provided additional information and illustrations of DSM use.

Griffing (1956) tied together the two concepts of combining ability and diallel crossing. The main elements were two combining abilities, four diallel systems, and two assumptions. The combining abilities were the familiar GCA and SCA, representing average and extraordinary performances, respectively, of lines in hybrid combination. The four diallel methods varied according to how many of the "players" are included: method 1 includes all, that is, parents, an F_1 set, and a reciprocal F_1 set; method 2 omits the reciprocals; method 3 omits the parents; and method 4 deals with one set only of F_1s. The two assumptions merely state whether the genotypes are random or fixed. Griffing thought that methods 3 and 4, where parents were absent, were likely the most useful in plant breeding.

Two cycles of random intermating in wheat and barley mass selection programs to increase tillering were shown to be very effective by Redden and Jensen (1974). Three populations were generated from one F_2 population in wheat and one in barley. One population (hybrid series) involved two cycles of random mating among plants selected before anthesis for tiller number; thus the matings were not in the truest sense random. A second population (selfed series) was mass-selected without intermating. A third population (control series) was selfed without being subjected to either mass selection or intermating.

When tested at an F_5 equivalent, averaging the hybrid and selfed series, mean selection response over two sites was 10.3%/cycle for wheat and 6.3%/cycle for barley. The results partitioned by site and type were for wheat (one site only), 22.6% hybrid and 18.5% selfed; for barley, site 1, 17.1% hybrid and 8.7% selfed; and for barley, site 2, 13.9% hybrid and 10.3% selfed.

These sizable advantages from both mass selection and intermating, but especially the latter, demonstrate the power of selection and intermating. Efficiency was increased by the fact that selection and intermating could be done sequentially in the same generation. One wonders if the selection advantage could have been increased through the use of directed mating, that is, applying DSM procedures to those individuals showing the highest expression of tillering rather than mating at random. Even so, the selection intensity was high for the selected group used in the random matings. The authors discussed plant breeding possibilities in recurrent selection using male sterility where it is available.

The DSM system is a flexible way of getting more from the arsenal of breeding weapons every breeder has. It is more than the rigid "recipe" format, de rigeuer in a scientific paper, of my 1970 article. A sequel in 1978 illustrated the supplementary nature of the DSM to established bulk methods. One of the important things the DSM allowed me to do was to design populations, involving substantial germ plasm input, that pointed toward specific goals. This was the "objective bases" approach, which built a breeding bulk hybrid population around a breeding objective.

REFERENCES

Griffing, B. 1956. Concept of general and specific combining ability in relation to diallel crossing systems. Aust. J. Biol. Sci. 9:463–493.

Jensen, N. F. 1970. A diallel selective mating system for cereal breeding. Crop Sci. 10:629–635.

Jensen, N. F. 1978. Composite breeding methods and the DSM system in cereals. Crop Sci. 18:622–626.

Redden, R. J. and N. F. Jensen. 1974. Mass selection and mating systems in cereals. Crop Sci. 14:345–350.

17

GENETIC MALE STERILITY

Suneson (1940) reported the 1936 discovery of a male sterile barley plant in the seed rows of C.I. 5368, a selection from a composite cross. Tests showed this to be a true male sterile, recessive in nature. Suneson was fully aware of its value to plant breeding, subtitling his notice of discovery, "A new tool for the plant breeder."

Stephens (1937) reported the finding of a genetic recessive male sterile plant in 'Texas Blackhull' in 1935. Stephens suggested ways of using the character in sorghum breeding.

Riddle and Suneson (1944) presented results of crossing and seed set studies with male sterile barleys. The effects of parent proximity, weather, and type of population structure were discussed in relation to the field use of male sterility in breeding.

Suneson (1945) enlarged upon the uses of the male sterile trait in barley improvement. He based suggested breeding procedures on the series of Composite Cross populations available to breeders through the USDA and California collections. These schemes involved making initial crosses on male steriles, harvesting and bulking the seeds from the male sterile F_2 plants, repeating this two or three times, then converting to bulk or pedigree methods for eventual extraction of lines.

Suneson (1951) outlined an innovative proposal for the development of hybrid synthetic barley. The population was built around the hybrid seed producing capacity of *msms*, the male sterile gene, with the male parent being some form of broad-based composite. The use of the male sterile gene results in a 25% sterile F_2 population, or 50% sterile in a backcrosss. The prospects for a successful synthetic rested on the efficiency in a population of a mixture of steriles, which are necessary for continued hybrid input, heterozygotes (highest heterosis), and fertiles. An F_2 barley population with the recessive male sterile gene contains these characters in the proportions of 1 : 2 : 1, respectively; however, because the heterozygotes are indistinguishable from the homozygous fertiles the perceived ratio is 25% sterile : 75% fertile plants. The planned commercial use would embrace the F_2–F_4, which Suneson hoped would perform at about the F_2 level.

There is a negative drag on yield from the presence of male steriles in a population because their seed set is seldom as high as in heterozygotes or fertile plants. Therefore Suneson was confronted with an insoluble problem—how to increase male steriles to provide the continuing hybird vigor, without sacrificing the yield gain to the presence of sterile florets in the incompletely filled heads. A plentiful supply of viable pollen was needed to ease this problem but was difficult to obtain in practice, given the vagaries of weather and the profusion of variation in flowering patterns. Grading planting seed for larger size increased the percentage of male steriles; however, this was negated by generally low proportions of hybrid seeds, and a perplexing tendency for kernels on low-seed-set florets to be smaller and lighter than those produced on high-seed-set florets. The experimental synthetics produced were competitive but not superior in yield to established varieties.

From my experience with the multiline method and with operating a winter barley program exclusively based on male sterile facilitated recurrent selection (MSFRS), a synthetic barley may also be viewed as being created via *msms*, but not perpetuated by a necessity for a continuing hybrid input. That is, once a superior producing composite had been identified it could be released on its merits and allowed to run its course without regard either for the declining percentage of male steriles or for heterosis. If desired, the variety could be upgraded through roguing or a rebulking of selections prior to release. However, this would not be a hybrid synthetic but rather a heterogeneous synthetic slowly declining in heterozygosity.

Thurman and Womack (1961) reported on irradiation treatment of oat seed by X-rays and neutrons. The treatments caused increases in natural crossing as high as 10%, which was verified by F_1 appearance and F_2 segregation. Because genotypes differed in response, a range of treatments must be applied experimentally in order to choose the most efficient dose for the genotype in question.

These findings are important for several reasons: no suitable genetic male sterile is available for oats, oats are the most difficult cereal to work with and obtain hybrid seeds in crossing, and there is negligible natural outcrossing in the domestic crop. Stock or reserve seed of certain genotypes suitably treated could serve as germ plasm reservoirs for augmented crossing programs.

The plant breeding possibilities inherent in genetic male sterility and the implications for germ plasm preservation were well illustrated in Suneson and Wiebe's 1961 and 1962 papers. The entire world collection of spring barleys (6200 entries) was crossed onto *msms* plants to form a hybrid population designated "Composite Cross XXI."

Pugsley's sterile wheat was described by Suneson (1962). Recovery of the sterile in heterozygotes ranged from 0 to 30%.

Jain and Suneson (1963) studied the persistence of the *msms* gene in successive generations of two composite barley populations. The known amount of natural outcrossing in barley in the absence of *msms* is extremely slight, estimated at 0.01–0.02, whereas the presence of the *msms* gene raised the probability to 0.42 and 0.36, presumably in the F_2 (data from Jain and Allard, 1960). The persistence of *msms* in the populations over years was higher than would be expected without a selective advantage, especially since the *msms* individuals are poorer in seed set. The authors postulated heterozygote advantage at different loci to explain the higher than expected persistence. This may be another way of stating that the extra vigor of heterozygotes is translated into more seeds so that such an individual contributes above-average amounts of seed to the total harvest. The lower fertility of the male steriles eventually outweighs the unknown selection advantage, and there is a gradual disappearance of the gene from the population.

Suneson and Ramage (1963) reported that the proportion of male sterility in a natural population gradually becomes reduced. Poor seed set is the critical factor, and outweighs others such as heterozygote advantage. Nevertheless, male sterility is an easily handled technique in breeding. Sterility may be maintained at a high level, for example, through selective crossing and backcrossing, and through collection of seeds on male sterile plants to form or enrich populations.

Gilmore (1964) suggested reciprocal recurrent selection for self-pollinating species, using genetic and cytoplasmic sterility. His article was directed toward the development of source populations for commercial hybrids. There was brief mention near the end of the paper that the same techniques might be useful in special situations where commercial hybrid production was not a goal. I would prefer to think that no distinction need be made as to eventual use. The technique is quite sound and has been used as the principal breeding method for creating both conventional (pure line) and hybrid cultivars.

Jain and Suneson (1966) introduced the male sterile gene to form barley populations with enhanced outcrossing capabilities and compared them with "normal" populations, which, however, are known to generate variability, presumably through small amounts of outcrossing. It was found that the higher outcrossing made possible by the added male sterility visibly changed the recombination system, but this was not necessarily translated into better performance than the normal system. The reason was that the continually produced extremes were selectively inferior. One receives a mental picture of a population burdened with its load of segregation products and disruptive heterozygotic effects, functioning in essence as though it were taking two steps forward and one back. Jain and Suneson addressed the theoretical aspects underlying the differences between low and high outcrossing. They summarized their observations:

> Thus, one is led to the inevitable conclusion that the amounts of variability produced by rather low rates of natural outcrossing in the bulk-hybrid populations of barley are large enough to allow optimal *rates of* utilization under natural selection for adaptive improvement.

Sprague (1966) proposed that the rapid approach to homozygosity, which limits recombination in self-fertilizing species, be modified through the forming of random-mating populations using genic male sterility. This has proved to be very effective, as for example, in barley.

Ross and Shaw (1971) considered the problem of how male steriles might be maintained in continuing populations, for it is well known that they are at a disadvantage compared with fertile plants and gradually disappear from a population. Ross and Shaw confirmed the belief that monogenic male steriles are lost from a population unless they are twice as fertile as the hermaphrodite fertiles. This seems logical inasmuch as observations have shown that the fertility of male sterile plants encompasses a range from heads with no seeds set, to heads with complete fertility. I have difficulty picturing a male sterile plant twice as fertile as a fertile plant because I see it in the fitness sense, that is, ability to produce seeds; perhaps extra tillering is involved.

The importance of pollinator distance was shown by research with cross-pollination in male sterile wheat by Stoskopf and Rai (1972). The average seed set in six male sterile plots was 28.8%, 10.3%, and 6.6% when the pollinator was separated by 0.2, 1.5, and 3.0 m, respectively. The importance of this in male sterile breeding as contrasted to commercial hybrid production is that planned crosses require knowledge of *ms* segregation ratios so that males can be positioned properly about the female plants, in rows or hills, and efficient direction given to the breeding effort. More details on such procedures are furnished in Chapter 18.

Doggett (1972), discussing recurrent selection in sorghum, pointed out a general breeding problem with self-fertilizing crops. After positive gains from major genes for yield and quality have been incorporated into cultivars and better lines, a new situation emerges where future gains require larger populations and more precise measurements in order to identify and capture the smaller increments. Genetic male sterility offers a way out of this dilemma. The recycling of random mating populations makes unnecessary huge segregating populations. Other advantages are that many parents can be involved in forming populations, and linkage arrangements are subject to breakage pressure.

Doggett's procedures as he experimented with genetic male steriles in Africa were of great interest to me because they paralleled my own experiences with barley (Doggett cites barley as an example of success). The details of his study should be read for an appreciation of the tremendous flexibility male sterility gives to a breeding program, from the germ plasm input possible with many parents, the "spin-offs" to begin new populations, the recycling or funneling back of selected material into oncoming composites, the chance to work population subsets toward different goals, the opportunity for mass and recurrent selection, the breaking of linkages, and other possibilities.

The widening use of male sterility, both genetic and cytoplasmic, is indicated by Nordcross, Webster, Gardner, and Ross's (1973) release of three sorghum random-mating germ plasm stocks.

The broadening of intellectual and operational opportunities for plant breeding that occurs when genetic male sterility becomes available in a crop is illustrated in Brim and Stuber's (1973) paper on recurrent selection schemes in soybeans. The simple recessive involved (Brim and Young, 1971) is similar to the recessive genes used in barley and the application to breeding is also similar with one exception: pollen in soybeans is insect-transported as contrasted to windborne in barley. Even though the proximity of mating plants is important in barley, in general, pollen panmixis can be assumed in field areas of reasonable size. The situation in soybeans requires sampling of subareas, which can be accomplished by setting up formal or informal field grids, as C. O. Gardner did for corn.

Brim and Stuber presented a straightforward one-, two-, and three-generation scheme for moving from one recurrent cycle to the next. A section on modifications was appended; in this they alluded to the many breeding and selection options available through the use of the synthesized germ plasm population.

Knowles' (1977) study on recurrent mass selection in intermediate wheatgrass is of interest, not only to show the efficiency in gain in yield per cycle, which was 10.3% when open pollination was used, but also to show the profound influence of controlled pollination where the advance was 20.3% per cycle. In the open-pollinated procedure, seed from selected plants was bulked to form the next generation; the male contribution was a random one from the population. In the controlled population, the selected plants were moved to the greenhouse and intermated to form a new bulked seed lot; thus the male and female contributions were approximately equal and one-directional. These procedures also are integrally a part of MSFRS.

The use of the genetic recessive male sterile genes for improvement of the self-pollinating crops has been a common practice for three or four decades. In the cereal small grains, for example, they have been most useful in barley, increasingly so in wheat, and in oats not at all. Prompted by a recent discovery by Sasakuma, Maan, and Williams (1978) of an induced-mutant dominant male sterile gene in wheat, Sorrells and Fritz (1982) presented background information and suggested procedures for the practical use in breeding programs of these alleles. Different applications for recurrent selection and backcrossing were shown. They pointed to two segregation characteristics of a genetic dominant male sterile that are different from a genetic recessive male sterile: namely, all fertile plants breed true, and male sterile (hybrid) progenies segregate 1:1 immediately (this the backcross situation for a genetic recessive male sterile). These advantages, now available for wheat breeding, are important, and make a significant procedural supplement to the recessive genetic MSFRS procedures long in use in barley breeding.

In plant breeding practice the gradual loss of male steriles in a population is of little consequence, except in long-range research projects. In planned breeding projects utilizing male steriles, the critical populations are the early-generation ones, particularly the F_2, where one-fourth of the plants are male steriles. The bulk of the F_2-harvested seed can be retained as a working reserve stock. Portions may be used to start successive annual populations for selection in later generations. But it is the F_2 population, beginning with directed hybridization on the field, where spin-off, backcross, recurrent selection, intermating, and outcross programs are arranged. These are illustrated in Chapter 18, but an obvious early move in the F_2 is phenotypic selection of male sterile plants and bulking their seeds to begin a second intermating population. If an older program does suffer too great a loss of male steriles, an infusion of backcross material that will segregate 1:1 fertiles and male steriles can be made. The introduced material can be chosen so as to advance the goals in the basic intermating population.

The plant breeding aspects of increased outcrossing using the male sterile under breeder selection pressure are exciting. In practice natural selection is subordinated to a secondary role while the breeder applies pressure toward desired objectives. Whole breeding programs have been converted to the MSFRS system because of its marvelous efficiency. In practice, male sterile breeding programs usually allow two or three extra generations beyond the standard F_4–F_6 line removal in order to take advantage of increased recombination and increased homozygosity. In fact, if a breeder has a promising MSFRS population, rigid termination dates need not be applied, and the population may be continued as an annual line source. These populations cross the boundary into the longer evolutionary breeding method populations, gradually shedding their male sterile plant component. One particular thing that breeders often do is to remove the newly set hybrid seeds in the field, usually by stripping or cutting heads of visually selected female phenotypes, and use these bulked seed lots to start new populations having a high frequency of male steriles.

The special case of the MSFRS system is the subject of the next chapter.

REFERENCES

Brim, C. A. and C. W. Stuber. 1973. Application of genetic male sterility to recurrent selection schemes in soybeans. Crop Sci. 13:528–530.

Brim, C. A. and M. F. Young. 1971. Inheritance of a male-sterile character in soybeans. Crop Sci. 11:564–567.

Doggett, H. 1972. Recurrent selection in sorghum populations. Heredity 28:9–29.

Gilmore, E. C., Jr. 1964. Suggested method of using reciprocal recurrent selection in some naturally self-pollinated species. Crop Sci. 4:323–325.

Jain, S. K. and R. W. Allard. 1960. Population studies in predominantly self-pollinated species, I. Evidence for heterozygote advantage in a closed population of barley. Proc. U.S. Natl. Acad. Sci. 46:1371–1377.

Jain, S. K. and C. A. Suneson. 1963. Male sterility for increased outbreeding in populations of barley. Nature 199:407–408.

Jain, S. K. and C. A. Suneson. 1966. Increased recombination and selection in barley populations carrying a male sterility factor. I. Quantitative variability. Genetics 54:1215–1224.

Knowles, R. P. 1977. Recurrent mass selection for improved seed yields in intermediate wheatgrass. Crop Sci. 17:51–54.

Nordcross, P. T., O. J. Webster, C. O. Gardner, and W. M. Ross. 1973. Registration of three sorghum germplasm random-mating populations. Crop Sci. 13:132.

Riddle, O. C. and C. A. Suneson. 1944. Crossing studies with male sterile barley. J. Am. Soc. Agron. 36:62–65.

Ross, M. D., and R. F. Shaw. 1971. Maintenance of male sterility in plant populations. Heredity 26:1–8.

Sasakuma, T., S. S. Maan, and N. D. Williams. 1978. EMS-induced male-sterile mutants in euplasmic and alloplasmic common wheat. Crop Sci. 18:850–853.

Sorrells, M. E. and S. E. Fritz. 1982. Application of a dominant male-sterile allele to the improvement of self-pollinated crops. Crop Sci. 22:1033–1035.

Sprague, G. F. 1966. The challenge to agriculture. African Soils XI: 23–29.

Stephens, J. C. 1937. Male sterility in sorghum. Its possible utilization in production of hybrid seed. J. Am. Soc. Agron. 29:690–696.

Stoskopf, N. C. and R. K. Rai. 1972. Cross-pollination in male sterile wheat in Ontario. Can. J. Plant Sci. 52:387–393.

Suneson, C. A. 1940. A male sterile character in barley. J. Heredity. 31:213–214.

Suneson, C. A. 1945. The use of male-sterile in barley improvement. J. Am. Soc. Agron. 37:72–73.

Suneson, C. A. 1951. Male-sterile facilitated synthetic hybrid barley. Agron. J. 43:234–236.

Suneson, C. A. 1962. Use of Pugsley's sterile wheat in cross breeding. Crop Sci. 2:534–535.

Suneson, C. A. and R. T. Ramage. 1963. Natural selection studies with barley populations featuring genetic male-sterility. Adv. Frontiers Plant Sci. 7:181–186.

Suneson, C. A. and G. A. Wiebe. 1961. A world composite cross of barley. USDA, ARS, Crops Res. Div. Publ., ARS 34–26:1–2.

Suneson, C. A. and G. A. Wiebe. 1962. A "Paul Bunyan" plant breeding enterprise with barley. Crop Sci. 2:347–348.

Thurman, R. L. and D. Womack. 1961. Percentage natural crossing in oats produced from irradiated seeds. Crop Sci. 1:374.

18

MALE STERILE FACILITATED RECURRENT SELECTION METHOD

The earlier discussion on male sterility can serve as an appropriate background for the MSFRS system, a highly efficient breeding and selection method especially suited in the cereals to barley and wheat improvement. The idea traces to the discovery in barley of the genetic recessive male sterile trait by Suneson in California. The name and acronym are generally credited to Eslick of Montana. Male sterile techniques were developed and polished by many scientists, among them Wiebe, Ramage, Reid, Eslick, and Hockett.

My association with the developing use of male sterility began when the cereal project at Cornell University started a two-decade-long period of cooperative research with the Arizona experiment stations at Tucson and Mesa, through the kindness and generosity of Tom Ramage, Rex Thompson, and others. Dr. Ramage, a USDA cytogeneticist, headed a barley research project that is internationally recognized. The presence in Tucson of Wiebe and Reid and the nature of the Arizona research provided a stimulating atmosphere for plant breeders.

As a cereal breeder I learned from these colleagues. Because the use of male sterility is closely tied to population breeding methods, I saw that my Cornell barley program could be converted to a bulk hybrid population MSFRS system. It already was a bulk hybrid population system, but the population origin procedures could be converted from conventional manual crossing to a sort of panmixis based on a judicious use of male sterility. By 1966 23 of my winter barley hybrid populations were founded on MSFRS procedures, and 10 years later the entire program involving almost 40 populations had come under the system.

Winter barley in New York is not a high-priority program, and at the time of my retirement there had not yet been time enough for the development of a new variety. However, Dr. Sorrells (1984), the present director of the program, released a cultivar, 'Willis', from the system in 1984.

I believe that my experience developing and operating a practical MSFRS breeding program may be interesting and helpful to breeders contemplating a change, and it is with that in mind that the following discussion has been added.

A basic distinction that separates a male sterile population from a non-male sterile one is that new hybrid seeds from each harvested crop are fed into the population every generation. Thus instead of a population proceeding inexorably towards homozygosis, a male sterile population after a few generations contains all F stages, that is, F_1, F_2, F_3, and so forth. This alone makes it desirable to delay line selection beyond the otherwise practical F_4–F_6 for non-male sterile populations. First withdrawal from male sterile populations becomes realistic beginning in F_7 or later (although the Willis line was taken out in the F_6). In the same vein, an MSFRS population is not discarded after the first line withdrawal, as is usually the case with a non-male sterile breeding population. It is kept as a reserve parent stock and can be grown again to make selections in a different environment, for example, at other locations or in different years, or the breeder may use portions in population building projects. Attrition gradually reduces the frequency of the male sterile genotypes in the population, the most likely cause being lower seed set on the male sterile heads. Countering this attrition is the obvious advantage of hybrid vigor in F_1 plants from the new hybrid seeds produced on the male sterile plants. An MSFRS population therefore occupies an extended position on the filial generation time scale, just beyound the standard short-range F_5 or F_6 line withdrawal point. A promising MSFRS population may be kept in seed stock reserve status indefinitely even though not used every year.

The promise held out by the MSFRS system is such that breeders need to concentrate on cross-pollinated crop breeding lore and methodology. Breeders are in the unique position of managing a self-pollinated crop, up to the point of line selection, with cross-pollinated crop procedures. Selections, if homozygous fertile (the test is that their progeny do not segregate male sterile individuals), are entered into the routine cereal nurseries.

Intermating is an inherent and continuous feature of MSFRS. Stam's (1977) finding that, if a single insertion of intermating is contemplated, it should be delayed (not F_2 but preferably F_4 or F_5) for maximum results, elicited his comment that this simply represented a re-starting of the program at a higher parental level in terms of reaching the selection goal. It is useful to think of MSFRS in these terms; however, I should comment that MSFRS changes the situation because natural intermatings occur in every generation. Nevertheless, Stam's remarks about each new formulation being a fresh start are on the mark, for actual breeding practice wth MSFRS involves the opportunity to begin new populations based on hybrid seeds obtained following an MSFRS intermating.

A breeding population is begun by pollinating heads on a male sterile plant. The double recessive male sterile plants are readily obtained from a variety of heterozygous sources, with sterile plants appearing at a 25% frequency in F_2, or a 50% frequency in a backcross. Especially useful are the backcross-derived sterile isolines of many cultivars and valuable genotypes of both spring and winter barleys. There are many ways of obtaining the desired starter crosses. For example, one could manually pollinate the male steriles. Another procedure from my own barley program was to mix the female seed stocks, heterozygous for male sterility, with the male seed stocks in such proportions that the barley field population would contain male sterile plants at a planned low level, for example, 2%. At this level the incidence of sibbing among the female component of the mixture would be very low and outcrossing very high. At harvest time the desired stock is obtained by picking heads from male sterile plants. From one such crossing at Cornell I collected 2.7 lb of hybrid seed in about three hours from a 2% male sterile mixture. The average seed set per head was 21 kernels, or 35%. In another instance involving a 4% planned male sterile participation, a yield of 4.6 lb was obtained; the mean seed set again was 21 kernels per head, or 36%. Seed set is given on a head basis, but on a plant basis it is much lower usually because of late or immature tillers that fail to receive pollination or to mature seeds. It is not uncommon, however, to find individual male sterile plants showing very high seed sets on all tillers. The seed sets given above were obtained from counts on 670 plants involving about 40,000 flowers capable of being pollinated.

I observed a specific case of seed set on a single male sterile plant that grew in the very corner of an F_2 barley field. Because of its prominent and isolated position I followed its growth during the season and at maturity counted the seeds. There were 1047 kernels on 2136 flowers. This amounted to 30.8 kernels per head or a 49% seed set from a plant that was not optimally situated for windborne pollination.

The Willis winter barley variety ($MS22F_1S_6-1$) provides an illustration of the procedures in an MSFRS system. The first recorded male sterile crosses in the Cornell project were made in 1960. By 1964 the project contained 16 male sterile hybrid bulk populations. One of these was a cross of a Cornell winter barley stock, estimated to contain 16–18% of male steriles, by an equal-seed composite of 10 spring malting quality barleys (Kindred, Traill, Trophy, Larker, Manchuria C.I. 2947, Swan, Forrest, Barbless, Moore, and Mars). In the Arizona environment, spring and winter types can be matched for flowering times and there is no appreciable loss of either type. The male and female seed stocks were planted in alternating rows with two dates of planting. It was decided to obtain all seed production from hand crosses using the twirl method; therefore no roguing of the fertile plants from the female rows was done, and there was no collection of open-pollinated seeds.

The hand crossing over a period of approximately two weeks resulted in about 2500 crossed heads and 4–5 lb of hybrid seeds numbering perhaps 1 million seeds. I named this stock 'Tucson Malt'.

The reader may have wondered about the no-roguing procedure on the female rows. Ordinarily, roguing is a necessary precaution against sibbing, but it was not needed here because of the way we planned subsequent procedures. We planted the hybrid seeds obtained in the spring at Ithaca. Spring habit is dominant in crosses between winter and spring barleys and the population would be expected to behave essentially as a late spring barley. However, if there were sibbed seeds from the female winter barley parent rows, they would behave as a winter barley planted in the spring, that is, a ground-hugging nonflowering rosette. About 5% of sibs were noted and they automatically eliminated themselves.

The base stock was completed by planting the F_2-embryo seeds in the fall at Ithaca, New York. The spring-type segregants were eliminated through winter kill, leaving a thin but ample winter-hardy population composed of 25% male sterile and 75% fertile plants. The fertile plants, visually indistinguishable, were genetically one-third homozygous fertile and two-thirds heterozygous for male sterility and fertility.

From this point forward, the stock was used in a variety of ways: (1) a portion was annually planted for further recombination and approach to homozygosis; (2) second and higher crossing cycles were started, using manual pollination of sterile plants, with other winter barley objectives; and (3) backcrossing cycles to spring malting types were added in order to increase the proportion of malting-quality genes. It was understood that later cycles of crosses to hardy winter barleys would be necessary to maintain good winter hardiness. At this time we already had achieved, from conventional procedures without use of male steriles, a malting quality winter barley in the NY 6005 Series. This selection later became the 'Wintermalt' cultivar, and this and sister lines were fed back as parents into the MSFRS system.

The next step was the MS22 cross whose pedigree was "1966 4% Tucson Malt F_2 ms × B: Dover, Hudson, Dutchess, Schuyler, NY5618–33, and NY5809–1." This cross also was made at Tucson by planting a mixture of 4% Tucson Malt stock segregating for *ms* plants with 96% of the winter barley variety bulk. The Tucson Malt component was so small that, practically, sibbing on male sterile plants in the large field could be ignored. In addition to natural pollination, desired seed set on the male steriles was augmented by hand pollination from adjoining plantings of the desired males.

The MS22 population was thereafter grown in New York under fall-sown winter barley conditions. $MS22F_1S_6$–1 first appeared as a parent in the 1976 register of crosses. Selection as a parent is usually a good omen of potential and a tentative recognition of candivar status.

The cross register for 1976 tells something also of the merging of MSFRS with parent building and composite building: among 71 crosses registered,

4 were new male sterile populations, 5 involved new *ms* crosses, 4 were standard single crosses, and 48 were crosses involving *MSFRS-derived lines*. The italics highlight the recurrent selection aspect of MSFRS, where outstanding fertile selections from the bulks are returned to the building stage.

A second illustration of MSFRS procedures was the two-cycle preparation of C.C. XXVI and XXVII winter barley stocks in the mid-1960s. Under the leadership of Dr. David Reid and the cooperation of Dr. Tom Ramage and myself (Reid, Jensen, and Ramage, 1964). C.C. XXVI began under irrigation at Tucson in 1963. The objective was to combine genetically the 1295 world winter and semi-winter barleys into one interbreeding population using male sterility. Female and male stocks were sown in alternate rows. The female stock was a heterozygous male sterile stock blended from seeds from a background of about 70 different winter barley cultivars and strains; the male parent was the population formed from the bulking of the 1295 world winter barleys. During the month-long pollinating season the principal activity was the daily roguing of fertile plants in the female rows to minimize sibbing. The 5 lb of seed obtained was increased the following year to yield 850 lb of F_2-embryo seeds, designated C.C. XXVI.

The second cycle, designated C.C. XXVII (Reid, Ramage, Jensen, and Thompson, 1966), came from planting a seed aliquot of C.C. XXVI the following year to obtain further recombination through pollination on the segregating male sterile plants. About 65 lb of hybrid seeds was collected from about 8000 male sterile plants that had been tagged at flowering time.

No human selection was practiced in either of the two cycles and natural selection pressure was minimal given the benign desert climate and the favorable irrigated conditions. The stocks therefore have value for worldwide use.

Hockett and Reid (1981) published an updated list of available genetic male sterile barley stocks. The entries are parental in nature, designed to be used for the development of hybrid composites or any other genetic or breeding use. The list included 176 spring and 147 winter barleys. All these stocks were available on request. A modified system of stock maintenance has been established. This will record accession data such as origin and genetic status, and make arrangements for perpetuation of seed viability.

Cultural procedures for growing male sterile populations are simple, generally involving only standard bulk population handling techniques. Ideally, a rectangular or square shape of the plot would be desirable for pollen dispersion and adequate isolation. It is desirable to orient the plot shape in consideration of the prevailing winds of an area. Isolation could be important if several populations were being grown in a concentrated area. For special objectives, isolation might be provided; however, in general it is not often a critical element in male sterile breeding. The transfer of genes through the cross drift of pollen would be expected to exert only a minor influence upon the composition of a population from the F_3 onward, and it may easily be argued that this is a favorable phenomenon.

As to size of plots after the F_2, which might be a square field plot arrangement, I have used drill-width plots of varying length. For efficiency in harvesting, the tedious cleaning of the combine harvester between plots can be eliminated by discarding the grain from the first part of each plot.

Head selection may begin about the F_7, and in a good composite, selection may continue annually for a few generations without a definite termination date. The generous amounts of harvested seeds also makes possible the growing of the same hybrid population at several locations within the same year. Line withdrawal procedures are identical with those I have described for the short-range bulk population method (Jensen, 1978). Careful attention to the management procedures described in that article for the line withdrawal stage is imperative. Line selection is a critical stage and these field line selection procedures ensure that the breeder will see the range of individuals in the bulk population. Selection of lines cannot be done properly from a solid-sown population, where the result is little better than random.

The selected heads go into head rows and are handled exactly the same as though they came from the bulk hybrid population method, or for that matter, any breeding and selection system. Any head row showing a male sterile plant should be reselected or discarded.

Twenty years of experience with MSFRS populations suggests there may be an average net growth to the program of about two composites per year until an equilibrium growth level is reached. Therefore, there is little reason to fear that the longer retention time for the exploitation of an MSFRS population in the program will result in uncontrolled growth.

I believe MSFRS occupies a special role in commercial barley breeding operations. Almost by definition such businesses strive to be efficient. For them, MSFRS is an ideal system: the combination of trained professionals and the ready availability of numerous male sterile population stocks can get an operation off to a fast start and efficiently produce superior cultivars.

REFERENCES

Hockett, E. A. and D. A. Reid. 1981. Spring and winter genetic male-sterile barley stocks. Crop Sci. 21:655–659.

Jensen, N. F. 1978. Composite breeding methods and the DSM system in cereals. Crop Sci. 18:622–626.

Reid, D. A., N. F. Jensen, and R. T. Ramage. 1964. A composite cross of world winter barleys. U.S. Agric. Res. Serv., ARS 34–67, 3pp.

Reid, D. A., R. T. Ramage, N. F. Jensen, and R. K. Thompson. 1966. C.C. XXVII—a second cycle of composite cross XXVI of world winter barleys. U.S. Agric. Res. Serv., ARS 34–86, 3 pp.

Sorrells, M. E. 1984. Release of 'Willis' winter barley. 1984 Barley Newsletter 28:51.

Stam, P. 1977. Selection response under random mating and under selfing in the progeny of a cross of homozygous parents. Euphytica 26:169–184.

19

METHODS SHAPED BY REQUIREMENTS OF TECHNIQUES

Four breeding methods are so fundamentally attached to techniques, both by origin and procedures, that it seemed useful to group the chapters in this part of the book.

MUTATION OR RADIATION BREEDING METHOD

Because mutation breeding is so little used in the United States I have not planned as complete a treatment of this method as I have given others. The several abstracted and referenced articles have been chosen for their breadth of subject and number of citations.

Gustafsson (1951) discussed mutations from the standpoint of viability relative to environments. He made the point that mutations are predominantly deleterious, harmful to the organism. Perhaps only one or two mutations per thousand reach homozygous stability in terms of surviving with the parent population. Many more exist in the heterozygous state. For use in plant breeding, Gustafsson recommended the bulk population method with yearly radiation of samples for planting the next generation. After a protracted period of time, final evaluation of selected lines might be made.

Plant breeders considering mutation breeding need to know the parameters of genetic response to radiation treatment. Krull and Frey (1961) published a worthwhile review of existing information and added to it a study on the effect of radiation on the variability of quantitative characters in oats. They used seeds from two spring oat varieties and the F_2 of their cross. These three categories each had matching nonradiated and radiated halves, giving six classes in all. Seed numbers were large, as was the whole testing program, which was moved to the field after isolated greenhouse increase generations. The plants were scored for chlorophyll and fatuoid mutants, heading date, nonheading plants, and height. It should be noted that oats is an allohexaploid, and the visibility of mutants would not be as obvious as in diploid barley, for example, although meiosis in oats is quite diploid in actuality.

Considering the two parents, evident and greater divergences were found for every trait in the radiated sector as compared to the nonradiated parent plants with one exception: neither "Clintland" group had any nonheading plants. The larger differences occurred between the hybridized (nonradiated) and the hybridized (radiated). In the latter group the variances for the various traits were higher than the hybridized nonradiated group by the following percentages: heading date 35%, segregating heading date 13%, fatuoids 2%, height differences, tall 6%, short 217%, and segregating 7%. In addition there was one fatuoid and one chlorophyll mutant (none in the hybridized-only group).

These results indicate that radiation treatment added a new dimension to the variance of a hybrid population. It is likely that there are also changes that were not visible given the hexaploid nature of the oat, the unknown internal effects of the radiation treatment, and the greater complexity of quantitative inheritance. Radiation treatment as a means of opening up diversity and widening genetic variance seems promising. It fills a role complementary to breeding, not competitive or mutually exclusive. One might expect it to be useful in "difficult" crosses where linkage prevents significant genetic advance.

The detection of mutations in the self-pollinated crops has been confounded by the likelihood of increased cross-pollination owing to sterility in the irradiated generation. Bhatia, Swaminathan, and Gupta (1961), in an informative review and research study on mutation for stem rust resistance in wheat, pointed out that the genetic basis of resistance to a particular race, that is, the number of genes involved, would influence the time and manner in which detection of resistance could be made. In careful tests involving a susceptible wheat and three stem rust races, and using X-rays and fast and thermal neutrons, they found only one M_3 family resistant to race 40 among hundreds of M_2 and M_3 families and thousands of seedlings. This was clearly a mutant and not a chance out-cross. The only other radiation-induced aberrations found were four *albina* mutants. The control population showed no offtypes.

Using four types of radiation treatment of oat seeds of two cultivars and exposing them to cross-pollination situations involving a breeding line containing marker genes, Thurman and Womack (1961) were able to demonstrate a remarkable increase in natural crossing over the essentially zero level of the checks. 'Arkwin' showed a range of 4.50 to 10.40%, and 'Ferguson 560' showed 1.43–9.60% natural crossing.

Wallace and Luke (1961) took seeds of 'Victorgrain', a successful oat cultivar which had succumbed to attacks of the pathogen, *Helminthosporium victoriae*, radiated them with cobalt-60 gamma rays, grew them to maturity, treated the X_2 with *H. victoriae* toxin, and selected resistant seedlings. This is an example of the specific goal case, where a mutation program not only can be successful but may do a better job than traditional methods. The study also emphasized the need for professional attention: several environmental and dosage variances were involved.

Suneson, Ramage, and Hoyle (1963) suggested the use of mutant populations for long-term evolutionary breeding programs.

The effects of four mutagen treatments on a pure line oat cultivar, "Clintland 60", were described by Abrams and Frey (1964). General observations were that fatuoid-type mutations were common and no sterility increase was noted in the M_3 population. Specifically, there was a shift toward later heading date and shorter height, but none in mean weight per 100 seeds. Genetic variability increased from the M_2 to the M_3. Ethyl methanesulfonate (EMS) treatment was deemed superior to radiation. Overall, the results did not seem encouraging unless specific goals and genetic source material were matched.

The paper by Khadr and Frey (1965) is an excellent review of the general problem of generating increased genetic variability via mutation-inducing agents, and of the consequences that might be expected. Their general conclusion, that mutation breeding is worthwhile, was qualified by these observations:

1. Genotypes differ in their efficiency as vehicles for generating variability.
2. Radiation treatment of hybrid material produced results below independently predicted additive expectations.
3. The first radiation treatment was more effective than the second.
4. Even so, recurrent radiation treatment expanded variability.
5. Selection between recurrent treatments is probably not cost effective.

Suneson, Murphy, and Petr (1965) used a chemical mutagen to create sterility, with a resultant increase of outcrossing in oats. The population to be treated was made by bulking about 10,000 entries from the World Collection of Oats. The project was deemed successful and seed lots of the M_2 and M_3 were offered to breeders. Continued recycling by radiating seeds from the treated populations would seem desirable. Alternating different mutagens, for example, thermal neutrons or EMS, might induce different effects in the population.

Gonzalez and Frey's (1965) paper dealt with the genetic variability of quantitative characters induced by thermal neutron treatments in five oat genotypes. The limited number of genotypes included one diploid, one tetraploid, and three hexaploids. In general, the hexaploids exhibited approximately as much induced variability (an average expression covering all characters) as did tetraploids or diploids. This is a favorable aspect because commercial cultivars are hexaploids. The authors made one other observation: hexaploid and diploid oats produced similar types of mutations.

In the absence of a useful male sterile for oat breeding, any research showing increased outcrossing in populations from mutagen-treated seeds is intriguing. Using seeds treated with thermal neutrons and EMS, Grindeland and Frohberg (1966) found that mean overall outcrossing increased three times, from 1.4% found in the checks to 4.0% in the treated populations.

Gustafsson, Hagberg, Persson, and Wiklund (1971) brought together a summary of accomplishments in barley improvement traceable to induced mutations. The list included seven barley varieties in Sweden. The authors also cite the production of translocations and their value to commercial hybrid barley (cf. Ramage, 1965). The review was prompted in part by a desire to lift the negative pall that has hung over the subject of induced plant mutations since its inception.

Ethyl methanesulfonate has been one of the more successful mutation-inducing agents in the small grain cereals. Arias and Frey (1973) reported on research specifically directed toward the use of this mutagenic agent to produce grain yield mutations in 16 oat cultivars. After EMS treatment in M_1, the M_2 plants were screened to remove all non-normal plants, for example, chlorophyll deficiencies. The M_2-derived line progenies, all with normal appearance relative to their parent cultivars, were then compared for yield with check lines drawn from each parent cultivar. The criterion for a yield deviant was defined as either higher or lower than the highest and lowest check line yield, respectively.

The results from 1141 lines were that there were 160 negative deviants and only one positive deviant. The positive deviant was 12% higher yielding than the highest check line.

These results add to the general belief that most mutations are deleterious, and they certainly raise questions about efficiency and whether a breeding search would be justified. Finally, it is not clear that the variants were indeed mutations. Doubt was cast on such an assumption by the finding of significant variation for yield among the *check* lines.

Two lines of descent established from a common F_3 bulk source were grown for 10 generations by Fatunla and Frey (1974). One stream was radiated with thermal neutrons or X-rays for four generations, followed by bulk propagation. It was found that the mean regression stability indices for grain yield appeared to be converging toward unity, with the nonradiated line decreasing from 1.17 to 1.02, and the radiated line increasing from 0.82 to 1.04.

Of further breeding interest, the use of mutagens to increase outcrossing in oats where useful male sterility genes and gametocides are lacking initially showed that yields and regression stability indices declined. However, following the five generations of bulk propagation the levels had been restored. It is normal for radiated populations to exhibit damage, but the hope is that with generation advance positive recombinations will be found. The authors recommended "the use of physical mutagens to enhance outcrossing among plants in a bulk population of an autogamous species."

The article by Yonezawa and Yamagata (1975) deals with certain practical cost and efficiency aspects of handling radiated plant material.

Gardner (1973), investigating the value of mass selection on radiated corn populations, concluded that mass selection was an effective method to improve corn. Paradoxically, the merit of the radiated population was evident not *en masse*, for treated populations performed as well as nontreated controls, but as it showed itself to be the source of superior and potent inbred lines.

Jalani, Frey, and Bailey (1979) compared equal numbers of oat lines from three related populations, two of which involved mutagen induction, for the magnitude of released genetic variance involving growth rate (GR) and harvest index (HI). The three populations each traced to the pure line, 'C.I. 7555'. Check lines, "C," were drawn from C.I. 7555, "M" represented M_2-derived lines emanating from EMS treatment, and "O" lines were F_2-derived lines from a cross of $M_1 \times$ C.I. 7555.

The results were that almost all (98%) of the variation for grain yield was due to the variation found in GR and HI. The variation was greater in M and O than in C, proof that the mutagen had worked, but the mutations resulted in lowered GR and HI levels, the opposite of breeding goals and therefore deleterious. Because HI remained in what had earlier been found to be an optimal range (40–50%) for Corn Belt oats, it seemed obvious that the low grain yields of the M and O populations traced to the effect of deleterious mutations for GR.

A dwarf mutant bean discovered by Nagata and Bassett (1985) showed promise of being an aid to plant breeders through enhanced cross-pollination in the highly self-pollinated common bean. Vulnerability to cross-pollination in the mutant was traced to delay in anther dehiscence, opening the door for insect pollination with foreign pollen.

MUTATION BREEDING: LIMITED DEVELOPMENT IN THE UNITED STATES

The apparent increase in additive genetic variance reported by Lonnquist, Cota A, and Gardner (1966) in a thermal neutron-irradiated corn population suggests a means of increasing the genetic variance of important hybrid populations. For example, two streams of the same cross population, one normal and one radiated, might be advanced for selection. Later, a third stream from repeated radiation treatment could be carried forward. Increasing variance within hybrid oat populations would be especially desirable in the absence of efficient male sterile stocks.

A good case can be made for the use of radiation in oat breeding. The unique floret and spikelet arrangement in the oat panicle is much less conducive to outcrossing than the spike arrangement in wheat or barley. Without male sterility genes, and with uncertain results from gametocides, the oat breeder faces yet another hurdle: oats are more difficult to hybridize and to obtain numbers of hybrid seeds than are the other small grain cereals, barley and wheat. In barley or wheat, crossing is usually a mass procedure on a whole head basis using the twirl method, but in oats there is not a comparable panicle situation, unless it is the "approach" method proposed by Hamilton (1953). Seed set in oats occurs through individual floret pollination. Thus it is more difficult with oats to carry through backcross or recurrent selection programs or any stepwise outcrossing program. Radiation provides an alternative way to create genetic variability and recombination opportunities. Nevertheless, plant breeders recognize that much of the increased genetic variation is undesirable.

In the late 1940s and early 1950s, I became interested in the possibilities of mutation breeding in the Cornell projects, principally in oats. We had excellent cooperative arrangements with the Brookhaven National Laboratories. The project was abandoned after about two years. The reasons for discontinuing the work were that I had no clear objective in mind, and the space and labor requirements quickly rose to a level approaching that of the entire oat program. The radiated seeds had to be space planted, and the tillers and panicles of each plant were harvested, threshed, and later sown separately in head rows. I understand that in Europe the radiation procedures have been refined and tailored to meet the breeder's desires, and that there is a close monitoring and attention to follow-up practices. Nevertheless, I came to the general conclusion that mutation breeding was suitable in my projects only for special cases, and even then an alternative procedure to radiation, for example, EMS, might be preferable. Sustained interest in mutation breeding has not been a characteristic of U.S. breeding programs.

REFERENCES

Abrams, R. and K. J. Frey. 1964. Variation in quantitative characters of oats (*Avena sativa* L.) after various mutagen treatments. Crop Sci. 4:163–168.

Arias, J. and K. J. Frey. 1973. Grain yield mutations induced by ethyl methanesulfonate treatment of oat seeds. Radiat. Bot. 13:73–85.

Bhatia, C., M. S. Swaminathan, and N. Gupta. 1961. Induction of mutations for rust resistance in wheat. Euphytica 10:379–383.

Fatunla, T. and K. J. Frey. 1974. Stability indexes of radiated and nonradiated oat genotypes propagated in bulk populations. Crop Sci. 14:719–724.

Gardner, C. O. 1973. Evaluation of mass selection and of seed irradiation with mass selection for population improvement in maize. Genetics 74:88–89. (abstr.)

Gonzalez, C. and K. J. Frey. 1965. Genetic variability in quantitative characters induced by thermal neutron treatment of diploid, tetraploid and hexaploid oat seeds. Radiat. Bot. 5:321–335.

Grindeland, R. L. and R. C. Frohberg. 1966. Outcrossing of oat plants (*Avena sativa* L.) grown from mutagen-treated seeds. Crop Sci. 6:381.

Gustafsson, A. 1951. Induction of changes in genes and chromosomes. II. Mutations, environments and evolution. Cold Spring Harbor Symp. Quant. Biol. 16:263–281.

Gustafsson, A., A. Hagberg, G. Persson, and K. Wiklund. 1971. Induced mutations and barley improvement. Theor. Appl. Genet. 41:239–248.

Hamilton, D. G. 1953. The approach method of barley hybridization. Can. J. Agric. Sci. 33:98–100.

Jalani, B. S., K. J. Frey, and T. B. Bailey, Jr. 1979. Contribution of growth rate and harvest index to grain yield of oats (*Avena sativa* L.) following selfing and outcrossing of M_1 plants. Euphytica 28:219–225.

Khadr, F. H. and K. J. Frey. 1965. Recurrent irradiation for oat breeding. Radiat. Bot. 5:391–402.

Krull, C. F. and K. J. Frey. 1961. Genetic variability in oats following hybridization and irradiation. Crop Sci. 1:141–146.

Lonnquist, J. H., O. Cota A, and C. O. Gardner. 1966. Effect of mass selection and thermal neutron irradiation on genetic variances in a variety of corn (*Zea mays* L.). Crop Sci. 6:330–332.

Nagata, R. T. and M. J. Bassett. 1985. A dwarf outcrossing mutant in common bean. Crop Sci. 25:949–954.

Ramage, R. T. 1965. Balanced tertiary trisomics for use in hybrid seed production. Crop Sci. 5:177–178.

Suneson, C. A., H. C. Murphy, and F. C. Petr. 1965. Fostering the recombination of oats with a mutagen. Crop Sci. 5:176.

Suneson, C. A., R. T. Ramage, and B. J. Hoyle. 1963. Compatability of evolutionary and mutation breeding methods. Euphytica 12:90–92.

Thurman, R. L. and D. Womack. 1961. Percentage natural crossing in oats produced from irradiated seeds. Crop Sci. 1:374.

Wallace, A. T. and H. H. Luke. 1961. Induced mutation rates with gamma rays at a specific locus in oats. Crop Sci. 1:93–96.

Yonezawa, K. and H. Yamagata. 1975. Practical merit of delayed selection after single and recurrent mutagenic treatments. Part 2. Optimum generation for selection. Radiat. Bot. 15:169–184.

20

HETEROSIS AND COMMERCIAL HYBRIDS

One of the early studies on heterosis in wheat was by Rosenquist (1931). He used his "approach" method (Rosenquist, 1927) to make a series of 26 crosses involving about 1600 hybrid plants and studied these over a period of three years in Illinois and Nebraska. Because a planned spacing of about 4 in. in rows was altered by the effects of differential winter killing, additional studies at approximately 1.5 in. in rows were organized. Parents used in crosses had distinguishable trait differences, such as awns versus no awns, or brown versus white chaff, so that selfs could be eliminated. Eight characters were monitored. The results were as follows:

1. *Grain Yield:* of the 26 crosses 14 were higher in yield than the high parent, 7 were intermediate, and 5 lower than the low parent. Overall, only slightly more than half showed heterosis for yield.
2. *Density:* hybrids developed relatively better under wider in-row spacing than narrower, but in general the results paralleled each other.

The number of crosses whose F_1s yielded lower than the low-yielding parent is disturbing. The study provided little promise for any projected use of heterosis. Interpretation could have been affected if more information on the growth conditions had been furnished: nothing was stated regarding the level of soil fertility or similar factors.

THEORETICAL ASPECTS OF HETEROSIS

Jinks (1955) compared two principal hypotheses as to the genetic basis of heterosis, using a method of analysis applicable to diallel crosses. He applied this to data from several species. The first hypothesis was based on the accumulation in the hybrid of dominant favorable genes of both parents. The second was that heterozygosity itself was responsible for heterosis (this has been extended by Hull's "overdominance" hypothesis). Jinks pointed out the great practical implications to the breeder: if the first is correct, maximum improvement should follow maximum accumulation of dominant favorable genes in the homozygous condition. If the second hypothesis is correct, then maximum productivity can be achieved only through maximum heterozygosity and will dissipate with approach to homozygosity.

Jinks's paper must be read for details, but a general finding was that a clear distinction could not be found between the two hypotheses. Wherever overdominance was found, nonallelic interaction was found also. Removing crosses with significant nonallelic interaction resulted also in a reduction or disappearance of the apparent overdominance. Furthermore, specific combining ability was always associated with nonallelic interaction, and general combining ability with ordinary dominance.

Another interpretation of heterosis was suggested by Sakai and Utiyamada (1957): vigorous heterotic growth may be due to independent competitive ability arising from overdominance in the heterozygous state.

Grafius (1959) proposed the novel view that genes for the complex trait, yield, do not exist—there are only genes for components of yield. He proposed a geometric model for yield, where the volume of a parallelepiped with three edges corresponds to the component traits, heads per plant, seeds per head, and average seed weight. Grafius provided data showing that changes in yield, indeed heterosis, can be explained by epistatic interactions among the components. To the extent that these are additive, they are fixable in inheritance. This corresponds to what we know about progress in breeding for yield. Grafius' views are important and I can see the relationships to selection pressure on components in his model. For example, with intense pressure on any component one gets a picture of a malleable box being pulled by the corners as one side lengthens and the box becomes flatter and reduced in volume.

Jain and Jain (1962) studied the within-family variances of F_4, F_5, and F_6 plant progenies of barley, using height and spike length as the measured criteria, and compared variances with those of the crosses' parents. Variances were found to be of the same magnitude as those of the parents for height, but not for spike length where the selections showed an increased range of variability. The authors attributed this to heterozygote advantage, which would tend to perpetuate bursts of heterozygosity, releasing variance via segregation into the population.

Although genetic male sterile barleys have been found readily, female sterile types are rare. Harvey, Reinbergs, and Somaroo (1968) reported the finding of such types as mutants in a colchicine-treated hybrid population. Sterility was governed by a single recessive gene. The authors explained that a female sterile, bred into the proper background, could become the pollen donor for hybrid barley seed production. It would be mixed with the female parent (male sterile) seed for field planting and would essentially disappear after serving its pollination function. A small amount of selfed seed could be ignored or removed mechanically, inasmuch as it likely would be different in size. The probability is that the seeds would be smaller, though there may be reason to expect they might be larger, given a better genetic background for the female sterile and less competition for seed development in the mostly sterile head.

Using *Arabidopsis thaliana*, Griffing and Zsiros (1971) studied heterosis in genetically similar F_2 populations of a cross of two races. The F_2 was used because of F_1 seed production limitations. The interaction of heterosis with three environments was studied. The controlled environments were three levels of temperature, two levels of nutrients, and two levels of density.

When the various closely controlled combinations of environmental components were analyzed, it was found that there were varying ranges of heterotic response. For example, changing the nutrient level when temperature was held constant resulted in a change of heterosis from negative to positive. Although the authors stressed the necessity of testing and corroborating the results on a field crop basis, they extracted the following generalizations (all comparisons were relative to best parents):

1. Heterosis is unlikely under moderate temperatures and low nutrient levels. There is a suggestion here that densely planted crops would create or aggravate such a situation.
2. Paradoxically, the same conditions but with higher prolonged temperatures are more favorable to heterosis.
3. Optimal nutritional status is always favorable to heterosis.

The plant breeding implications seem to be that caution must be used in interpreting results from single environment tests, and it would be desirable to have controlled environment facilities to create differing environments for testing.

Zevan (1972) provided an excellent review on the effect of spacing on heterosis, but his own experiments led him to the surprising conclusion that "the expression of heterosis for yield or for one of its components is not affected by density."

Moreover, Zevan stated that this unexpected conclusion was supported by the results in five published papers (Briggle, Cox, and Hayes, 1967), (Briggle, Petersen, and Hayes, 1967), (Fonseca and Patterson, 1968), (Knott and Sindagi, 1969), and (Bitzer, Patterson, and Nyquist, 1971), and was contravened only by one paper, that of Rosenquist (1931), who observed heterosis for certain characters at low density but not when grown close together.

Zevan then undertook the study that is the subject of this paper. He termed it a "laborious work" and indeed it was, as may be seen from its scope and detail and not least from its impressive list of nine cooperating private and public institutions having a common interest in hybrid wheat.

The research was done separately with spring and winter wheat; however, spring wheat was dropped from the study when the results (from 60 F_1s, presumably different crosses) showed that all F_1s yielded less than the highest-yielding pure line variety. This uniform failure of any spring F_1 to surpass the best variety is interesting and raises questions without answers.

In winter wheat the situation was otherwise: several F_1s were superior in yield to their parents and to the highest-yielding varieties. Six F_1s involving a total of 10 parents, along with some F_2s and F_3s, were chosen for the study. One thousand F_1-embryo seeds were produced for each combination. This was enough seed to plant replicated trials of four-row plots 1 m long, with within-row spacing to create seeding densities of 32, 64, 128, and 256 plants/m^2. Survival counts in the plots indicated that these densities were achieved approximately. Data were taken on grain yield per row and eight other yield component traits.

The results showed that per-row yields increased with increasing density. The important finding was that when the yield of F_1 was expressed as percentage of P_1 and P_2 the figures were very alike across the four densities, and statistical analysis showed there was no density effect. The same conclusion was reached after an analysis of each of the components studied. A further finding of importance to hybrid wheat prospects was that the yield per row of the F_1 populations was statistically equal to that of the F_3 populations (although the latter actually yielded slightly less).

Zevan himself stated that his findings were contrary to expectations. Most hybrid vigor studies are made under spaced density and restricted research conditions with results that are often difficult to corroborate or sustain under wheat and barley field conditons. If these findings continue to be borne out it would mean that a distinction exists between all neutral-density F_1s and the testing of other generations where spacing seems to exert profound influence.

MEASUREMENT OF HYBRID VIGOR

Suneson and Riddle (1944) compared seven fertile barley cultivars crossed with male steriles for heterosis in their hybrids. Male parent differences were observed in the hybrids whose yields generally followed the yielding capacity of the cultivars. The general heterotic advantage averaged approximately 20%.

Pawlisch and Van Dijk (1965) studied heterosis for both grain and forage yields in crosses of four barley varieties of intermediate winter type. They found forage increases ranging from 8 to 31%, and grain yields ranging from 8 to 37% above the better parent. It seems logical that forage objectives would involve the range between spring and winter varieties. Reid (1965) presented information that showed the genetic relationship of growth habit in spring and winter barley.

Knott (1965) recorded the results of a study of hybrid vigor in seven crosses of spring wheat. The 14 parents used represented an unusual diversity of germ plasm, displaying broad differences in origin and commercial life span. The hybrids and parents were evaluated in two tests in rows of different density, 5 in. apart ("rows") and 12 in. apart ("spaced"). The yield results in rows were that four of the seven hybrids exceeded parental means and three of these showed heterobeltiosis. As competition for commercial use, the comparisons with 'Thatcher', a then-standard pure line variety, were more meaningful: no F_1 hybrid yielded significantly more. The yields of the widely spaced plants were compared only with Thatcher; five yielded more, two less, and none was statistically significantly different.

Knott found a high correlation (.79*) between the yield of F_1 hybrids and the yields of their parents. This conveys information of general interest on the subject of choice of parents in plant breeding. There was some indication that wide spacing favored heterosis from the fact that yields ran higher in the spaced tests; otherwise the conclusions drawn were not favorable to commercial hybrid prospects.

McNeal, Baldridge, Berg, and Watson (1965) failed to find significant heterosis in three wheat crosses and attributed this to the close relationship of the parents, which indeed were high in shared common germ plasm.

Upadhyaya and Rasmusson (1967) evaluated heterosis in eight spring barley cultivars and their 28 possible F_1 and F_2s. Five traits were studied: yield, kernels per head, heads per plant, kernel weight, and height. Average heterosis for yield was 21.5% over the midparent and 9.1% above the high parent. For the other characters, heterosis figures (midparent base) were kernels per head 7.1%, heads per plant 7.6%, height 3.2%, and kernel weight 5.9%. Inbreeding depression for yield measured in the F_2 was 26.1%. Substantial superiority was shown over the highest-yielding parent cultivars, 'Traill' and 'Liberty'.

A possibly significant factor in this study is that the research was done under spaced-plant conditions (1 ft apart). Generally, spaced plant results are not corroborated by later crop density experiences. In this respect, however, the experiences of Ramage (personal communications) with commercial carefully spaced plantings of hybrid barley hold promise of a direct relation between test and crop results.

Briggle, Cox, and Hayes (1967) reported a three-year study of heterosis in spring wheat under irrigation in Idaho. A salient feature was that the research on the F_1, F_2, F_3, and parents was done at five population levels, a crucial area of concern in relating levels of heterosis found in research to commercial field heterosis.

Heterobeltiosis (yield above high parent), measured over all population levels was, by years, 19.5%, 19.2%, and 16.5%.

Maximum yields were obtained with seeds ½ in. apart in the row; however, there was little penalty if the spacing widened to 1 in. apart. These represented, respectively, sowing rates of 6½ and 3¼ pecks/acre. It should be noted that rows were 1 ft apart whereas standard field seeding row widths are customarily 7–9 in. apart. The authors reviewed other research on this subject.

Another 1967 paper by Briggle, Petersen, and Hayes (1967) was a research report on heterosis in winter wheat, a companion to that on spring wheat by Briggle et al., just reviewed. The format was similar, and irrigation was used throughout at the Idaho location. Although slight discrepancies in consistency occurred, the heterosis advantage (above the higher-yielding parent and at all population levels) of the F_1s was large: by years, 33.9%, 28.9%, and 6.5%.

Populational level effects were important. As with the spring wheats, levels 1 and 2, more widely spaced, were lower yielding. There were no real differences in yield among levels 3–5, which meant that seed costs could participate in the spacing decision.

Nettevich (1968) studied heterosis in 48 hybrid combinations of a broad range of Russian wheats. He concluded, "Only if high yielding parental forms are crossed it is possible to obtain high yielding hybrids." He added that hybrid wheat breeders must also work to develop higher-yielding wheats for use as parents in hybrids.

Carleton and Foote (1968) looked for heterosis in the components for the two complex traits, yield and leaf area, in 12 hybrids of two- and six-rowed barleys. The plants were grown at wide spacings (18 in.). The authors anticipated that interaction of components would occur and that heterosis of the complex trait would be enhanced thereby. The results were disappointing; in fact, heterosis ranging from negligible to negative was found. Inasmuch as the growing conditions were excellent, the authors were left with the feeling that optimal conditions may not foster the expression of heterosis.

Using different but adapted winter wheats (soft and hard reds, whites) Gyawali, Qualset, and Yamazaki (1968) found that 10 of 21 hybrids showed heterobeltiosis, and the average yield of all hybrids was 24% greater than the better parent. Soft red × soft white hybrids generally gave lower heterotic values.

In studies by Fonseca and Patterson (1968) on heterosis in wheat grown at different plant spacings it was shown that greater heterosis occurred in hill plantings than in nursery plantings when standard crop density rates were used.

Severson and Rasmusson (1968) showed that the spacing of hybrid plants profoundly affected the expression of heterosis. Average estimates of heterosis for yield (midparent criterion) of five hybrids (eight parents) were 3.2%, 16.7%, 17.3%, and 22.5% for the expanding within-row spacings of 2.5, 7.5, 15.0, and 22.5 cm, respectively. There were specific individual hybrid responses to spacing; nevertheless this inverse relationship between amount of heterosis and spacing (for meaning, read low heterosis at crop density) cannot be ignored because hybrid varieties are customarily grown at crop density spacings. This practice, however, is always subject to new interpretations and to change, based on continuing research into capturable heterosis. Practically, yield per area must also be considered along with plant spacing.

Knott and Sindagi (1969) obtained disappointing results relative to heterosis in hybrid hard red spring wheats. The material used came from 15 diallel crosses of six productive cultivars and genotypes, with the F_1 and parent populations grown at two seeding rates. Heterosis indications were low. Average yield increases of the F_1s over the better parent were 4%, with two hybrids significantly better. Several hybrids yielded more than the highest-yielding parent in the test but none significantly more. A positive note was that parental yields were moderately good predictors of hybird yields (correlation of 0.541, just below significance).

Experimental winter wheat hybrids from hand-crossed seeds sown at crop densities in Ontario, Canada, gave promising results in a study reported by Rai, Stoskopf, and Reinbergs (1970). Yields from the best were 26% above the better parent and 19% above 'Genesee', a standard commercial cultivar. Two other hybrids outyielded the better parent by 5% and 15%.

Grant and McKenzie (1970) studied heterosis in hybrids of spring × winter wheats, two groups that have experienced a significant amount of breeding isolation. They found that under dryland conditions the hybrids outyielded the spring parents by 35%, 40%, and 2%, the last figure not significant, in three crosses, and under irrigation all three hybrids significantly outyielded the spring parents by 36%, 25%, and 19%.

Thirteen of 15 F_1 hybrids from a six-parent diallel cross of soft winter wheats (five red and one white) produced yields superior to the higher-yielding parent in research reported by Bitzer, Patterson, and Nyquist (1971), but the differences were not statistically significant.

Busch, Lucken, and Frohberg (1971) estimated heterotic effects in F_1 hybrids of three hard red spring wheats and compared their yields with F_5 lines from random F_3 lines drawn from the same crosses. None of the F_1 hybrids significantly outyielded its better parent, although two outyielded it. In each cross F_5 lines were found that exceeded and performed significantly better than the F_1 hybrid. One cross was "difficult" in that it failed to produce many superior lines; nevertheless it is notable that one outstanding line was found, emphasizing the point that generalizations about the worth of a cross do not always hold true.

Yap and Harvey (1971) made six crosses of four barley cultivars and evaluated six agronomic traits under two spacing environments, namely, widely spaced and densely spaced. Heterosis was essentially absent in all six traits in all crosses under all densities, although there was a tendency for the majority of hybrids to exceed the midparent value. Interestingly, observed heterosis was higher in the dense, more competitive seedings than in the spaced plantings. Number of heads per area showed a close relation to grain yield.

Bitzer and Fu (1972) surveyed diallel crosses of six southern U.S. soft red winter wheats for heterosis. Three F_1 hybrids produced statistically significant higher yields than their better parent, averaging 9.7%, with one ranging as high as 25% greater. The heterosis level of all F_1s over their midparent yields averaged 17%. 'Blueboy' was the highest-yielding cultivar and produced the highest-yielding F_1s. An unexpected result was the performance of 'Coker 67–14' which, although the lowest-yielding of the cultivars, contributed nevertheless to the greatest average yield increase of the hybrids. This tendency of Coker 67–14 to produce such results is reminiscent of the high–low cross syndrome. These two varieties, respectively, had the highest GCA values.

Fedak and Fejer (1975) found no agreement between F_1 performance data in spaced and in solid plantings. Actually, heterosis was shown in spaced trials but not in solid plantings. In a general way these findings reflect the experience of many scientists, particularly those wrestling with the challenge of commercial hybrid wheat and barley. Levels of heterosis found in research trials are infrequently matched under field conditions. There is another thought here concerning breeding and selection: can hybrid cultivars be selected to match different planting rates?

The problems associated with possible commercial production of hybrid rice were assessed in Davis and Rutger's (1976) paper. Attributes of rice that currently depress hybrid rice prospects are difficulties in producing large amounts of seed, inconsistency in heterosis findings, and erratic yield interactions with plant spacings. The authors' conclusions were that the prospects for hybrid rice were less favorable for direct seeding, high-density production than for transplanted production at wide spacings.

In the experimental part of the study concerned with heterosis at different plant spacings, the widely spaced (30 × 30 cm) hybrids showed significant heterosis over high parent in only about one-fourth of the 41 hybrids tested. However, an additional discouraging note was that only two of these were significantly higher yielding than the best cultivar in the test. In the 15 × 15 cm experiments with only four hybrids, the best hybrid came only to within 92% of the yield of its high parent.

The authors also looked for residual heterozygosity in F_2 and F_3 hybrids. In 12 two-parent mixtures none yielded significantly more than its high parent.

Many of the characteristics of an F_1 hybrid are similar to those of a pure line cultivar, for example, the unvarying replication of an identical gene package among individuals of each group. However, one might ask whether hybrids have similar genotype × environment interaction and stability responses. This was the problem addressed by Johnson and Whittington (1977). They created 16 barley genotypes for test in 16 environments. The genotypes were eight commercial cultivars, and the eight hybrids from crossing each with a single female, the latter being a male sterile of the balanced tertiary trisomic type developed and described by Ramage in his 1965 paper. The 16 environments were created by different combinations in two years of fertilizer applications, disease control, and dates of planting. The Finlay–Wilkinson model was used to measure stability.

The results were that parents and hybrids performed similarly across the gamut of conditions that produced distinctly different yield levels in the two years.

All the hybrids except one outyielded their male parent (the female male sterile parent could not be scored; however, it was assumed the adapted males were the better-yielding parents). The yield advantage was not great, especially in view of the finding that only one hybrid exceeded the best cultivar in yield, by only 1.3%. Admittedly, the use of only a single female, scarcely adapted to European conditions, restricted the range of responses; nevertheless the results were not favorable for hybrid barley prospects.

Cregan and Busch (1978) compared F_1, F_2–F_5 bulks, and F_5 lines produced by a 28 population diallel of eight spring wheat parents, seeking to learn more about heterosis and pure line performance. Their study illustrated the complexity surrounding these matters, for parts of the findings were positive to hybrids and part favorable to pure lines, yet overall there existed a degree of uncertainty. The F_1s showed credible heterosis on the midparent basis, generally not considered a critically favorable level, with a range from negative to positive (41%) heterobeltiosis, and compared with the highest-yielding pure line from each cross, a range of 75–119%. But the mean yields of best lines and F_1s over all crosses were similar. Of course, averages are somewhat irrelevant because the breeder would look for the highest F_1 response. Nevertheless, in this study no F_1 line was found that significantly outyielded the best pure line, leaving the authors with an uncertain choice between pure line or F_1 hybrid programs. They pointed out that the parents used had a moderate degree of genetic relatedness.

For purposes relating to the production of commercial soybean hybrids, it would be helpful if hybrid seeds could be distinguished and separated from selfed seeds. Kilen (1980) examined the premise proposed by Bradner (1975) that the use of a large-seeded male parent in cross-pollination was conducive to the production of larger hybrid seeds. Using large- and small-seeded cultivars differing in seed weight by a minimum of 30 g/100 seeds, Kilen was unable to find a significant difference in hybrid seed weights between hybrids whose male parent showed a heavier seed weight advantage over the female parent and self-pollinated seeds, provided the effect of pod numbers at nodes was removed.

From the premise, "If genetic effects, including additive, dominance and epistatic, were known, then the performance of the F_1 could be predicted," Bailey, Qualset, and Cox (1980) addressed the very important problem of predictability of F_1 yields, that is, to estimate the future performance of a commercial wheat hybrid, and also to gauge the amount of heterosis in three-way wheat hybrids. They used four spring wheat parents and prepared their 6 F_1s (diallels), 6 F_2s, and 12 three-way F_1s of the single crosses × the parents not in the single crosses (triallels).

The genetic model of Eberhart and Gardner (1966) was used. Although average all-entry yields in the first year were more than double those of the second year owing principally to late planting in the latter year, the heterosis results were very similar. Additive effects dominated and there was little evidence of epistatic effects. An interesting observation was that the average value of midparent heterosis was much higher in the second lower-yielding year, for single crosses two and one half times larger, and for three-way crosses four times larger. Also, the average heterotic values of single crosses were almost double those of three-way crosses.

The plant breeding implications of this study are that predictions can be made on the basis of F_1 hybrid performance when the single cross data are unavailable. The rankings of predicted single cross means matched well with observed means. The model also identified one hybrid that showed heterobeltiosis for both years, and also one three-way cross showing high parent heterosis.

COMMERCIAL HYBRID VARIETIES

The principal basis for hybrid wheat, heterosis, was covered in a broad review by Briggle (1963).

Johnson and Schmidt in 1968 published the comprehensive review, "Hybrid wheat." It provides an excellent means of grasping quickly all the elements of this relatively new breeding and selection methodology. A long article, its breadth can be gauged by the main subject headings: cytoplasmic sterility, fertility restoration, heterosis, milling and baking quality, seed production, and agronomic and economic considerations.

A detailed procedure for hybrid seed production in barley was given by Ramage (1965) using specialized chromosome arrangements known as balanced tertiary trisomics, which carry a recessive male sterile gene. The breeding behavior and the handling of special seed production fields are complex. This article should be read in its entirety by those with a special interest in the subject.

For commercial success of hybrid wheat the magnitude of the heterotic yield effect is crucial, so much so that the effect on other traits has been overshadowed. It is not surprising, therefore, that Parodi, Nyquist, Patterson, and Hodges (1983) found a significant hybrid response to cold resistance in winter wheat. Parents and F_1s of a diallel cross were subjected in the seedling stage to varying severe cold exposures for four hours and allowed to recover. After 10 days the F_1s averaged 26% and 30% more regrowth tissue and crown weight, respectively, than their better parents.

Performances such as these could be as important to productivity as yield itself; indeed, cold resistance might be considered a component of yield. The authors expressed only one reservation: that they had insufficient data to gauge the effect of the possibly larger seed size of the F_0 seeds.

An ingenious proposal for hybrid wheat seed production (Driscoll, 1972) was modified in 1985 by Driscoll (1985). The XYZ system now became, in effect, the YZ system by the elimination of the need for the X-line. The new system is also simpler to implement.

The system can best be described in Driscoll's own words:

> The XYZ system of producing hybrid wheat... involves three lines: i) the Z-line contains the normal 21 pairs of wheat chromosomes and is homozygous for a recessive male-sterile mutant on one of the chromosomes; ii) the Y-line is the same as the Z-line except it contains an additional 'alien' chromosome that bears a compensating gene for male fertility; and iii) the X-line is the same as the Y-line except that it contains a pair of the alien chromosomes. The alien chromosome, e.g., a chromosome of rye (*Secale cereale* L.), does not pair meiotically with the wheat chromosomes; and of the two types of pollen produced on the Y-line, the 21-chromosome type is largely favored in fertilization.
>
> Seed production using the XYZ system has three steps. The X-line pollinates the Z-line to produce a Y-line which in turn pollinates the Z-line to produce an enlarged Z-line. The Z-line is used as the female block, and normal wheat is used as the male block to produce the hybrid seed.
>
> The modified XYZ system differs from the original system in two ways: first, no X-line is involved; and, secondly, the progeny of a selfed Y-line is used instead of a Y-line.

A "cornerstone" of the XYZ system is the 'Cornerstone' male sterile mutant (Driscoll, 1977), which became the Z-line.

The known effects on wheat seeds resulting from cross pollination on *T. timopheevi* Zhuk. cytoplasm male steriles are relatively lower test weight, and higher protein percentage and kernel weight than in self-pollinated seeds. Research by Johnson and Lucken (1986) compared selfed seeds and hybrid seeds produced by using A, B, and R lines and verified these effects. Their interpretations offered hope for improving these traits in seed production for commercial hybrid use.

Lower seed set on male sterile plants, a partially controllable factor, affected not only the seeds produced but also their progeny performance. The authors concluded that higher seed production levels resulted in improved seed, seedling, and field performance. Cross-pollinated seeds showed low initial heterosis which, however, was overcome as the vegetatively generated growth appeared. Final performance showed expected high yield differences in favor of hybrid seeds. This does raise the question as to whether final yields might be further enhanced if the initial seed-generated growth could be improved. The hybrid wheat breeder must concern himself with two aspects of test weight, namely, that of the seed to produce hybrid wheat and that of seeds of the commercial crop. Of the two, the latter is the more important because of its direct relation to economic value. Improving both aspects, however, may involve separate breeding considerations.

I have devoted little attention to hybrid wheat and certainly have not given it the consideration that it deserves. The massive efforts directed to commercial hybrid wheat research during the past few decades not only illustrate, I think, the complexity of the problems facing the breeders but also explain why ordinary breeding programs cannot enter the competition. The promise of yields higher than those of pure line cultivars has been difficult to achieve. I think the following assessment by Sneep and Hendriksen (1979) has about the proper balance:

> A cytoplasmic male sterility and restorer system is available in wheat. In producing F_1 hybrids, breeders have encountered difficulties in maintaining pure male sterility, in restoring full fertility, in synchronizing the times of stigma receptivity and pollen shedding in the parents and in ensuring adequate cross pollination, especially under unfavourable weather conditions. Nevertheless, progress is being made in exploiting heterosis in this important cereal.

Included is a good discussion on the prerequisites for successful hybrid wheat production and on the progress that is being made to meet these requirements.

REFERENCES

Bailey, T. B., Jr., C. O. Qualset, and D. F. Cox. 1980. Predicting heterosis in wheat. Crop Sci. 20:339–342.

Bitzer, M. J. and S. H. Fu. 1972. Heterosis and combining ability in southern soft red winter wheats. Crop Sci. 12:35–37.

Bitzer, M. J., F. L. Patterson, and W. E. Nyquist. 1971. Hybrid vigor and gene action in a six-parent diallel cross of soft winter wheat. Can. J. Genet. Cytol. 13:131–137.

Bradner, N. R. 1975. Hybrid soybean production. U.S. Patent 3,903,645. Off. Gaz. U.S. Patent Office 938:480.

Briggle, L. W. 1963. Heterosis in wheat—a review. Crop Sci. 3:407–412.

Briggle, L. W., E. L. Cox, and R. M. Hayes. 1967. Performance of a spring wheat hybrid, F_2, F_3 and parent varieties at five population levels. Crop Sci. 7:465–470.

Briggle, L. W., H. D. Petersen, and R. M. Hayes. 1967. Performance of a winter wheat hybrid, F_2, F_3 and parent varieties at five population levels. Crop Sci. 7:485–490.

Busch, R. H., K. A. Lucken, and R. C. Frohberg. 1971. F_1 hybrids versus random F_5 line performance and estimates of genetic effects in spring wheat. Crop Sci. 11:357–361.

Carleton, A. E. and W. H. Foote. 1968. Heterosis for grain yield and leaf area and their components in two- and six-rowed barley crosses. Crop Sci. 8:554–557.

Cregan, P. B. and R. H. Busch. 1978. Heterosis, inbreeding, and line performance in crosses of adapted spring wheats. Crop Sci. 18:247–251.

Davis, M. D. and J. N. Rutger. 1976. Yield of F_1, F_2 and F_3 hybrids of rice (*Oryza sativa* L.). Euphytica 25:587–595.

Driscoll, C. J. 1972. XYZ system of producing hybrid wheat. Crop Sci. 12:516–517.

Driscoll, C. J. 1977. Registration of Cornerstone male-sterile wheat germplasm. Crop Sci. 17:190.

Driscoll, C. J. 1985. Modified XYZ system of producing hybrid wheat. Crop Sci. 25:1115–1116.

Eberhart, S. A. and C. O. Gardner. 1966. A gerenal model for genetic effects. Biometrics 22:864–881.

Fedak, G. and S. O. Fejer. 1975. Yield advantage in F_1 hybrids between spring and winter barley. Can. J. Plant Sci. 55:547–553.

Fonseca, S. and F. L. Patterson. 1968. Hybrid vigor in a seven-parent diallel cross in common winter wheat (*Triticum aestivum* L.). Crop Sci. 8:85–88.

Grafius, J. E. 1959. Heterosis in barley. Agron. J. 51:551–554.

Grant, M. N. and H. McKenzie. 1970. Heterosis in F_1 hybrids between spring and winter wheats. Can. J. Plant Sci. 50:137–140.

Griffing, B. and E. Zsiros. 1971. Heterosis associated with genotype–environment interactions. Genetics 68:443–445.

Gyawali, K. K., C. O. Qualset, and W. T. Yamazaki. 1968. Estimates of heterosis and combining ability in winter wheat. Crop Sci. 8:322–324.

Harvey, B. L., E. Reinbergs, and B. H. Somaroo. 1968. Inheritance of female sterility in barley. Can. J. Plant Sci. 48:417–418.

Jain, S. K. and K. B. L. Jain. 1962. The progress of inbreeding in a pedigree-bred population of barley. Euphytica 11:229–232.

Jinks, J. L. 1955. A survey of the genetical basis of heterosis in a variety of diallel crosses. Heredity 9:223–238.

Johnson, G. F. and W. J. Whittington. 1977. Genotype–environment interaction effects in F_1 barley hybrids. Euphytica 26:67–73.

Johnson, K. M. and K. A. Lucken. 1986. Characteristics and performance of male-sterile and hybrid seed produced by cross-pollination in hard red spring wheat. Crop Sci. 26:55–57.

Johnson, V. A. and J. W. Schmidt. 1968. Hybrid wheat. Adv. Agron. 20:199–233.

Kilen, T. C. 1980. Paternal influence on F_1 seed size in soybean. Crop Sci. 20:261–262.

Knott, D. R. 1965. Heterosis in seven wheat hybrids. Can. J. Plant Sci. 45:499–501.

Knott, D. R. and S. S. Sindagi. 1969. Heterosis in diallel crosses among six varieties of hard red spring wheat. Can. J. Genet. Cytol. 11:810–822.

McNeal, F. H., D. E. Baldridge, M. A. Berg, and C. A. Watson. 1965. Evaluation of three hard red spring wheat crosses for heterosis. Crop Sci. 5:399–400.

Nettevich, E. D. 1968. The problem of using heterosis of wheat (*Triticum aestivum*). Euphytica 17:54–62.

Parodi, P. C., W. E. Nyquist, F. L. Patterson, and H. F. Hodges. 1983. Traditional combining-ability and Gardner–Eberhart analyses of a diallel for cold resistance in winter wheat. Crop Sci. 23:314–318.

Pawlisch, P. E. and A. H. Van Dijk. 1965. Forage and grain production of four F_1 barley hybrids and their parents. Crop Sci. 5:135–136.

Rai, R. K., N. C. Stoskopf, and E. Reinbergs. 1970. Studies with hybrid wheat in Ontario. Can. J. Plant Sci. 50:485–491.

Ramage, R. T. 1965. Balanced tertiary trisomics for use in hybrid seed production. Crop Sci. 5:177–178.

Reid, D. A. 1965. Inheritance of growth habit in barley (*Hordeum vulgare L.* Emend. Lam). Crop Sci. 5:141–145.

Rosenquist, C. E. 1927. An improved method of producing F_1 hybrid seeds of wheat and barley. J. Am. Soc. Agron. 19:968–971.

Rosenquist, C. E. 1931. Hybrid vigor in wheat. J. Am. Soc. Agron. 23:81–105.

Sakai, K. and H. Utiyamada. 1957. Studies on competition in plants. VIII. Chromosome number, hybridity and competitive ability in *Oryza sativa* L. J. Genet. 55:235–240.

Severson, D. A. and D. C. Rasmusson. 1968. Performance of barley hybrids at 4 seeding rates. Crop Sci. 8:339–341.

Sneep, J. and A. J. T. Hendriksen (Eds.). 1979. Plant breeding perspectives. Centennial publication of Koninklijk Kweekbedrijf en Zaadhandel. D. J. Van der Have. 1879–1979. Centre for Agricultural Publications and Documentation, Wageningen.

Suneson, C. A. and O. C. Riddle. 1944. Hybrid vigor in barley. J. Am. Soc. Agron. 36:57–61.

Upadhyaya, B. R. and D. C. Rasmusson. 1967. Heterosis and combining ability in barley. Crop Sci. 7:644–647.

Yap, T. C. and B. L. Harvey. 1971. Heterosis and combining ability of barley hybrids in densely and widely seeded conditions. Can. J. Plant Sci. 51:115–122.

Zevan, A. C. 1972. Plant density effect on expression of heterosis for yield and its components in wheat and F_1 versus F_3 yields. Euphytica 21:468–488.

21

THE SINGLE SEED DESCENT METHOD

In 1939 C. H. Goulden (1939) proposed a rapid and broad germ plasm approach to homozygosis in segregating populations that has since become generally known as the *single seed descent method* (SSD). He proposed reaching homozygosity in as many lines as possible, reducing progenies of single plants to one or two, and utilizing two winter greenhouse crops to supplement the summer field planting.

ORIGIN AND MODIFICATIONS

Some time elapsed before Goulden's suggestions were seized upon by others; however, it is clear that the procedures were used locally as evidenced by growth chamber experiments by Gfeller and Goulden (1954), and particularly by Gfeller and Svejda (1960), where in a cross between 'Cascade' and 'Renown' spring wheats three seeds were taken from each of 400 F_2 plants. At harvest time one plant of the three from each line was harvested. This procedure was repeated until the F_6 when the seeds from the three plants in each line were bulked to produce the F_7 lines.

In 1961 Kaufmann (1961a) proposed his random method of oat breeding, and followed this with a second paper in the same year (Kaufmann, 1961b). Referring to the Gfeller and Svejda procedure, he proposed a concise seven-step method which he termed the "random method" of oat breeding. Although this term has given way to the name *single seed descent*, his short descriptive note embodies the first adaptation of Goulden's proposal by a plant breeder. Kaufmann's proposal grew out of his experience in oat breeding where the negative association between early maturity and lower yield mitigated against selecting for higher-yielding lines. He was searching for a method whereby large numbers of lines could be advanced to yield trials without preliminary selection. His seven steps follow:

1. Make the cross and grow the F_1.
2. Grow the F_2 under favorable conditions and retain the seeds of each plant.
3. "Follow procedure. . .random sampling. . .segregating generations (to F_6) described by Gfeller and Svejda. Essentially this is propagating one plant derived from each F_2 plant in each generation. A large increase from each F_6 plant is desirable."
4. Grow the plant rows in F_7, with alternating parents or standard varieties every fifth row. Discard lines that have obvious serious defects. Harvest remaining lines.
5. Sow the F_8 in multiple-rowed plots, with alternating parents or standards every third plot. Record agronomic and yield data. Retain lines outstanding in at least one feature.
6. Advance selected lines to replicated yield trials.
7. "Go back and propagate the F_3 of lines that show excellent promise in all but one feature and attempt to either select that feature in the F_3 or F_4 or carry through each generation by random sampling."

Kaufmann emphasized the random, unbiased nature of generation advance; in fact, selection was virtually absent. He indicated that this method would work in the field, greenhouse, or growth chamber. Disadvantages were that it was difficult to work with many crosses, and inferior lines were carried along with promising lines to homozygosis.

Grafius (1965) proposed cycling plant populations on a larger scale by utilizing greenhouse benches and controlling plant size, maturity, and seed set through changes in density and the withholding of nutrients. He found that 8000 cereal plants could be grown in a bench 1 × 8 m in size.

Brim (1966) also credited Goulden for his modified pedigree method of selection in soybeans. Brim attempted to bring the technique out of the plant growth chamber and adapt it to field conditions. He stressed rapid and efficient cycling, giving the example of going from a field F_2 population one season, utilizing two greenhouse generations, and returning to the field the next season with F_5 seed. As is the practice of most breeders, Brim used visual selection throughout the progeny sequences.

Compton (1968) extended the single seed descent system by proposing to base it on F_1s from different crosses of self-pollinated crops. Thus the breeder would be advancing lines (SSD) among crosses rather than within a cross. The top group of lines after yield testing would be recombined through random crossing. The scheme is based on hand pollination but obviously could be adapted to male sterile use. The program might run through two or three cycles and would have the twin goals of selecting superior lines and of creating a recurrent selection population with high genetic variability. Compton's paper is important for its contributions to the understanding of methodology.

Allen and Kaufmann (1981) traced the history of SSD in oat breeding at Lacombe, Alberta, where four varieties, 'Random', 'Cavell', 'Athabasca', and 'Cascade', had been licensed. An interesting development was the use of SSD to evaluate crosses for further exploitation. This assessment began with the seed increase of lines in F_7 and usually was completed in one season. The details of their prediction method were given as follows:

> Fifty lines from each of 5–8 crosses are seeded in blocks of 5 m rows 45 cm apart with one line from each cross in a block along with one or two controls. The average yield, maturity, lodging and growth rate are obtained for each cross and control and the CV calculated. From this information the crosses that we wish to work with are chosen.

The 50 lines apparently represented only a sample of the total lines in each cross, and the remaining lines were brought back into test only if a cross were chosen for further exploitation. Allen and Kaufmann stressed the point that the SSD system produces random, that is, unbiased lines that are suitable for genetic, correlation, or prediction studies. Their method has interesting possibilities for selecting among crosses.

Cisar, Howey, and Brown (1982) adapted procedures from the laboratory to field conditions in a modified SSD arrangement for oats. Their proposal was built upon the stress environment concepts of Grafius (1965) and Tee and Qualset (1975), wherein the limitation on seed production per plant was imposed by forces of the environment as opposed to the rigid harvest of one seed per plant. In consequence, the system moves subtly from a pedigree to a bulk approach.

The authors were concerned with finding an optimal density that would produce one or two seeds per plant, and determining whether losses were random or selectively imposed by the environment (one indicator of stress observed was barren plants). The principal criterion for the study was the number of seeds produced per plant. Tests were done using both cultivars and hybrid segregating material.

The results showed that a density of 120–150 seeds per 6 in. diameter pot produced an optimal number of single-seeded plants; the loss of the population at this density was 20–30%. The authors' analysis of the selective or nonselective elimination of genotypes led them to conclude that elimination was random, not differential. This is not to say that valuable genotypes might not have been lost; there was no evidence as to that, but the losses were similar in populations of greatly contrasting densities. The combination of answers from the two questions asked signified that the SSD stressed-environment approach was a feasible one. The next step would be to take the procedure to the field with a blending of SSD and bulk hybrid population procedures.

It would be interesting to find out whether these stress–culture procedures could be adapted to separate F_2 genotypes into two classes for later evaluation. It was found that 120–150 seeds per plot produced an optimal number of single-seeded plants. In the process a mortality of 20–30% was observed—this might be considered a one-directional penalty. The other directional extreme would be represented by the surviving plants that had more than one seed per plant, and it is possible that there were some plants with several seeds. The question is, taken as a group do these multiple-seeded plants represent a genetically different group from the single-seeded plant group? If these two groups of seeds were separately bulked and grown the following year, and bearing in mind that all were produced under extreme competitive conditions, what would be the differences in mean yields between them?

COMPARISONS OF SSD WITH OTHER METHODS

Empig and Fehr (1971) compared four methods of generation advance in three bulk hybrid soybean populations. In essence, SSD was compared with three cross-bulk formulations derived from different samplings of selected 200-plant maturity groups (E, M, L) to form 600-plant populations within each of the three crosses. The three cross-bulks were as follows:

1. *Restricted Cross-bulk (RCB).* Two hundred plants for each maturity group harvested in a bundle. A 10-cm sample from the center of each bundle was cut and threshed. The three samples from each cross were bulked and 600 seeds taken at random to plant the F_3.
2. *Maturity-Group Bulk (MGB).* Two hundred plants in each maturity group in each cross were threshed in bulk, following which 200 random seeds from each maturity group were planted in separate groups.
3. *Cross Bulk (CB).* Two hundred seeds from each maturity group in a cross were composited and the 600-seed bulk was grown in F_3.

Procedures established in the F_2 population were used to advance the populations two generations, F_3 and F_4. The F_5 populations were space-planted and 45 plants were selected at random from each equivalent-sized method group. Field tests at two locations were made on replicated bordered hill plots of the 45 lines, two parents of each cross, and all methods, plus three check varieties.

The study showed that for yield there was no significant difference among the four methods at two locations; however, SSD and MGB (selection maintained in maturity groups) had the best record of large numbers of superior lines and small numbers of inferior lines compared with cross means. For maturity, it was evident that natural selection favored late maturity. SSD and MGB had higher proportions of early maturing lines. SSD had the highest means and estimates of genetic variance for seed size. It was twice as effective in maintaining large-seeded genotypes. Height and lodging were highly correlated; SSD showed little difference from the other methods. Although natural selection caused shifts in maturity, seed size, height, and lodging, SSD showed good stability; that is, it maintained its proportion of early, large-seeded, and short plants. MGB was similar to SSD in forced maintenance of the original proportions of the populations. Although in efficiency, SSD was the third, possibly fourth, most time-consuming, it adapted best to environmental changes; that is, it was resistant to shift. Empig and Fehr concluded: "Therefore, maximum genetic gain per unit time, attained by growing multiple generations per year in artificial environments, can best be achieved without loss in efficiency with the use of SSD."

Tee and Qualset (1975) compared SSD, bulked each generation with pedigree not maintained, to the standard random bulk population method in two spring wheat populations. Hybrids I and II involved 1195 and 1196 seed-producing F_2 plants, which were the basis for the SSD and BP populations. Comparable lines from the various methods were brought together for testing in common nurseries at two locations. Information was developed on response to selection for different traits, but the general conclusions concerning the two methods were that they were remarkably similar in all points relating to an efficient germ plasm development scheme. Because of savings in labor, the BP method showed a slight advantage in situations where competition effects were important.

The authors commented on a practical consequence of using the SSD: unavoidable attrition over time will steadily decrease population size. This would seem to be more of a hazard in field than in greenhouse plantings.

Knott and Kumar (1975) simulated breeding program conditions in a series of tests comparing early generation yield testing (YT) with SSD procedures in two spring wheat crosses. A 2500-plant F_2 population was stress-screened, using the rusts and agronomic traits, to a selected group of 300 F_2 plants from each cross. The two procedures were conducted during the same time period, though not under similar conditions (e.g., greenhouses were used for SSD and a winter increase in Mexico for the YT). The procedures generated 232 SSD and 180 YT lines for the first cross, and 190 and 222, respectively, for the second. The SSD lines were one generation older than the YT lines. The researchers applied slightly greater selection pressure to the YT lines—that is, more were eliminated—than to the SSD ones.

The authors found that the YT lines significantly outyielded the SSD lines in both crosses. The correlations between F_3 and F_5 YT line yields were

.29** and .14**, deemed not high enough for practical adoption. When the yield of the F_5 YT lines, selected on the basis of the prediction data from the top 20% of the F_3 lines, was compared with the top 20% of the F_6 SSD lines, the results favored the SSD procedure. Overall, the SSD procedure showed to advantage.

Knott and Kumar have proposed a practical program to modify and utilize the SSD procedure in wheat breeding programs. A number of options for use at different phases of the selection process were specified. It would have been instructive to have the genetic identity of each plant line carried through. Even though the SSD selects different subsamples of germ plasm, a knowledge of identity would help to answer the important question as to whether different selection procedures, operating on a common germ plasm pool, act to select identical or similarly performing lines. With genetic pedigree information, it would be possible to determine whether superior lines from both methods might have traced to the same F_2 plant.

Casali and Tigchelaar (1975), working with tomatoes, compared genetic advance within pedigree selection, SSD, and combination procedures. In pedigree selection, they found cross effects to be greater than generation differences, "indicating selection of parents to be of greater importance than selection within segregating populations."

They found that the combination of

> ...pedigree selection through F_3 followed by winter generation advance to F_6 appears to optimize advantages of both breeding methods. Initiating SSD following selection in F_2 did not take full advantage of potential gains from early generation improvement from pedigree selection, whereas initiating SSD at F_5 added one additional year before line selection began. In a crop such as tomato, where space imposed severe limitations on the number of lines which can be maintained, a combination of stringent pedigree selection in F_2 and F_3 for characteristics of high heritability and advance of F_4 pedigree lines to F_6 via SSD, minimized the number of undesirable lines retained and still maintained a reasonable genetic base for selection among F_6 SSD lines for characteristics of lower heritability.

The authors pointed out that these combination procedures resulted in considerable saving of time and in the ability to develop a larger number of advanced lines.

Boerma and Cooper (1975a) compared three selection procedures in four soybean crosses. The procedures were modified early-generation testing (EGT), pedigree selection (PS), and SSD. All procedures were applied to the same F_2 populations derived from the four crosses. There were 943, 1001, 431, and 243 plants in the four crosses. The testing procedures were applied under field conditions to produce an equal number of "best" lines from each procedure within each cross for F_8 yield testing. The final line groups were grown in triple lattice experiments at two locations in Illinois.

The general results from all four crosses failed to indicate any consistent superiority among the three testing procedures; therefore attention was directed to other aspects of the methods. The PS procedure retained more yield variability than the other two. This is a finding at variance with other research, which shows SSD to be superior in retaining germ plasm breadth within a population. The EGT procedure produced the latest maturing lines. When efficiency or the amount of effort required was considered, the EGT and PS procedures were at a distinct disadvantage owing to large yield testing and pedigree evaluation requirements. The authors concluded that the SSD procedure was the most efficient because it used the least selection effort, bypassed yield testing until late generations, and allowed a rapid generation advance.

In a second paper the same year, Boerma and Cooper (1975b) compared the efficiency of developing heterogeneous lines (HL) from F_2 plants of four soybean crosses (without further within-line selection) by early-generation testing, to that of creating pure lines (PL) developed by PS and SSD procedures. The interesting comparison here is between the PS–PLs and the SSD–PLs, a point of secondary importance to the authors: the results showed PS–PLs to have a slight advantage over SSD–PLs. In two of three crosses (no data from the fourth) the mean of two or three lines favored PS–PL, and in three of four crosses the highest-yielding line was a PS–PL. The eight highest-yielding PLs from the four crosses at Urbana and Dekalb, Illinois were divided five PS–PL to three SSD–PL. Overall, the eight-line mean yield for crosses I–III was 3017.5 kg/ha PS–PL to 2982.5 kg/ha SSD–PL. The mean of the highest lines was 3160.0 PS–PL to 3070.0 SSD–PL. These differences, amounting to a percent or two, are slight and do not suggest an advantage of the pedigree over the SSD procedure. In fact, SSD as the untried "challenger" appears to have done surprisingly well.

Also in a second paper in 1975, Casali and Tigchelaar (1975) employed a complex computer simulation program to compare five selection methods, "pedigree, bulk, SSD, and combinations of early generation pedigree and bulk followed by SSD," for breeding progress measured in terms of best F_6 lines. Four heritability levels, 75%, 50%, 25%, and 10%, were assumed. The results showed that "single seed descent was inferior to pedigree or mass selection at all heritabilities when equal numbers of F_6 lines were compared for each method."

However, significantly greater variability to F_6 was maintained. The authors believed that for characters above 50% heritability, larger SSD numbers cannot compensate for pedigree and mass selection progress; below this, however, compensation through use of greater numbers is effective.

The pedigree or mass selection followed by SSD options appeared attractive. Although pedigree to F_4 followed by SSD was not better than pedigree selection alone, there was a saving of two years. Mass selection to F_4 followed by SSD did not offer similar advantages because this was inferior to mass selection alone except at the 10% heritability level.

Snape (1976) compared the theoretical efficiency of the methods of diploidized haploids and of SSD in producing inbreds in barley to best exploit the recombinant possibilities of a cross between two inbred lines. He found that in the absence of linkage the genetic expectations were similar. He concluded:

> In a plant-breeding situation the breeder is generally interested in maximizing the variation from a cross, which involves the recombination of genes dispersed between the parents. In this case SSD would appear to be a more efficient method than haploid production due to the opportunity for recombination over more than one generation.

Park, Walsh, Reinbergs, Song, and Kasha (1976) compared the field performance of doubled haploid (DH) barley lines with those developed using pedigree (PD) and SSD procedures. They found that all three methods produced lines that were equally good agronomically. The PD technique was useful in that it eliminated poor yielding lines ("...the lower limit of the yield range for PD lines was slightly higher than for DH or SSD lines"), but it did not produce significantly higher-yielding lines than the DH or SSD methods. Their conclusion:

> There was no difference in grain yield, heading date and plant height between the DH population and the populations derived by the other two breeding methods. Similar means and ranges, genetic variances and frequencies of desirable genotypes were obtained in the populations produced by the three breeding methods for grain yield, heading date and plant height. The mean grain yields of superior lines were similar for all three methods.

Jinks, Jayasekara, and Boughey (1977) in a previous study had used family selection to isolate the four combinations of high–low mean performance and high–low environmental sensitivity. When in this 1977 study they compared random SSD with the conventionally selected (F_7) they found SSD to be clearly superior. Not only did the SSD lines better meet the selection criteria, but also they were more evenly distributed. The authors stressed the relative ease and economy of the SSD procedures.

Peirce (1977) compared SSD with pedigree and with SSD following one cycle of pedigree selection. He found that "SSD produced generally inferior fruit size, earliness and total yield among progenies" in three tomato populations. SSD did produce some superior lines. The author concluded that despite a poor comparative showing, SSD provided major advantages to the breeder in rapid advance of generations with a relatively small space requirement.

Riggs and Snape (1977) used computer simulation to test the effects of linkage in inbreds initiated and developed by the DH and SSD methods. They found, as had been shown before, that in the absence of linkage there was no difference in the means and variance of the quantitative trait, but when linkage was present "recombination was more frequent in the SSD populations." From theoretical considerations the SSD was superior, although the authors recognized that other factors would help dictate the choice of method.

The objective in Ivers and Fehr's (1978) research was to evaluate the pure line family (PLF) method of selection in comparison with three other methods, namely, BF (bulk family), PS (pedigree selection), and SSD. They found that even though SSD sampled twice as many F_2 families as PLF (80 versus 40), PLF selected a greater number of superior F_5 lines. The authors implied a preference for the SSD procedure, conveyed through their listing of four disadvantages of the PLF method. Basically, the comparison rests on the known relative simplicity of the SSD method in contrast to the rather laborious procedures inherent in the PLF. The PLF requires the harvesting and threshing of all F_2 plants and bulk maintenance for several generations until acceptable homozygosis is reached. Following that, one or more individual plant lines are selected and tested and the results used to predict the value of the F_2 family. Presumably, this information would be used for reselection within the superior family lines.

Wright and Thomas (1978) compared the SSD method with their normal field pedigree system for wheat in which selection is practiced in the F_2 and F_3 and lines are established by F_4. They found the results to be similar even though the SSD population was smaller. However, they noted two objections to the SSD method; namely, pedigree selections were markedly more resistant to leaf rust and mildew, and the pedigree operations were less demanding of labor. They concluded that "single seed descent should not become a major breeding technique."

Knott (1979), in a sequel to the 1975 Knott and Kumar paper in which it was concluded that F_3 yield testing was of doubtful value because SSD-derived lines appeared to be as high yielding as YT lines, prepared new SSD lines from the two spring wheat crosses and placed them in a yield trial with the best lines from the earlier F_3 yield tests. A preliminary yield test had been held to select the highest-yielding F_4 SSD lines; thus the best lines from both methods appeared in the common yield trial. The results showed the YT lines to have an average 1.5–3.8% yield advantage over the SSD lines. The YT lines were also later in maturity, an undesirable feature. Because of the small differences in yield and the lower labor requirement of the SSD, Knott favored the SSD procedure over the YT method.

Doubled haploid lines from an F_1 hybrid are considered to have status equal to random homozygous lines developed from the same hybrid population. Research by Schnell, Wernsman, and Burk (1980) raised doubts that this was always so. Anther-derived tobacco dihaploids from an F_1 were compared with an equal number (50) of random F_8 lines developed by the SSD method. Final direct comparisons were made on plot plants for seven traits. Economic yield in tobacco is based on leaf quantity and quality.

The overall results decidedly were in favor of SSD over DH lines; in almost every category SSD was superior. Despite a wider range of variability for leaf yields the mean of DH was lower than that of SSD. SSD also did relatively better in relation to the two parents. Specifically, "The DH population yielded 10.6% less cured leaf, possessed inferior leaf quality, was later flowering, produced reduced leaf numbers, and its cured leaf was lower in total alkaloids than the SSD population."

The yield depression observed in the anther-derived DH populations leaves little choice but to accept the conclusion that selection by SSD provides greater genetic advance than by DH procedures.

AMONG-LINE AND WITHIN-LINE VARIANCES

Some information on the rapidity with which yield variance narrows in F_2-derived lines was provided by Weiss, Weber, and Kalton (1947). They studied 51 pairs of sister F_3 lines and 51 pairs of lines from sister F_4 selections in soybeans. In the F_3 line pairs, 15 differed significantly, six at the 5% and nine at the 1% level; in the F_4 pairs only four differed significantly, all at the 5% level of significance for yield. They concluded that retention of one line from the F_4 was sufficient. For the SSD situation, the variance of lines derived from a single F_2 still must be examined to determine the magnitude of possible loss through limiting progeny to the random selection of a single seed. Nevertheless, their findings emphasized greater variability among than within lines.

In an excellent review of plant breeding problems, Shebeski (1967) raised questions about the balance of effort applied to selection in the early generations. He believed that more emphasis should be given the F_3 generation of F_2-derived lines. Specifically, he recommended larger numbers of seeds per line, "from 50 seeds to 750," more attention to between-plot competition (spacing adjustments), and especially, placing control plots next to every line. By following these concepts, Shebeski was able to show near-perfect correlation between F_3 and F_5 line yields.

I have brought Shebeski's ideas into the SSD discussion because the two proposals illustrate two extremes of line variance. On the one hand, the SSD promotes between-line variance, because except for losses through attrition all lines are perpetuated from generation to generation, and rapidly reduces within-line variance. On the other hand, Shebeski's proposal expands within-line variance through increased family seed numbers without necessarily curbing between-line variances. Both views are subject to modification, for example, by two seeds or a head per line in the SSD (see below), and by selection and early discard of whole lines in Shebeski's method.

Sneep (1977) examined the problem of selecting for yield in early generations. On the assumption that selection for yield in the F_2 is unreliable, he concluded that the SSD method was not a "proper way to select populations originating from parents which differ for many gene pairs." His principal objection was to the genetic channeling that automatically follows the taking of only one seed from each F_2 plant and eventually leads to an F_4 population the same size as the original F_2. He believed that the strict population control forced by the SSD method would not give a sufficiently large F_4 "smallest appropriate population" to include the potential best yielder.

Muehlbauer, Burnell, Bogyo, and Bogyo (1981) conducted a computer simulation of what might happen in situations of major concerns to breeders. They described these as follows: "Genetic variability and opportunities for selection in advanced generations can be reduced in SSD by plant loss during generation advance while these factors can be reduced in BP by differential fecundity rates and inadequate sampling."

They found that in bulk population methods survival was highly correlated with fecundity, with low seed number individuals being eliminated through competition. There is not, however, a logical correlation between the degree of fecundity of the individual and the success of a cultivar in its own autocompetitive population. Furthermore, the restrictive sampling procedure used to advance BP generations also acts to limit the pass-through of genotypes and is a factor involving a rapid narrowing of germ plasm. By comparison, in the SSD a significant danger of a narrowing germ plasm base (genetic variance) could be found in the loss of plants from generation to generation, as through occasional inviability of the single seed used to propagate each individual. The SSD variance was predominantly between-line, whereas the BP variance was predominantly within-line. No clear choice was evident between the two methods but the effect of the study was to put a qualified stamp of approval upon the newer SSD method.

EFFECT OF SSD ON PLANT CHARACTERS

Roy (1973) dealt with the problem raised with the SSD where random selection of wheat seeds results in populations having large and small seeds. His aim was to study the effect within genotypes of interplant competition on yield components when density is very high and tillering virtually absent. His study had eight treatments: two varieties × two seed grades (large and small) × two seeding systems (mixed and pure), with three replicates in randomized blocks. The large seeds were 45–47 mg and the small ones 32–35 mg. The varieties were sown alone for the pure stand treatment. Plantings included 10 seeds per row, 1.7 cm apart, in 10 rows per box, 3 cm apart. All borders were excluded at harvest. The SSD method was used for generation advance. Measured were plant height, ear characters, and yield, all from 10 plants taken at random from each replication of each treatment.

The conclusions from the study were that under SSD procedures with the environmental conditions of close spacing and favorable moisture and nutrient factors, compensation of some kind within the plants produced seeds of uniform and similar weights throughout the population regardless of the size or weight of the sown seed. Whether the seeds sown averaged 47 or 32 mg, the seed weights harvested were in the range of 37–33 mg. The importance of this is that there would be no carryover to the next generation of seed size differences.

Snape and Riggs (1975) examined by computer simulation the genetic consequences of the SSD method on quantitative characters. Genotype values were assumed for different genetic situations. An important finding was that "in the absence of dominance, the distribution of the means of F_6 lines is similar to the genotypic distribution of the F_2 individuals," but if heterosis is shown, the F_6 lines fell below the F_2 expectations and "a greater genetic advance would be expected by using the normal pedigree method."

The authors chose the SSD over the pedigree method because pedigree selection procedures are tedious and costly, and selection in the F_2 is also possible with SSD procedures.

Grace, Jenkins, and Roffey (1976) attempted to adapt SSD procedures to the growing of winter barleys in a southern hemisphere nursery during a northern winter period. These barleys had a range of vernalization requirements. They found genotype × treatment (density, vernalization) interactions to be complex, and concluded that heterozygous populations (for maturity, vernalization requirements, height, or head type) might be expected to sustain serious genetic shifts over several generations under SSD. They suggested an optimal density and vernalization guide for winter barley populations.

Inasmuch as the situations described would require similar adjustments to any breeding and selection technique chosen, it would seem that the SSD would be neither favored nor discriminated against. Under the same conditions of competition, similar adjustments in density and vernalization treatments would be required for pedigree or composite procedures.

Roy (1976) studied intergenotypic plant competition in wheat under SSD breeding. He tested four genotypes grown in pure stand and diallelic combinations of two in alternate rows under density pressure. Certain genotypes in mixtures showed a marked reduction in survival (8–14%) compared to survival in pure stand. Concerning SSD, Roy concluded that the "main aim should be to avoid competition severe enough to reduce plant survival. Loss of non-competitive plants could mean elimination from the population of genotypes some of which are likely to be high yielding."

The SSD method was used successfully by Gale and Gregory (1977) in an efficient extraction of desired dwarf genotypes in an F_2 hybrid population of a tetraploid wheat cross. Presumably the method could be applied to hexaploid wheats. The procedures involved isolating plants homozygous for the dwarfing genes, *Rht. Rht*, by their gibberellic acid (GA) insensitivity; all others were removed. The SSD method facilitated removal, and required only one seed per plant for continuing the population into the F_3.

Martin, Wilcox, and Laviolette (1978) wished to test whether the relatively high mortality of soybean plants experienced with the SSD method affected gene frequencies of different qualitative and quantitative traits. By increasing planting density they artificially encouraged higher seedling losses. In the low density (three plants per pot) populations plant losses totaled 26%, 18%, and 15%, respectively, of the original populations of the three crosses over the four generations, F_3–F_6. The corresponding losses for the high density (nine plants per pot) populations were 62%, 58%, and 47%. It was found that two qualitative traits, flower color and pubescence color, agreed with expected values at low density in all three crosses; however, the deviation from expected was significant in each cross at high density.

Later evaluation for other traits showed that progenies from the high-density populations were later in maturity, showed greater lodging, and were taller than plants advanced from the low-density populations. The authors concluded that high density, which increased losses, significantly altered gene frequencies for qualitative characters and at least certain quantitative characters.

Although Bravo, Fehr, and Cianzio (1980) had indicated the possibility that width of mature pod in soybeans was an effective method for indirect selection of seed weight, a number of questions remained to be answered before the method could be considered reliable for field use. Frank and Fehr (1981) undertook to answer three questions: what is the effect of pod seed numbers on pod dimension parameters, what is the relation of measurements on green and mature pods, and which of the three pod dimensions, length, width, or thickness, would provide the most effective measurement? The study was made on F_4 plants of the same six hybrid populations of two-way and three-way crosses used in the Bravo, Fehr, and Cianzio study. Results were as follows:

1. It was found that pod length was significantly influenced by number of seeds per pod. The key to reliable use of pod length was to use pods with the same number of seeds. On the other hand, pod width and thickness were little affected by seeds per pod.
2. Either green or mature pods were effective for both pod length or width; however, both should not be used in the same study because of the confounding effect of pod shrinkage with maturity.
3. Pod width measurement was favored over length, area, or volume, because it did not require screening for number of seeds per pod. It was also a trait that could be measured more quickly than pod length.

The advantage gained by indirect selection for yield by using green pod width, a preanthesis trait, is that hybridization of selected plants can take place within the same generation. For use of pods in SSD field procedures, however, it still would seem necessary to screen and make adjustments for number of seeds per pod.

WAYS IN WHICH SSD HAS BEEN USED

Baker, Bendelow, and Kaufmann (1968) used the SSD technique to advance lines from F_2 to F_6 in a study of inheritance in common wheat.

Bliss and Gates (1968) used computer simulation to compare mass selection with SSD as a means of generating populations from a hybrid. They found mass selection to be superior to SSD in selecting the best individuals from a population; however, with small numbers and tight linkage there were no significant differences in genetic gains achieved by the two methods. They emphasized the substantial saving in time through use of the SSD; in fact, they suggested,

> It should be possible in some crops to proceed through two or three recurrent cycles (i.e., from the F_2 to the F_5–F_6 for each) of crossing the best individuals followed by selection among pure lines in the same length of time required for one complete cycle of mass selection.

Pesek and Baker (1969) found the SSD technique useful in concept to generate 600 F_6 lines from the F_2 of a cross between two inbred lines (computer simulation) to study tandem and index selection techniques.

Kaufmann (1971) reported on the random method of oat breeding for productivity (essentially the random method is the SSD in the same way that Brim's modified pedigree method is SSD). Kaufmann moved to the SSD because he believed a yield plateau had been reached in oats. The pedigree method had produced no selections better than the parents in several years of use. The SSD method was found to be effective in advancing oat segregants through the F_2–F_5 generations. Transgressive segregants for yield were produced with as much as an 8% yield advantage over the parents.

Kaufmann realized that the absence of selection during the generation advance to homozygosity called for some form of ongoing screening system. He tried taking the dry weight of five panicles per plot and found that the correlation with yield was consistent and positively significant. However, it was so low that he thought it was not useful as a selection tool.

Baker (1971) presented guidelines to help the breeder estimate the validity of F_6 selections out of SSD programs. He wondered how representative they were of the base population. If heritability of a certain trait were known, Baker's table of coefficients of variation of response to modified pedigree selection allows the breeder to estimate the number of lines that need to be tested in order to select the best performing one.

Baker's study might be termed a "procedural validity" test in that it is a theoretical consideration of the variance of response to a particular selection procedure, in this case Brim's modified pedigree method. The study envisaged a population proceeding through a plant breeding program as a sample of an unknown base population with its possible variances and biases. Two types of sampling, random and conditional, were considered. The SSD, through generations, continues an original random sample; however, the sampling is not random once choosing and testing begin. Baker pointed out that if the selection program (same cross, same base population) were repeated, the observed responses would show their own variances. Equations covering these situations were presented. The theoretical considerations led to useful estimates of numbers of lines necessary to ensure the appearance of a desired genotype.

The theoretical nature of Baker's 1971 study in some ways raises it above the practical level suggested by its attachment to the modified pedigree method. With regard to repeatability, it ignores the very subjective nature of nonrandom selection, especially as it pertains to the breeder. Indeed, it would be difficult to factor breeder differences into an equation. Variance in response is to be expected, but repeatability is not something that occurs often in conventional cereal breeding. It is discouraged by the availability and pressure of large numbers of crosses "waiting in the wings," as it were. This is a feature of aggressive programs that countenance a "one-passage-per-cross' procedure.

There are, of course, sound exceptions justifying the repeating of another sample of a base population. Examples are:

1. Repeating a cross that showed outstanding performance characteristics
2. Repeating selected crosses used in a small-scale screening or prediction test
3. Repeating the same cross at different locations in the same year or in subsequent years to sample more environments
4. Repeating a bulk mixture of many crosses using combinations of years or locations.

The number of lines out of a population necessary to include various desirable gene combinations in one individual, if it could be known, is a point often subordinated to the exigencies of field decisions. It has been my experience in working with advanced generation material that even in the year of line selection, that is, when selection is being made out of the population for heads or plants, there might be a range in numbers of plants selected per cross ranging from zero to a thousand among several hybrid populations. Even after years of growing and observing a hybrid population, it is still possible that when the plants are seen as individuals, the breeder may decide to discard the population without any selection. At the other extreme, we might cite the oat hybrid at Cornell that produced both 'Orbit' and 'Astro'. This cross had such a promising appearance during early generations that several thousand heads were selected as lines. The head row nursery of lines from this one outstanding cross occupied a separate small field. Plant breeders do recognize that populations of any cross they might work with are only finite samples of an infinite base population, but their concern with the amount and kind of genetic variance within any one population is relatively slight compared with their concern for differences among the many finite samples that make up the whole breeding program. Expressed another way, breeders first look for the between-cross differences, discard the poor crosses as early as possible, and then turn their attention to the within-cross differences of the retained crosses.

Jinks and Pooni (1976) employed the SSD method to generate homozygous lines from which were obtained the necessary genetic components, and the family means and variances that allowed prediction of inbred line distribution in the general population. The availability of these data would provide the breeder with a picture of the merits of a cross, and would indicate whether it would be profitable to exploit the hybrid population beyond the F_2. The tests, although useful, are time-consuming. The role of the SSD is to shorten time requirements and reduce labor demands.

Riggs and Hayter (1976) discussed practical aspects of the SSD method for barley breeding.

Hurd (1977) reported that a survey of Canadian wheat breeders indicated that one-third used the SSD method in their breeding programs, a tribute to the high regard given the method.

Pooni, Jinks, and Cornish (1977) used SSD techniques to establish 80 F_{11} lines of a cross between varieties of *Nicotiana rustica* in order to test the method's predictive capabilities for estimating the distribution of pure-breeding lines under two conditions of nonnormality, namely, epistasis and genotype × environment interactions. It was found that predictions of line properties that assume normality are satisfactory if a correction to the expected means of the distribution is made equal to the epistatic component. The following year, Pooni and Jinks (1978) used the same 80 lines derived by SSD to predict the properties of lines selected for two or more characters simultaneously.

Needing to choose between SSD and bulk population (BP) methods for a lentil breeding program, Haddid and Muehlbauer (1981) compared the two methods in a greenhouse study using three hybrid populations for three generations, F_2–F_4. On the basis of the test results they concluded that the SSD was superior to the BP method for lentil breeding. The perceived SSD advantages were as follows:

1. Space requirements in the greenhouse were only about 13% of that required by BP.
2. Labor requirements were less because the seed pod to be advanced could be chosen earlier.
3. Genetic variation and the number of transgressive segregants were higher.

The authors noted a loss of plants, amounting to 29% in one population, attributable to seedling mortality. Losses can also occur in BP populations owing to seedling mortality, competition, and sampling omissions. To counter this loss they proposed advancing two seeds from each F_2 plant followed by single seeds in subsequent generations. Although this doubles the starting population, the advantages for within-line variability and protection against whole line losses are obviously great. They also considered the merits of adding a certain amount of selection for identifiable characters during the generation-advance period.

The authors stressed the point that the method chosen must act to retain the "most desired plant genotypes" in the segregating population. Crop or plant morphology thus can play an important part in selection of a breeding system. In this case, SSD was not superior in advancing all desired traits, but the SSD populations did show more erect types and high lowest-pods, characters that are needed for mechanical harvest.

These results were conducted under greenhouse conditions, arguably more favorable to SSD than BP because a practical SSD program might reasonably operate from a greenhouse, whereas BP is more appropriate to field operations because of space and equipment considerations.

Valentine (1984) proposed an "accelerated pedigree system" or APS as a modification of the pedigree system. I call attention to it here because he also compared it throughout with the SSD method, citing three advantages, namely, lower resource requirements, avoidance of high mortality, and early evaluation of lines. Also, without selection APS maintains higher levels of between-line and within-line variability.

A direct test of the SSD method against pedigree and bulk procedures applied to early-generation yield testing was made by the team of Ntare, Aken'ova, Redden, and Singh (1984). All procedures began from the same F_2 plants selected from two cowpea hybrids. They succinctly stated the requirements for success in early generation selection as "the ability to distinguish differences between genotypes in early generation and the persistence of these differences in later generations."

Basic details of the methods were these: 500 F_2 plants from each cross were grown; 100 from each cross were selected visually and harvested, and seeds of each were divided to produce the comparison streams. SSD procedures for F_3 and F_4 were orthodox, except that five seeds per pot instead of one were planted in F_4, and one pod per pot in F_5, and the numbers thinned to one plant. The F_5 was field sown as line progenies for yield testing, and 40 "agronomically superior" lines chosen visually for the F_6 yield test of the methods.

The combined pedigree and bulk procedures involved an F_3 yield test of the 100 F_2-derived lines. Data for yield were taken; however, the authors also scored a field visual rating of yield. From the yield data two types of bulks were formed; the first type was 10 bulks made from each of the 10 highest-yielding lines. The second type was 10 bulks made from each of the 10 lowest-yielding lines. A portion of each bulk was held in reserve, and another portion was repeatedly grown to produce seed for the F_6 yield trials of the methods. There was the additional variation of selecting two plants from each F_4 bulk to furnish the needed total of 40 F_2-derived lines for the F_6 trials. F_6 plots were visually scored to estimate yields. The final trials revealed these points:

1. The SSD method produced lines with grain yields equivalent to those from the combined pedigree–bulk early-generation yield testing. The numbers of superior lines was about the same in each method.
2. F_3 yields were good predictors of later line performances (high significant correlations, F_3 and later Fs).
3. Visual discrimination was proved successful, as shown by significant positive correlations between F_3 and F_6 yields.
4. High and low selections from the F_3 yield test maintained their separate ranges throughout, indicating the effectiveness of the F_3 data criterion.

The high input of visual judgment deserves mention. To begin with there was an 80% discard in the basic F_2 population; from 500 plants only 100 were saved. It is remarkable that such large high : low differences remained in the selected 100. Visual discrimination was used at other procedural junctures, for example, for the ratings in F_3 and F_6. Ntare et al. noted the positive correlations between visual scores and the actual F_3 and F_6 yield data. Visual scores seemed to do as well in identifying promising lines as the actual yield data. In view of the fact that visual ratings cost little to make, they could well be used as a supplement to yield data.

DISCUSSION OF SINGLE SEED DESCENT

Nearly half a century has passed since Goulden's 1939 suggestion of the outlines of what since has become known as the SSD method. The slow evolution of a breeding and selection method may be seen in the papers we have just reviewed. There were no papers in the decade following Goulden's talk, and only one in the 1950s. There were approximately 10 papers in the 1960s and 30 in the 1970s. This suggests that the ultimate form of the SSD as a method has not yet been completed. It is also clear that SSD now rates the status accorded an established breeding and selection method. It is evident also that the SSD method has interesting properties that make it a useful research tool for biometricians and theoretical geneticists.

Single seed descent developed as a response to two needs of plant breeders. Goulden referred to the first in his talk entitled "Problems in plant selection," in which he raised the question as to the desirability of simultaneously practicing selection for superior lines while moving the population toward homozygosis. Inseparable within this question is whether selection for quantitative characters is meaningful at all when applied to nonhomozygous individuals of a population. This question is so basic to an understanding of plant breeding problems that Goulden's comments, from the abstract of his talk, are worth repeating:

> In wheat breeding there are two main objectives. The first is to obtain true-breeding lines by selection from the hybrid progenies; the second is to select from these lines, those that have the desired combination of characteristics. The methods of breeding that are now employed are based on the principle that progress in obtaining the lines having the desirable characters can be made at the same time as the lines are being selected for homozygosity. . . .if it were possible to separate the two phases of breeding, the first step would be to reach homozygosity in as many lines as possible. The progenies of single plants could be cut down to a minimum of one or two plants, and two generations could be grown in the greenhouse during the winter and one in the field during the summer. The F_6 generation could be reached in two years, whereas under the system being used it takes at least five years.

The second need of plant breeders associated with the SSD method was the development of the controlled environment chamber, or growth chamber. Goulden was one of the scientists who pioneered the growth chamber and I have always believed that this developmental work, which took place parallel in time with his views on breeding and selection, contributed significantly to the birth of the SSD method. I saw an early working model of a reach-in chamber filled with barley on a visit to Ottawa in the early 1940s. Goulden and Fred Gfeller together explained to me that it was possible to grow five crops of a spring crop such as barley within one calendar year. The only way a finite-size container could accommodate successive generations of a hybrid population would be through an SSD procedure. It is probable that this parallel research on equipment and breeding techniques influenced Goulden's views on selection and led to the SSD method.

The record of the SSD when compared with older selection methods is a good one. Its strong points are rapid generation advance, maintenance of an unbiased broad germ plasm base, labor and time efficiencies, ability to handle large numbers, and amenability to modification. Its measured pace of development has moved it out of the growth chamber into the greenhouse and fields. Applications for its use are still being found, for example, in theoretical and statistical studies. It is used with a wide range of crops even in cases where rapid cycling of generations is not feasible. Very importantly, it has been matched with other systems, for example, pedigree followed by SSD. It has proved a useful research tool beyond plant breeding, in genetics, biometry, and computer simulation. It may be used with directed selection. It appears adapted to bulk population methods; for example, Grafius' 1965 suggestions are close to a workable bulk method. It can be blended with the pedigree method utilizing record-keeping and reserve F_2 stocks. An emerging use is for the prediction of merit among hybrid populations. DePauw and Clarke (1976) corroborated and successfully modified a method of inducing immature wheat seeds to germinate (see Mukade, Kamio, and Hosada, 1973), thereby making possible four to six generations per year, and suggested its application to SSD procedures.

In my opinion the strengths of SSD outweigh its weaknesses. Its major flaw is a too narrow exploitation of the F_2 generation. Although every individual of the F_2 is sampled, thus advancing a broad germ plasm representation of the population, only one seed represents the genetic base of each F_2 plant. Sneep in 1977 worried about this point and concluded that the genetic drift over generations was too large a penalty to pay for the use of the SSD method. The question is, is one seed per plant enough? Basically, what is the degree of genetic similarity for yield among the seed progeny of a single F_2 plant? The answer to this is not clear; there are suggestions that consanguinity is high (Ivers and Fehr, 1978; Thorne, Weber, and Fehr, 1970). The practical consequences of this theoretical deficiency of the SSD method could be important, as illustrated by the following hypothetical but routine breeding case. The anticipated outcome of a breeding project, of course, is a superior variety or genotype. However, when such a genotype has been created using SSD procedures, it is already too late to do other than speculate as to the relative merits of the numerous sister seeds abandoned in its ancestral F_2 family line. What we would know in this case is only that one seed, randomly picked from an F_2 plant, spawned a superior genotype. To reverse a parable, what might have been the value of the lost ninety and nine?

Seitzer and Evans (1978) provided some information on this potential loss. First they found between-family variances much greater than within-family ones. They cited one instance where 15 superior wheat lines were obtained by the extensive testing of 60 lines each from the F_2-derived families: 14 of the superior lines came from *one* family. This case highlighted the importance of selecting that F_2 plant, but nevertheless it is easy to see that an unmodified SSD system would have carried forward only one of the 14 superior lines. It is difficult to assess fully the loss because the 14 lines might have been similar sibs.

Because this weakness in the pure SSD method is unique to the F_2 generation it may be softened or corrected through various modifications. One way, mentioned by several breeders, is to retain the reserve seed stocks of F_2 plants and return to them for further exploitation if the initial SSD screening identifies the population as one containing superior lines. This modification carries a penalty in that the SSD is now a modified pedigree system requiring seed storage, record-keeping, and so forth; that is, the efficiency of the pure SSD is lowered. Nevertheless it is an attractive and powerful option. Other breeders have found attractive a pedigree system through the F_3 followed by SSD procedures.

In my small grains breeding project at Cornell University, I circumvented this problem by a single head descent procedure (SHD). This was a field technique for which hybrid populations were thinly planted. Individual plants were selected, one head retained from each, and the seeds from all bulked for the next generation. Future options were to continue SHD, continue as a bulk population, or for rapid advance, grow head populations in the greenhouse under a high-density, low-nutrient regime. The advantage of the SHD procedure is the perpetuation of 30–50 seed units, in the case of wheat, from each F_2 plant, rather than the single seed advancement in the pure SSD version. Variations of this technique are in use elsewhere; for example, Qualset (1977) stated:

> In the wheat breeding program at Davis we continue to use bulk populations extensively, but in many instances we deviate from the random bulk and single seed descent procedures in a very simple way, using the single-spike selection procedure that Wade Dewey of Utah State University has successfully used for a number of years.

A second concern about the SSD system is the attrition loss suffered cyclically in the population because of the nonviability or loss of the single seed per line. Enough has been written to emphasize the importance of planting at least two seeds, preferably more, to ensure that the number of progeny lines is carried forward undiminished.

Grafius' (1965) suggestions for high-density, controlled-nutrient plantings in the greenhouse are intriguing and could lead to an efficient, fully bulk population type of operation. If the density and control of nutrients could be "fine-tuned" to a reliable point, that is, low mortality or sterility and at least one seed per plant, then the entire population could be bulk harvested and replanted under the same conditions for the next generation. I would view the production of more than one seed (perhaps two to five) by some plants as a valuable addition to the population because, even accepting dominance and heterosis situations, there are possibilities for natural selection for vigor and yield genes. Adequate control of growing conditions at the required low level of seed production would perpetuate an approximately constant-sized population.

The advent of the SSD method, beginning with Goulden in 1939, can be seen emerging in Gfeller and Svejda's 1960 paper. Grafius in 1965 and Brim in 1966 were instrumental in publicizing the SSD and modifying it for uses outside of the growth chamber. It should not be lost sight of, however, that Kaufmann had set forth clear procedures for its use in a practical breeding program several years previously, in 1961. Perhaps his use of a less widely known vehicle for publishing and his choice of a general term, "random method," were factors obscuring his contribution.

A relatively recent development in the use of the SSD in plant breeding is concerned with its merits as a predictive tool for estimating the merits of hybrid populations to produce best lines. Plant breeders need assistance to choose *among* populations so that scarce resources are not wasted on inferior germ plasm. Kaufmann in his 1961 paper emphasized the retention of F_2 parent lines and in his Step 7 advised, "Go back and propagate the F_3 of lines that show excellent promise in all but one feature and attempt to either select that feature in the F_3 or F_4 or carry through each generation by random sampling."

However, he was beyond the F_8 at this stage and the discriminatory function, as to judgment of merit differences between populations, could scarcely be termed efficient at this point.

Jinks and Pooni in 1976 apparently were the first to combine the SSD method with prediction concerning the distribution of the inbred lines of a hybrid population, although the SSD method appeared to be incidental to the thrust of their research because it merely reduced the labor of handling generations. The authors state, "There is no reason why we need ever go beyond the F_2 of an inbreeding programme without a fairly clear idea of the final outcome."

Although intriguing to a practicing plant breeder, the proposed procedures are actually fairly laborious and time-consuming, because they involve high F generations, time in years, and additional hybridizing (triple test cross). When it is considered that in some of the self-pollinated crop breeding programs as many as 50–2000 hybrid populations may be formed each year, it would not seem feasible to allocate the resources apparently required by this method. Perhaps if there were a reliable preliminary method to screen for elite crosses it then would be feasible applied to the elite crosses.

While in Australia in 1979 I became acquainted with the research of I. A. Rose and K. S. McWhirter, which employed SSD procedures in a population screening procedure to predict elite 'Founder' hybrid populations on which major allocation of resources could be directed. McWhirter was using derived procedures in a soybean breeding program at the University of Sydney. The method required that reserve seed stocks be placed in storage while the analysis was being done; however, the time delay was not longer than two years. This could be absorbed in a breeding project, especially if these preliminary operations were made part of a routine first phase of the entire project. Once the initial lag had been absorbed into the project operations there would be no further apparent delay, because annual inputs would appear in normal succession.

McWhirter's procedures required taking two seeds from a small sample of F_2 plants from each potential Founder hybrid population. These were advanced by SSD procedures through two generations, F_3 and F_4, followed by field observation and evaluation in F_5 three-plant hill plots replicated once. Notes and data were taken on yield and other traits of interest, such as days to flower, maturity, height, seed size, protein, and oil content. The potential Founder populations were then arrayed on an information base sheet for each important trait, for example, seed yield. Each cross population was represented on this sheet by a simple line graphic which showed, by its length and designated tick marks, the variation in character expression via the mean, the predicted genotypic variance, and the parent positions for the trait. The crosses could then be ranked in order of h_2, which merely indicated the repeatability of character expression values. This information should give the breeder a comparative base for choosing among populations those most promising in which to search for superior lines for the traits of interest. By combining the information from the sheets for different traits, estimates may be deduced as to the possibilities for genetic advance for two or more traits selected simultaneously.

Aside from the screening of populations for later intensive effort, McWhirter's procedure has the useful property of also producing a large number of homozygous lines immediately ready for field evaluation. Even though a "discard" decision may be made on a particular cross population, one can still have a look at a representative sample of its progeny, an important check on one's judgments.

On balance, the SSD selection method, with its possible modifications, must be considered seriously by all breeders of autogamous crops. It has performed better, has shown more promise, and has more intriguing possibilities than I at first credited. It neatly accomplishes what Goulden proposed to do, namely, separate the early generation advance to homozygosity procedures from the selection among and within lines. In the process of SSD there is the further advantage that the progenies preserve a facsimile of the F_2 breadth of genetic variance uncontaminated by human selection and, to a lesser degree, natural selection. Kaufmann, I believe, recognized this in his concept of "random."

REFERENCES

Allen, H. T. and M. L. Kaufmann. 1981. Single seed descent—two decades. 1980 Oat Newsletter 31:28.

Baker, R. J. 1971. Theoretical variance of response to modified pedigree selection. Can. J. Plant Sci. 51:463–468.

Baker, R. J., V. M. Bendelow, and M. L. Kaufmann. 1968. Inheritance of and inter-relationships among yield and several quality traits in common wheat. Crop Sci. 8:725–728.

Bliss, F. A. and C. E. Gates. 1968. Directional selection in simulated populations of self-pollinated plants. Aust. J. Biol. Sci. 21:705–719.

Boerma, H. R. and R. L. Cooper. 1975a. Comparison of three selection procedures for yield in soybeans. Crop Sci. 15:225–229.

Boerma, H. R. and R. L. Cooper. 1975b. Effectiveness of early-generation yield selection of heterogeneous lines in soybeans. Crop Sci. 15:313–315.

Bravo, J. A., W. R. Fehr, and S. R. de Cianzio. 1980. Use of pod width for indirect selection of seed weight in soybeans. Crop Sci. 20:507–510.

Brim, C. A. 1966. A modified pedigree method of selection in soybeans. Crop Sci. 6:220.

Casali, V. W. D. and E. C. Tigchelaar. 1975a. Breeding progress in tomato with pedigree selection and single seed descents. J. Am. Soc. Hort. Sci. 100:362–364.

Casali, V. W. D. and E. C. Tigchelaar. 1975b. Computer simulation studies comparing pedigree, bulk, and single seed descent selection in self-pollinated populations. J. Am. Soc. Hort. Sci. 100:364–367.

Cisar, G., A. E. Howey, and C. M. Brown. 1982. Optimal population density and random vs. selective elimination of genotypes under a modified single-seed descent method with spring oats. Crop Sci. 22:576–579.

Compton, W. A. 1968. Recurrent selection in self-pollinated crops without extensive crossing. Crop Sci. 8:773.

DePauw, R. M. and J. M. Clarke. 1976. Acceleration of generation advancement in spring wheat. Euphytica 25:415–418.

Empig, L. T. and W. R. Fehr. 1971. Evaluation of methods for generation advance in bulk hybrid soybean populations. Crop Sci. 11:51–54.

Frank, S. J. and W. R. Fehr. 1981. Associations among pod dimensions and seed weight in soybeans. Crop Sci. 21:547–550.

Gale, M. D. and R. S. Gregory. 1977. A rapid method for early generation selection of dwarf genotypes in wheat. Euphytica 26:733–738.

Gfeller, F. and C. H. Goulden. 1954. The effect of the intensity of artificial light on the growth of cereals. Can. J. Bot. 32:318–319.

Gfeller, F. and F. Svejda. 1960. Inheritance of post-harvest dormancy and kernel colour in spring wheat lines. Can. J. Plant Sci. 40:1–6.

Goulden, C. H. 1939. Problems in plant selection. *In* R. C. Burnett (Ed.), Proc. 7th Int. Genet. Congr., (Edinburgh). Cambridge University Press, pp. 132–133.

Grace, R. M., G. Jenkins, and A. P. Roffey. 1976. Single seed descent studies in winter barley. 1975 Barley Newsletter 19:69–70.

Grafius, J. E. 1965. Short cuts in plant breeding. Crop Sci. 5:377.

Haddid, N. I. and F. J. Muehlbauer. 1981. Comparison of random bulk population and single-seed-descent methods for lentil breeding. Euphytica 30:643–651.

Hurd, E. A. 1977. Trends in wheat breeding methods. Can. J. Plant Sci. 57:313 (abstr.).

Ivers, D. R. and W. R. Fehr. 1978. Evaluation of the pure-line family method for cultivar development. Crop Sci. 18:541–544.

Jinks, J. L. and H. S. Pooni. 1976. Predicting the properties of recombinant inbred lines derived by single seed descent. Heredity 36:253–266.

Jinks, J. L., N. E. M. Jayasekara, and H. Boughey. 1977. Joint selection for both extremes of mean performance and of sensitivity to a macroenvironmental variable. II. Single seed descent. Heredity 39:345–355.

Kaufmann, M. L. 1961a. A proposed method of oat breeding for Central Alberta. Cereal News 6:15–18.

Kaufmann, M. L. 1961b. Yield-maturity relationships in oats. Can. J. Plant Sci. 41:763–771.

Kaufmann, M. L. 1971. The random method of oat breeding for productivity. Can. J. Plant Sci. 51:13–16.

Knott, D. R. 1979. Selection for yield in wheat breeding. Euphytica 28:37–40.

Knott, D. R. and J. Kumar. 1975. Comparison of early generation yield testing and a single seed descent procedure in wheat breeding. Crop Sci. 15:295–299.

Martin, R. J., J. R. Wilcox, and F. A. Laviolette. 1978. Variability in soybean progenies developed by single seed descent at two plant populations. Crop Sci. 18:359–363.

Muehlbauer, F. J., D. G. Burnell, T. P. Bogyo, and M. T. Bogyo. 1981. Simulated comparisons of single seed descent and bulk population breeding methods. Crop Sci. 21:572–577.

Mukade, K., M. Kamio, and K. Hosada. 1973. The acceleration of generation advancement in breeding rust-resistant wheat. Proc. 4th Int. Wheat Breeding Genetics Symp., Missouri Agric. Exp. Stn., Columbia: 439–441.

Ntare, B. R., M. E. Aken'ova, R. J. Redden, and B. B. Singh. 1984. The effectiveness of early generation (F_3) yield testing and the single seed descent procedures in two cowpea (*Vigna unuiculata* (L.) Walp.) crosses. Euphytica 33:539–547.

Park, S. J., E. J. Walsh, E. Reinbergs, L. S. P. Song, and K. J. Kasha. 1976. Field performance of doubled haploid barley lines in comparison with lines developed by the pedigree and single seed descent methods. Can. J. Plant Sci. 56:467–474.

Peirce, L. C. 1977. Impact of single seed descent in selecting for fruit size, earliness and total yield in tomato. J. Am. Soc. Hort. Sci. 102:520–522.

Pesek, J. and R. J. Baker. 1969. Comparison of tandem and index selection in the modified pedigree method of breeding self-pollinating species. Can. J. Plant Sci. 49:773–781.

Pooni, H. S., and J. L. Jinks. 1978. Predicting the properties of recombinant inbred lines derived by single seed descent for two or more characters simultaneously. Heredity 40:349–361.

Pooni, H. S., J. L. Jinks, and M. A. Cornish. 1977. The causes and consequences of non-normality in predicting the properties of recombinant inbred lines. Heredity 38:329–338.

Qualset, C. O. 1977. Population management for efficient methods of breeding and genetic conservation. Plant Breeding Papers, 3rd Int. Congr. Soc. for the Advancement of Breeding Researches in Asia and Oceania (SABRAO). Aust. Plant Breeding Cong., Canberra, Feb. 1977.

Riggs, T. J. and A. M. Hayter. 1976. Practical aspects of the single seed descent method in barley breeding. *In* Proc. Third Int. Barley Genetics Symposium, Munich, pp. 708–717.

Riggs, T. J. and J. W. Snape. 1977. Effects of linkage and interaction in a comparison of theoretical populations derived by diploidized haploid and single seed descent methods. Theor. Appl. Genet. 49:111–115.

Roy, N. N. 1973. Effect of seed size differences in wheat breeding by single seed descent. J. Aust. Inst. Agric. Sci. 39:70–71.

Roy, N. N. 1976. Inter-genotypic plant competition in wheat under single seed descent breeding. Euphytica 25:219–223.

Schnell, R. J., II, E. A. Wernsman, and L. G. Burk. 1980. Efficiency of single-seed descent vs. anther-derived dihaploid breeding methods in tobacco. Crop Sci. 20:619–622.

Seitzer, J. F. and L. E. Evans. 1978. Yield gains in wheat by the pedigree method of selection and 2 early yield tests. Z. Pflanzenzuecht. 80(1):1–10.

Shebeski, L. H. 1967. Wheat and wheat breeding. *In* K. F. Nielson (Ed.), Proc. Can. Centennial Wheat Symp. Modern Press, Saskatoon, pp. 249–272.

Snape, J. W. 1976. A theoretical comparison of diploidized haploid and single seed descent populations. Heredity 36:275–277.

Snape, J. W. and T. J. Riggs. 1975. Genetical consequences of single seed descent in the breeding of self pollinating crops. Heredity 35:211–219.

Sneep, J. 1977. Selection for yield in early generations of self-fertilizing crops. Euphytica 26:27–30.

Tee, T. S. and C. O. Qualset. 1975. Bulk populations in wheat breeding: comparison of single-seed descent and random bulk methods. Euphytica 24:393–406.

Thorne, J. C., C. R. Weber, and W. R. Fehr. 1970. Character fixation and its relationship to the F2-derived line method of soybean breeding. Iowa State J. Sci. 44:271–278.

Valentine, J. 1984. Accelerated pedigree selection: an alternative to individual plant selection in the normal pedigree breeding method in the self-pollinated cereals. Euphytica 33:943–951.

Weiss, M. G., C. R. Weber, and R. R. Kalton. 1947. Early generation testing in soybeans. J. Am. Soc. Agron. 39:791–811.

Wright, G. M. and G. A. Thomas. 1978. An evaluation of the single seed descent method of breeding. New Zealand Wheat Rev. No. 13. 1974–1976.

22

DOUBLED HAPLOID METHOD

Haploids long have been known in plants, but they have not been useful in plant improvement because the technology of creating haploids in large numbers was not available.

Nei (1963) presented a general overview of the efficiency of haploid versus diploid methods of plant breeding. This preceded the important discoveries in barley, although work had been done in corn and potatoes. Without considering linkage, Nei pointed out that the probability of obtaining a desired genotype involving *n* loci segregating independently is much higher in the haploid than the diploid method in view of the $(\frac{1}{2})^n$ versus $(\frac{1}{4})^n$ relationship, respectively. The efficiency of the haploid technique rises with the number of genes involved. Nei pointed out that selection is favored in the diploid method, but only for qualitative traits. Of overall importance is the saving of time in the haploid method.

The lower variability of doubled haploids (DH) in comparison with conventional hybrid-derived lines, which has been reported in tobacco, also has been found in maize according to Chase and Nanda (1965).

The problem of sufficient numbers appeared solved when Kasha and Kao (1970) reported on techniques to produce haploids in the cereals in large numbers. In earlier experiments with tetraploid *H. vulgare* and *H. bulbosum* it was noted that most of the progeny were diploid with 14 chromosomes, the expected gametic number. To see if this phenomenon held with diploid forms of the two barley species, such crosses were made and, indeed, haploid plants were obtained instead of the expected diploids. A decline in vigor of the developing seeds after 10 days necessitated resorting to embryo culture, which was successful in bringing the haploid plants to maturity.

Unfortunately, these plants were sterile. The application to cereal breeding was presented in the following way:

> These observations lead us to suggest that haploidy could be used in barley breeding programmes. By selecting haploids in the first generation (F_1) from a cross and subsequently doubling their chromosome number, homozygous lines could be obtained within two generations. Present conventional barley breeding methods require six to seven generations of selfing in order to obtain an adequate level of homozygosity. The success of the application of haploidy to barley breeding programmes depends on efficient embryo culture techniques and the achievement of doubling of the haploids into homozygous diploid lines.

Doubled haploids are pure lines par excellence, with essentially nil within-population genetic variability; thus they lend themselves well as material for genetic and plant breeding research. Meredith, Bridge, and Chism (1970) reported comparisons of DH cottons and their parent varieties, F_1 and F_2 hybrids, and their populational characteristics such as variety × location interactions and stability. They found that doubled haploids showed no significant differences from their parents, nor was their value as parents in hybrids any different. However, there was an indication that the parents were less homozygous than their DH progeny.

Haploids of cotton are readily obtained through attention to the phenomenon of semigamy, which is fertilization where there is incomplete fusion of the sperm with the egg nucleus. Background for this phenomenon may be found in papers by Meyer and Justus (1961) and Turcotte and Feaster (1969). In the present 1973 article, Feaster and Turcotte (1973) compared the yields and stabilities of three American Pima cotton cultivars with a doubled haploid from each, and a composite of the three doubled haploids. The results showed yields and stabilities of the parent cultivars and their derived double haploids to be similar. The importance of this rests on the differences in homozygosity and heterogeneity of the two. Cotton, although classified as a self-pollinator, is nevertheless subject to considerable cross-pollination. Consequently, a cultivar could be quite heterogeneous. On the other hand, doubled haploids are unusually homozygous. Given these differences, it was surprising to find that the parents and derived doubled haploids were similar in stability. The authors concluded that doubled haploids would be acceptable as commercial cultivars on the basis of stability. A composite of the three doubled haploids showed even greater stability.

Walsh, Reinbergs, and Kasha (1973) strongly recommended that "original" seeds, those produced on the original DH plants, be avoided for use in tests for yield because their smaller size and lowered plumpness produced a population often markedly different in emergence and stand from lines out of conventional breeding programs, which had several generations of normal seed increase behind them. Instead they recommended that all DH plants go through one generation of seed increase before being used in comparative tests.

Subrahmanyam and Kasha (1975) attacked the problem of getting an efficient chromosome doubling technique for the haploid system. It was noted in the haploid controls that about 3% natural doubling occurred. They experimented with nitrous oxide, colchicine, and dimethyl sulfoxide (DMSO) agents, singly and in combinations. The most effective treatment to induce doubling was colchicine plus DMSO.

Burk and Matzinger (1976) evaluated 46 anther-derived dihaploid (colchicine-doubled haploids) tobacco families obtained from an extremely inbred strain of 'Coker 139', which was self-pollinated via a single plant for 15 generations. The dihaploids showed generally reduced vigor and growth inferior to 'Coker 139'. On the other hand, considerable variation was shown in other characters, especially increased percentages of total alkaloids and reducing sugars. The range of differences among the dihybrids was as great as often found in segregating generations of conventional cultivar crosses. It was not clear whether the variability might be due to mutation or residual heterozygosity in the inbred line.

Park, Walsh, Reinbergs, Song, and Kasha (1976) reported on a comparison of DH barley lines with pedigree-derived and SSD-derived lines. The DH lines were obtained by colchicine doubling of haploids emanating from the interspecific hybrids between *H. bulbosum* and the F_1s of two crosses from the breeding program. An equivalent number of lines was drawn from these two program crosses to establish the pedigree (PD) lines and the SSD lines. These lines were advanced to the F_6 for the field comparison.

The nursery trials showed that lines from the three different methods were essentially alike in means and ranges for yield, but they showed slight differences in heading date and height that might be accounted for by slight procedural differences in moving the lines to homozygosity. Both DH and SSD lines surpassed PD lines in genetic variance. Little evidence of transgressive segregation was noted; however, DH (and SSD) produced more transgressive lines than PD, compared on the basis of grain yield equal to the high parent, which was a control. These results were consistent over the four environments in which the tests were conducted. Park et al. concluded that the DH method was as good as PD or SSD, and additionally had the advantages of shorter time to homozygosity and of being available for use at any stage of the breeding process. I interpret this to mean that F_1s of any desirable hybrid within the breeding project can be crossed to *H. bulbosum* as convenience dictates.

Doubled haploids differ from pure lines developed from conventional breeding methods in their absolute homozygosity; so-called pure lines likely are less homozygous. Reinbergs, Song, Choo, and Kasha (1978) compared the yield stability of 38 doubled haploid, high-yielding lines from six crosses, with 13 cultivars that included all of the *H. vulgare* parents. Tests were conducted for two years at five locations. Eberhart and Russell's 1966 stability model was used for comparisons. They found the stability of the doubled haploids equivalent to the parental cultivars, regardless of the haploids' origin, or different testing environments.

Serious doubts as to the suitability of dihaploid line production from anther cultures as a principal breeding method in tobacco improvement were raised in research reported by Arcia, Wernsman, and Burk (1978). Using the same parents two F_2 populations were prepared, one of conventional hybrid origin, the other dihaploid derived. The latter was created by producing 10 dihaploids from each parent and randomly crossing between the separately derived groups to produce 10 dihaploid F_1s, followed by selfing to F_2.

When the two F_2 groups were examined for differences it was found that they differed significantly. First, the dihaploids from the parents yielded significantly less (15%) than the parent varieties: they grew more slowly and were agronomically inferior. Therefore it is no surprise that the dihaploid F_1s and F_2s were also inferior to their conventional counterparts. Residual heterozygosity in the parents was ruled out because of their known maintenance history. The authors theorized that the imposition of a mutagenic treatment in the forming of the dihybrids, or the anther culture technique of haploid sporophyte production, somehow may have resulted in the production of inferior dihaploid lines.

In tobacco, where leaf yield and quality represent economic yield, anther-derived DH line populations were clearly inferior to F_8 SSD-developed populations in research reported by Schnell, Wernsman, and Burk (1980). The haploid lines (later doubled) were taken from F_1 plants of the same population that produced the F_8 lines via selfing. The difference was attributed not to the selection methods, but to some effect from the way the dihaploid lines are obtained, possibly connected to the mutagenic treatment.

The doubled haploid breeding method is a powerful instrument for barley breeding. I am attracted to it partly because of the leverage exerted by using a *Hordeum vulgare* hybrid ($2n = 14$) as one of the parents, the other parent being the $2n = 14$ version of *H. bulbosum*. The use of a hybrid that is already segregating as one parent sets up an infinitely variable series of possibilities. Technically, this makes a three-way cross within which chromosomes are selectively eliminated, leaving a production of haploid seeds carrying a complement of segregating genes, which when doubled as seedlings give rise to completely homozygous diploid individuals.

The procedure also has a high technological requirement: adherents must be prepared to set up a factory-like production line in order to generate enough lines to justify a selection program. In this sense, it may not be a procedure for all breeders. It is questionable whether small-scale operations would be efficient. On the other hand, it is conceivable that the DH system would mesh well with conventional systems if the small-scale production line of doubled haploids, which might be a laboratory operation, could be put on a sustaining basis. Then the common elements of all programs would include greenhouse and field facilities, crossing operations, nursery and testing procedures, and the feedback and weaving of new lines into the crossing pattern and into MSFRS schemes.

REFERENCES

Arcia, M. A., E. A. Wernsman, and L. G. Burk. 1978. Performance of anther-derived dihaploids and their conventionally inbred parents as lines, in F_1 hybrids and in F_2 generations. Crop Sci. 18:413–418.

Burk, L. G. and D. F. Matzinger. 1976. Variation among anther-derived doubled haploids from an inbred line of tobacco. J. Hered. 67:381–384.

Chase, S. S. and D. K. Nanda. 1965. Comparison of variability in inbred lines and monoploid-derived lines of maize (*Zea mays* L.). Crop Sci. 5:275–276.

Feaster, C. V. and E. L. Turcotte. 1973. Yield stability in doubled haploids of American Pima cotton. Crop Sci. 13:232–233.

Kaṣha, K. J. and K. N. Kao. 1970. High frequency haploid production in barley (*Hordeum vulgare* L.). Nature 225:874–876.

Meredith, W. R., Jr., R. R. Bridge, and J. F. Chism. 1970. Relative performance of F_1 and F_2 hybrids from doubled haploids and their parent varieties in upland cotton, *Gossypium hirsutum* L. Crop Sci. 10:295–298.

Meyer, J. R. and N. Justus. 1961. Properties of doubled haploids in cotton. Crop Sci. 1:462–464.

Nei, M. 1963. The efficiency of haploid method of plant breeding. Heredity 18:95–100.

Park, S. J., E. J. Walsh, E. Reinbergs, L. S. P. Song, and K. J. Kasha. 1976. Field performance of doubled haploid barley lines in comparison with lines developed by the pedigree and single seed descent methods. Can. J. Plant Sci. 56:467–474.

Reinbergs, E., L. S. P. Song, T. M. Choo, and K. J. Kasha. 1978. Yield stability of doubled haploid lines of barley. Can. J. Plant Sci. 58:929–933.

Schnell, R. J., II, E. A. Wernsman, and L. G. Burk. 1980. Efficiency of single-seed descent vs. anther-derived dihaploid breeding methods in tobacco. Crop Sci. 20:619–622.

Subrahmanyam, N. C. and K. J. Kasha. 1975. Chromosome doubling of barley haploids by nitrous oxide and colchicine treatments. Can. J. Genet. Cytol. 17:573–578.

Turcotte, E. L. and C. V. Feaster. 1969. Semigametic production of haploids in Pima cotton. Crop Sci. 9:653–655.

Walsh, E. J., E. Reinbergs, and K. J. Kasha. 1973. Importance of seed source in preliminary evaluations of doubled haploids in barley. Can. J. Plant Sci. 53:257–260.

SECTION III

BREEDING AND SELECTION STRATEGIES

The four major chapters of Section III address the roles of visual selection, selection indices, harvest index, and stability. They are a necessary precursor to Section IV, which covers the important related subjects of germ plasm, cross quality level, predicting and selecting parents, predicting and selecting crosses, predicting and selecting lines, and crossing techniques.

23

ISSUES IN SELECTION—A PREAMBLE

In discussions bearing on variances and heritabilities, Hutchinson (1940) highlighted the importance of separating environmental and genetic variances during selection. He considered some of the ways of minimizing or measuring environmental variance, for example, progeny row testing and plot design. This is the breeder's perennial dilemma. If environmental variance were entirely eliminated, the genetic variance would be unmasked and true genetic differences exposed. But this cannot be the ultimate goal because cultivars are not grown in a perfect environment, but are successful in their given environments because of the adjustments they make as a result of their genotype × environment interactions. Interactions and average performances are important. If environmental variance could be eliminated at a site the results could be interpreted only as representing one sample of location, year, microclimate, local hazards, and many other factors. It would seem in such a case that the breeder would require many more tests in order to obtain reliable data on average performance. Palmer (1952) also makes this point. Granting that rigid control of environment would increase selection efficiency in that environment, he stated, "However, because the products of selection are grown in a diversity of ever changing environments, rigid control, while increasing the apparent effectiveness of selection, would limit its usefulness."

Recognizing this, a breeder can select a cultivar for a known type of environment, including the clustering of, and testing in, similar environments. He may arrange his nursery entries so that the cultivars also are similar. Or he can breed for general adaptability. In the final analysis, breeders must have the information on genotype × environment interactions, but because environmental variance is often extreme they must think always of the need to reduce environmental variation to its practical minimum so that the interaction, read in the yield figures, represents a reasonable and meaningful response. Variation can be damped by paying attention to the items mentioned above, and by making certain that the planting seed is uniform, appropriate field sites are chosen, and that sound statistical designs are used.

The goal of partitioning variances between what is heritable (G) and what is not (E) has provided a basic thrust underlying genetic and breeding research in both plants and animals. A foundation paper was provided by R. A. Fisher (1918). Later, Lush (1945) proposed the term *heritability* for the ratio of genetic to total variance.

Jenkins (1940) found limited segregation for yield in inbred corn lines. This was an early indication that variability should be sought for among lines rather than within lines.

Sprague (1946) presented the arguments for early-generation testing of maize inbreds for yield prepotency. Early testing not only has great predictive value, but tremendously speeds up the evaluation of inbreds as compared with earlier visual selection and testing, concurrent with selfing, before evaluating for combining ability. It had already been determined by Sprague and others that relative variability was immeasurably greater among inbred lines than within lines. Sprague's research showed that there was a consistent pattern wherein yield prepotency of inbreds could be traced through top cross progenies, selfing generations, within-line segregations, relations with commercial performances, and consistent differences in between-line yields. In other words, early testing was successful and reliable for yield. He stressed that the value of early testing would be less in cases where important characters first might be selected visually in the very early stages of a breeding program, for example, the first selection of plants from an open-pollinated variety.

What I especially want to draw from Sprague's paper is the universality of genetics and methodology that underlies all breeding endeavors. Regardless of species, type of pollination, or other differences among crops, there are common themes and problems and solutions. The matter of determining yield prepotency of inbreds in corn is a good example, and similar analogies may be found in autogamous species in parent selection for crosses, and in early-generation testing. In corn, the common practice was to choose inbreds largely by visual selection, and after selfing the continuing lines for several years, to obtain a value assessment by top crossing; after further discard, single crosses were made and tested to predict the best double crosses. Early-generation testing was needed to shorten this long drawn-out process. Research in this area showed that early testing, by outcrossing to a tester simultaneous with first selfing, made possible the immediate identification, after performance testing of top cross progeny, of families showing the greatest potential. This is what breeders of self-pollinators try to do using data from parent means, F_1 performance, and F_2 families.

I believe I am correct in saying that corn breeders came earlier to the recognition that there was greater variance for yield expressed *among* open-pollinated plants (the source of inbreds) than there was *within* the segregation progenies derived from inbreds. This is crucial knowledge to the breeder of self-pollinated crops and we are only now coming to its full realization. It tells us that selection methodology has pinpointed, in the F_2 or F_3 family line, *between*-line variance as being more important than *within*-line variance for yield. Moreover, between-line selection always supersedes and precedes within-line selection, which follows the discard of the unwanted F_2 "between liners."

It is not to be construed from these comments that within-line variance is unimportant. In both cross-pollinators and self-pollinators continued inbreeding fine-tunes, and testing identifies, the best of the within-line segregates. This says something, too, about the unmodified SSD procedure, which methodically casts aside all save one random survivor of its within-line variance while retaining a full complement of between-line entities.

Frey and Horner (1955) used four traits in two barley crosses to compare gains predicted from heritability studies to gains actually realized. They reviewed three methods of calculating heritability. One, based on variance within an isogenic line, was excluded because it was not applicable to this situation and also produced broad sense estimates. The two methods used in the comparisons both produced narrow sense estimates, based on additive genetic effects, and might be expected to yield similar values. The methods were parent–progeny regressions (F_4–F_5) and variance component analysis of the F_4.

Frey and Horner projected a general belief that parent–offspring regression would be the more realistic because "it more nearly represents what plant breeders practice when selecting within segregating populations." However, heritability estimates prepared by the two methods were in close agreement in only two of the seven comparisons, and in the others the regression method gave lower heritability percentages.

Predicted versus actual gains, calculated using the components of variance method, were in good agreement in all comparisons except for test weight in one cross. This method was judged superior to the parent-regression method.

Breeders would like a simpler, less laborious way to use heritabilities for prediction purposes. Is it possible to combine heritability experiences for traits and species to produce average heritability tables? For example, knowledge of an average heritability figure for yield in barley might serve a breeder's general needs in many cases. After all, separate calculations of heritability for each case require many assumptions (Frey and Horner listed five needed for creating a heritability formula). Breeders add many compromises of their own with the environment, for example, in the Kindred ×

Bay cross only stem rust-resistant lines were used. In many ways, each separate case we examine can be said to be unique, and the derived heritability figure becomes merely a point in a range of possible points.

Could confidence be placed in an average heritability figure for a trait in a given species? For an example, I have taken the test weight and data in Frey and Horner's Table 3. Combining the four heritability figures in Table 1 for the two crosses gives an average figure of 0.72. Applying this against average *D* figures of 1.4 and taking a percentage of the average population mean (34.0) gives a predicted figure of 4.1, which compares with average actual gains (6.7 + 1.7) of 4.2. Doing the same with plant height produced a predicted figure of 5.5, which compared with their actual gain figure of 4.8. Although these manipulations might transgress rules of averaging, they do get to the point as to whether the pooled experiences of several breeders with several crosses within the same species could provide a meaningful figure for general use. An average figure per trait might eliminate the need for the next-generation actual gain test. Its usefulness to the breeder would be in sorting among crosses for those showing the greatest predicted advance. In any case, I feel sure breeders would find useful a convenient table of experienced heritabilities classified by species and trait.

The paper by Clayton, Morris, and Robertson (1957) is concerned with quantitative genetic theory in animal breeding, but has relevance to plant breeding, especially for cross-pollinating species and crops in which male sterility is used. An odd pattern, for which the authors had no explanation, occurred in bidirectional intensity-of-selection experiments for abdominal bristles in the fruit fly: "In both directions the ratio of observed to predicted response is highest in the 20/100 lines and declines continually to the 20/25 lines."

This is to say that prediction is best, with better agreement between observed and expected, at the highest intensity of selection. I don't know how to interpret this; perhaps the phenomenon cannot be generalized. In any case a breeder cannot raise selection intensity too high because of the risk of losing some good material.

Finney's (1958) paper is well worth reading because of its extensive and rounded look at breeding and selection, with emphasis on the intensity and the integration of selection procedures. He uses the theme of yield, interpreted as a general index value of a crop. He separated breeding procedures into the preliminary years of lines production and the later main testing period of the selections, devoting his attention to the latter. Throughout, various modifications were considered; for example, if numbers were too large in routine truncation, he suggested that a random discard might be employed. Discussing "fairness" in selection, he pointed out that fairness to farmers is more relevant than fairness to genotypes. Both of these examples subtly let the breeder know that he or she is in charge of the program.

Donald's (1968) classic paper, "The breeding of crop ideotypes," begins by characterizing plant breeders' activities as belonging to one of two categories, "defect elimination" or "selection for yield." It is also clear that the breeder of a successful cultivar may not be certain always of just why the new cultivar is more productive. Donald proposed a third category which involves the use of models to design the ideal plant. He acknowledged that theoretical models eventually must produce prototypes subject to testing as is the practice in industry. He termed his biological models *ideotypes*.

Before specifying details of model building, Donald discussed the principles of design and the general assumptions concerning the ideotype and its relation to its surroundings. He assumed resources to be nonlimiting ("the community as a whole must press on total resources to maximum degree"). If resources are limited it would be easier to face that specific problem later. He saw the environment, in its relation to the model plant population, as being of two kinds, namely, (1) the simplest environment having maximum available resources, such as high fertility, and (2) an environment that must be changed to meet the needs of the ideotype, for example, by irrigation or fertilizer. Because the model plant would be homozygous and surrounded by itself it must be a weak competitor, but at the same time aggressive relative to its environmental resources.

A critical element of design relates to harvest index, the ratio of economic yield to biological yield. Donald asserted it is not enough to measure and record harvest index; plant design requires a deeper search for the morphological and physiological processes behind it.

In summing design principles, Donald stressed three areas where theory and experiment would be crucial:

1. Photosynthesis in cereal communities
2. The cereal ear as a sink for photosynthates
3. The role of plant competition in crop communities.

Following these general considerations, Donald listed the model characters that he felt would be basic to a wheat ideotype: a short strong stem, erect leaves, few small leaves, a large ear, an erect ear, awns, and a single culm. Agronomic considerations to match the ideotype included planting density, fertilizer application, row spacing, weed control, and water supply.

It is not possible to do justice to this classic paper in a review. It should be read in the original. The most intriguing and as yet unproven feature of Donald's ideotype is the single culm per plant.

Reviews are difficult to review: the excellent paper on selection by Walker (1969) attempts to bridge the gap between the plant breeder's experience and practice and the statistical geneticist's experiments with their assumptions. In addition to the breadth of subject covered, the article contains several pages of references. Walker's economy and simplicity of expression are illustrated in his comment that there are under any conditions but three basic principles of selection: "First, it [selection] operates because some individuals are favoured in reproduction at the expense of others;

secondly, it acts only through heritable differences, and thirdly, it works upon variation already present in the organism."

Disturbing findings about heritability estimates for grain yield in oats were reported by Pfahler (1971). The study was conducted on homozygous lines from 94 oat populations grown over a six-year period. The heritability estimates were erratic and inconsistent, ranging widely over years from negative to positive. Surprisingly, selection, which was carried out (hypothetically) at three pressure levels, did not increase heritability estimates as years accumulated; therefore it was concluded that selection in any one year would be as effective as selection over a combination of years.

Valentine's (1979) research concentrated on the F_2 as the critical pressure point for selection in self-pollinated crops, in this case, barley. He asked whether the credibility of F_2 individual plant selection could be enhanced by indirect selection of characters related to yield. Furthermore, he examined ways of modifying the planting of the F_2 in order to improve selection efficiency.

The plan incorporated four population arrangements, the use of seeds of uniform size, three varieties, pure and mixed stands, and the measurement of six primary traits (from which six additional characters were derived by calculation). The three pure line varieties were used to simulate plants of the F_2, because being homozygous, they were amenable for replication of individual plants (the homozygosity, however, introduced other considerations into the milieu; that is to say, it could be argued that a population of homozygous individuals cannot represent fairly a population of F_2 individuals showing maximum heterozygosity). The summarized results follow:

1. Heritability of quantitative characters, as determined from F_3 plots, was markedly higher than heritability derived from the F_2, which is to say that selection between families would be more efficient than selection between individuals. This statement is general; there were variations according to trait, and certain traits were unresponsive or neutral to conditions.
2. Certain components of yield suggested that greater genetic advance for yield could be obtained through their selection than selection for grain yield itself. To illustrate, grain number per tiller (h = 21.9%) and 1000-kernel weight (h = 11.2%) had higher heritability at the single plant level than did heritability for yield itself (5.3%). Harvest index showed very low heritability, about 1%.
3. Close spacing competition did not change character values in pure stands. Because this was intragenotypic rather than intergenotypic, this would seem to be a specific case where the use of a homozygous population might not represent realistically the responses of a segregating population.

The author suggested that in selecting, breeders should look for and use components of high heritability, and pay attention to uniformity in seed size and sowing methods. His final statement, I believe, sums the meaning for methodology:

> The present evidence suggests that in spring barley breeding, the efficiency of single plant selection in the F_2 may be sufficiently high to allow the use of indirect selection for yield in the F_2. The F_2 generation can then be regarded, both in theory and practice, as the first and vital step in the sequence. There seems little reason not to begin positive selection for yield in the F_2 of the other inbreeding small grain cereals, notably wheat and oats.

Does it sometimes happen that a variety masquerading under a high yield mean is nevertheless well-adapted to a low-yielding environment? Armed with Finlay and Wilkinson's 1963 suggestion that a high mean yield in association with a below-average regression coefficient, less than 1, would indicate such a genotype, Walker and Cooper (1982) examined 10 Maturity Group II soybean cultivars and found that indeed there was such as genotype, 'Amcor'. Amcor's ability to increase yield, relatively, over other cultivars in a stressed low-yield environment was traced at least partially to its added height and its already proven yielding ability. It might be said that the height of Amcor represents a hazard in very favorable environments because it was conducive to lodging, a feature whose yield "drag" would be removed in less favorable situations. Although breeders do not wish to contemplate breeding for low-yielding environments, which are usually low economic level situations, a dual-purpose variety such as Amcor has obvious attractions.

According to Bridge and Meredith (1983), a puzzling decline in cotton lint yields began about 1961 in Mississippi and other parts of the South. The reason for this is unknown, but Bridge and Meredith were able to show that the decline was not due to a decline in genetic gains in yields. These were shown to be continuous from 1910 to about the present time.

ASSUMPTIONS

Without dealing with the substance of Sokol and Baker's (1977) paper concerning the place of assumptions in interpreting analyses of quantitative characters via diallel experiments, I thought their view (that diallel experiments are of little value in self-pollinating species) novel in that, because all material under test has been selected by the breeder, "random" has little meaning, and a new connotation or definition is needed for models designed to analyze diallels in selected material. This was also the view of Eberhart and Gardner (1966).

Baker (1978) touched an area of biometry that has been of concern to me as a practicing plant breeder, namely, the validity of assumptions used, specifically here with reference to diallel analysis. Baker subdivided diallel analysis into its "critical issues"; for background he reviewed five kinds of analysis that have been proposed. In his coverage of diallel analysis Baker projects the following positions (my interpretations):

1. Independent distribution of genes in parents is not a realistic assumption and, if invalid, dominance levels are likely overestimated, and it is not clear whether combining ability components of variance have been biased.
2. Assumptions about epistasis are so diverse and rest upon such varied bases, for example, independent gene distribution, sufficient number of parents to represent average population parameters, and so forth, that no credible assumption position is possible.

Baker is not sanguine about the value of general use of diallel analysis for genetic interpretations except for situations where totally random pan-mating is possible. General use might best be restricted to estimating general and specific combining ability effects.

Sidwell, Smith, and McNew (1978) made estimates of heritability and genetic advance for several wheat characteristics, using two kinds of selection units, single plots (SRS) and multiple replications (MRS). They found, as would be expected, that MRS gave higher heritability estimates and lower standard errors than single replicates. However, the bias in favor of MRS was less when genetic advance was examined, for "as the reference unit changed from MRS to SRS, the predicted genetic advance decreased less than the heritability estimates."

Of the traits considered, namely, grain yield, plant height, and heading date, the authors selected increased kernel weight as the character most subject to direct selection and positively related to increases in grain yield.

COMPONENTS OF YIELD

Although I have not seen the paper by Engledow and Ramiah (1930), it is referred to as an early study of components of yield.

In a study of yield components in barley and oats, Stoskopf and Reinbergs (1966) found that tillering, grains per head, and grain weight characterized yield per plant but were not reliable indicators of yield per area, the growers' chief index of cultivar value. Of the three components, grains per head most closely described yield. All components were variable, showing interaction with varieties, seeding rates and dates, years, and soil fertility levels.

Stoskopf and Reinbergs proposed approaching higher grain yields by selecting for favorable morphological features, such as leaf angle and size and plant height, and by altering the vegetative and reproductive developmental phases.

The problems of interpretation may be illustrated by Stoskopf and Reinbergs' results on yields from seeding rate studies conducted over two years. Results were diametrically opposite: in 1956 the highest mean yield of three cultivars was at the greatest seeding rate, declining steadily to the lowest rate; the next year the highest yields were from the *lowest* seeding rates, increasing with the seeding rate. Under crop density conditions the tillering rate decreased to approximately one tiller per plant, lending support to Donald's concept of plant ideotype. Later dates of seeding also caused tillering rates to decline.

Fonseca and Patterson (1968) studied heritabilities of components of yield in progenies of a seven-parent diallel cross of winter wheat. The three components of yield were very highly correlated with the complex trait of grain yield itself. Because each component showed direct effects on yield, the authors suggested that selection for components might be hampered by the negative correlations also found between number of spikes and number of kernels per ear. These suggested compensating reactions were taking place, leading to the necessity for accepting compromises in the degree of expression of component traits. It was noted incidentally that none of the seven parents in the diallel contained top expressions for all three yield components.

Rasmusson and Cannell (1970) compared the progress made in selection for yield in F_4 family lines from two barley crosses by yield itself and by the three components, heads per hill, kernels per head, and kernel weight. They selected the five high and five low lines for each selection category (about 4% selection intensity) and compared progeny performances the following year. In both crosses selection was effective for yield and kernel weight, but differed for kernels per head and number of heads, where selection was effective in one cross but not in the other. Kernel weight was very effective as a selector for yield, providing large gains—so much so that one feels that the seed size–yield relationship must not be overlooked in the selection, processing, testing, and preparation of seed lots for planting. Especially, uniformity must be stressed for a particular favorable kernel weight lot.

The authors pointed out that the keys to use of yield components for selection are heritability and a positive association with yield. Their conclusions reflected an acknowledgment of the fluctuating and inconsistent results obtained when using the same procedures on two different-origin populations. A firm conclusion was that selection for high kernel weight ought to be advantageous in any environment. I think the confused status of selecting for yield by components of yield is accurately stated by the authors: "While the literature indicates an active interest in the components of yield, we suspect that they are generally ignored by plant breeders as selection criteria except when visual selection is practiced."

Grafius and Okoli (1974) studied the components of yield in the 28 F_1s of an 8 × 8 diallel cross of barley. The negative correlation values found between yield and kernel weight and between seeds per head and kernel weight implied that increases in yield in heads per plant and seeds per head would be at the expense of seed weight. They recommended that selection for seed weight be stabilized at an acceptable economic point that allows for higher selection pressure on the other two components.

The authors derived shape parameters representing yield from the relationships of the components to each other. These shapes varied according to influences, for example, of environment or yield group. Grafius and Okoli point out that spaced- or drilled-row conditions would be expected to influence the shape of the yield parallelepiped. The assumption, attributed to Okoli, that the highest-yielding F_1 would have the most efficient arrangement of yield components, seems to me a powerful clue to the relation between yield and components of yield.

McNeal, Qualset, Baldridge, and Stewart (1978) acknowledged that historically the subdivision of yield into component units had been largely unsuccessful as a means of improving selection efficiency for yield. Their study was one of the most comprehensive treatments of the subject. They followed the path of grain yield per plant and the four characters, spikes/plant, kernels/spike, kernel weight, and spikelets/spike, for seven generations in one spring wheat cross, then brought all together with parents, for a comparison with grain yield per area (kg/ha).

Grain yield was increased over that of both parents by indirect selection for kernels per spike and for kernel weight. Both have possibilities for use in an index or in conjunction with harvest index. Selection for spikelets per spike produced yield at about the midparent level, that is, the effect was indifferent or neutral. Grain yield was decreased below that of parents when selection was for grain yield per plant, or spikes per plant, both of which might influence visual selection decisions in the F_2 and early generations.

The two components showing the most negative relationship to grain yield (kg/ha), namely, grain yield per plant and spikes per plant, would seem to be obvious candidates for elimination as yield criteria. I am reminded, however, of Dr. R. T. Ramage's experiences with developing hybrid barley. (Ramage is a USDA cytogeneticist with outstanding qualifications as a plant breeder.) Ramage reasoned that 10,000 barley plants per irrigated acre, each producing 1 lb of grain, would produce an acceptable yield of 5 tons/acre. This allowed 4.4 ft^2 of space per plant and, importantly, it was not unusual to find space-planted barleys that produced one pound or more of grain—they typically had about 100 heads. Ramage space-planted acreages of specially formulated hybrid composites, then hired student or nonprofessional labor to find and tag plants that met his criteria. His criteria were simple and required perhaps an hour of instruction: learn to identify visually 100-head plants, and tag all of these that meet a height requirement, for example, below waist, belt, or pocket. After initial selection, Ramage and his staff followed routine selection procedures to screen for seed quality, disease reactions, and so forth. The relevance of this to the studies on components of McNeal et al. and others is that it absolutely turns around frames of reference: the two most negative components in "standard" selection methodology, namely, grain yield per plant and spikes per plant, become in Ramage's case the two most positive components.

An interesting aspect of the study by McNeal et al., which they commented upon, concerns the wide yearly swings in trait expression. This could be attributed to environmental influences, but also may be what Adams (1967) and Adams and Grafius (1971) referred to as "oscillatory" expression, and Jain (1961) termed "environmental slippage." This projects the likelihood of continued wide swings, seen in the figures of McNeal et al. into the future, perhaps even at F_{20}. Although this is serious enough for one trait, it becomes more complex when one is dealing with several related traits experiencing component compensation. The best one could expect might be a line fitted to least squares.

The reduced genetic variance with advancing generations provides incentive to place greater reliance on fewer, earlier generations, for example, F_2–F_5. The authors suggested, however, that relaxation of the selection intensity (1–2%) to perhaps 20–40% would foster wider latitude of genetic variance and a more reliable outcome expression of selection. They also suggested the desirability of intermating at about F_5 to start a new recurrent selection cycle (in this connection more crosses, or a broader parent base, would aid a recurrent selection program). Overall, the study suggested the use of a restricted selection index.

Hanson, Jenkins, and Westcott (1979), working with barley, found visual selection of plots to be the best predictor of yields, superior to nine other components that were evaluated for indirect selection. They favored delaying selection until F_4.

A surprising but helpful discovery of the relationship between pod cavity width (suture to suture) and seed weight in soybeans came as the result of studies by Bravo, Fehr, and de Cianzio (1980). In fact, pod width was more effective in increasing seed weight than selection for seed weight itself. This is explained by the very rapid growth of pods after pollination, reaching their maximum length and width within 20 days, whereas final seed weight is subject to hazards of the longer maturation period. A valuable side effect of using pod measurements is that this may be done on lower pods, and crossing of selected plants done on the same plant (a suggested breeding alternative would be to have two dates of planting). It was further found that a relatively few pods per plant, and a relatively few plants per plot, were needed for reliable measurements.

It is difficult not to agree with Cavalieri and Smith's (1980) statement that "study of historically important cultivars may provide information on physiological traits that have been changed during selection for yield." They studied 21 commercial hybrids and one open-pollinated maize cultivar that had been released over a period of 50 years. The most significant finding contributing to increased yield was the lengthening of the grain filling period, accomplished surprisingly without changing relative maturity; in other words, physiological maturity was more efficient, occurring at a lower water content.

SELECTION STRATEGIES

Frey (1968) used an ingenious scheme to simulate performances of lines under three selection arrangements. He first obtained extensive field data, then drew portions of the data to test what would happen if those lines had been under a particular selection scheme. The ancestor of this concept could be the uniformity trial, where an entire field would be grown to one genotype under a given format, for example, rod rows. Once all the data had been obtained the breeder could ask questions about any plot or replicate arrangement on the field, and get the answers by going to the data. From the data much information could be derived, such as efficient check placement, most efficient size of plot, field fertility patterns, and so forth. Cornell University's Professor Love, with his interest in statistics, was an avid exponent of uniformity trials and they were still being exploited when I came to Ithaca in 1939.

Frey's simulated selection methods involved six traits in two crops, namely:

1. *Random F_3 System:* lines were chosen without regard to their F_2 source.
2. *Stratified F_3 System:* lines were chosen on the basis of one line from each F_2.
3. *Tandem System:* the best 20% of evaluated F_2-derived lines were chosen, and three lines per F_3-derived family tracing to each F_2 were tested.

Overall results showed the tandem method to be superior in producing the largest genetic gains, whereas the random and stratified schemes were essentially equal. Frey determined that the superiority of the tandem system was related to the heritability percentage of the selected trait; that is, at 100% heritability the methods provided equal gain, but as heritability declined the advantage of the tandem system quickly became evident.

Frey and Huang (1969) studied the relation of seed weight to grain yield of oats and proposed an intriguing selection scheme that involved identifying promising individuals from their position in a scatter diagram of yields and seed weights. The procedures reflected both indirect selection for yield (seed weight) and index selection; that is, floors were established for both yield and seed weights.

The materials used were 150 F_8 lines from a broad-based heterogeneous bulk population. These were graphed in a scatter diagram with yield of grain and weight per 100 seeds as axes; a significant correlation of .22 was found, but as the authors stated, it had little value for predictive purposes. What they did was to divide the scattergram into six unequal parts, giving it somewhat the appearance of a tic-tac-toe diagram, of two yield classes × three seed weight classes. The yield classes separated into an upper group having few individuals and a lower group with many individuals, established by the line of mean yield plus one standard deviation. The seed weight class chosen was the center segment, between 2.75 and 3.10 g/100 seeds, from a total range of about 2.0–3.6. The number of individuals falling within this one chosen segment was 18, slightly more than 10%.

This partitioning is designed to channel and facilitate mean population improvement. Additionally, a breeder might wish automatically to save any individuals that were above the yield floor and of heavier seed weight. These would be relatively few; in this study there were only two plants.

A thoughtful article by Hurd (1969) illustrates many of the problems that face breeders in different environments: his problem was how best to breed wheats for yield in a drought-prone environment. Hurd reviewed a large number of research papers and concluded in general that few reliable guides existed for the plant breeder and that breeding for drought resistance was tantamount to breeding for yield under semiarid growing conditions. The research cited here was an experiment to search out and evaluate direct ways to produce high-yielding wheat varieties for these conditions. Although Hurd's intention in this paper was not a study of breeding methods, there are interesting things that relate to our interest in metholodogy. For example, the parents used in the wheat cross were two genetically different lines discarded from previous elite tests. Obviously of good breeding, they had the additional merit of being adapted to the area and having complementary quality traits. They illustrate the value residing in lines that are not quite good enough to be released as varieties, and show that developmental work need not be discarded if the objective of a cultivar is not reached, but may be recycled.

Hurd incorporated certain general principles into his selection methods. One example was a decision to discard profusely tillering plants on the grounds that tillering was inimical to highest production in semiarid conditions. In fact, the principle of emphasis on discard was rigidly adhered to: "In this work selections were the plants left when all the ones with undesirable traits were discarded."

Yield tests began in F_4 with 1462 lines, which were reduced to 47 lines for F_7 testing. At this point testing at multiple sites began and the number of lines was reduced to 20. This number was slowly reduced as testing continued over sites and years.

The results clearly showed the method to be successful: in 465 comparisons, 275 were significantly higher yielding than the best check variety, 'Thatcher', despite the fact that none was suitable for release as a cultivar for reasons of minor quality defects. Hurd believed that the choice of sites with heavy clay soils was important because of its inherent heavy moisture stress. His paper is most relevant for the insights it provides into a breeder's train of thought while analyzing and deciding on how to approach the special problems posed by the environment.

Briggs and Shebeski (1971) found that four selections imposed on three different wheat hybrid populations during F_3 and F_4 resulted in significant yield advances in F_5, based on comparisons with unselected F_3 lines (at F_5), commercial cultivar controls, and percent of control yields. The first selection was for yield (actual data and control comparisons), the second tested the selected lines for breadmaking qualities, the third restricted retained lines to the upper group of high yielders, and the fourth was a visual selection for desirable agronomic traits.

Bhatt (1972) proposed a limited selection approach in single plant selection in order to create a heterogeneous population, that is, one with a residue of heterozygosity rather than a homozygous population. However, limiting the selection to F_4 or F_5 as suggested would not be a radical move because there are cases of hybrid populations that require no selection for uniformity. Bhatt and Derera (1973) expanded on the idea of limited selection. Although I agree with Bhatt and Derera's thoughts, I believe an advancement in this area would have to focus on the F_3, and possibly the F_2 of certain crosses, if the goal is to effect a change in cultivar heterogeneity.

In an attempt to widen, by crop, the use of Gardner's grid and investigate its use in selection, Verhalen, Baker, and McNew (1975) reported on its use in cotton. Based on bidirectional selection for fiber length, they found that gridding "reduced phenotypic variation by 22%; lowered selection differentials by 11 to 14%; produced 20 to 35% greater selection responses; and estimated 'realized' (i.e., narrow-sense) heritabilities 41 to 52% higher than did the identical selection procedure without grids."

Boerma and Cooper (1975) established nine heterogeneous parental lines (HPL) chosen from four soybean crosses after three years of selection under decreasing intensity pressure for yield. Following this, 30 pure lines (PL) from each HPL were selected, apparently at random, the seeds were increased one year, and the yield testing was done the following year.

If the 30-unit sample size were representative, it would be expected that their mean yields would approximate the yield of the parental HPL. This did not occur. In four of the nine comparisons (HPL versus PL), PL mean yields were significantly lower than their ancestral HPL. This suggests overcompensation for yield in the HPL (a phenomenon that would be welcomed in multiline breeding). In the lowest-yielding HPL, its PL mean yields were significantly higher (undercompensation). Differences in yield were nonsignificant in the other four cases.

It is difficult to interpret these results. The implication could be that heterogeneous bulk populations provided unreliable indications as to the value of their line components. The authors did not so conclude but took the positive stand that selection within heterogeneous bulk lines is worthwhile in the search for improved line performance.

Seitzer and Evans (1978) compared wheat yield gains realized from three methods, namely, pedigree and two early-yield tests, one based on control plots, the other on replicated hill plots. On the whole, small differences exacerbated by uncertainties were found. They proposed instead delaying yield tests to F_4, thereby at least partially bypassing problems traceable to insufficient seed amounts, heterosis, dominance effects, and the like. The testing unit would be a "selected bulk" made up of seeds from 5 to 10 plants from an F_3 family.

The authors' other findings and comments on selection seem to be of central importance to methodology. First they found between-family variances several times greater than within-family. Examples were given showing that superior lines often traced to the same families. The following excerpt is a profound statement of importance to selection methodology (italics mine):

> The fact that one family in Cross I yielded 14 of the 15 superior lines despite the extensive testing of 60 lines for each, points to the importance of the F_3 family, the F_2 plant. Similar results were obtained in Cross III: nine of the 13 lines exceeding the check by one standard deviation came from one family. *Thus the key to progress in selection seems to be the F_2 plant.*

Without restricting for maturity, Miller and Fehr (1979) formed a breeding population, cycle 0 (C_0), by a four-stage crossing procedure where each of 12 high-protein soybeans was separately crossed with one of 12 high-yielding soybeans, then the hybrids were crossed in a diallel and bulked, followed by two additional hand-pollinated intermatings. This base population was the source of two cycle 1 populations aimed at selecting for increased protein directly (C_1-HP), and indirectly selecting for low oil (C_1-LO). These were formed, respectively, by analyzing for protein and oil and separately intercrossing the 10 highest-protein lines and the 10 lowest-oil lines.

The comparative results showed that protein increased from 43.1% in C_0 to 44.6% (C_1-HP) and 43.9% (C_1-LO). Carbohydrate was reduced in the HP but not in LO. Oil was reduced in both; thus indirect selection for protein by selecting for oil was not a practical procedure especially in view of the heavy laboratory cost. However, there were problems also with direct selection for protein—there was an unacceptable increase in later maturity (also found in the indirect selection). The association of carbohydrate with yield was also considered, a positive relationship being assumed.

These considerations led to evaluation of three selection methods, namely, maturity class selection, linear regression adjustment, and selection index. Fundamentally, all such methods are analogous to selection index proceedings, attempting to partition and reassemble the influencing factors into a more meaningful result. In this case they were each successful, and as the authors pointed out, each exhibited trade-offs in advantages. The advantages of linear regression and selection index turned out to be slightly higher mean performance and adaptability to computer processing. Furthermore, it was found that the lines selected under each were the same or similar. The advantages of the maturity class selection were that it was a familiar breeding procedure and was found to assure a wide range of maturities among the selections, whereas the other methods were maturity-neutral.

Bos (1983) suggested refinements to grid selection in order to improve its efficiency. His argument was based upon the state of uniformity in a selection field. If uniformity is high, the environmental influence becomes more of a constant, and the breeder sees in the phenotypes a truer estimate of the genotypic value. Thus in a highly uniform field, gridding would be less effective and might not be needed at all. His reasoning led to this conclusion: "Therefore, a necessary requirement for G-selection to give better results than T-selection is the observation that there are significant differences among the grids." ("T" refers to truncation, a general overall reduction of the population.)

Bos cited cases of negative results of the application of grids to support his thesis. He suggested two possible reasons for negative or ineffective results from gridding, namely, a fixed number of selected plants per grid and the arbitrary size and shape of grids.

The issues raised by Bos are stimulating and have implications beyond gridding. For example, if it were argued that complete uniformity in a selection field would lead to selection for truer 1 : 1 phenotype : genotype values, then gridding might not be necessary. However, such selection would be for *that* specific environment and not for a series of different environments such as would be sampled in a properly gridded but nonuniform field. Which is better? One could say the question is moot because gridding was first used by Gardner to deal with nonuniform areas. At the worst, gridding in a uniform field where it might not increase selection efficiency would entail only the penalty of wasted effort.

I find Bos's second reason more persuasive than the first, that is, that size, shape, and orientation of grids are important. Picture, for example, a rectangular field with rows running across the shorter dimension. Suppose this field had a strip the length of the field along one side that was markedly superior in productivity for reasons of better soil or drainage. If selection were made with the field as one unit, it is obvious that most selections would be taken from the productive strip with very few from the much larger but less productive area. If the area were gridded with each selection row being considered as one grid unit, it is again obvious that there would be great differences within each grid, but not among the grids, and a disproportionate number of selections still would come from the productive areas within each grid. These objections could be met, or at least the situation ameliorated, by a serpentine selection path along the rows, based either on a fixed number, for example, selecting the 10 best out of every 100 plants, or a fixed distance of any length other than the exact cross-field row length. Another solution would be to arrange the gridding in certain geometric shapes, so that selections would be made in areas located at random, thus sampling areas of low-, variable-, and high-quality environments. Gardner used areas rather than rows; I have used 100-plant grid units in rows following a serpentine back-and-forth route across the selection field.

It is interesting to think that the outlines of the gridding technique must have been evident long ago from studies on sampling for yield in cereal grain crops. In 1929, Clapham (1929) showed that the random location of sampling units was superior to systematic procedures, and that more and smaller units, for example, 1 m-length of drill row, gave more reliable estimates of yield, that is, lower standard errors, than fewer but larger units. He stated that "it is of great advantage to divide the area to be sampled into a small number of parts within each of which an equal number of samples is taken."

Thus random sampling, which may be seen as selection for plants as well as for yield determination, can be assured for a whole area, that is, Gardner's whole field. An effective distribution of samples or plants is also assured.

A further discussion point in connection with efficiency of gridding procedures turns on the details of the actual selection. Visual discrimination is recommended wherever it is suitable, and it is known to be suitable for many characters, for example, height, disease resistance, lodging resistance, and vigor. Bos, however, suggested a certain amount of data accumulation in his discussion, for example, adjusting for the grid mean, in order to select variable rather than fixed numbers among grids. I think the decision as to whether to use Bos's augmented procedures rests on the nature of a breeder's overall program: the additional effort can be absorbed if the hybrid populations are few, but is less feasible in a larger program. Another solution is a two-stage gridding procedure: I used this in a program to select for high protein in wheat. In the first, or field, stage we simply selected agronomically best plants from 100-plant groups using the serpentine arrangement described and a liberal selection intensity that would provide 1500 plants. The labor requirement was a one-person half-day effort. The second stage was a laboratory protein analysis of all lines; this was more tedious but unavoidable, and it resulted in the identification of 46 high-protein individuals (about 3%). The important point is that gridding was used both in the field and in the final selection, where the highest individuals were chosen from each 100-plant grid unit rather than the 46 highest from the total 1500 plants.

Johnson, Helsel, and Frey (1983) compared direct selection for increased grain yield, that is, harvested grain weight, with three indirect methods under which selections were made on the basis of another character. All selections were completed before grain yields were known. The indirect methods and the procedures were as follows:

1. *Harvest Index:* grain yield divided by biological yield.
2. *Vegetative Growth Rate:* biological yield minus grain yield, divided by days to heading.
3. Harvest index plus vegetative growth rate.

The plant material consisted of 678 F_2-derived lines from 12 hybrids. No selection was indicated prior to F_4 and F_5 examinations. Trait measurements were taken on all lines from eight replicates divided over two sites in Iowa. Selection from the 678 lines, at 10% intensity, was made on the basis of these measurements, providing about 67 lines within each criterion group. A random group of 68 lines from the overall population was chosen to serve as a comparison base; that is, their grain yield means represented the population base from which genetic gains or losses of the selections would be scored. Heritability was estimated in three ways. The selected lines were again tested and measured at two sites in 1980.

The results showed actual gains in grain yield from the four criteria in the following order, highest to lowest: vegetative growth rate 8%, grain yield 7%, harvest index plus vegetative growth rate 6%, and harvest index 4%, all statistically significant. The indirect effects were deduced from measurements of genetic correlations and direct comparisons of the means of the selected versus random groups.

The plant breeding derivations from this study by Johnson et al. are many. First, grain yield (direct) was near the top in effectiveness, and had an efficiency advantage over the highest-ranking procedure, vegetative growth rate, which required measuring biological yield and days to heading (however, some programs may routinely take these measurements, in which case only calculating time would be involved). Second, a breeder would wish to know from genotypic correlations what positive or negative associations accompanied each selection procedure. These were determined, but because of differences between calculated and observed results I use only the observed results to characterize the different selection criteria:

Grain Yield Directly: The associations were with later-heading, taller plants, greater biological yields, higher vegetative growth rates, and lower harvest index.

Harvest Index: The associations were with earlier heading, shorter plants, equivalent biological yield, and a higher sum from harvest index plus vegetative growth rate.

Vegetative Growth Rate: The associations were with later heading, taller plants, greater biological yields, and lower harvest index.

The authors discussed these results in terms of their meaning to the end-product cultivar. Economic weights must be considered for certain traits; for example, early heading and maturity are necessary, a priori, to combat late-season heat, drought, and diseases; taller plants must be balanced against increased lodging; and so forth.

I am left with two thoughts concerning additional alternatives: (1) to follow the obvious indications suggesting a selection index, or some way of combining the separate criteria, and (2) to adopt a course involving large numbers, early visual selection for suitable heavy-economic-weight traits such as height and lodging resistance, use of disease screens, dividing into two maturity groups, and finally, concentrating intensive yield efforts on the selections remaining.

Sebern and Lambert (1984) tested another version of stratified early selection in soybeans. The selections in F_2 for two hybrid populations were made in this way:

1. Visual selection of 100 well-podded plants from each cross, reduced to the sixty having the highest seed amount.
2. Analysis of seeds of each plant for percent protein and oil.
3. Retention, based on protein results, of the 10 highest, 10 lowest, and 10 intermediate F_2-derived lines in each population. These F_4 lines were yield-tested and further reduced to the five highest (protein) lines from the high group, five lowest of the low group, and five intermediate from the intermediate group. Line samples were advanced for F_6 yield tests.

The results showed a significant correlation between early and later generations for percent protein, showing that early selection was a useful practice. As might have been expected the correlation between percent protein and percent oil and seed yield was negative. The most significant aspect of the paper, as I read it, was the finding that despite the negative correlations, individuals with reasonable combinations of these three traits were found, and most came from the *intermediate selection group*. Apparently, selecting for an intermediate measure of the target character, protein, acted in a manner analogous to index selection; that is, by the breeder forswearing the maximum extreme gain in one character, lesser gains may be made in others.

Soil salinity is an important and continuing problem in arid areas, notably in irrigated areas where salts tend to accumulate in the top soil layer. Kingsbury and Epstein (1984) concentrated their efforts on ways to identify and select for salt-resistant or more tolerant wheat genotypes. They screened 5000 world collection accessions by germinating them in 85% seawater adjusted for nutrient balance, and they recovered 312 promising lines. Their article includes a useful list of references to previous work on this subject.

Earlier conflicting studies on the relative value of estimators of yield prompted Openshaw (1985) to subject five corn hybrids, common to 123 trials over two years, to analysis using eight estimators including the arithmetic mean, median, and Trim 10 and Trim 20 (mean calculated after trimming these percentages of data from each tail of the distribution). The results showed the three best estimators to be the mean, Trim 10, and SRO (rank order weighted mean), and the worst to be the median. The overall conclusion favored the arithmetic mean based on familiarity, accuracy, stability, simplicity, and ease of comprehension.

SEARCHING FOR LOCKS AND KEYS

The incredible adaptation shown by the stem rust organism has kept plant pathologists and breeders striving to attain a consistent level of control over this pest. Somewhere there is—must be—a lock to which the organism cannot fashion a key. The history of controlling rust has been a seesaw battle between scientists and the rust organism with notable gains on each side. In Australia, beginning about 1950, the outlines of a successful strategy began to emerge. The strategy continued to employ vertical resistance using specific genes but with this difference: single genes were abandoned for a combination of genes to achieve a broad-based resistance.

The first consideration in breeding for stem rust resistance (the situation is not so clear in leaf rust) in New South Wales (Watson and Butler, 1984) was to ensure that hybrids received genetic diversity for resistance genes. Second, the genes had to be established in a suitable agronomic background to obtain stability of resistance. A serendipitous result has been occasional protection better than expected: a cited example is 'Timgalen', which had remained field-resistant in Australia for 20 years even though rust strains capable of attacking its individual genes were present.

A similar situation was experienced with stem rust of oats in New York State. Prior to the 1950s, commercial oat cultivars carried no known genes for stem rust resistance. The ravages of stem rust were severe during many of these years and were visually shocking—fields of dead immature oats on soils orange-red from fallen spores of both crown and stem rust. Some protection was obtained by incorporating a single gene for resistance, for example, in the Canadian cultivar 'Rodney'. But when two genes for resistance (known then as A and B-C) were inserted, notably in the Canadian cultivar, 'Garry', essentially complete protection was obtained. These genes were incorporated into all subsequent Cornell cultivars, and in my experience, stem rust ceased to be a major problem in New York State (although this was not true with leaf or crown rust).

Watson and Singh (1952) proposed a new term, *resgenes*, for resistance genes. Resgenes provided an approach to an ingenious lock system for which a new pathogen race (stem rust of wheat) would be unlikely to create a virulence key that would fit simultaneously the two locks of the two resgene loci in the wheat cultivar. Some cooperation among breeders would be required because the system could be breached by introducing a cultivar containing only one of the resgenes. The scheme assumes an absence of a sexual stage, that is, absence of barberry bushes, but mutation cannot be excluded.

Flor (1956 and his earlier reports on flax rust) established the well-known "gene-for-gene" theory: "For each gene conditioning resistance in the host there is a specific gene conditioning pathogenicity in the parasite."

The concept implicitly states that studies of rust reaction necessarily involve the gene systems of both host and parasite. Flor presented an argument for combining two or more resgenes:

> As virulence is usually recessive, a race can attack a heretofore resistant variety only if it is homozygous for the mutant gene. . . . Because the chance that a race will become homozygous for two or more pairs of mutant genes is much less than that it will do so for one pair, varieties with two or more genes for rust resistance are less apt to succumb to new races.

Flor expanded on this concept (1971) and prepared 20 lines, each containing three to five different resgenes.

The type of generalized resistance known as tolerance, a form of horizontal resistance, has been explored, notably by Caldwell, Schafer, Compton, and Patterson (1958) for wheat, and Simons (1966) for oats, but has not received the same attention as other methods. Tolerance may be a "fringe" form of resistance lacking a simple firm genetic base but founded on a marshaling of a plant's resources that is inimical to the epiphytotic dynamics of virugenes (genes for virulence), thus creating a slowing reaction of disease buildup that is analogous to resistance. Tolerance is a promising approach to crop protection yet to be fully developed and understood.

Flor's (1971) paper, discussed in Chapter 13 on multilines, is a good example of a search for a new lock. Flor raised the ante in the host–pathogen battle by developing backcross lines with not one, but two and three resgenes (flax rust) to oppose the pathogen's onslaught through its rampant breeding aided by natural selection. The objective was to present an impenetrable barrier to the pathogen, in any case to slow down its buildup. Left unanswered is a consideration of the strategic and tactical options available to breeders within a region as to whether and how one or more of the 20 two- and three-resgene flax lines might be deployed as cultivars. The multiline-possibilities are intriguing. A search for a new lock was behind the research for the multiline cultivar, as I have related elsewhere.

A full range of choices in breeding and selection strategy was presented by McIntosh (1976) in an important paper on breeding for stem and leaf rust resistance in wheat. Because of the high likelilhood of mutation as the source of new virulent strains of rust in Australia, and through the knowledge of the gene-for-gene relationship that prevails, pathologists and breeders have initiated the novel approach of anticipating the virulence gene combinations that will likely occur in the natural course of events. This was done by artificially creating mutants—in the greenhouses, for precautions against release must be enforced. This innovative approach takes from the pathogen some of the advantages of initiative and surprise it previously had held. It is not expected that the created mutants will be exactly the same as may later occur in nature. However, some will be similar, and these procedures permit the scientists to search for and stockpile new resgenes of possible future value.

The general incorporation of useful single gene and multigenic resistances in the world's wheats is not without risk. Genes used in multigenic resistance patterns are often used elsewhere as single gene resistances, so that as McIntosh wrote, "Ladders are being provided. . .for the pathogen to overcome those hurdles by cumulative single-step changes."

The paper by Frey, Browning, and Simons (1977) is a fascinating tale of the search for locks and keys in the oat–crown rust confrontation.

The University of Sydney (New South Wales, Australia) through its Narrabri station has been in the forefront of research on preharvest sprouting of wheat. Derera, Bhatt, and McMaster (1977) found that the levels of sprouting resistance in known sources were so low that no one parent could provide adequate protection for the white wheats. Consequently, they proposed a program of crossing to several partially resistant sources in an attempt to build up resistance through an additive combination of components.

Marshall and Pryor (1979) (See chapter 13) showed how the pattern of introduction of resgenes into a multiline might be used to fashion a lock difficult for a pathogen's virulence genes to unlock.

The phenomenon of slow leaf-rusting has intrigued plant pathologists and breeders for decades. Not as dramatic as hypersensitive resistance to leaf rust, and often overlooked, it has been found not only in wheat but in oats and barley; indeed, certain old varieties, for example, 'Red Rustproof' oats, owed much of their persistent longevity to slow rusting resistance. Many scientists have worked on slow rusting, but it will always be associated in my mind with Dr. Ralph Caldwell of Purdue University, who made it a lifelong research interest in wheat breeding.

Slow rusting has been shown to have at least three components, namely, length of latent period and size and number of pustules. Together they act to lengthen each infection cycle and thus reduce the number of cycles—as is well known, it is the saturation buildup in later cycles that results in major damage as a crop approaches maturity. Kuhn, Ohm, and Shaner (1980) found that two genes controlled longer latent period in slow leaf-rusting 'Suwon 85' and partially controlled smaller pustule size, both favorable to greater resistance. The authors suggest what I interpret to be index selection for all three components. The use of slow rusting resistance does not preclude combining it with vertical, horizontal, or other polygenic types of resistance. There is a challenge here to forge a lock to which the pathogens may have a key but are foiled and foreclosed each year by the maturing crop.

REFERENCES

Adams, M. W. 1967. Basis of yield component compensation in crop plants with special reference to the field bean, *Phaseolus vulgaris*. Crop Sci. 7:505–510.

Adams, M. W. and J. E. Grafius. 1971. Yield component compensation—alternative interpretations. Crop Sci. 11:33–35.

Baker, R. J. 1978. Issues in diallel analysis. Crop Sci. 18:533–536.

Bhatt, G. M. 1972. Methods in plant breeding. III. Limited selection approach in breeding crop plants. *In* T. M. Varghese and R. K. Grover (Eds.), Vistas in Plant Science, Vol. 2, Hissar, India: Hissar International Bio-Science Pubs., pp. 23–48.

Bhatt, G. M. and N. F. Derera. 1973. Heterogeneity in relation to performance in bread wheat (*Triticum aestivum* L.). Proc. 4th Int. Wheat Genet. Symp., Missouri Agric. Exp. Stn., Columbia, Mo.

Boerma, H. R. and R. L. Cooper. 1975. Performance of pure lines obtained from superior-yielding heterogeneous lines in soybeans. Crop Sci. 15:300–302.

Bos, I. 1983. About the efficiency of grid selection. Euphytica 32:885–893.

Bravo, J. A., W. R. Fehr, and S. R. de Cianzio. 1980. Use of pod width for indirect selection of seed weight in soybeans. Crop Sci. 20:507–510.

Bridge, R. R. and W. R. Meredith. 1983. Comparative performance of obsolete and current cultivars. Crop Sci. 23:949–952.

Briggs, K. G. and L. H. Shebeski. 1971. Early generation selection for yield and breadmaking quality of hard red spring wheat. Euphytica 20:453–463.

Caldwell, R. M., J. F. Schafer, L. E. Compton, and F. L. Patterson. 1958. Tolerance to cereal leaf rusts. Science 128:714–716.

Cavalieri, A. J. and O. S. Smith. 1985. Grain filling and field drying of a set of maize hybrids released from 1930–1982. Crop Sci. 25:856–860.

Clapham, A. R. 1929. The estimation of yield in cereal crops by sampling methods. J. Agric. Sci. 19:214–236.

Clayton, G. A., J. A. Morris, and A. Robertson. 1957. An experimental check on quantitative genetical theory. I. Short-term response to selection. J. Genet. 55:131–151.

Derera, N. F., G. M. Bhatt, and G. J. McMaster. 1977. On the problem of pre-harvest sprouting of wheat. Euphytica 26:299–308.

Donald, C. M. 1968. The breeding of crop ideotypes. Euphytica 17:385–403.

Eberhart, S. A. and C. O. Gardner. 1966. A general model for genetic effects. Biometrics 22:864–881.

Engledow, F. L. and K. Ramiah. 1930. Investigations on yield in cereals, VII. A study of development and yield of wheat based upon varietal comparison. J. Agric. Res. 20:265–344.

Finney, D. J. 1958. Plant selection for yield improvement. Euphytica 7:83–106.

Fisher, R. A. 1918. The correlation between relatives on the supposition of Mendelian inheritance. Trans. Royal Soc. Edinburgh. 52:399–433.

Flor, H. H. 1956. The complementary genic systems in flax and flax rust. Adv. Genet. 8:29–54.

Flor, H. H. 1971. Flax cultivars with multiple rust-conditioning genes. Crop Sci. 11:64–66.

Fonseca, S. and F. L. Patterson. 1968. Yield component heritabilities and interrelationships in winter wheat (*Triticum aestivum* L.) Crop Sci. 8:614–617.

Frey, K. J. 1968. Expected genetic changes from three simulated selection schemes. Crop Sci. 8:235–238.

Frey, K. J. and T. Horner. 1955. Comparison of actual and predicted gains in barley selection experiments. Agron. J. 47:186–188.

Frey, K. J. and T. F. Huang. 1969. Relation of seed weight to grain yield in oats, *Avena sativa* L. Euphytica 18:417–424.

Frey, K. J., J. A. Browning, and M. D. Simons. 1977. Management systems for host genes to control disease loss. Ann. N. Y. Acad. Sci. 287:255–274.

Grafius, J. E. and L. B. Okoli. 1974. Dimensional balance among yield components and maximum yield in an 8 × 8 diallel of barley. Crop Sci. 14:353–355.

Hanson, P. R., G. Jenkins, and B. Westcott. 1979. Early generation selection in a cross of spring barley. Z. Pflanzenzuechtg. 83:64–80.

Hurd, E. A. 1969. A method of breeding for yield of wheat in semi-arid climates. Euphytica 18:217–226.

Hutchinson, J. B. 1940. The application of genetics to plant breeding. 1. The genetic interpretation of plant breeding problems. J. Genet. 40:271–282.

Jain, S. K. 1961. Studies on the breeding of self-pollinated cereals. The composite cross bulk population method. Euphtyica 10:315–324.

Jenkins, M. T. 1940. The segregation of genes affecting yield of grain in maize. J. Am. Soc. Agron. 32:55–63.

Johnson, S. K., D. B. Helsel, and K. J. Frey. 1983. Direct and indirect selection for grain yield in oats (*Avena sativa* L.). Euphytica 32:407–413.

Kingsbury, R. W. and E. Epstein. 1984. Selection for salt-resistant spring wheat. Crop Sci. 24:310–315.

Kuhn, R. C., H. W. Ohm, and G. Shaner. 1980. Inheritance of slow leaf-rusting resistance in Suwon 85 wheat. Crop Sci. 20:655–659.

Lush, J. L. 1945. Animal breeding plans. Collegiate Press, Inc., Ames, Iowa, pp. 90–102.

McIntosh, R. A. 1976. Genetics of wheat and wheat rusts since Farrer. J. Aust. Inst. Agric. Sci. 42:203–216.

McNeal, F. H., C. O. Qualset, D. E. Baldridge, and V. R. Stewart. 1978. Selection for yield and yield components in wheat. Crop Sci. 18:795–799.

Marshall, D. R. and A. J. Pryor. 1979. Multiline varieties and disease control. II. The "dirty crop" approach with components carrying two or more genes for resistance. Euphytica 28:145–159.

Miller, J. E. and W. R. Fehr. 1979. Direct and indirect selection for protein in soybeans. Crop Sci. 19:101–106.

Openshaw, S. J. 1985. An evaluation of estimators of location for maize yield comparisons. Crop Sci. 25:938–939.

Palmer, T. P. 1952. Populations and selection studies in a *Triticum* cross. Heredity 6:171–185.

Pfahler, P. L. 1971. Heritability estimates for grain yield in oats (*Avena* sp.). Crop Sci. 11:378–381.

Rasmusson, D. C. and R. Q. Cannell. 1970. Selection for grain yield and components of yield in barley. Crop Sci. 10:51–54.

Sebern, N. A. and J. W. Lambert. 1984. Effect of stratification for percent protein in two soybean populations. Crop Sci. 24:225–228.

Seitzer, J. F. and L. E. Evans. 1978. Yield gains in wheat by the pedigree method of selection and 2 early yield tests. Z. Pflanzenzuecht. 80(1):1–10.

Sidwell, R. J., E. L. Smith, and R. W. McNew. 1978. Heritability and genetic advance of selected agronomic traits in a winter wheat cross. Cereal Res. Commun. 6:103–111.

Simons, M. D. 1966. Relative tolerance of oat varieties to the crown rust fungus. Phytopathology 56:36–40.

Sokol, M. J. and R. J. Baker. 1977. Evaluation of the assumptions required for the genetic interpretation of diallel experiments in self-pollinating crops. Can. J. Plant Sci. 57:1185–1191.

Sprague, G. F. 1946. Early testing of inbred lines of corn. J. Am. Soc. Agron. 38:108–117.

Stoskopf, N. C. and E. Reinbergs. 1966. Breeding for yield in spring cereals. Can. J. Plant Sci. 46:513–519.

Valentine, J. 1979. The effect of competition and method of sowing on efficiency of single plant selection for grain yield, yield components and other characters in spring barley. Z. Pflanzenzuecht. 83:193–204.

Verhalen, L. M., J. L. Baker. and R. W. McNew. 1975. Gardner's grid system and plant selection efficiency in cotton. Crop Sci. 15:588–591.

Walker, A. K. and R. L. Cooper. 1982. Adaptation of soybean cultivars to low-yield environments. Crop Sci. 22:678–680.

Walker, J. T. 1969. Selection and quantitative characters in field crops. Biol. Rev. 44:207–242.

Watson, I. A. and F. C. Butler. 1984. Wheat rust control in Australia: national conferences and other initiatives and developments. The University of Sydney, 79 pp.

Watson, I. A. and D. Singh. 1952. The future for rust resistant wheat in Australia. J. Aust. Inst. Agric. Sci. 18:190–197.

24

VISUAL SELECTION

Clearly, observing is an inseparable part of the plant breeding process. It is the "art" of the art and science of selecting, and is a practice that gains in credibility when one bears in mind that scientific measurements too are subject to errors and misinterpretations. Small wonder that a successful breeder is said to possess the plant breeder's eye. But what is it the plant breeder sees when he looks at his plants? Can he trust his vision? Are his interpretations valid? And what about the "bottom line," the soundness of decisions that flow from repeated observations and ponderings? The purpose of this examination of the art of observing is enlightenment—of the prospective plant breeder. Information in the form of knowledge of others' experiences is valuable, indeed necessary, because the breeder has no choice but to resolve to do a more intelligent and better job of it. Fortunately, he does not have the option of eliminating eye input and judgment.

We should not expect this information to be of a pattern. The use of vision in decision-making is so complex and so pervasive throughout plant breeding operations that the breeder is usually faced with a series of contextual situations; that is, the breeder's frame of reference will *depend on the context of the plant material in its environment.* Suppose the breeder is observing an F_2 nursery. Here the visual framework is backstopped by knowledge of the parents used, biases in vigor differences, profile ranges of characters, thoughts about heritability, appraisals of environmental effects, selection cutoff points, and so forth. Now when the breeder moves on to take notes on a yield nursery the whole visual frame of reference changes, for here he or she is rating homozygous lines with a history of performance behind each. All the way up the chain, from the selection of the best 2000 lines from a 15,000-entry F_4 head row nursery to the selection of a variety among seven good lines, breeders have no choice but to trust their own eyes.

To begin with, breeders quickly learn elementary caution. They learn the fake signals transmitted by heterozygotes. Immer (1942) said: "It seems clear that the yield of an F_2 plant is determined very largely by factors of the environment." He could as well have substituted "appearance" for "yield."

Heterosis in early-generation plants has been the cause of much early enthusiasm; in fact, I found a sizable proportion of my cereal selections at Cornell University to be based on the transient expression of hybrid vigor even as late as the F_5.

Selection for qualitatively inherited traits can be practiced successfully in early generations, particularly if one is familiar with the heterozygous expressions of the character. But productivity, performance or *yield*, is the touchstone, a complex quantitatively inherited trait severely buffeted by high genotype × environment forces. Sprague and Milller (1952) said about corn, "Yield, being the most complex from a genetic standpoint and most affected by environmental variations, is probably influenced least by visual selection," and, "This suggests that the selection practiced was not effective in modifying combining ability."

CHARACTERISTICS OF VISUAL SELECTION—HOW IT WORKS

One of the earliest discussions of visual selection occurred in Engledow and Wadham's (1923) paper. The use of "eye-judgment" arose in a description of the pedigree system. The authors acknowledged deficiencies in the use of eye-judgment, but made it clear that there existed no alternative to its employment by breeders to aid in the truncation selection necessary to reduce numbers. They further explained that eye-judgment is based on the evaluation of sound observation, and its use improves with the experience of the breeder.

The most serious objection, which planted the label of "guessing" on eye-judgment, concerned estimation for yield. In the end the authors reiterated their belief that breeders cannot avoid the use of eye-judgment because the alternatives were too costly and inefficient. They expressed the need for an "index of the desired kind. . .concerning the salient characters of the cereal plant," in retrospect, an early call for a selection index.

Engledow and Wadham did cite one instance of a successful indicator for yield: the "migration coefficient" used by Beavan before 1914, which was defined as the proportion of grain weight to total produce weight. This ratio today is known as the harvest index. The authors also were pioneers in the partitioning of yield into its component parts.

The problems inherent in preparing a standard reference scale for use in visual discrimination ratings of leaf and plant leaf area of potato plants were discussed by Bald (1943). It was found that the eye or mind did not judge variations in leaf area linearly but made adjustments that were farther from the mark at certain points on a scale; for example, at large increments in leaf area the eye continued to measure on a basis of proportionality but reduced the scale. A scale was constructed that recognized and corrected for this. It had 21 ratings, covering a range of area from 0.7 to 129.0 dm^2.

The rating scale was used in this way: with leaf prints of the various ratings in hand, an observer spent an initial period of time calibrating his eye-judgment against the prints. Conversion was made to a plant size scale, so in effect the observer was making a tandem judgment. When preliminary guesses matched the standards, the observer was ready to set aside the scales and visually judge and call out ratings during a walk through the field. The important points from this study were the need for a set of reference standards and for the calibration of the observer's visual judgment to these standards.

Byth, Weber, and Caldwell (1969) reported that early lodging in soybeans was the most effective predictor of yield among nine traits studied, including yield, although it was found that actual yield advances required the selection of yield itself. It was not understood why early lodging should be a predictor of yield (inverse relationship, negative correlation) but it was thought that it was related to light penetration into the canopy at pod filling. It seems logical, too, that lodged plots yield less, thus guaranteeing the higher yields to the upright populations. The authors suggested that visual classification of canopy shape during pod development might be an effective selection screen for identifying high-yielding genotypes. This would be analogous to a visual estimation for harvest index, a feat not yet accomplished in the cereals.

In research by Briggs, Faris, and Kelker (1978), the effectiveness of early-generation visual selection of single barley plants was found to depend on the characters observed. Using randomly spaced sets of single plants of 12 different barley cultivars, they examined and measured seven characters. Variance in the mixed rows and single-cultivar control rows as well as estimates of heritability were determined. Analysis indicated that the seven characters were grouped in three categories, and in only the first group could the desired genotypes be successfully selected for. Group I characters were seeds per head, 1000-kernel weight, days to ripe, and height. The group II characters, yield per plant and seeds per plant, showed low or nonsignificant variance, heritabilities, and performance consistencies. The single group III character, heads per plant, was intermediate between I and II.

Of the seven characters, only three, namely, days to ripe, height, and heads per plant, were susceptible to true visual selection, that is, by estimation without measurement. These three characters found in groups I and III were confirmed as credible early-generation selection possibilities. The data for the remaining four characters indicated that early-generation selection from measurements on single plants were possible for two characters, but not for the other two. This is useful information because it tells the breeder where applied effort is worthwhile. Especially, the findings showed that early-generation single-plant selection for yield was unreliable.

The breeder does receive and process other information in visual discrimination, which functions as a crude form of index selector. The whole plant, its vigor, a morphological profile, head size differences, and potential weaknesses are all subconsciously matched in the breeder's mind to a host of past successful cultivars.

Another part of this paper was of interest to selection methodology, especially for situations where visual assessment was judged to be of high value. An example was the authors' selections of the "desired genotype," which was defined for each character as the best mean expression of that trait among the 12 cultivars. For example, 'Gateway 63' had the most seeds per plant, 'Centennial A' was the shortest, and so forth. After measurements had indicated the three best plants for each trait, the plants were identified by cultivar from their known in-row positions. From this it could be determined how often one or both (each cultivar appeared twice in a row) of the desired cultivars were chosen in each selected three-plant character group per row. Because the three plants for each character were chosen from data, the possibilities of bias from visual recognition of any of the 12 cultivars could be discounted.

The results showed that under 12.5% selection intensity (3 of 24), the number of rows having at least one correctly identified genotype among the three plants was higher than chance expectations, and was especially high for the characters that were suitable for visual selection only. This suggests that, for characters selectable by visual means only, a more liberal selection intensity would ensure against serious discard of valuable genotypes. If visual selection were applied generously on the single-plant basis in F_2, and data taken on other characters in selected F_3 head rows (after additional discard), the measurement effort would be lessened appreciably. Some modification in this direction seems desirable, because without it the single-plant data might as well be incorporated into an early-generation selection index procedure applied to each plant. This paper showed where character selection boundaries might be drawn: visual selection first on some traits, followed later by yield (and component) determinations from data.

Wilcox and Schapaugh (1980) stated that limited plant-to-plant variability in certain crosses of equal-quality superior parents could well be the reason for unsuccessful selection of single plants for yield. Accordingly, they conducted a research trial comparing the effectiveness of visual selection in soybean progenies tracing to two distinctly different hybrid formulations, namely, crosses among superior soybean cultivars and crosses between these and plant introductions (PI). The hypothesis was that the latter might produce greater plant-to-plant variability. Three cultivars and nine PIs (three each from Manchuria, Central China, and Korea), all of group II maturity, were the parents. Twelve crosses were made: three among the superior cultivars, and nine combinations made by mating each cultivar with one PI from each geographical origin. In each cross the main control was an unselected F_2 bulk retained from the F_1 plants. The comparison treatments were bulks formed from visually selected-for-yield plants (and lodging resistance) in successive generations, F_2– and F_4. Selection preference was approximately 7% for the F_2 populations, and 5% for the F_3 and F_4. After a year of seed increase in a common environment, yield trials were conducted for two years. The results were as follows:

> Visual selection for high yielding plants from heterogeneous populations for three generations had no effect on seed yield (Table 2). Yields of selected F_4 bulk populations were no different from the yields of the unselected F_2 populations in crosses among cultivars or in crosses between cultivars and plant introductions.

This was true for both of the two different hybrid formulations mentioned earlier. Visual selection was found to cause a shift toward taller, later-maturity plants with an apparent reduction in lodging resistance.

Physiological maturity is that point when maximum dry weight of grain seed is reached. It has several practical applications and is very useful in physiological research. It has long been common practice for farmers to judge when a crop is ready for harvest by noting the ripening changes in color and so forth—even judging grain hardness by chewing a handful of kernels. In recent years, with the advent of the combine harvester, growers of seed grains have been interested in the earliest possible harvest in order to avoid costly hazards such as sprouting and storm losses if the crop is left for extra days in the field. They need to know physiological maturity, and thus they rely heavily on grain moisture tests with the expectation of harvesting early and artificially drying the crop.

Hanft and Wych (1982) examined 13 visible plant characters in a search for a reliable visual indicator of physiological maturity in wheat, and also one that would give an earlier signal, at about 95% of maximum kernel weight. They found that the 95% maximum kernel weight date was about seven days before physiological maturity. The visual indicators deemed useful were the following:

1. About one day before physiological maturity, a strand of brown to black pigment appeared in the kernel crease, visible as a spot in cross-section.
2. At approximately the same time all green color disappeared from the glumes.
3. Complete loss of green color from the flag leaf signaled the 95% maximum kernel weight and coincided with a sudden drop in grain moisture.

All these changes except for the drop in moisture content could be observed visually in the field.

Copeland and Crookston (1985) sought visible markers for physiological maturity in barley. They detected three indicators that could be observed in the field, namely, loss of green color from the glumes and peduncle (more striking in the peduncle), loss of green color from the flag leaf blade, and the appearance of pigment in the kernel crease.

THE INFLUENCE OF THE SELECTOR

Hanson, Leffel, and Johnson (1962) reported an interesting study on visual evaluation in soybeans. The plant material was 45 groups, each involving 20 random F_3 lines taken from 45 diallel design crosses, respectively. These were planted as 20 subgroups replicated twice, each subgroup containing one line from each of the 45 crosses. The test was repeated the following year with a second set of lines. Three experienced observers considered the population of 45 lines as belonging to five groups of nine each. Using a save/discard concept they charged themselves to choose the top ⅕ and bottom ⅘, and the top ⅖ and bottom ⅗ according to estimated yield. Discrimination was based on the total phenotype of the plants including the specifically measured characters of seed yield (number of pods), maturity, lodging, and height. The results were discouraging: the three observers correctly identified only 22% of the plants in the top ⅕, and 60% of the plants in the bottom ⅗. They concluded, "When considering that the observers attempted to identify only the top ⅕ or the top ⅖ phenotypically, one must conclude that the power of individuals to discriminate yield potential of a plot is low."

They discovered that visual discrimination worked better on the poorer yielders, whereas "the higher yielding plots tend to be merged on the visual scale."

Shebeski (1967) reported an experiment wherein four plant breeders each selected the 10 most vigorous wheat plants in each of 11 large F_2 populations. These 440 best plants were taken from a total population estimated at 110,000. In the F_3 the progeny of each plant was grown adjacent to an unselected plant (from the same population) as a control. The results were that "approximately 50 percent of the lines yielded more and 50 percent of the lines yielded less than their corresponding controls." The visual selection therefore was no better than random selection.

In a follow-up, McGinnis and Shebeski (1968) reported on the findings of three selectors choosing the best F_2 spring wheat plants for future yield potential from a nursery of 8000 plants. The general consensus was that there was little correlation between the yields of the F_2 plants and their F_3 line yields. Selectors' choices performed no better than random, but it was noted that the selection of well-tillered F_2 plants did raise the general yield capacity of F_3 lines.

The researchers stated that "no difference was found between those selected for high yield and those taken at random in F_2."

Even though the mean of the F_2 plant yields of the random group were lower by as much as 10 g/plant, one would have to conclude that they were not random, and that they were better than random; they were closer to the selected group because low-yielding plants were excluded from the random group. Therefore, the fact that the selected plots consistently outyielded the control plots indicated that selection in the F_2 was effective.

There are interesting things in this study that are pertinent to what one sees in selection. In their "Methods" they noted:

> Just prior to harvest, three breeders went through the nursery and made over 100 selections each strictly on a visual basis of what they believed to be the high yielding plants with superior qualitative agronomic characters. These plants were tagged using a different colored tag for each selector.

Because these were independent evaluations it may be assumed that one breeder went first, one seond, and one third. But were the colored tags placed on the plants as they were selected? If so, only the first selector could be considered as having made an independent selection, because the presence of a tag on a plant could scarcely be ignored by the next selector. From the results in the paper, the first selector chose 6 of the 10 highest-yielding plants; the second, 9; and the third, 10. If the numbers assigned to the selectors also indicated the order in which they made selections it could be argued that the last selector benefited from knowledge of the choices of the earlier selectors. Success could also be related to the number of selections chosen by the selectors: selectors 1, 2, and 3 chose 95, 162, and 157 plants, respectively. Perhaps the number of best selections is a function of numbers of selections taken: to make the point clear, assume that all 8000 plants were chosen by a selector—obviously that group would contain the best 10. The first selector simply gave himself fewer chances at the best 10. This discussion is speculative because it is not known in what way the colored tags were actually used, but the methodological relevance is clear.

On the basis of what has been shown to this point, there seems to be profound meaning in the fact that the selections chosen by all selectors—there were 43 such plants—were 25% better than random selections. But there was disillusionment when the F_3 yields of the selected lines were compared with the control: there were no differences between selections of different selectors, of selectors as a group, or randomly selected lines. However, the "best" yielding 48 F_2 plants, which yielded 50 g or more, all were 7–8% above the F_3 control yields. It is indeed difficult to reconcile these findings, and we must return to the authors' conclusion that selection was effective because the F_3 selected plots consistently showed a 7–8% yield advantage over the random bulk plots. They also observed that well-tillered plants seemed to correlate with high yields.

Kwon and Torrie (1964) used three observers to make visual yield assessments in the F_2–F_5 of two soybean populations. The agreement among the observers was good, but their overall assessment was that visual evaluation was only 50% or so as efficient as selection based on plot yield. They could visually discriminate extremes successfully when yield differences were large but not when they were small. Visual judgment was distracted by uneven stands as well as by height, maturity, and lodging differences. In conclusion, they suggested that the breeder should advance lines in early generations "based upon performance of traits other than yield."

According to Briggs and Shebeski (1970), Krull, Zapata, and Lopez (1966) "visually evaluated 3,274 wheat lines for yield on the basis of agronomic characteristics, using a scale of 0 to 4. More than 1,700 lines were eliminated because of poor appearance prior to harvest, but a very close correlation was found between the actual and estimated performance of the remainder."

Briggs and Shebeski conducted an especially revealing study of visual selection in wheat. Their goal was to evaluate the ability of selectors "to visually identify superior lines in a breeding nursery." The nursery involved 828 F_3 spring wheat lines grown under optimal conditions at one site. Every third plot was sown to 'Manitou' check. Fourteen individuals served as selectors. Each was instructed to make a visual selection for estimated crop yield and to rate 90 plants by choice from the 828 available according to the following structure:

Group 1	Top 10 selections
Group 2	Next best 20
Group 3	Next best 20
Group 4	Next best 20
Group 5	Poorest 10

Thus there were 80 positive selections to be made (about 10% selection pressure) because even group 4 would be an attempt to get the top 80 in the nursery, and 10 negative selections, because here the opposite selection extreme would be tried. In addition, for comparison four sets of random selections were obtained for each of the five groups.

The results led to several interpretations and conclusions. First, the overall positive selection pressure on the 80 lines produced a significant improvement in yield over random selection. Second, the mean yields of selectors' choices were in the order of groups 1–5. Third, the ability of selectors visually to identify actual high-yielding plots was "limited." Fourth, as might be expected, the ability of selectors varied; for example, two selectors were especially good at picking high yielders and one was good at picking low yielders. Here is a sampling of findings by the authors:

> Of the 80 plots selected for high yield by selector 1, three were in the 10 highest yielding plots and 11 were in the 25 highest yielding plots. For the 10 highest

yielding plots the mean number chosen by the 14 selectors was 2.6 and for the 25 highest yielding plots, 6.6, compared to 1.0 and 2.4, respectively, expected by random selection. Three of the 10 highest yielding plots were not identified by any of the selectors, nor were 6 of the 25 highest yielding plots.

A clear signal from this study is that a selection proportion higher than 10% of the population would be needed to identify all, or nearly all, high-yielding lines. The question is, how high? The fact that 3 of the 10 highest-yielding plots were not identified by any selector is of great interest. Although there may have been logical and important reasons for their rejection, for example, weak straw or extreme height, it would still be important to know what were the visible morphological characteristics of these plots.

Townley-Smith, Hurd, and McBean (1973) found essentially no difference in the abilities of experienced or inexperienced selectors visually to discriminate for yield among early-generation wheat lines.

Tai (1975) distinguished two levels of efficiency of visual evaluation in potatoes that would seem to exert an important influence on how visual capabilities might best be used in potato breeding. He found the efficiency (correlations) low when individual clones were evaluated when separated by seasons or different practices. However, when the selectors expanded their scrutiny to the value of different crosses, their visual score correlations rose significantly. The author's explanations seem logical: "the influential trait had poor correlations between experiments" (which I interpret to mean that G × E interactions confounded, changed, or masked the trait's expression), and the pattern of a cross population of clones forced the selectors to see a broad range of traits. For whatever reason this mental index computation seemed to represent a more stable and reproducible value. The authors concluded that visual discrimination might be concentrated on between-cross values rather than on individual clones within a cross.

Salmon and Larter (1978) worked with F_3 triticale lines from eight diverse populations sown in three-row plots with checks every seventh plot. Ten selectors made estimates in 1973 and 11 in 1974. Selectors were described as "experienced" (breeders), "novice" (graduate and postdoctoral students), and "inexperienced" (summer laborers). The charge to the selectors was to identify visually lines whose yields would equal or exceed the nearest check plot. In addition in 1974 they were required to attempt to identify the 20 highest-yielding lines in each of the eight populations.

The results of the study showed that most selectors, even the least experienced, were able to identify lines that were better than the 'Rosner' check more frequently than random expectations. However, efficiency was highly correlated with the following:

1. Selector group, in the descending order of experienced–novice–inexperienced.

2. Individuals, who showed great personal differences. Groups overlapped; for example, one novice did as well as selectors in the experienced group. The most efficient selector might not be necessarily the most successful breeder.

Overall, the study gave support to the view that visual selection, expecially by experienced breeders, was effective in identifying high-yielding lines:

> The mean yield of those lines selected visually by plant breeders (experienced selectors) did not deviate significantly from the mean yield of the best 20 selected by actual yields in populations 1, 3, and 4. In population 2, two of the three experienced selectors selected lines whose mean yields did not deviate significantly from the mean yield of the best 20 selected on the basis of yield.

Hanson, Jenkins, and Westcott (1979) were able to compare visual selection for yield in a population of F_3 single plants, and among F_5 drilled plots, of a barley cross, in both cases measured against random performance. Four breeders were each assigned one-fourth of the area and asked to choose the best high-yielding plants. The number of selections of each ranged from 90 to 148 plants. A "random" group was grown separately and remained unselected.

When the mean performance of the breeders' selections were tested for yield in the F_5 and compared with performances of other breeders and the random group, it was found that only one breeder's selections had a significantly higher mean value (at the 5% level) than that of the random group. Analysis of plant traits showed that this breeder's group of plants was taller and earlier than random. Of the others, "it was of interest that selection by the other three breeders did not result in a significant shift in population mean compared with RS values for any character measured."

The authors made a distinction between mean yield level and the number of potentially highest-yielding lines in each breeder's selection group: there were no significant differences, and numbers were proportional to breeder group and random group sizes. The conclusion reached was that selection for yield at the single plant stage was ineffective, but was effective for certain qualitative traits such as maturity and height.

The visual selection for yield of drilled plots was done on a series of random plots. The assessor "groups" were three: two were breeders operating singly, and the third represented two technicians acting jointly. Scores were assigned over a range of 1 to 9 (high yield), and were applied to a single replicate by the breeders, and to three replicates by the technician pair. When the data were collated and scores assigned to high- and low-yield groups, it was seen that the mean yield of selections scored into the high-yielding groups were dramatically higher than those of the low-yielding groups (I detected an approximate 18% difference: the high group was about 7% higher than random, and the low group was about 10% below random). Furthermore, these visual measurements could be compared with mean yields resulting from selection for nine other traits (which may be thought of as indirect selection for yield; for example, selection for high 1000-kernel weight produced a certain yield figure). The authors concluded, "Selection by eye of drilled plots therefore seems to be a better method of selecting for yield than any measurements taken in the previous season."

Turner, Hayman, Riches, Roberts, and Wilson (1953) suggested that an observer in effect operates a computer that uses a weighted index system. In such an analogy the depth of a breeder's knowledge and experience would markedly affect the success of the system.

There appears to have been little standardization of observers in the past, although some research has been done. Young, Turner, and Dolling (1960), working with fleece scoring in sheep, stated, "Where more than one observer took part in a sampling or scoring operation, standardization of observers was carried out."

Williams (1954) observed observers:

> Five observers estimated the leaf area of an average 7 leaves on 15 tomato plants. Compared with the actual measured areas, all observers underestimated the (total) area, with greater bias on the smaller leaves. However, all observers were generally successful in placing the plants in the correct size sequence.

According to Williams, Bald (1943) found empirically that the mind judges differences in the areas of individual leaves on a basis of proportionality, and he set up a 12-point scale with which the logarithmic interval was constant at 0.105. Young (1955) analyzed and suggested standardization procedures for observer differences in sheep breeding experiments.

EFFICIENCY OF VISUAL SELECTION

High efficiency of visual selection of wheat plants and plots was shown in a report by Boyce, Copp, and Frankel (1947). The selections of three observers were pitted against unselected populations for yield. The findings were as follows:

1. Visual selection was as successful as selection by grain weight in raising the average yields of progenies.
2. Selection of plants was more successful than selection of plots.
3. Visual selection of plants increased in efficiency the greater the intensity of selection.
4. Differences in selectors were evident.
5. Combining selectors' top choices raised the efficiency of selection.

The authors referred to the saving of labor of the visual over a weighing procedure.

Grafius, Nelson, and Dirks (1952) employed visual selection in the F_2 for an F_3 plant row progeny test, and were able to show yield advances that they attributed to increased homozygosity (thus additive) and larger plot size. Any further relation to visual selection efficiency was not explored inasmuch as this was not part of the research study objectives.

Visual selection for three generations from 11 bulk hybrid barley populations by Atkins (1953) showed a yield advantage for visual selection over unselected bulk populations in eight evaluated populations as follows: 1948 F_4, 1.9 bu/acre; 1949 F_3–F_5, 3.1* bu/acre; 1950 F_4–F_6, 1.0 bu/acre.

"Selection was directed toward criteria that might be expected to have greatest effect on yield"; however, the author commented that visual criteria were not particularly evident. There was a shift in the selected bulks toward taller, later plants; no conscious selection was made involving these characters. Twenty F_7 lines drawn from each of the bulk and selected bulks of four crosses were compared for yield, and the lines from the selected bulks were significantly higher yielding than those from the unselected bulks.

Wellhausen and Wortman (1954), working with maize, concluded, "In the Mexican program visual selection and early testing for combining ability among S_1 lines has been very effective, but further inbreeding plus visual selection within high combining S_1 families appears to offer limited opportunities for greater improvement in yield."

Lupton and Whitehouse (1955) believed that it was necessary for wheat breeders to supplement eye judgment in selection with early-generation yield trials of F_2 plant progenies; otherwise yielding capacity might be overlooked and valuable families or lines discarded. Visual selection could not by itself recognize the low-yielding or poor crosses, which nevertheless might contain some high-yielding segregants.

Lupton and Whitehouse (1957) based the beginning stages of their two selection proposals, the F_2 progeny method and the pedigree trial method, on visual selection. Based on the analyzed results of these two methods, they considered the F_2, F_3, and F_4 visual selection for apparently superior-yielding plants to have been very effective.

Weber (1957) made a visual selection for yield from three bulk F_6 and three bulk F_3 soybean populations in drilled (1-in.), 4-in., and 8-in. spacing arrangements. He concluded that final yield differences from visual selection were small and inconsequential.

Osler, Wellhausen, and Palacios (1958) found that visual selection for yield resulted in improvement in 57% of corn hybrid combinations studied (76 out of 134).

McKenzie and Lambert (1961) practiced divergent (good–poor) selection using visual discrimination for two years in two spring barley crosses. The paired good and poor F_6 lines, each derived from an F_2 plant, were compared for yield and seven other characters in a replicated yield test. The results showed that the visual selection within families essentially had no effect on improvement in yield. Surprisingly, and especially in the 'Montcalm' × 'Feebar' cross, there was a preponderance of statistically significant gains in all of the other six characters—surprising because they were not selected for directly by data but solely by visual means.

Correlations were also calculated between the F_3 and F_6 generations for eight characters in the two crosses. Correlations significant at the 1% level were found for all characters in both crosses except for resistance to lodging in one cross and visual yield ratings in both crosses.

These results under the conditions of this experiment are very negative regarding the usefulness of visual selection, but it must be remembered that these were within-family comparisons.

One part of the procedure was not clear to me. Seeds from 96 F_2 plants from each cross were sown in a three-replicate F_3 row test where "on the basis of yield, 'visual rating,' kernel weight, and percent extract, the twenty highest and ten lowest F_3 lines from each cross were retained for further study." If the initial F_3 selection of lines that became the family and selection streams were unwittingly preselected by being based on a confounding of actual yield and visual selection, this would have implications as to variance in the continuing family selections and to the validity of the test as a measure of visual discrimination. This is puzzling because the authors were aware of this, citing reports of other investigators of small variances attributed to a high degree of fixation, found in F_4 lines derived from a single F_3 line.

Frey (1962) made "good," "random," and "poor" plant selections in F_2 oat populations, increased the seed in F_3, and then tested the F_4 and F_5 plot performances. He found that single-plant evaluation was inadequate; however, the further visual evaluation of "good" progeny rows identified lines superior to those from "random" and "poor," and substantially higher than the F_2 lines from which they were derived. Thus one could conclude that the yielding capacity of oat lines was detectable visually, but only when viewed as populations, not individuals.

Green, Luedders, and Monaghan (1971) examined the seed from three generation lines of six soybean crosses for five visually rated seed quality traits and compared the heritability estimates thus obtained with field emergence percentages (which should reflect the quality worth of the seed). The five seed traits were defective and wrinkled seed coats, shriveled and green cotyledons, and overall appearance. In general, correlations with field emergence were low, but overall they were equal to any other rating available. The visual ratings of estimates of heritability were generally higher than for field emergence. Consequently, the authors suggested that the greatest advantage in selecting for high field emergence would be attained by selecting for seed quality by a visual rating backed by a laboratory early germination test.

Reminiscent of Harrington's mass pedigree selection method was the suggestion that "an alternative method is to concentrate on improving visually rated characters by selection when environmental conditions cause adequate expressivity."

Striving for uniform environmental conditions, Knott (1972) accurately space-planted large (800–1450 plants per cross) F_2 populations from eight diverse parent wheat crosses. The "best" plants were chosen as measured by visual appearance. For comparisons, a random sample was chosen, as well as a smaller number of "poorest" plants. In all, 958 F_3 lines were generated for a yield test that featured a Thatcher check every fifth row. The yield of each F_3 line was expressed as a percentage of its neighbor checks.

From the data it was obvious that F_2 visual selection was effective, for example, taking the three crosses in which poor, random, and good F_2 selections were made, and considering random as 100%, F_3 poor lines yielded 94% as much and good lines 108% as much (my calculations). There were, of course, overlaps in the ranges, but the point is made clearly that visual selection of F_2 plants under carefully prepared conditions is effective.

As regards any particular plant selection and line, the whole system seemed less reliable, and Knott thought it would be difficult to evolve a practical system given the nature of the first selection and the F_3 overlapping found among groups.

Meredith and Bridge (1973) found that the correlation coefficient for lint yield in cotton between the F_2, visually selected for high yield and F_3 progenies, was 0.48 and not significant. However, this is not as negative for visual discrimination as it appears at first glance. Of the 240 F_2 plants selected for the test, one-third (80) were random (R) draws. The remaining 160 came from reducing a series of five-plant draws, based on a visual assessment for high yield, from each of 10 crosses and eight replications (initial total 400), to two plants in each category (total 160). One plant (Y) was judged the highest yield (visual judgment only) and the other plant (S) was chosen by measurement as having the highest fiber strength. Therefore, a judgment was made that one plant was potentially a higher yielder than the other. This was, in fact, borne out by Y outyielding both S and R, which were equal in yield, by 5.7%.

DePauw and Shebeski (1973) corroborated the experience of several scientists that visual selection of F_2 plants for potential yield was ineffective. From a nursery of 10,000 F_2 plants of a spring wheat cross, 528 lines were advanced to an F_2 yield test that included bulk control plots. The mean yields of the two categories were not significantly different.

Visual selection played a significant role in an early-generation procedure suggested for soybeans by Thorne (1974). The original truncation for yield, reducing 3100 F_4 lines to 658 for further yield testing, was done visually. A modification proposed a tandem visual and testing selection done in the same season; presumably the visual-only selection procedure would be merged with a small replicated yield trial and the twin results used for final selection. This modification expands the effort required, although it is possible to have a two-stage approach in the same season, whereby only the visually selected lines would be harvested; this would advance yield information by one year.

Thakare and Qualset (1978) found "no disadvantage, and possible gains, by visual selection of spaced plants..." Indeed, there appeared to be sizable gains traceable to visual selection. They compared individual plant selection to three forms of family selection, and within each kind they contrasted gains made by random and visual means. The best strategy was a combination of family and within-family selection, which showed the highest gains when done visually. Ways of incorporating visual selection of plants with a multistage family selection based on means of yield were suggested.

Mundel (1972), cited by Salmon, Larter, and Gustafson (1978), found that visually selected lines from wheat F_3 yield trials gave small increases in yield.

Salmon, Larter, and Gustafson worked with triticale at Canadian and Mexican sites, and made 100 plant selections from each of four hybrid F_2 populations. On the basis of an F_3 line yield test, three bulks were formed in each population made up of the 10 highest (HY), the 10 lowest (LY), and 10 random lines (RS). Concurrently, a head row nursery of selected F_3 plants of the same origin furnished from each population a visual bulk (VS) composed of five superior plants from the 10 best head rows. The four bulks from each population were yield-tested in F_4 at two locations in Manitoba and one in Mexico. Comparisons were made as follows: VS with HY, VS + HY with RS, and LY with HY, RS, and VS. Although a variety of results occurred in populations and locations, an overall assessment indicated that "no significant difference occurred between the VS and HY bulks (Table 3). At all locations, the RS bulk was lower-yielding than the VS and HY bulks combined. Similarly, the LY bulks across all populations produced the lowest yield at all locations."

In sum, visual selection in F_3 was equal to F_3 yield testing in identifying high-yielding lines, and was better than random. Salmon et al. believed also that visual selection could be instrumental in choosing *among* populations.

Nass (1978) referred to unpublished data from a study that showed that F_2 visual head selection in spring wheat sown at crop density seeding rates resulted in higher average yields of F_4 and F_5 lines than were obtained from selection based on head weight. However, this finding was muted by the fact that the high head weight selections produced more lines yielding above the highest parent mean than did those from random or visual selection.

Qualset and Vogt (1980) evaluated four different selection strategies applied to F_3 plants. Each strategy included a part that used visual selection. Regardless of strategy, that is, in every case, "there was a numerically better selection response when visual selection was supplied." As a result Qualset and Vogt relied on visual selection of plants in their pedigree system.

Field visual ratings of cowpea lines from two crosses for yield were almost as useful as actual F_3 yield data in selecting promising lines, as shown by Ntare, Aken'ova, Redden, and Singh (1984). Visual discrimination was used initially in the F_2 to achieve a major population truncation. Later evaluation of the field ratings taken at maturity showed highly significant correlations between visual ratings and actual yields of F_3 and F_6 lines, .40** and .52** for populations A and B, respectively, in F_3, and .46** and .56** for F_6.

Frederickson and Kronstad (1985) used visual selection for early and late heading in an F_2 wheat population to test the efficiency of two breeding methods. The two methods produced essentially equal results, but gains in both maturity directions were obtained from the visually chosen base populations.

VISUAL SELECTION PARTICULARLY EFFECTIVE FOR DISCARD

Atkins (1964) visually selected F_3 plants for yield in a barley cross, using the categories, "good," "poor," and a random set. Yields were obtained in the F_5–F_7. He found the mean yields to be in the order of good–random–poor; however, the differences were small, only about 1 bu/acre. Visual discrimination was more effective for the low range. He concluded the practical value of visual selection to be negligible.

Luedders, Duclos, and Matson (1973) found that visual selection aided pedigree and early generation testing methods and was effective for eliminating low-yielding lines of soybeans. Pedigree selection, of course, relies heavily on visual stimuli.

Knott and Kumar (1975) visually rated F_3 and F_5 wheat lines for yield, and correlated the ratings with actual yield. They found the values too low and inconsistent to be useful; however, they felt that visual discrimination would be helpful in eliminating lines, which would increase efficiency by easing harvest problems.

Park, Walsh, Reinbergs, Song, and Kasha (1976) made a comparison of the field performance of their doubled haploid barley lines with lines from pedigree and single seed descent procedures. The pedigree selections were advanced by field visual selection for yield from the F_2 to the F_6. In general, they found no differences in grain yield, heading date, and plant height among the lines attributable to breeding and selection procedure; all were equally good agronomically. The pedigree method did not produce significantly higher-yielding lines; however, pedigree lines did have a higher lower limit of the yield range, suggesting that visual selection was effective in eliminating poor yielding lines.

Stuthman and Steidl (1976), working with four oat populations, made high and low visual selections among F_3 lines and compared yield results with those of randomly selected lines. In the "high" group of selections, highs yielded more than random in three populations, and in one population the yields were equal to those of random selections. Although the direction of movement was favorable, the authors concluded that visual selection, particularly when involving elimination, should be used with caution.

Ivers and Fehr (1978) used visual selection in conjunction with pedigree selection (PS) in soybeans. They found, as have many others, that "the PS method was effective in eliminating some of the low yielding F_5 lines."

In talks with my graduate students on the subject of early-generation discarding of lines, I have said in effect, "Don't dwell on possible mistakes in judgment. There is no second guessing on discarded material." This is logically and factually correct—I know of no case where rejects have been collected and evaluated against their selected peers. But a by-product of recent research by Sage, Roffey, and Stanca (1984) has provided information, a second look as it were, at discarded genotypes. This came about from a study of the adaptations of selections made from the same hybrid barley material grown simultaneously at two European locations, England and Italy, selected independently by two teams of plant breeders. The authors were interested in the adaptation shown by the surviving selections.

Three suitable crosses of two-row winter barley were chosen and the two teams working together in England selected 200 F_2 plants from each cross. The seeds harvested from each plant were divided into two lots, one for planting in Italy and the other in England. From this point forward each team made its own selections among the F_3 lines at its respective site, harvested five plants from each line, divided the seeds into two portions, and sent one lot to the other location. Each team furnished the other team the list of its selections, which were then harvested at the other location, and the seed divided and returned. Thus at both locations, each generation to F_6 included all selections made by both teams, with identical five-line groups growing side by side at each location. Visual discrimination was the criterion for selection during the early generations. Finally, yield trials were conducted after seed increases had been made.

The revealing results showed that simulataneous selection of the same lines at both sites was initially low: 16% in F_3, 13% in F_4, and 43% in F_5, the latter representing more homozygous material more readily judged visually. Further, it was seen that both teams freely selected rows from families chosen in the other environment. The authors stated that "Both teams were selecting rows from families which they had discarded in the previous season and which they only had available to them because they had been selected in the alternative environment."

The plant breeding implications are profound. It seems clear that discard decisions on the same plant material can differ according to the environment, the time, and of course, the selector. Indeed there may be value in discarded material. A correction should not go so far as the reexamination of discarded material, but I see suggestions here of positive measures that might be taken within a breeding program:

1. Use two independent selectors instead of one and accept the total number of selections.
2. Use multiple sites, selecting from the same material using divided seed lots (this would have to be F_3). Dates of planting may be superimposed on any location to add additional diversity.
3. Share seed with other breeders. For example, note the several

varieties that have been selected by different breeders from some of the Composite series of barley bulks.

4. Plant the same material for selection in successive years.

A comment on the main objective of the research of Sage et al.: visual early-generation selection had little to do with adaptation, which could be recognized only after performance trials. Therefore, any breeding system might have been used to bring material to this testing stage. Three levels of adaptation were found among the selections, namely, a highly stable group adapted to the lower-yielding environment (England), a below-average stability group associated with the higher-yielding environment (Italy), and an in-between stability group that yielded well in either environment.

DISCUSSION OF VISUAL SELECTION

Any judgment of the efficiency of visual selection is unavoidably a judgment of all breeding and selection procedures. The integration of "vision decision" with precise measurement tools is inseparable. Earlier it was stated that the breeder does not have the option of distrusting his vision; he is commanded only to do better, which he can begin to do by learning from history. To start, there are three steps that recommend themselves for adoption:

1. The breeder, because he lacks an alternative, must retain his confidence and view himself as a competent observer.

2. The breeder must test and calibrate himself to understand visual strengths and weaknesses. The test is simple: choose a replicated nursery of any crop. A nursery of promising new lines rather than established cultivars, which might be recognized, is ideal. Working "blind" without the field plan, he must score every plot in the nursery with his estimation of its yield. He should use actual yield measurements for his estimates. Before beginning to record data he should spend a few minutes in any part of the nursery getting agreement within himself as to yield levels and changes. He should not worry that estimated yield levels may be higher or lower than the actual yields that will be recorded later—these can be adjusted. He should do this at the time yields are most visually evident, probably shortly before maturity and harvest.

Field calibration will help in making visual judgments. When correlated with actual yields it allows one to compare estimation with reality. It alerts the breeder to recognize upward and downward yield swings. Repeatability may be checked from replication data. Calibration will make one a closer observer of detail, with an enhanced ability to make overall decisions.

3. The breeder should explore ways to incorporate visual discrimination steps into the wealth of breeding and selection options available. For

example, visual procedures can be adapted to control or reduce numbers in an F_2 population or in a SSD operation.

An examination of breeders' visual experiences leaves one curious as to what and how one sees. Frey (1962) referred to visual "elements." Several authors have referred to phenotypic appearance as a whole in judging yield, and have alluded to factors that confound yield estimates, such as differences within the genotype population of maturity, height, or canopy meshing.

Feekes (1952) referred to the "breeder's eye," which he characterized as the plant breeder's insight and intuition, based on his interest, talents, and experience. Historically, the term predated the era of scientific evaluation of genotypes. Apparently, Feekes felt not all plant breeders have the breeder's eye, but a gifted few are able to use this faculty to make not only preliminary selections but deeper assessments of a genotype's worth. The question remains, is this a trait capable of general development or is it simply a highly subjective ability that certain persons possess and others do not?

A startling revelation from Briggs and Shebeski's 1970 paper was that *not 1 of 14 individuals* identified 3 of the 10 highest-yielding plots, nor 6 of the 25 highest. What was there about the appearance of these high-yielding plots that masked their productivity? Much could be learned about "seeing" if these genotypes were subsequently observed in detail.

Crop reporting is done in a myriad of ways, but some industrial grain interests maintain a roving observer in a region to make independent crop production estimates that the firm might use, for example, in advance of federal crop estimates. I have known one or two of these road reporters and I believe their judgment on yield is sound. Here it is easier to visualize the mind's computer in action; indeed, we may essentially have a mean of estimated yields put together from over-the-fence glimpses of many fields.

King and Jebe (1940) explored techniques in preharvest sampling of fields for estimation and forecasting wheat yields. Their route sampling method relied on detailed analysis of data and heads collected from a standard 1/10,000 of an acre (24 in. × 26.14 in.), and represents an enhancement of "car window" visual estimates.

The computer analogy brings to mind the subject of creating an index. Every breeder has wished for an index with which the merits of each individual genotype or line could be reduced to a single number in a field book or computer pad with columns for every trait, to be scored on a 1–10 or percentage basis with each trait weighted according to its relative importance. This problem was of intense interest to me at one time, but I was never able to develop a working index. I came to the conclusion that a successful index would be very user specific—no two breeders would agree on weights. For example, the value of stem rust resistance in wheat might be one breeder's first priority and of no importance at all to a breeder in a different environment.

There are a number of conclusions to be derived from this look at visual competence in breeding and selection. First, because it is well-known that the performance of a single early-generation plant is a poor predictor of progeny performance, it is no surprise that visual estimates of early-generation plants may also be unreliable. Does this apply also to single plants of a pure line? Yes, but with a difference: the heterozygous, hybrid vigor component is gone, there remains the potential for high genotype × environment interaction, and replication is available as a check on variance. This suggests applications relative to small plots, spaced plants, and hill plots. It also suggests a window to the understanding of what to look for and "see" when observing single plants.

One way to measure and understand the worth of a single plant would be to look at individual plants of a successful cultivar (Jensen, 1967). The objective would be to examine and describe the characteristics of such single pure line plants. These characteristics were probably never seen by the breeder, because in most cases, potential cultivars are not identified early enough in the generation sequence. The cultivar's success is the result of a population of plants; nevertheless the single-plant profile of released varieties would be interesting to contemplate. If breeders do not look back, it is unlikely that they can recognize the relationship between the appearance of the single plant and its later mass population. Thus they cannot project this knowledge forward into their selection "vision." Such an "exploded" study of successful cultivars has not been done so far as I know.

From the comments of several writers, it seems clear that visual discrimination works better for the discard option than it does for retention. This means that cutoff levels for discard can be raised with a lower risk of losing valuable material. Levels of 50% have been suggested.

Visual discrimination *among* hybrid populations appears possible with emphasis on the discard option; that is, concentration would be focused on discarding populations rather than on the merits of the better ones. This might be thought of as a three-category situation. There are always populations in which the decision to save or discard is clear-cut. Doubtful populations might be carried and observed for an additional season. Also, selection on the basis of vigor seems positive, bearing in mind the transient nature of early generation hybrid vigor.

Earlier I alluded to ways of incorporating visual discrimination into plant breeding operations. The ambiguous results from visual assessments on yield suggest that other ways to use it efficiently should be considered. A sound operating procedure would be to screen first for agronomic type, disease resistance, seed quality, or in fact, any trait of a qualitative nature. Often these are simply inherited, and even the heterozygous phenotype may be recognizable. Assuming that yield genes are evenly distributed, this is an excellent way to pare down population sizes or line numbers.

This type of bypass selection procedure has broader breeding implications beyond visual considerations. It is a deferral of the yield selection option to a later generation because of the difficulties inherent in visual selection for yield and parent–progeny correlations in early segregating generations. I recognize now that the predominantly bulk population breeding system used in the Cornell University project was a bypass procedure that enabled me to pregroom populations for later individual plant selections among highly homozygous material. In the same sense the SSD is also a bypass procedure.

The subject of prediction in all its aspects is discussed elsewhere, but visual selection among hybrid populations is a viable option, particularly for the experienced breeder. The population form in which the F_2 or F_3 is viewed is not critical. Population yield data can be helpful but is not critical—the eyes alone should be able to trigger a sizable discard portion from the total number of populations examined. For the simplest illustrations consider the extremes of appearance, and the type of observable segregation in the plots; for example, some populations may be too tall, or too weak.

The question of choosing among hybrid populations is one of the most critical facing the plant breeder. Breeding programs can generate more hybrid populations than a policy of equal attention can justify; thus it is necessary to begin immediate reduction through the discard of whole populations at once. Indeed, the process of paring begins earlier: in my wheat breeding project at Cornell, 2000 hybrids might be reduced to 400 populations making an initial entrance to field operations. By the F_5 these would be pared down to about 50 for individual plant selection. This reduction was accomplished primarily be eye with some aid from performance and yield notes of the composites. Prediction can rest on a more secure base than visual discrimination only; nevertheless, visual discrimination remains a necessary and efficient tool for the early generations. Needless to say, the plant breeder must strive for objectivity and the elimination of bias in forming his ratings.

REFERENCES

Atkins, R. E. 1953. Effect of selection upon bulk hybrid barley populations. Agron. J. 45:311–314.

Atkins, R. E. 1964. Visual selection for grain yields in barley. Crop Sci. 4:494–496.

Bald, J. G. 1943. Estimation of the leaf area of potato plants for pathological studies. Phytopathology 33:922–932.

Boyce, S. W., I. G. I. Copp, and O. H. Frankel. 1947. The effectiveness of selection for yield in wheat. Heredity 1:223–233.

Briggs, K. G. and L. H. Shebeski. 1970. Visual selection for yielding ability of F_3 lines in a hard red spring wheat breeding program. Crop Sci. 10:400–402.

Briggs, K. G., D. G. Faris, and H. A. Kelker. 1978. Effectiveness of selection for plant characters of barley in simulated segregating rows. Euphytica 27:157–166.

Byth, D. E., C. R. Weber, and B. E. Caldwell. 1969. Correlated truncation selection for yield in soybeans. Crop Sci. 9:699–702.

Copeland, P. J. and R. K. Crookston. 1985. Visible indicators of physiological maturity in barley. Crop Sci. 25:843–847.

DePauw, R. M. and L. H. Shebeski. 1973. An evaluation of an early generation yield testing procedure in *Triticum aestivum*. Can. J. Plant Sci. 53:465–470.

Engledow, F. L. and S. M. Wadham. 1923. Investigations on yield in the cereals. Part I. J. Agric. Sci., Cambridge 13:390–439.

Feekes, W. 1952. Some aspects of wheat breeding. Euphytica 1:77.

Frederickson, L. J. and W. E. Kronstad. 1985. A comparison of intermating and selfing following selection for heading date in two diverse winter wheat crosses. Crop Sci. 25:555–560.

Frey, K. J. 1962. Effectiveness of visual selection upon yield in oat crosses. Crop Sci. 2:102–105.

Grafius, J. E., W. L. Nelson, and V. A. Dirks. 1952. The heritability of yield as measured by early generation bulked progenies. Agron. J. 44:253–257.

Green, D. E., V. D. Luedders, and B. J. Monaghan. 1971. Heritability and advance from selection for six soybean seed-quality characters. Crop Sci. 11:531–533.

Hanft, J. M. and R. D. Wych. 1982. Visual indicators of physiological maturity of hard red spring wheat. Crop Sci. 22:584–588.

Hanson, W. D., R. C. Leffel, and H. W. Johnson. 1962. Visual discrimination for yield among soybean phenotypes. Crop Sci. 2:93–96.

Hanson, P. R., G. Jenkins, and B. Westcott. 1979. Early generation selection in a cross of spring barley. Z. Pflanzenzuecht. 83:64–80.

Immer, F. R. 1942. Distribution of yields of single plants of varieties and F_2 crosses of barley. J. Am. Soc. Agron. 34:844–849.

Ivers, D. R. and W. R. Fehr. 1978. Evaluation of the pure-line family method for cultivar development. Crop Sci. 18:541–544.

Jensen, N. F. 1967. Agrobiology: specialization or systems analysis? Science 157:1405–1409.

King, A. J. and E. H. Jebe. 1940. An experiment in pre-harvest sampling of wheat fields. Iowa State College Agric. Exp. Stn. Res. Bul. 273:624–649.

Knott, D. R. 1972. Effects of selection for F_2 plant yield on subsequent generations in wheat. Can. J. Plant Sci. 52:721–726.

Knott, D. R. and J. Kumar. 1975. Comparison of early generation yield testing and a single seed descent procedure in wheat breeding. Crop Sci. 15:295–299.

Krull, C. F., B. M. Zapata, and O. R. Lopez. 1966. Effectiveness of visual selection for yield in wheat. Rev. Inst. Colomb. Agropec. 1:33–36 (abstr.).

Kwon, S. H. and J. H. Torrie. 1964. Visual discrimination for yield in two soybean populations. Crop Sci. 4:287–290.

Luedders, V. D., L. A. Duclos, and A. L. Matson. 1973. Bulk, pedigree, and early generation testing breeding methods compared in soybeans. Crop Sci. 13:363–364.

Lupton, F. G. H. and R. N. H. Whitehouse. 1955. Selection methods in the breeding of high yielding wheat varieties. Heredity 9:150–151 (abstr.).

Lupton, F. G. H. and R. N. H. Whitehouse. 1957. Studies on the breeding of self-pollinating cereals. Euphytica 6:169–185.

McGinnis, R. C. and L. H. Shebeski. 1968. The reliability of single plant selection for yield in F_2. Proc. 3rd Int. Wheat Genet. Symp., Canberra, Aust:410–415.

McKenzie, R. I. H. and J. W. Lambert. 1961. A comparison of F_3 lines and their related F_6 lines in two barley crosses. Crop Sci. 1:246–249.

Meredith, W. R., Jr., and R. R. Bridge. 1973. The relationship between F_2 and selected F_3 progenies in cotton (Gossypium hirsutum L.). Crop Sci. 13:354–356.

Mundel, H. H. 1972. An evaluation of tillering, grain protein, and other selection criteria, as early generation indicators of yield and quality in hard red spring wheat. Ph.D. Thesis, University of Manitoba, Winnipeg, Canada.

Nass, H. G. 1978. Comparison of selection efficiency for grain yield in two population densities of four spring wheat crosses. Crop Sci. 18:10–12.

Ntare, B. R., M. E. Aken'ova, R. J. Redden, and B. B. Singh. 1984. The effectiveness of early generation (F_3) yield testing and the single seed descent procedures in two cowpea (*Vigna unuiculata* (L.) Walp.) crosses. Euphytica 33:539–547.

Osler, R. D., E. J. Wellhausen, and G. Palacios. 1958. Effect of visual selection during inbreeding upon combining ability in corn. Agron. J. 50:45–48.

Park, S. J., E. J. Walsh, E. Reinbergs, L. S. P. Song, and K. J. Kasha. 1976. Field performance of doubled haploid barley lines in comparison with lines developed by the pedigree and single seed descent methods. Can. J. Plant Sci. 56:467–474.

Qualset, C. O. and H. E. Vogt. 1980. Efficient methods of population management and utilization in breeding wheat for Mediterranean-type climates. Proc. Third Int. Wheat Conf., Madrid, pp. 166–188.

Sage, G. C. M., A. P. Roffey, and A. M. Stanca. 1984. Simultaneous selection of segregating two-row winter barley material in England and Italy. Euphytica 33:187–198.

Salmon, D. F. and E. N. Larter. 1978. Visual selection as a method for improving yield of triticale. Crop Sci. 18:427–430.

Salmon, D. F., E. N. Larter, and J. P. Gustafson. 1978. A comparison of early generation (F_3) yield testing and pedigree selection methods in Triticale. Crop Sci. 18:673–676.

Shebeski, L. H. 1967. Wheat and breeding. *In* K. F. Neilsen (Ed.), Proc. Can. Centennial Wheat Symp. Modern Press, Saskatoon, pp. 249–272.

Sprague, G. F. and P. A. Miller. 1952. The influence of visual selection during inbreeding on combining ability in corn. Agron. J. 44:258–262.

Stuthman, D. D. and R. P. Steidl. 1976. Observed gain from visual selection for yield in diverse oat populations. Crop Sci. 16:262–264.

Tai, G. C. C. 1975. Effectiveness of visual selection for early clonal generation seedlings of potato. Crop Sci. 15:15–18.

Thakare, R. B. and C. O. Qualset. 1978. Empirical evaluation of single-plant and family selection strategies in wheat. Crop Sci. 18:115–118.

Thorne, J. C. 1974. Early generation testing and selection in soybeans: association of yields in F_3- and F_5-derived lines. Crop Sci. 14:898–900.

Townley-Smith, T. F., E. A. Hurd, and D. S. McBean. 1973. Techniques for selection for yield in wheat. Proc. 4th Int. Wheat Symp., Missouri Agric. Exp. Stn., Columbia, Mo. pp. 605–608.

Turner, H. N., R. H. Hayman, J. H. Riches, N. F. Roberts, and L. T. Wilson. 1953. Physical definition of sheep and their fleeces for breeding and husbandry studies, with particular reference to Merino sheep. C.S.I.R.O. Aust., Div. Anim. Health Prodn., Rep. No. 4 (Ser. S.W.-2).

Weber, C. R. 1957. Selection for yield in bulk hybrid soybean populations with different plant spacings. Agron. J. 49:547–548.

Wellhausen, E. J. and L. S. Wortman. 1954. Combining ability of S_1 and derived S_3 lines of corn. Agron. J. 46:86–89.

Wilcox, J. R. and W. T. Schapaugh, Jr. 1980. Effectiveness of single plant selection during successive generations of inbreeding in soybeans. Crop Sci. 20:809–811.

Williams, R. F. 1954. Estimation of leaf area for agronomic and plant physiological studies. Aust. J. Agric. Res. 5:235–246.

Young, S. S. Y. 1955. The importance of observer differences in some criteria used in sheep breeding experiments. M.Sc. Thesis, University New South Wales, Sydney, Australia.

Young, S. S. Y., H. N. Turner, and C. H. S. Dolling. 1960. Comparison of estimates of repeatability and heritability for some production traits in Merino rams and ewes. I. Repeatability. Aust. J. Agric. Res. 11:257–275.

25

SELECTION INDICES

My thinking about the role of selection indices in selection is conditioned by a belief that breeders employ fundamental index selection concepts without conscious realization that they are doing so. They are forced to operate in this fashion because the end-product commercial cultivar must meet reasonable and accepted standards across a range of traits; for example, a new variety must be productive but at the same time it cannot be too tall, or too low in quality. Selection index theory makes formal what breeders have done empirically. Furthermore, it establishes technical standards and procedures that help the breeder channel his trait selections into reasonable compromises. Selection indices serve as a form of insurance to the breeder, making it more certain that selection will result in genotypes with traits balanced according to a desired goal.

EARLY RESEARCH AND DEVELOPMENT

H. F. Smith (1936) was a pioneer in early discussions on index selection, and he believed in a "linear function of observable characters which will be the best available guide to the genetic value of each line." Although I agree with this, I think it is only proper to add that economic value must be incorporated to get the whole view of genotypic value.

Referring to efficiency in selection, Hutchinson (1940) by implication supported index selection, stating that changes sought under selection should be coordinated changes, "not merely intensification of a single character." The background to this remark involved selection among components of yield in cotton.

Hazel and Lush (1942) described and compared three of the simple methods of selection, namely, tandem, total score, and independent culling (tandem is a somewhat misleading term inasmuch as it describes a sequence of one-trait selections). For selection efficiency a breeder must know a trait's economic value and heritability, and its genetic and environmental covariance with other traits which are being selected. The authors showed that simultaneous selection (total score) for several traits is the most efficient of the three, independent culling second, and tandem method last. This paper is good background reading for an understanding of basic index selection.

Hazel's (1943) paper is one of the classics in the developing literature of index selection. Theory and practice have developed separately but concurrently in both plants and animals. Hazel's index selection, or "yardstick" as he termed it, evolved out of the animal field (swine). His method, widely used, is built around the following constants necessary to construct an index:

1. *Economic Values for Swine:* growth rate ⅓, market suitability 1, and litter size 2.
2. *Phenotypic Needs:* standard deviations and correlations.
3. *Genetic Factors:* heritability and genetic correlations.

Hazel considered progress from index selection might be 30–40% of that possible with a "perfect" index—the latter unobtainable because of the effects of insoluble phenotype–genotype confounding interactions. His comments that local indices might lack general application seem as appropriate to plants as to animals.

An early use of a primitive analog computer to give an immediate single index number incorporating the weighted values of several plant properties was described by Pate, Ewald, and Duncan (1962) for cotton breeders.

There exists a premise that the widely used Smith (1936) and Hazel (1943) selection indices are optimal measures in the sense that "optimal" implies the existence or need of a high degree of accuracy of estimates of genetic variances and covariances, and that such conditions are difficult to meet in some cases owing to the sampling size variations in different populations. Williams (1962) reviewed and analyzed this situation, which in practical terms can lead to an overstatement of expected genetic gain. From several alternatives discussed by Williams, one, the base index, has received attention as a supplement or adjunct to other index procedures.

Elston (1963) addressed the problem of interpreting rankings of individuals obtained by scoring and summing more than one trait. This problem encompasses the common breeding observation of negative relations between component traits when selection for one denigrates the other. Index selection attempts to mediate the differences, usually resulting in slower advancement of one and compromise for both.

Elston's solution is an (economic) weight-free index that is multiplicative in nature, with the line values obtained from phenotypic deviations for each trait. An important feature is that the index becomes operable without the necessity of calculating estimates of genetic variances and covariances.

Caldwell and Weber (1965), using a group of seven traits for their indices, compared general, average, and specific indices in tests for selection efficiency in F_4 and F_5 populations of soybeans. The justification for these tests rested on the different effort levels required to prepare each type, that is, establishing the parameter estimates for traits and the actual construction of the index. These differences are readily apparent when the parameter estimates required are matched to the type of index:

General: obtained from unrelated populations.

Average: obtained from related populations.

Specific: obtained from a particular population.

The results showed that very small gains in selection advance were attained through specific indices as compared to average or general ones. Furthermore, the general index was favored because of its simpler construction and wider potential use. The authors pointed out that if a series of closely related populations were to be used, the preparation of a specific index on one population for use on all could be efficient.

An interesting observation was the good showing of selection on yield performance alone (involving two replications)—it was almost as efficient as specific index selection.

Byth, Caldwell, and Weber (1969) applied general, average, and specific index selection to two soybean populations grown in three environments. Definitions of these three indices were based on the source of the parameter estimates used; that is, "general" represented all other populations, "average" represented pooled data from various crosses being studied, and "specific" represented data from a particular cross population. Seven important soybean traits including yield were the indices. Weights for phenotypic traits were based on genetic and phenotypic variances and covariances established in a previous study.

Specific indices predicted greater genetic advance, and in fact, actual advances exceeded those of average indices. However, there were a number of situations where one or more of the six indices studied exceeded the specific indices (considering the environment and generation options).

The study developed information of interest about two phenomena:

1. Although the greatest predicted advances were in nonstress environments, the actual realized genetic gain was found to be higher in a low-yield, moisture-stressed environment than in an irrigated high-yield environment. This is the opposite of certain findings in other crops (the option should be held open that crop species may respond differently).

2. Index selection may not be the best indicator or selector for yield. Some of the traits may have inverse relationships for yield. The authors found that the most reliable yield indicator, surprisingly, was early lodging, "a character describing canopy shape and uprightness during pod filling and presumably related to light penetration into the canopy at that stage."

Pesek and Baker (1969a) compared two-stage tandem selection with index selection in a computer simulation study which involved a number of assumptions as to the genetic situation in a population of 600 individuals tracing by single seed descent to 600 F_2 individuals. Selection was practiced on the F_6 and F_7.

Two-stage tandem selection involves the alternate selection of two traits. After a selection is made for trait X, the selected individuals are selfed and selection made for trait Y. The rapid reduction in variance in this system is obvious. In reading this I wondered why a two-population scheme similar to reciprocal recurrent selection might not be employed: in one population Y would be the first trait selected, in the other population X would be first.

Index selection involves the creation of a single number index from the assembly of various parameters of interest such as phenotypic, economic, and genetic. These measurements are used to score and select individuals from each generation.

In the simulation study it was possible to use the same F_6 and F_7 populations to compare the two selection methods. An arbitrary selection intensity of 40% per generation provided an effective 16% final save after two years. Factored into the study were varying levels of linkage intensities, environmental variances, heritabilities, and economic values of two traits. Especially, a negative correlation between the two traits (linkage, repulsion) was inserted. The selected populations were compared according to their "genetic worth," a calculation that expressed the genetic gain for each trait in terms of its economic value. The results, from all aspects, showed that index selection was superior to tandem selection, actually by a substantial margin. The authors concede that index selection requires more effort from the breeder.

The importance of Pesek and Baker's study is not, I think, in the results themselves, but in the insights that lead to an understanding of these very complicated biological situations. After all, a simulated study is not "real life" but only an approximation guided by several assumptions. The simulation tells us what would happen if the assumptions were correct and comprehensive. Nevertheless, the superiority of index selection is impressive, and plant breeders must ask themselves whether they can afford to ignore it.

Pesek and Baker (1969b) showed a recognition of plant breeders' problems with index selection when they noted that selection index principles are limited by the problem of affixing relative economic weights to the quantitative traits involved. They proposed a modified formula for index selection that substitutes "desired gains" for the economic weights, believing that a breeder can express more readily what amount of improvement he would accept in a particular cycle of selection. This is an attractive thought. The breeder is, in fact, constructing a framework of compromise when he builds an index, and the modified theory permits an unfettered consideration of true value relationships of the various traits.

In another paper, Pesek and Baker (1970) made an important point about selection indices that is often not appreciated by breeders: their principal use is application not to homozygous already selected lines under nursery testing, but rather to recurrent selection programs. In this paper they developed an intriguing modification of index selection that emphasizes "desired genetic gains" in contrast to other methods that employ economic value weights (even though in some cases the weights are set at zero for secondary traits). The problems associated with the assignment of economic weights led to their concept of desired genetic gains which requires no economic weights at all. The concept may be placed in the evolutionary development framework of index selection as following Kempthorne and Nordskog's (1959) "restricted selection" and Tallis's (1962) "optimum genotype."

Line material for the study was generated from the unselected F_9 of a spring wheat cross. Pesek and Baker discussed the requirements of quantitative data and estimates of genetic parameters, which were extensive. Once these had been obtained, the authors' illustration was helpful: a wish list specifying desires for each trait is stated, for example, "a decrease of 4.83 days (5%) in days to ripen." These were used in conjunction with estimates of genetic variances and covariances in linear equations that solved for index coefficients; for example,the index coefficient for days to ripen was -14.14. The index value of a line was obtained as the sum for all characters of the index coefficient times its average; in this case, the single character of value days to ripen was obtained by multiplying $-14.14 \times$ average days to ripen. The breeder may select the parents as x lines from a ranked order of index values.

The procedures are attractive; however, the work load is daunting. It helps to think of this in terms of elite combinations in a recurrent selection program where the effort is related to choosing the parents that will generate the next cycle. Intermating should be helpful, and of course is not separable from the recurrent selection approach. Pesek and Baker recommended taking more than two lines for intercrossing to start a new cycle for selection. The diallel selective mating scheme would be particularly appropriate for this situation.

Eagles and Frey (1974), in a paper incidentally notable for its recognition of a second trait of economic value in oats (the value of the straw is often ignored in computing economic return), compared five selection methods of increasing the economic value of the oat crop via the components, grain and straw yields. The five methods were as follows:

1. Smith's 1936 index
2. William's 1962 base index
3. Independent culling levels (Young and Weiler, 1960)
4. Grain alone
5. Straw alone.

The authors' goal was to find the best procedure for selecting cultivars having the greatest economic return per acre.

Whether breeders will adopt or adapt some form of index selection in their breeding programs will depend to a large degree on their estimate of the ratio of expected gains to added costs. Generally, what breeders do now is to select for grain alone and straw alone and mentally assign these to a judgment procedure that weighs a number of observed things, such as straw height and lodging, tillering compensation with short straw, heavier weighting to grain than straw, straw strength and yield, and so forth. Usually the result is a cultivar acceptable to grower and industry; however, it would be difficult to know what tandem gains in grain and straw were made or what gains might have been possible. Eagles and Frey found that selection alone for either grain or straw was the least efficient of the five methods tested. The two indices, Smith's and William's, were about equal in efficiency of selection for genetic gain. This would favor the use of William's base index on the basis of its simplicity, because the weighting and estimates of population parameters required for Smith's index (and independent culling levels) are not needed.

Eagles and Frey noted the option of using combinations of selection methods applied as appropriate throughout the life cycle of the crop. In the Iowa program, independent culling levels calibrated to check cultivar standards were used during the growing season prior to harvest, and an index was applied to grain and straw yields for selection after harvest.

The need for continuous updating of indices because of changes in product requirements is illustrated in an article by McClintic (1986). Recent technological improvements in yarn-making machines required cotton with stronger fiber regardless of length. The stronger fiber is genetically available but difficult to combine with acceptable high yields. The use of tandem or higher index selection methods in breeding cotton might be helpful.

RESEARCH ON THE NATURE OF INDEX SELECTION

Association between characters is important knowledge to the breeder because it gives additional approaches to improvement of a character. The ideal would be an association between a visible qualitative character and a difficult quantitative one. Johnson, Robinson, and Comstock (1955) reported on correlation coefficients, both genotypic and phenotypic, in a range of 24 soybean traits in two populations. The evaluation was done in the F_4 on family lines tracing to individual F_2 plants.

The researchers found general agreement in two situations; namely, correlations were more credible (consistent) when measured on a plot rather than plant basis, and genotypic correlations were higher than phenotypic ones.

The important quantitative character, yield, was positively correlated with long fruiting period, lateness, heavy seeds, resistance to shattering, low protein, and high oil. Surprisingly, these offered good individual opportunities for increasing yield because of their high heritabilities (actually higher than that for yield). The known negative relationship between oil and protein was confirmed. High oil was associated with high yield and with long fruiting period, both favorable to yield improvement. Their Tables 3 and 4 showed expected progress in yield and oil percentage, respectively, when selection was for designated characters.

Perhaps because of the large number of characters studied it is difficult to ignore their relation to index selection. The authors considered building selection indices based on their findings. Their Table 5 put together expected advances in yield expressed as percentages from selecting for yield alone, and Table 6 did the same for oil. There seems little question that index selection could be very effective; however, analysis showed that in some cases, notably oil percentage, selection for oil alone was as effective as and simpler than using an index of characters genetically highly correlated with oil.

Johnson et al. dwelt on the breeding implications of genotypic and phenotypic correlations between characters, pointing out that selection for an indicator trait is valuable to the extent that it improves the desirable target character. Positive correlations may identify an otherwise negligible character as being of use to advance improvement in another. They suggested a general approach of first selecting for important characters other than yield, which requires more extensive testing. In the process, yield might receive some benefit through an awareness of indicator traits correlated with it. The same awareness for negative correlations must be kept in mind. Specifically related to soybeans, the authors pointed out that character associations are generally favorable to the breeder's objectives if oil (correlated positively with yield) retains an economic emphasis over protein.

Is linkage the cause of some of the correlations? The authors said it cannot be ignored, but there was enough evidence of random assortment of genes and indications of other restraining plant processes to believe that linkage would not be a major influence.

The complexities and uncertainties connected with constructing and operating an index are well illustrated in the experiences of Brim, Johnson, and Cockerham (1959). They built indices using 15 combinations of six characters in soybeans: weight of oil, protein, and seed, seed yield, lodging resistance, and fruiting period. The relative value of oil and protein was weighted according to the historic average price ratio, approximating 1 : 0.6. The two populations they examined did not respond in a similar manner. In general the authors were cautious about the value of indices. This caution seemed to turn on the *b* values (the weights assigned to the various traits). In many cases the *b* values must be estimates or have a fluctuating movement, such as market price. Thus it would to possible to calculate genetic advance in several different ways depending on choice of *b* values. They pointed out that "in some crops, a single character such as yield largely determines worth, and in others the definition of worth is very complex."

A further uncertainty lay in the observation that differences in populations, as to their variances and covariances might invalidate a "universal" index and require indices to be population-specific. Nevertheless, the authors pointed out that the "alternatives, such as independent culling levels, visual appraisals or mental thumbrules, do not overcome any of the objections to indexes."

Thoday (1960) dispensed with confusing terminology with the following statement: "Stabilising selection is selection for the mean. Disruptive selection is selection away from the mean in both directions in the same population."

He compared the two in populations of *Drosophila melanogaster*, the fruit fly. The principal findings were that stabilizing selection reduced variance and was favorable to the establishment of repulsion linkage complexes, whereas disruptive selection increased variance, so much so that a polymorphism trend was noted, and coupling phase linkages were favored.

It is easy for a plant breeder to understand the role of stabilizing selection; it is, in fact, somewhat akin to index selection. Disruptive selection, on the other hand, is more like independent culling, or tandem selection, where cutoff levels play a part, and where, to complete the analogy, the extreme selections for the different traits are tossed back into a common population. An interesting thought is the nature of the population that would come from intermating the extreme selections from independent culling or tandem selection, in other words, using indices to choose parents by concentrating trait expression.

Young (1961) reaffirmed Hazel and Lush's 1942 assessment that a well-constructed index provided a higher expected genetic gain than independent culling levels, which in turn were more precise than tandem selection. One cannot read Young's paper without being aware of the myriad of detail involved in index analysis, ranging from environment to economics, heritabilities, weighting considerations, genotypic and phenotypic correlations, number of characters, the selection pressure or intensity, assumptions, method to choose, market value, accuracy of estimates, and so forth.

In an obscure way the body of literature and theoretical activity surrounding the subject of optimizing selection [e.g., see Jain and Allard (1965) for one discussion on this subject] is related to index selection in the practical arena of plant breeding. Optimizing selection is based on the presumption that under natural selection there is a predilection for phenotypically intermediate individuals to be favored or more fit. On occasion breeders have observed this phenomenon, teleologically almost as though the population shunned the rarefied air of extremes and drew toward the comfortable middle. Index selection recognizes the necessity for compromises between the "selfish" level attainable by concentration on one character and the modest shared gains possible through compromise.

The suggestion by Thoday (1960) that disruptive selection showed a developing trend toward polymorphism in the fruit fly was independently observed and corroborated by Doggett and Majisu (1968) from an extensive hybrid index survey based on plant characters of the cultivated and wild members of *Sorghum* in Africa. The species and subspecies formed distinct populations, which, however, could be fitted into a polymorphic frequency distribution complete with "tails" that connected the wild and domestic extremes. Disruptive rather than directional selection was postulated because both natural and human selection favored different phenotypes in the populations. Gene flow between wild and domesticated types was relatively unimpeded.

Without attempting a detailed analysis, I nevertheless want to record the existence of several papers authored by Adams (1967), Grafius (1969), Grafius and Thomas (1971), and Thomas, Grafius, and Hahn (1970, 1971). The topic was stress and the correlation of sequential traits, for example, components of yield. It seems that the first trait in a series is more likely to be subject to genetic control, whereas later developing traits have lost, in a sense, degrees of freedom and are more subject to prior restrictions and the influence of environment, thus producing an oscillatory pattern. Teleologically speaking, a successful variety adopts a strategy for high yield in the manner in which it meets the strictures of available resources (actually, it was selected by the breeder because these responses had been observed).

The statistical properties of small indices, for example, treating a ratio of two traits as though it were a unit, was examined by Ikehashi and Ito (1971). They referred to their quotients as plant type indices that show the relationship, correlation, regression, and heritability of significant components of type or yield. The concept is familiar, for example, in harvest index, a ratio that has been treated as an entity and examined for associations and heritability as though it were a single character. In this study the statistical derivations were the focus of emphasis. The practical application of the formulas is illustrated in a rice cross in which five traits were studied in F_3 lines. One interesting result was that the heritability of the quotient of two traits is sometimes found to be greater than either of the components; for example, the respective heritabilities for panicle weight and total weight were 0.47 and 0.50, whereas the heritability of the ratio quotient was 0.85. The authors viewed such cases as efficient selection indices.

Singh and Bellman (1972) dealt with the problem of an index's breadth and suitability. They postulated that indices are population-sensitive, sometimes tailored to "specific" populations and environmental situations, and thus likely to be efficient. In order to get an efficient "general" index, the sample base must be enlarged and the parameters broadened to correspond to general populations. To illustrate, an efficient specific index might be designed around the parameters of a known population to be grown in a known environment, but might be totally unsuitable if the population, or species, or environment were changed.

Subandi, Compton, and Empig (1973) summarized the important characteristics as follows:

1. *Construction Procedures and Details Needed for Indices Regarding Traits:* (a) relative economic values, (b) phenotypic and genotypic variances, (c) phenotypic and genotypic covariances per pair of traits.
2. *Advantages of Indices:* (a) indices are generally superior to other selection procedures, and (b) they are more efficient than selection based solely on trait.
3. *Disadvantages of Indices:* (a) high genetic correlation of traits are needed for efficiency, (b) there are problems in assigning economic importance to traits, (c) obtaining reliable estimates of variances and covariances is difficult, and (d) extensive computations are required.

The authors tested five indices in corn populations and found one, a multiplication type that involved yield × % lodging × % dropped ears, better than three others and equal to Smith's 1936 index.

Fatunla and Frey (1976) explored the degree of repeatability of regression stability indices and were surprised to find it low in view of earlier positive indications that production stability was a heritable, selectable trait, as noted by Perkins and Jinks (1968), Bucio Alanis, Perkins, and Jinks (1969), Finlay (1971), and Frey (1972), and as implied by knowledge of individual and populational buffering.

A positive interpretation of these findings concerns the merits of using the set of all entries [Finlay (1963 and 1971) and Eberhart and Russell (1969)] or a set of separate check cultivars (Freeman and Perkins, 1971) to determine environmental productivity indices. Fatunla and Frey found that the use of either set made no difference in the relative ranking of their experimental strains. However, the use of a single standard set of cultivars would make comparisons possible over a series of years.

MODAL SELECTION

Modal selection as proposed by Manning (1951 and 1955) was conceived as a stabilizing selection to further his goal of a stable check variety that might be used to measure future breeding gains. When I first encountered modal selection I saw it as a form of index selection that represented a compromise reached by tolerating gains less than those obtainable for any one trait in order to produce acceptable gains for all traits. Manning, however, took the route of minimum deviations from the mean, a procedure that should have assured constancy and stability. On the contrary, he found a slow but steady increase in yield over the years, raising questions about stability.

Elements of the similarity of modal selection to index selection are found in Manning's 1956 paper, in which index selection was used for yield improvement in cotton. One of his checks or standards was a modal bulk constructed from the same population. Means and standard deviations were calculated for the three traits lint per seed, lint length, and seed weight per seed successively, so that after each individual trait calculation those individuals deviating by more than one standard deviation were removed, and calculations made for the next trait were applied to the remaining individuals. This resulted in the elimination of slightly less than half of the original individuals.

Both the index selection and modal selection took place in the 'BP52', variety which had originated as a single-plant selection and had been maintained as an open-pollinated bulk with some selection for more than a decade. The yield advance attempted was on the basis of a selection index involving three traits, namely, total lint yield, seeds per boll, and lint per seed (lint index). The yield advances realized were based on large-scale yield trials and were compared against projected modal yields that had built-in escalator figures based on Manning's earlier estimates of per-year gains; thus yield advances were net gains after subtracting the modal bulk figures (actually the total gains would be the sum, as the author indicated).

The results were astounding, given the presumed homogeneity of the base BP52 variety, amounting to an estimated gain of 38% over six years, or about 6%/year. These results are difficult to reconcile without postulating a cause for the large variability found in the population; perhaps the answer lies in the fact that the population had been maintained in open-pollinated bulks for 10 years.

Walker (1964) reported on further investigations concerning the unexpected yield advance. Lint length increased for five generations, after which the mean then stabilized. Heterzygote advantage appeared to be a strong force, with extremes eliminated but variance remaining high through breeding. This implies heterozygosity with low heterosis. Lint length increase was attributed to partial dominance, and yield increase to heterosis and nonallelic interaction. A very important finding was that the order of selection of traits sensibly affected the truncation outcome. This is a phenomenon on which more research is needed.

As background, the amount of outcrossing in cotton in the research region (Uganda) was of the order of 20–30%, but it is implied that the population used, BP52, was closely maintained so that outcrossing would be largely sib crossing—still a force in generating intravarietal variability. Walker's studies convinced him that reported yield gains from modal selection were overestimated and scarcely great enough to inspire adoption of the program for yield improvement. Walker's studies showed that Manning's first goal, stabilization, was in fact achieved. There is ambiguity as to the status of yield improvement, notably as to the causes of the small gains. Where sampling is involved, in this case field and laboratory, there are the possibilities both of unknown biases and probable causes: Manning suggested three possibilities, but Walker concluded that heterozygote advantage, partial dominance, and pleiotropism were the more likely causes. Although not powerful enough as a breeding method, modal selection could be useful in maintaining seed stocks. In that sense it could be applied to more completely self-fertilized crops, but there would then arise the question of whether the tedious examinations would be superior to the more usual techniques of visually selecting and examining lines from a cultivar and recombining to form a new seed stock.

In modal selection, mean values of a population become more important than extremes. Modal selection is imposed on bulk populations, proceeding from an initial large sample of selected plants. Laboratory analysis is then made of the attributes of several important traits on each plant, always taken in the same order, and truncation occurs through the discard of individuals that deviate by more than one standard deviation in either direction. This process is continued for several generations. Readers may be puzzled, as I was, as to the basis for deciding which characters should be selected first. Is it not possible that the order of trait selection might be associated with the unexplained yield increases?

The original concept was that modal selection would act to stabilize populations (varieties, stocks, and checks). Instead it was found that this had not occurred for yield, which in the early attempts showed an average 2–3% gain per generation. Because of this Manning (1963), who had done the early research on cotton, suggested that modal selection might be used as a breeding method even though the reasons for its producing yield gains were not fully understood.

The plant breeding implications are several even though the populations dealt with were different. There are the interesting situations of both genetic advance and stabilization, maintenance of high genetic variance, and the implied opportunity for selection under genotype × environment situations. Walker concluded that modal plant selection was not suitable for single plants because of the experienced large variability. I do not fully comprehend the point that modal selection favors heterozygotes that adhere to midparent values, unless it is that the constant elimination of extremes gradually reduces the heterotic range. Another explanation could be traced to Sakai and Gotoh's (1955) finding that F_1 barley hybrids showed heterosis for characters but not for competitive ability.

Riggs (1967) cast doubt on the value of modal bulk selection in cotton breeding even though extensive testing showed approximately a 1%/year improvement in yield. Modal selection originally was seen as a way to impose a stabilizing influence on a population that could thereafter be used as a comparison base. This was accomplished by examining and measuring a sufficient sample of plants and saving all those whose characters closely matched the norm or mode. This ought to create a stable population with time. That it does not always do that—improving slightly in yield and harboring individuals that later exceed the permissible deviation from the mode—is believed due to heterozygote advantage and bias, perhaps unintended, in the sampling and selection activities. Also, natural adaptation of populations to the sites where they are grown plays a part. For example, all selections at one site produced a bulk population maturing much earlier than populations from other sites. Riggs believed that the yield gains were too small to characterize modal selection as a useful breeding technique.

THE ROLE AND IMPORTANCE OF ASSUMPTIONS

Lupton and Whitehouse (1957), in discussing selection schemes, alluded to difficulties if a selection method presumes "knowledge of the mode of inheritance of a large number of characters. This is an ideal situation which is unlikely to be attained in the cereals, particularly when considering such complex characters as yield, strength of straw and milling and baking quality, where it is probable that polygenically controlled systems are involved. Even for characters such as disease resistance, in which inheritance is more likely to be simple, the picture is often complicated by polyploidy so that in a particular cross the number of factors determining the inheritance of any character will rarely be known in advance."

This situation relative to assumptions is also applicable to the problems of constructing a selection index.

Looking at correlations in plant breeding, Gilbert (1961) came to two conclusions: that any correlation, no matter how small, between two characters (in this case, between seedling and adult characters) can be effective in selection and that small correlations require increases in progeny size. Concerning index selection, the question comes to mind as to whether there might exist a correlation between an unnoticed trait and a desirable trait, which if known and the minor trait selected for, might be used to achieve gains in the desirable character.

Heidhues (1961) examined the problem of the accuracy of selection indices and concluded from a computer simulation study that accuracy depended on errorfree estimates of the genetic parameters. Accuracy increased with each increment of knowledge about the genetic structure of a population. The use of assumptions without logical bases was discouraged.

To a breeder unacquainted with the use of indices, the prospect of using or constructing one is daunting. Sprague (1966) and others have pointed out that a credible selection index requires considerable preknowledge: for each character, its genetic and phenotypic variances and correlations, as well as the economic weights to be assigned, must be known.

APPLICATIONS OF INDEX SELECTION

Robinson, Comstock, and Harvey (1951) concluded that a form of index selection would be useful for selection in maize. This was indicated from a consideration of the fact that grain yield, although of primary concern, is only one of a few characters having economic importance. For example, lodging affects the harvesting quality and collection of grain yield, and other plant parts are valuable as livestock fodder. Their paper dealt with the construction of indices with various combinations, and assigned weights for eight characters.

Caldwell, Weber, and Byth (1966) used a six-trait weighted-average index in soybeans—the construction of an average index was reported previously (Caldwell and Weber, 1965). The traits were removed, one by one, from the index in the study. Actual gains were higher than predicted, but trends were similar. When yield was removed from the index, predicted yield dropped 38.9 lb/acre, but yield values remained remarkably stable in all cases where other attributes were eliminated. Yield and oil, yield and protein, and yield, oil, and protein were necessary, separately, to advance oil, protein, or both, respectively. Yield alone was not bad except that it discriminated heavily against advancing oil gains.

Andrus and Bohn (1967) used index mass selection successfully in cantaloupe breeding. Scores were taken on 18 characters, on a scale of 1–3 not acceptable and 4–5 acceptable, to produce a total acceptability score. It was quite possible to have high total scores that included low individual component scores. The authors believed fine-tuning of weighting factors would improve the worth of the procedures.

In a review of selection and stability in plant breeding, Peirce (1968) favorably considered index construction as a possible aid in selection problems.

A tandem or indirect selection scheme based on partitioning selections for both yield and seed weight via a scattergram was proposed by Frey and Huang (1969). The partitioning resulted in an approximate 10% selection.

Watkins and Spangelo (1971) used selection indices to measure genetic progress under three breeding procedures in strawberries. A high-quality successful commercial cultivar, 'Redcoat', was used as a standard for comparison. The three procedures, all employing index selection for the same seven traits, were:

1. Selection based on additive variance
2. Selection based on total genetic variance
3. Progeny selection followed by individual selection.

The third procedure was deemed the most promising. The progress was greatest for six of the seven characters, only berry appearance failing to show progress. This was not considered a serious defect inasmuch as it was understood that Redcoat exemplified the highest expression of appearance and represented a visible comparison goal. An important consideration in choosing progeny selection followed by individual selection was that the index predicted 10% genetic progress—almost as great as that predicted for yield—therefore, it appeared possible to improve yield without sacrificing advances in other traits. Furthermore, the method was amenable to modification to suit the breeder; for example, both progeny and individual selection intensities might be altered. Another illustration was that the between-line selection could be broadened by raising the progeny selection and lowering the within-line individual selection (or vice versa to reverse direction).

Watkins and Spangelo noted the higher cost of index selection and suggested for higher efficiency that a prescreening for simply inherited major genes of economic importance would be helpful; for example, the cross's genetic base could be improved by backcrossing for disease resistance.

Several of the authors' comments seem appropriate and helpful to understanding the why of index selection. Selection for a single character, for example, yield, tends to neglect other characters important to breeding objectives. If it is shown that index selection works then the breeder has an excellent chance of devising indices suitable to his own concepts of weights and so forth. Experience has shown that a range of weights, if not extreme, results in the selection of similar plants. Thus once begun, indices may be modified with experience.

Campbell and Davis (1972) reported that a correlation of .74** was found between predicted and actual fresh cut sweet corn yield rankings among 28 single crosses. A "total score" was calculated for each single cross based on a five-element index applied to the inbreds. By using cutoff points in the predicted rankings, the index could be used to predict the highest yielding hybrids and to discard others.

The need for compromise in the use of indices is shown in the research study by Rosielle and Frey (1975). Unrestricted selection for oat grain yield showed impressive gains, but these were accompanied by an unwanted increase in height and days to heading. This situation could be avoided by earlier culling for acceptable height and maturity; however, this runs the risk of loss of high-yielding lines. Using rigid and restricted selection on height and heading date, and following the models modifying Smith's index (Kempthorne and Nordskog, 1959; Tallis, 1962; Pesek and Baker, 1970), the authors found that efficiency was 57% of the unrestricted rate.

Among several variations of indices tried, one combination involving harvest index raised the efficiency to 70%. This index kept the weights for height and heading dates low, and rated harvest index above grain yield.

The unrestricted selection for grain yield in oats described by Rosielle and Frey suggests other interesting methodological approaches. The high yields of the selected plants could be "locked in" in a new hybrid population formed by a diallel crossing scheme using these individuals as parents. The culling for unacceptable extremes of height and maturity could then be applied to the F_2 of the new population. This scheme would be more feasible in wheat or barley than in oats because of greater crossing difficulties in oats.

Radwan and Momtaz (1975) evaluated selection indices based on the three economically important flax characters, straw yield (A), seed yield (B), and earliness (C). These were arbitrarily assigned economic weights of 13 for each gram of straw, 65 for each gram of seed, and 10 for every day to first flower. The seven indices included and evaluated each character alone plus all combinations. The F_2 data provided all phenotypic and genotypic variances for the three characters. These were applied over the F_2–F_5 populations, inclusive, of a two-parent hybrid population. Yield and earliness trials were conducted from F_3 to F_5. Entries included the parents, a bulk, and best families as scored by each of the seven indices. A breeding nursery paralleled the trials to provide more statistical detail and parameters of the indices.

The results were favorable to the use of index selection. Inasmuch as plants were scored by a succession of indices, there was considerable overlap of plants and progenies being retained (no plant, superior in even one trait, was discarded). In this sense, early use of a simple selection index, although it would increase selection numbers, would protect the breeder from discarding potentially valuable "one-merit" germ plasm. Genotypes of this kind often have value as parents in further intercrossing.

The means of lines scored by any of the indices were significantly higher in all generations than the means of the control parents and bulks. The ABC index was the most effective for advancing seed and straw yields, and did as well in maintaining earliness, which is an accomplishement in view of the positive relationship between yield and lateness. The authors stated that "the relatively low heritability estimates of the characters justified the use of selection indices."

Following up earlier research on choosing parents and predicting the best source of progenies, Sampson (1976) explored a range of options and concluded with observations on the construction and use of selection indices. Sampson worked with 28 oat hybrids derived from diallel crosses of eight cultivars. F_1s, F_2s, and F_3s were examined at different spacings and under different moisture regimens. Nine traits were followed that provided the opportunity for index studies involving four traits. The study provided information and insights which led to the revision of some earlier-held expectations.

Sampson's approach involved the following:

1. Obtaining general combining ability (GCA) effects of each parent to suggest the breeding value for separate traits.
2. Evaluating the progenies for combinations of desired characters.
3. Making a study of the correlation of these traits.
4. Applying the use of index selection to obtain the multiple traits. In

addition Sampson rhetorically asked whether one could obtain the same results by predicting performance from the parents themselves.

GCA parental effects were derived from the deviation of the mean of seven progenies, tracing to a particular parent, from the overall mean of the 28 progenies. Selection indices were constructed, that is, weighted for goals, for four traits (Pesek and Baker's 1970 method) using extracted components of variance and covariance from the parents and progenies. In an earlier study Sampson had used control cultivars as the goals base. The basis of progeny yields was yield per area rather than primary components of yield.

The results for the arrayed GCA effects quickly identified the best parents for individual traits studied. At this point, Sampson considered the correlations found among traits and was faced with the necessity for compromises, because the nature of the correlations and heritabilities precluded the probability of getting all desirable traits at their highest expression. Thus Sampson set up two indices to represent different goals: the first goal stressed plant type at the expense of grain characters; the second maximized yield and 1000-kernel weight at the expense of short stature and thick stems (lodging resistance). From an examination of the top progenies in the two indices over two years, the 28 progenies could be stratified for rating high in all indices; 15 progenies fell within the top one-fourth of all indices and of these, two crosses (progenies) stood out from the others by falling in the top quarter both years. Furthermore, when Sampson examined the parents of the 15 progenies, he found that the same parents reappeared in many of the crosses; for example, the leading variety, 'Astro', appeared six times. Astro and 'Clintford' were rated the best overall parents.

Sampson applied the index goals to the 28 midparent figures in an attempt to find out whether the parent plants could have furnished the same predictive values. Correlating these with the progenies he obtained nonsignificant values. Furthermore, calculations of GCA effects produced conflicting indications. On the matter of progeny performance over generations (F_1, F_2, and F_3), Sampson found good agreement between the first year's F_1 test and the third year's F_3 test, but not with the second year's F_2 test. However, an important relationship was seen here: the first and third years were alike in that no moisture stress was present whereas there was stress in the second year. Therefore he concluded that "cage-grown F_1 plants can be predictive of field performance under optimum conditions of soil moisture." In addition, Sampson stated on the basis of a high correlation between midparents and F_3 progenies for yield that "midparent values can be reliable predictors in non-stress environments."

Rosielle, Eagles, and Frey (1977) added the value of oat straw and constructed 12 selection indices (four unrestricted and eight restricted) to measure the selection potential for economic value of biological yield. General conclusions were that heading date and height restrictions affected straw yields more than grain; plant weight and grain yield were highly correlated; short, early cultivars having high yields of both straw and grain would be difficult to select; and plant weight alone is a good indicator of economic value.

Lin (1978) gave detailed guidelines for calculating parameters of index components.

Miller, James, and Lyrene (1978) constructed different selection indices for sugarcane. As expected, the greatest genetic advance for yield was on metric tons of cane per hectare (not an index), but of course the objective of an index is to advance several traits and avoid losses in attempts at specific genetic gains. An index based on stalk length, stalk diameter, and stalk number resulted in yield gains of 89% of selection for yield alone. When the emphasis was put on yield of sucrose, the addition of Brix (% soluble solids in juice) to the index allowed yield gains of sucrose of 92% of that expected from selection for sucrose alone. An interesting conclusion was that data from indices could be obtained more easily and at less cost than the current procedures of collecting, hauling, and milling.

A merit competition to sort out the relative efficiencies of nine forms of selection indices and single-trait selection, applied to three components of cold tolerance in two corn populations, was conducted by Crosbie, Mock, and Smith (1980). The two corn populations were part of a recurrent selection program to advance cold tolerance. Mock and Eberhart (1972) proposed simultaneous improvement in three components of cold tolerance, and Mock and Bakri (1976) evaluated economic weights assigned to index traits and pointed out that incorrect weights might distort results because they force trait estimates out of proper balance. In Crosbie, Mock, and Smith's study, the cold tolerance characters were percentage and rate of emergence, and seedling dry weight. These single-trait indices were considered to represent 100 (%), and selection differentials were expressed as percentages of their figures. Additionally, a base index was constructed of equal weights for each trait. Not all indices required the information; however, phenotypic and genotypic variances and covariances and heritabilities were calculated for the three component traits, for up to four cycles of recurrent selection. The indices compared were:

Index	Ref.
SH, Smith–Hazel	Smith, 1936; Hazel, 1943
DG; Desired gain	Pesek and Baker, 1969

Because desired gains for the components were difficult to ascertain, four forms (DG_1–DG_4) were included to present different weights (BASE represented equal weights; see Williams 1962).

Index	Ref.
EWF; Elston weightfree	Elston, 1963
RSI; Rank summation	Mulamba and Mock, 1978
BSD; Linear	Baker, 1974

The results fell into two categories. Among the multiple trait indices the authors chose RSI, EWF, and BSD as the best procedures to improve the composite character, cold tolerance (or presumably, any similar objective involving ambiguous or nonlogical economic weights). These indices were simple to use and resulted in excellent selection differentials and predicted gains; of paramount interest was that they did not require the estimation of genetic parameters—an attractive feature to the busy breeder.

The second category was the observation that single-trait selection for dry weight per plot was effective, through association, in advancing the seedling dry weight and percentage emergence. Rate of emergence gains were smaller, but in the right direction. An additional favorable point in this respect was that selection for dry weight per plot avoided the problem of genotype × environment interaction found among cold tolerance traits by Mock (1979). Easily measured, this trait can be assessed at different locations.

The paper on selection indices by St. Martin, Loesch, Demopulos-Rogriguez, and Wiser (1982) is an exhaustive study of selection indices for maize improvement and should be read by those having more than a passing interest in the subject. Eleven traits were considered for an index, the number subsequently being reduced to five because more than that were found to be of questionable value largely because they showed low heritability. Four types of recurrent selection designs were considered with the research concentrated on S_1 testing. Visual selection of plants for characters and disease resistance (ear rot) was practiced.

The most important traits were found to be yield, moisture, kernel hardness, lysine (Opaque-2 was the population to be improved), kernel weight, and kernel light transmission. Many traits showed negative correlations. The authors discussed the complex relationships involved, the confidence level resting on the parameters used, and suggested the need for a certain amount of subjective judgment input from the breeder.

The article raised questions, not about index selection, but about the base population that had already derived from a recurrent selection program. Was it the best base population on which to base index selection? Suppose all the measurements intended for index selection were made from a larger sample of the Opaque-2 population, and the 10 highest individuals in each category regardless of any relation to another trait were identified, and all selected individuals from all categories were brought together in a breeding plot. Would this population base, which concentrated genes for desired traits at the expense of overall variance, be more amenable to index selection?

Openshaw and Hadley (1984) compared three selection indices for efficiency in increasing protein in soybean. The indices were Kempthorne and Nordskog's (1959) restricted index, Williams' (1962) base index, and Pesek and Baker's (1969) desired gain index. Sugar and oil were used as secondary traits for selection in two hybrid soybean populations. The authors used seven test criteria (combinations of ways of forming indices) to assess efficiency. A brief summary of the salient findings included the following:

1. There was a showing of superiority for desired gains indices, which, however, were less useful in choosing the end products of selection for oil and protein. Because of high heritabilities, direct selection for either was simpler and more effective.
2. Using oil and sugar as secondary traits was not superior to direct selection for protein.
3. Selection for both protein and oil was slow and only moderately successful.

The paper brought out a further two points that were of interest, namely, the observed reasonable performance of direct selection for protein, and the fact that estimates of heritability were obtained from several researchers to establish a base for assumptions and for use in statistical computations.

The level of cutoff points in selection is a major concern of a breeder. The level of selection intensity may either result in the discard of valuable germ plasm or alternatively, represent inefficiency through carrying forward material that might better be discarded. Takeda and Frey (1985) noted that harvest index and growth rate, two traits that earlier had been shown to account for 95% of variation in oat grain yields, were negatively associated, that is, working in opposite directions where selection is involved. They stated, "Selecting too intensively for either of these traits may hamper yield improvement." The question arose as to what selection strategy would blend these two traits into a favorable combination to improve grain yield.

The answer seemed to lie with independent culling, selecting grain yield by harvest index associated with two manifestations of vegetative growth, namely, index and unit straw weight, the latter having been found to be closely correlated with the vegetative growth index. In all the binary comparisons the harvest index was selected first, the other trait, "vigor," second. Selection intensity throughout was 2%.

Intense harvest index (HI) selection resulted in higher HI but no increase in yield, and biomass increase was small. Intense selection for either vegetative growth rate or unit straw weight resulted in small grain yield improvement and low HI. Selection for intense grain yield alone resulted in increased biomass but little change in HI. Takeda and Frey concluded that biomass, more than harvest index, was important to increased grain yield.

The best that can be said here for harvest index is that it is a "good indicator of plant type." It is also seen that intense selection can reduce grain yield, because the other so-called vigor traits are negatively associated with it.

The selection strategy decided upon for high grain yield was described as one that maximized the vigor traits and retained a good HI. Because HI is a more accurate measurement it was selected first (this is a case where a reason is given for a first choice of character). Takeda and Frey's suggested procedures envisage independent culling for HI, discarding low HI readings (genotypes), and in a later generation selecting for vegetative growth rate or unit straw weight. The optimal independent culling levels were found to be 25% for HI and 8% for either of the vigor traits. Independent culling was thought to be similar in results to tandem selection.

A fitting conclusion to this chapter is the recent notice that Robert Baker, an acknowledged authority on selection indices, has published a book on indices in plant breeding (1986).

REFERENCES

Adams, M. W. 1967. Basis of yield component compensation in crop plants with special reference to the field bean, *Phaseolus vulgaris*. Crop Sci. 7:505–510.

Andrus, C. F. and G. W. Bohn. 1967. Cantaloup breeding shifts in population means and variability under mass selection. Proc. Am. Soc. Hort. Sci. 90:209–222.

Baker, R. J. 1986. Selection indices in plant breeding. 232 pp., CRC Press, Boca Raton, FL.

Brim, C. A., H. W. Johnson, and C. C. Cockerham. 1959. Multiple selection criteria in soybeans. Agron. J. 51:42–46.

Bucio Alanis, L., J. M. Perkins, and J. L. Jinks. 1969. Environmental and genotype–environmental components of variability. V. Segregating generations. Heredity 24:115–127.

Byth, D. E., B. E. Caldwell, and C. R. Weber. 1969. Specific and non-specific index selection in soybeans, *Glycine max* L. (Merrill). Crop Sci. 9:702–705.

Caldwell, B. E. and C. R. Weber. 1965. General, average, and specific selection indices for yield in F_4 and F_5 soybean populations. Crop Sci. 5:223–226.

Caldwell, B. E., C. R. Weber, and D. E. Byth. 1966. Selection value of phenotypic attributes in soybeans. Crop Sci. 6:249–251.

Campbell, W. M. and D. W. Davis. 1972. An inbred based predictive index for superior yielding sweet corn single crosses. HortScience 7(3):46 (abstr.).

Crosbie, T. M., J. J. Mock, and O. S. Smith. 1980. Comparison of gains predicted by several selection methods for cold tolerance traits of two maize populations. Crop Sci. 20:649–655.

Doggett, H. and B. N. Majisu. 1968. Disruptive selection in crop development. Heredity 23:1–23.

Eagles, H. A. and K. J. Frey. 1974. Expected and actual gains in economic value of oat lines from five selection methods. Crop Sci. 14:861–864.

Eberhart, S. A. and W. A. Russell. 1969. Yield and stability for a 10-line diallel of single-cross and double-cross maize hybrids. Crop Sci. 9:357–361.

Elston, R. C. 1963. A weight free index for the purpose of ranking or selection with respect to several traits at a time. Biometrics 19:85–97.

Fatunla, T. and K. J. Frey. 1976. Repeatability of regression stability indexes for grain yield of oats (*Avena sativa* L.). Euphytica 25:21–28.

Finlay, K. W. 1963. Adaptation—its measurement and significance in barley breeding. Int. Barley Genet. Symp. Proc. 1:351–359.

Finlay, K. W. 1971. Breeding for yield in barley. Int. Barley Genet. Symp. Proc. 2:338–345.

Freeman, G. H. and J. M. Perkins. 1971. Environmental and genotype-environmental components of variability. VIII. Relations between genotypes grown in different environments and measures of these environments. Heredity 27:15–23.

Frey, K. J. 1972. Stability indexes for isolines of oats (*Avena sativa* L.). Crop Sci. 12:809–812.

Frey, K. J. and T. F. Huang. 1969. Relation of seed weight to grain yield in oats, *Avena sativa* L. Euphytica 18:417–424.

Gilbert, N. 1961. Correlations in plant breeding. Euphytica 10:205–208.

Grafius, J. E. 1970. Stress: a necessary ingredient of genotype × environment interaction. Barley Genet. II:346–355.

Grafius, J. E. and R. L. Thomas. 1971. The case for indirect genetic control of sequential traits and the strategy of deployment of environmental resources by the plant. Heredity 26:433–442.

Hazel, L. N. 1943. The genetic basis for constructing selection indexes. Genetics 28:476–490.

Hazel, L. N. and J. L. Lush. 1942. The efficiency of three methods of selection. J. Hered. 33:393–402.

Heidhues, T. 1961. Relative accuracy of selection indices based on estimated genotypic and phenotypic parameters. Biometrics 17:502–503. (abstr.)

Hutchinson, J. B. 1940. The application of genetics to plant breeding. 1. The genetic interpretation of plant breeding problems. J. Genet. 40:271–282.

Ikehashi, H. and R. Ito. 1971. Statistical property of the selection by the plant type index given by quotient of two traits. Jap. J. Breeding 21:106–113.

Jain, S. K. and R. W. Allard. 1965. The nature and stability of equilibria under optimizing selection. Natl. Acad. Sci. 54:1436–1443.

Johnson, H. W., H. F. Robinson, and R. E. Comstock. 1955. Genotypic and phenotypic correlations in soybeans and their implications in selection. Agron. J. 47:477–483.

Kempthorne, O. and A. W. Nordskog. 1959. Restricted selection indexes. Biometrics 15:10–19.

Lin, C. Y. 1978. Index selection for genetic improvement of quantitative characters. Theor. Appl. Genet. 52:49–56.

Lupton, F. G. H. and R. N. H. Whitehouse. 1957. Studies on the breeding of self-pollinating cereals. Euphytica 6:169–185.

McClintic, D. 1986. Needed: stronger fiber. The Furrow 91(1):35–36.

Manning, H. L. 1951. Prog. Rep. Exp. Stn., Uganda, 1949–1950, Cott. Res. Corp., p. 7.

Manning, H. L. 1955. Response to selection for yield in cotton. Cold Spring Harbor Symp. Quant. Biol. 20:103–110.

Manning, H. L. 1956. Yield improvement from a selection index technique with cotton. Heredity 10:303–322.

Manning, H. L. 1963. Realized yield improvement from twelve generations of progeny selection in a variety of Upland cotton. Stat. Genet. Plant Breeding, NAS–NRC 982:329–349.

Miller, J. D., N. I. James, and P. M. Lyrene. 1978. Selection indices in sugar cane. Crop Sci. 18:369–372.

Mock, J. J. 1979. Investigation of genotype × environment interaction for cold tolerance of maize. Iowa State J. Res. 53:291–296.

Mock, J. J. and A. A. Bakri. 1976. Recurrent selection for cold tolerance in maize. Crop Sci. 16:230–233.

Mock, J. J. and S. A. Eberhart. 1972. Cold tolerance in adapted maize populations. Crop Sci. 12:466–469.

Openshaw, S. J. and H. H. Hadley. 1984. Selection indexes to modity protein concentration of soybean seeds. Crop Sci. 24:1–4.

Pate, J. B., P. R. Ewald, and E. N. Duncan. 1962. An elementary analog computer to aid cotton breeders in making selections. Crop Sci. 2:178–179.

Peirce, L. C. 1968. Selection and stability of phenotype in plant breeding. Hort-Science 3:250–252.

Perkins, J. M. and J. L. Jinks. 1968. Environmental and genotype-environmental components of variability. III. Nonlinear interactions for multiple inbred lines. Heredity 23:525–535.

Pesek, J. and R. J. Baker. 1969. Comparison of tandem and index selection in the modified pedigree method of breeding self-pollinated species. Can. J. Plant Sci. 49:773–781.

Pesek, J. and R. J. Baker. 1969. Desired improvement in relation to selection indices. Can. J. Plant Sci. 49:803–804.

Pesek, J. and R. J. Baker. 1970. An application of index selection to the improvement of self-pollinated species. Can. J. Plant Sci. 50:267–276.

Radwan, S. R. H. and A. Momtaz. 1975. Evaluation of 7 selection indices in flax *Linum usitatissimum*. Egypt J. Genet. Cytol. 4:153–160.

Riggs, T. J. 1967. Response to modal selection in Upland cotton in Northern and Eastern Uganda. Cott. Gr. Rev. 44:176–183.

Robinson, H. F., R. E. Comstock, and P. H. Harvey. 1951. Genotypic and phenotypic correlations in corn and their implications in selection. Agron. J. 43:282–287.

Rosielle, A. A. and K. J. Frey. 1975. Application of restricted selection indices for grain yield improvement in oats. Crop Sci. 15:544–547.

Rosielle, A. A., H. A. Eagles, and K. J. Frey. 1977. Application of restricted selection indexes for improvement of economic value in oats. Crop Sci. 17:359–361.

Sakai, K. J. and K. Gotoh. 1955. Studies on competition in plants. IV. Competitive ability of F_1 hybrids in barley. J. Hered. 46:139–143.

Sampson, D. R. 1976. Choosing the best parents for a breeding program from among eight oat cultivars crossed in diallel. Can. J. Plant Sci. 56:263–274.

Singh, R. K. and K. Bellman. 1972. Problems of generalization of selection indices. Theor. Appl. Genet. 42:331–334.

Smith, H. F. 1936. A discriminant function for plant selection. Ann. Eugen. (London) 7:240–250.

Sprague, G. F. 1966. Quantitative genetics in plant improvement. *In* K. J. Frey, (Ed.), Plant breeding. Iowa State Univ. Press, Ames, pp. 315–347.

St. Martin, S. K., P. J. Loesch, Jr., J. T. Demopulos-Rodriguez, and W. J. Wiser. 1982. Selection indices for the improvement of Opaque-2 maize. Crop Sci. 22:478–485.

Subandi, W. A. Compton, and L. T. Empig. 1973. Comparison of the efficiencies of selection indices for 3 traits in 2 cultivar crosses of corn. Crop Sci. 13:184–186.

Takeda, K. and K. J. Frey. 1985. Increasing grain yield of oats by independent culling for harvest index and vegetative growth index or unit straw weight. Euphytica 34:33–41.

Tallis, G. M. 1962. A selection index for optimum phenotype. Biometrics 18:120–122.

Thoday, J. M. 1960. Effects of disruptive selection. III. Coupling and repulsion. Heredity 14:35–49.

Thomas, R. L., J. E. Grafius, and S. K. Hahn. 1970. Genetic analysis of correlated sequential characters. Heredity 26:177–188.

Thomas, R. L., J. E. Grafius, and S. K. Hahn. 1971. Stress: an analysis of its source and influence. Heredity 26:423–432.

Walker, J. T. 1964. Modal selection in Upland cotton. Heredity 19:559–583.

Watkins, R. and L. P. S. Spangelo. 1971. Strawberry selection index components. Can. J. Genet. Cytol. 13:42–50.

Williams, J. S. 1962. The evaluation of a selection index. Biometrics 18:375–393.

Young, S. S. Y. 1961. A further examination of the relative efficiency of three methods of selection for genetic gains under less-restricted conditions. Genet. Res. 2:106–121.

Young, S. S. Y. and H. Weiler. 1960. Selection for two correlated traits by independent culling levels. J. Genetics 57:329–338.

26

HARVEST INDEX

Harvest index is the ratio of the economic yield (grain) to the biological yield (total dry matter). For practical reasons the root weight is excluded and biological yield represents harvested dry matter—grain and straw—theoretically, straw cut at ground level, but practically, cut at uniform mechanical height of nursery cutting.

ORIGIN AND HISTORICAL BACKGROUND

There is common agreement that harvest index (HI) had its origins in writings of the English barley breeder, E. S. Beaven (1914, 1920), who noted the high stability of the ratio and referred to it as the "migration coefficient." Beaven's use was based on taking data from a few thousand plants and using these data to provide one ratio figure for a unit area to describe a variety. Engledow and Wadham (1924) extended this by suggesting its use as an "index of yielding power for single plants." Their research, however, raised doubts as to its successful use for selecting individual plants, but did indicate that it might be helpful in the "first elimination of low-yielding lines" if based on a sufficient number of plants.

DEVELOPMENTS RELATED TO HARVEST INDEX

Love (1912) studied a number of plant characters and their correlations in a pure line of wheat. He found a high degree of correlation between height and yield, height and seed size or weight, number of kernels and yield, and number of kernels and average weight of grains. He stated,

> The data were taken on each culm separately and then averaged, and the averages are used to represent the plant. In all of the tables the plant has been used as the individual and not the culm. Other studies have shown that in general we may expect similar correlation when the culms are used as the individuals rather than the plants themselves.

This reinforces the question, apropos harvest index, as to whether choosing the primary culm alone might serve as a useful harvest index base for plant selection. Also, what information on yield potential would be given, for example, by a rapid visual measuring or estimating of a plant's total heads' length in inches or centimeters?

Arny and Garber (1918) found a positive association between plant height and yield, and number of culms and yield. This was suggestive, relative to harvest index, that increased yield of kernels was closely accompanied by an increase in total length of spikes. Today, in the semidwarf era, it would seem that this relationship of height and yield would have to be reinterpreted.

King and Jebe (1940) were concerned with techniques of estimating and forecasting wheat yields over large areas using route sampling (collecting data and material for analysis from two 1/10,000-acre samples per randomly chosen field). Because of the time required for laboratory analysis they searched for variables more easily obtained, thereby eliminating the best barometer variable measured, number of kernels per sample. The next best indicator of yield was number of heads per sample, followed by length of head.

King and Jebe were not concerned with harvest index, but breeders will see a connection to harvest index in the highlighting of kernels per sample and number and length of heads. Harvest index has one aspect for populations, but a special challenge in plant breeding is in selection at the single-plant level. It would be helpful if plant selection in the field could be done on the spot using visual and simple supplementary aids. Is single-plant, single-culm harvest index such an aid?

Moussouros and Papadopoulos (1935) concentrated on a study of the effect of maturity on yields of spring wheat in Greece. They stressed that earliness, singularly related to high yield, must be thought of in two ways: earliness in heading and earliness in maturity. The former is the more important; the latter is influenced by seasonal weather and more subject to disease damage. The difference between the two represents variation in the length of the vegetative period. The importance of early heading rests on the fact that a longer vegetative growth period before maturity—favorable to yield—can be exploited during the earlier, more favorable season gained in part by a genotype's inherent early heading. Thus within reasonable ranges, earliness of heading is desirable because it permits the longer favorable extended growth period before late season stresses appear.

Smith (1936) provided the answer to a philosophic question sometimes asked, whether any more could be learned about the phenomena involved if a ratio were used. Smith wrote:

> In general the derivation of these functions demonstrates that grain : straw ratio can never supply more information than can be given by a consideration of both grain and straw separately and can give as much information only if the coefficients of X_1 and X_5 are numerically equal and opposite in sign. It seems that the latter condition may arise only under exceptional circumstances. Since both grain and straw yields must be determined before their ratio can be evaluated, and since all the available information can be obtained from the two primary variates, it seems an unnecessary step to evaluate the ratio.

A startling difference in yield response of corn inbreds differing in height to population density was shown by Nelson and Ohlrogge (1957). Three different semidwarf mutants were compared for yield and other morphological traits with the normal inbred. The mutants were all either derived from or converted to the normal background genotype. When tested at levels of 13,000, 26,000, 52,000, and 78,000 plants per acre, the mutant line *compact (ct)* reached its highest yield (about 57 bu/acre) at 26,000 plants per acre and showed only modest declines thereafter. The normal inbred, higher yielding at 13,000 plants per acre, dropped to about 45 bu/acre at 26,000 and declined precipitously thereafter. Harvest index studies on material of this kind would be illuminating.

Watson, Thorne, and French (1958) studied the physiological cause of higher yields in certain cultivars of barley and concluded that the major cause was greater photosynthesis of the ear. There was some indication that larger size or weight of ears (total increased exposure per land area) might be responsible, but an alternative explanation could be higher photosynthetic efficiency.

In Britain, although there is a four-months difference in growing seasons, yields of spring and winter wheats are not markedly different, as they are in some other parts of the world. Watson, Thorne, and French (1963) used two varieties each (an older variety and a recent release) of spring and winter wheats to study growth analysis and physiological changes that might account for similarities in yield between spring and winter types, and for the higher yield of newer varieties.

In the first objective, similarity of yield, it was found that winter wheat showed "early foot," but its advantage was slowly overcome by the spring wheats owing to their LAI (leaf area index) increasing. At the same time the spring wheats had a higher proportion of their shoots that had formed ears beyond the time when the maximum already had been reached in the winter types. Also after ear emergence, LAI decreased at a slower rate than that of the winters. The result was a rapid closing of the yield gap between the two kinds of wheat.

For the second objective of the study, determining why newer varieties yield more than older ones, they found newer varieties placed more of their total plant dry weight in the ears, which also emerged earlier, with the consequence that the lengthened period of grain growth, plus greater photosynthetic area close to the grain, enhanced yield.

Robinson and Bernat (1963) found that panicles of grain sorghum (whole plots cut just below the lowest branch), dried for two weeks and weighed, gave almost perfect correlations (.95 and above) with bushels of grain per acre. Dry weight of panicles thus could be used to rank entries in a yield trial. The similarity to harvest index is obvious.

In studies on barley by Thorne (1963), the best estimates were "that 70% of the grain was contributed by ear photosynthesis."

Thorne's (1965) research revealed distinct differences in photosynthetic activity between barley and wheat. Differences in size of ear (larger in barley) were directly related to photosynthetic activity. On the other hand, wheat flag leaves had more surface than barley and contributed relatively more to grain dry weight. The influence of shading and presence or absence of awns apparently is substantial.

Welbank, French, and Witts (1966) found that three wheat varieties had grain yields proportional to their leaf area durations during grain development.

In a review of contributions of leaf area to grain production, Voldeng and Simpson (1967) cited six research papers where the estimated contributions to grain dry weight from photosynthesis above the flag leaf node ranged from 60 to 85%. Their hypothesis that yield per plant should be proportional to the photosynthetic area above the flag leaf node was confirmed when a close relationship was found among seven closely related lines of spring wheat. They suggested that large flag leaf and ear area might be a valid criterion for selecting high-yielding plants. It is not a big step from this to the question of whether cutting stems at the flag leaf node might serve as a convenient base for determining a modified harvest index.

This study was continued when Simpson (1968) conducted a more extensive and detailed examination in the greenhouse of 120 varieties divided into three 40-variety groups each of tall, medium, and short types. An impressive degree of linearity was shown in a graph of regression of yield per plant against above-flag leaf photosynthetic area. Tillers of tall plants have more photosynthetic area above the flag leaf than shorter plants; consequently, they yield more per tiller *but less per plant* than short wheats because the short plants produce more tillers. This is important where harvest index is contemplated on a single tiller or head basis—somehow plant height must be factored in. An easy solution would be to set an upper-level cutoff point for height and thus eliminate the problem. The study contains detailed information on the components of photosynthetic area.

Walton (1969) found a well-extruded wheat head, regardless of length, to be the element in the flag leaf and above area that contributed most to yield.

Reaching back to one of the earliest papers that explored the division of yield of cereals into its component parts (that of Engledow and Wadham, 1923), Hsu and Walton (1970) considered the list of components too limited and organized an experiment to evaluate simultaneously the contributions of several morphological characters to yield. The study was made on the parents and F_1s of a diallel cross among five spring wheat cultivars. The highest correlations with yield were shown by number of ears per plant, number of kernels per ear, and flag leaf sheath length. As shown earlier by Walton (1969), peduncle length and head extrusion were highly correlated and found here to be associated with increased sheath length. Dominance relationships were determined for the various characters, but in general additive gene effects prevailed. The authors suggested that more attention be give to morphological characters in breeding for increased yield.

Hsu and Walton (1971) studied the above-flag-leaf-node traits and related them to yield. The study included 13 characters, both components of yield and morphological, which is an indication of the complex nature of yield and its contributing factors. It is simpler to begin with what was *not* related to yield per plant: peduncle length and flag leaf length. The characters making the greatest contribution to yield were number of ears per plant and yield per ear, sheath length, and flag leaf breadth.

Several inferences may be drawn from these few selected findings. Number of ears per plant suggests that tillering has a desired association with high yields under favorable environmental conditions, for example, irrigation. Sheath length and flag leaf breadth suggest that photosynthetic area is important. The authors constructed an ideal ideotype, which incidentally is a reasonable description of many semidwarf commercial wheats. An excellent literature review is appended.

The area above the flag leaf node in cereals has long shown a positive association with yield, and is of further interest today because this same area is related to harvest index. Yap and Harvey (1972) found that spike and peduncle surface, flag leaf area, and erect portion of the flag leaf in barley were positively associated with and could be used as indicators of potential grain yield.

Alessandroni and Scalfati (1973) compared F_2 yields of plants and heads with F_4 progeny plot yields. They found a statistically significant association between F_2 yield per head and F_4 plot yield but not between F_2 yield per plant and F_4 plot yield. From the description it appears that yields per head were generated from head-bearing tiller counts and the yields of F_2 plants. The absence of data to show the variance of individual head yields makes it impossible to explore the relationship to harvest index and particularly single-head harvest index.

Baker and Dyck (1974) in a study on early-generation wheat noted a high correlation between weight of unthreshed spikes and grain yield, two characters that show a positive genetic association. They suggested this might be a screening procedure for yield: "Only those samples which yield a satisfactory level of unthreshed spikes would have to be threshed out for further study."

Considerable evidence exists to link morphological characters above the flag leaf to yield of the cereal plant. On the premise that increasing their size might also increase yield, Briggs and Aytenfisu (1980) studied the effect of different seeding rates and dates on changes in the morphological characters above the flag leaf node, and particularly, the interrelationships between the expressions of yield, that is, per unit area, per plant, per tiller, and per grain yield components.

The study involved seven spring wheats, three seeding dates, and seven seeding rates (from 30 to 180 kg/ha). The characters evaluated were grain yield per tiller, plant, and plot, number of plants, ears per plant, kernels per ear, 1000-kernel weight, ear length and extrusion, and areas of flag sheath and lamina.

The significance of this paper is not in the detailed results, which were influenced by myriad compensating interactions affecting the complex above-the-node traits, but in a particular finding and a commonsense conclusion. The singular finding related to the behavior of "extrusion length," which is defined as "that part of the culm between the tip of the flag leaf sheath and the base of the ear," that is, the peduncle. This was negatively associated with yield (that is, the shorter the extrusion the higher the yield) and was consistent in its response, suggesting that it might be used as a selection indicator for the associated character, yield. The important conclusion is, in effect, a call for "standard growing conditions" for nurseries in which selection for plant characters is practiced. These are needed in order to minimize the environmental interaction traceable to rates and dates of seeding. Briggs and Aytenfisu suggested choosing early dates and medium rates, comparable to commercial practices.

SELECTION, STABILITY, AND INHERITANCE RELATIONS

An important paper by Rosielle and Frey (1975a) adds to our understanding of harvest index. It is not enough to know that harvest index is the ratio of grain yield to total above-ground plant weight; the characteristics and powers for use in selection that are inherent in this new entity also become a focus for inquiries. In particular, a breeder might wish to know the relative merits of strategies that call for either direct selection for grain yield or some form of indirect selection for yield. It is within this context that Rosielle and Frey cast their studies.

Their materials were 1200 F_9 oat lines from a broad-based hybrid composite grown in six environments over two years. Data taken included plant dry weight, grain yield, height, and heading date. Heritability percentages were determined by variance component, standard unit, and realized heritability procedures (the latter at 1% and 10% selection intensities), relying heavily on theory and formulas devised for ratios by Turner (1959). Standard-unit heritability percentage procedures were considered best because there was less confounding of genotypic variation with genotype × environment interaction owing to the self-correcting influence of the multiple environments.

Important trait correlations (pooled means) found in the study were these:

Harvest index and grain yield	0.42
Harvest index and plant weight	0.07
Grain yield and plant weigh	0.88

In addition, relative selective efficiencies were calculated, with results as follows:

Indirect selection for grain yield by HI	43%
Indirect selection for grain yield by plant wt.	86%
Index selection combining yield and HI	100%

The first two of these showed negative association, meaning that "when harvest index was most efficient, plant weight was least efficient." Several possible checks made with these and other data convinced the authors that the Turner formulas were reliable in dealing with ratio traits.

What, then, do these findings mean to plant breeding? For one thing, the authors found an explanation for a seeming paradox: the high correlation (.88) between grain and plant weight suggests that unrestricted selection for grain yield—the believed common practice of breeders—would choose more high-plant-weight individuals than high-harvest-index ones (where the correlation of yield and harvest index was less, .42). This actually is not what happens in plant breeding because there have been improvements in harvest index rather than biological yield. The interpretation of this, which seems logical, is that breeders do not practice unrestricted selection for yield; they use a form of mental index selection, discarding lines on the basis of height, late maturity, and perhaps other factors associated with plant weight.

Rosielle and Frey (1975b) examined several variations in formulating restricted indices aimed at increasing yield in oats, while countering the tendencies to later heading and increased height. When height and heading date were held constant the gain in yield was (reduced to) 57% of the unrestricted gain. However, the inclusion of harvest index, weighted to greater importance than yield, resulted in raising efficiency to 70%. This suggests that harvest index can be used to advance yield beyond the point where direct selection and restricted indices become ineffective.

A comprehensive treatment of biological yield and harvest index was given by Donald and Hamblin (1976). Most of what follows has been derived from their section on "Biological yield and harvest index as criteria in cereal breeding." After noting that the harvest index of cereals has steadily risen without an increase in biological yield, the authors remarked that this has occurred without purposeful planning by breeders: "Any increase has been an unplanned secondary effort of breeding for grain yield, shorter straw, and earliness." This raises a question: can breeding for HI be a means of increasing grain yields (GY)? Their conclusion was that heritability of HI is high enough to suggest that this might be a promising approach.

Coupling biological yield (BY) with HI, the authors cited three ways in which they may be used:

1. *By Using Parents of Known High BY or HI.* High HI should be searched for even if BY and GY are low. Attempts should be made to combine high HI with high BY. These procedures, of course, require the measuring of all three traits.

2. *By Using High-Density Populations and/or High Fertility to Assess Lines for BY and HI.* The authors felt that this type of environment is necessary to understand the potential of breeding material: "The successful cultivar will have increased biological yield at high density or fertility together with a satisfactory maintenance or minimum decline of harvest index."

3. *By Using BY and HI in Early-generation Selection.* This category led to an in-depth discussion of breeders' practices and quandaries. The authors described three ecosystems in which cereal genotypes are grown, each differentiated by plant and population characteristics, namely, as spaced plants, as plants in mixed or segregating populations, and as plants in a dense monoculture. They visualized and described three different plant ideotypes, A, B, and C, each of which would be successful only in the matching ecosystem, respectively. A was termed the "isolation ideotype," B the "competition ideotype," and C the "crop ideotype." The plant breeder's enduring dilemma is that the objective, the crop ideotype, is scarcely recognizable in the confounding milieu of spacing, heterosis, competition, and other factors that have impact on the early generations.

Donald and Hamblin listed four criteria that breeders use in early-generation testing to gauge GY potential. These are listed with brief comments:

1. *Grain Yield Itself.* The problem is that the criterion of successful selection is field performance.
2. *Components of Yield.* The problem here lies with compensation among components.
3. *Vegetative Characters (Morphology).* Generally, these have not been

sufficiently explored, but are promising criteria because of certain known negative correlations between visible traits and GY.

4. *Harvest Index*. HI has a known positive *r* with GY on both a plant and single-culm basis.

Donald and Hamblin believed that the B competition environment will be dominated by a B genotype, which is likely to have a high HI. However, this genotype is different from the desired C genotype that would be successful in the C crop environment. Therefore, HI may not be a useful selection criterion in competition environments such as segregating populations. The plant form, however, is suggested as promising for the C environment. They summarized their views as follows:

> Harvest index may be a valuable criterion for early generation selection among spaced plants (Environment A) where performance per culm is carried forward to the crop situation (Environment C). On the other hand, it seems unlikely to be of value for the prediction of crop grain yields from a segregating population at crop density (Environment B) to the crop situation (C). Here vegetative characters may prove useful.

The authors' explanations for the good agreement between spaced-plant HI and commercial-crop GY are worth quoting in full:

> Single plants in pots or spaced plants in the field, as in Syme's and Fischer's experiments, were in a situation approximating to the isolation environment (A) of Table XVII, where free tillering genotypes are able to achieve high biological and grain yields—higher, relative to other genotypes, than they would in a crop situation. As a result, the relationship of grain yield of spaced plants to grain yield in crops is distorted and the correlation is low. On the other hand, harvest index, which depends much more on the performance of the individual culms, tends to be sustained across all culms of a genotype, whether it has many culms (as in spaced plants) or few culms (as in a crop). The harvest index of spaced plants therefore tends to show a lesser genotype/density interaction than does single-plant grain yield, and hence a better relationship to plot grain yield.

Jain and Kulshrestha (1976) observed plant character changes in four height groups of wheat, namely, tall and the three dwarf height groups conditioned by D_1 (semidwarf), D_2 (dwarf), and D_3 (very dwarf). Harvest index showed a continued positive increase with reduction in plant height—an inverse relationship. Accompanying this was an increase in number of effective tillers (I observed this phenomenon in Cornell's semidwarf wheat cultivar, Yorkstar). Thus there was compensation for biological yield in the shorter wheats through additional tillers. Grain yield also increased with reduction in height.

Of significance to breeders considering harvest index as a selection tool, biological yield was not significantly correlated with grain yield, nor any other character studied. However, harvest index was significantly positively correlated with grain yield and number of effective tillers (presumably head-bearing tillers) per area, and negatively with plant height. The authors projected an objective strategy for breeding and selecting higher grain yields in wheat that involved the maintenance of high harvest index while selecting for greater dry matter production (no negative correlation was found between these two attributes). These views, involving effective tillering capacity, are in contrast to Donald's 1962 concept of the ideal wheat ideotype, a density-controlled uniculm.

Bhatt (1976) published a significant paper on the nature of inheritance of harvest index (if, indeed, a ratio trait may be considered to be an independent entity; harvest index is easier to comprehend if one notes the very high correlation shown in his Table 3 between grain yield per plant and harvest index). Harvest index was treated as a normal quantitative trait and examined in six parents, two of which showed medium harvest index and were the testers crossed reciprocally to two high, one medium, and one low harvest index genotypes. The pattern of harvest index variations and distribution was remarkably uncomplicated. Reciprocal crosses showed little differences and were pooled. Midparent heterosis for harvest index was exhibited in all crosses and no high parent heterosis was evident. Segregation for harvest index followed a normal distribution in all crosses. There appeared to be partial dominance of high over low harvest index and both additive and nonadditive gene action was noted. In sum, the paper lends ecouragement to the use of harvest index as a selection tool.

Rosielle and Frey (1977) explored the manner of inheritance of harvest index in three yield groups of crosses designed to present a broad variation in harvest index, namely, high × high, high × low, and low × low. Stratification for similar heights and heading dates may have influenced the study although height variance was still substantial. The study elicited no clear evidence of the mode of inheritance of harvest index except that it seemed to be controlled predominantly by additive gene action.

Bhatt (1977) used bidirectional (high, low) selection for harvest index at 10% intensity in two wheat crosses and correlated this response with grain yield. Low harvest index F_2 plants gave low indices in the F_3; however, high index F_2 plants segregated high and medium index F_3s. Again a moderately high correlation was found between harvest index and grain yield. The possibilities of separately selecting for grain yield, biological yield, and harvest index, in order to raise grain yield, were discussed. If progress could be made in all three areas the selections would make interesting parents for an intermating program.

Important confirmation of the year-to-year stability of harvest index in spring wheat was provided by Bhatt and Derera (1978). Near-homozygous lines (F_5–F_7) and two cultivars were grown in hill plots, and harvest index and grain yield per hill recorded. From the 153 lines tested, the five highest, the five lowest, and five lines intermediate in harvest index were chosen.

These 15, plus the two cultivars, were planted in a replicated yield nursery the following year, and harvest index and grain yield again recorded. The correspondence between the harvest indices of the 17 entries over two years was remarkably close. The correlation between the harvest index in both years with grain yield in the second year, although not so high as between the two-year harvest index, was significant at the 1% level. The importance of these findings is that harvest index can be a reliable criterion for yield if it can be efficiently inserted into the selection procedures.

Schapaugh and Wilcox (1980), working with soybeans, reminded us of a facet of harvest index we (in cereals) may have forgotten: that harvest index has two faces, which they term "actual" and "apparent." One concession to convenience has already been made—we apply the ratio to the above-ground parts of the plant, ignoring the roots. The authors pointed out that plant loss of canopy can amount in the extreme to 40% of biological yield in soybeans. It is not feasible to monitor and assess this loss, which is part of the "actual" harvest index. Their study was designed to evaluate the relationship between actual harvest index and the readily obtained "apparent" figure, which is based on what remains of seed and plant at harvest. The point is particularly important in soybeans because of determinate–indeterminate growth types found throughout maturity groups.

The main finding was positive: actual and apparent harvest indices were highly significantly correlated. Also, harvest index was not affected by growth habit differences. Unfortunately, harvest index and yield lacked a consistent relationship over seasons; however, as a balance to this, index means were more stable than yield means.

Whan, Rathgen, and Knight (1981) reported results from a rare comparison of correlations of four early generations, F_2–F_5 inclusive, made in *one* season. One of the traits measured was harvest index. The results were negative, and unfavorable to the use of harvest index in early generations. Measurements were taken in F-derived lines from random samples. The F_2–F_3 correlations were similar to those for yield, but there was no correlation between harvest index and total plot yield in the same or different generations (all grown in the same year). The explanation given is that low harvest index was associated with taller and later high-yielding types. Their conclusions: harvest index was not useful as a selection-for-yield criterion in early generations, and not as good as selection for yield itself considering the extra work required.

Determining the true selective value of harvest index might be said to be the research objective of Sharma and Smith (1986). They selected for low and high harvest index in three different winter wheat populations. They found harvest index-realized heritabilities to range around 0.50, and there was good predictability of the harvest index level from F_3 to F_4 (but harvest index was a poor predictor of grain yield!). It was not a surprise that high harvest index and shorter plants were correlated; high harvest index was also associated with earlier heading and lower total plant yield.

This amassing of information about harvest index is helpful in giving the breeder a sense of the trait's place in the overall selection picture. The known good correlation between grain and biomass yield cannot be ignored. The ratio relationship intrudes here: grain yield and harvest index often rise at the expense of biomass, which decreases with a desirable reduction in plant height. What is left out is the perhaps desirable goal of also raising biomass yield. The authors' conclusions lead to suggestions that this may be an ideal situation for the use of selection indices that embrace all of the elements and attempt to strike a balance for the final genotype product.

INFLUENCE OF DENSITY ON HARVEST INDEX

Fischer and Kertesz (1976) reported on a study in wheat of the effect of density on grain weight and harvest index. Three plot formations were used: large and small plots at normal densities, and spaced plants. The small microplots principally differed from large plots in overall size and edge exposure. The study involved 40 homozygous bread and durum wheats (two triticales). The results showed that grain yields and harvest indices were closely correlated in the larger, normal-density plots. The relationship for grain yield correlation did not hold in the spaced plants, there being a very weak association between grain yield per plant and grain yield per large plots. However, spaced-plant harvest index was high and correlated well with larger plot grain yields.

Fischer and Kertesz's paper is of special interest because they used a miniaturization of harvest index, which they called *shoot harvest index*. They described it thus: "A single central shoot with the lamina and sheaths of the four uppermost leaves intact was taken from each plant in the space planted treatment just before harvest. The shoots were bulked for each replicate, dried, weighed, and threshed."

First, it should be noted that Fischer and Kertesz found that spaced-plant yield is questionable as a predictor of potential yield in large plots, but that harvest index is population- or density-neutral. The surprising thing was that the best predictor was neither harvest index of plants nor of large plots, but shoot harvest index: the means of these three for the 40 genotypes were 45.2%, 42.4% and 46.6%, respectively.

My interest in a shoot harvest index stems from the fact that my principal selection procedure with any and all of the small grain cereals was the collection of a single culm with head, based on a plant's visual appearance. Although it was known that single-plant selection with harvest index was reliable, there have been only vague references to using the central or principal plant culm alone. Furthermore, if one considers the thousands of "bending overs" done by the plant breeder to take ground level and higher cuttings of a culm, necessary for shorter plant material, it will be appreciated that even more was wanted: a harvest index based on a cut at the top node, in other words, an above-flag-leaf-node harvest index. These are very practical considerations, and explain why Fischer and Kertesz's shoot harvest index is a step in the direction of efficiency. It should be noted, however, that their data were from dried replicate bulks of culms; research is necessary to establish harvest index for single fresh weight shoots and for above-flag-leaf-node material.

Admittedly, harvest index may manifest itself differently in different species. Buzzell and Buttery (1977) found harvest index in soybeans to have a negative association with maturity and yield when measured in hill plots. Furthermore, harvest index did not respond to changes in density within the hill plot. Though this could be seen as a favorable stability outcome, the aim of the authors had been to see if hill plots could be used to screen genotypes for the high density they would experience in commercial row culture.

Okolo's (1977) experiences with harvest index applied to F_2 spaced-plant populations of four wheat crosses (there was no significant correlation between F_2 plant harvest index and either F_3 or F_4 bulk yields) were cited by McVetty and Evans (1980a). They found similar results in spaced F_2 plants, that is, no significant correlation between harvest index and F_2-derived F_4 bulks. McVetty and Evans made a distinction involving origin and homogeneity of F_2 plants being tested for harvest index as a predictor of yield potential: harvest index may be successful when dealing with similar types having similar productivities, but unsuccessful when dealing with an F_2 population showing an array of types and productivities.

Nass (1978) used F_2 plant head weight as an indication of yielding potential in four spring wheat crosses sown at both high and low seeding rates, comparable to spaced and commercial densities. High head weight selection in F_2 produced a 50% increase in F_4 lines yielding in the top 15%. Also, the F_2 selection at high population density resulted in a 100% increase in high-yielding lines compared with that in low-population densities. Even so, the frequency of high-yielding lines coming from low head weights could not be disregarded if the selection were done at high density; for example, more such lines were found than when selection for high head weight was done at low density. Nass concluded that selection based on head weight has moderate prediction value, and selection at high population density is preferred over spaced density conditions.

Baker and Briggs (1982) found no significant effect of plant density on harvest index in a study of 10 barley cultivars measured in five density patterns ranging from a thin 1.6 plants to a dense 400 plants/m^2.

GENERAL STUDIES OF HARVEST INDEX

Hanway and Russell (1969) studied biological and grain yields, and harvest indices of different corn hybrids grown at high (52,000 plants/ha) and low (35,700) densities. Both biological and grain yields of all hybrids were higher at high densities but the increases were not proportional: 16% greater for biological yield but only 5.5% for grain yield. The disparity was due to a lowered harvest index. The real significance for selection was found in the performance of hybrids when separately examined. There were hybrids that showed huge increases in biological yields, but that were low in grain yield and harvest index. There were others with increases in all three attributes. For maximum genetic advance it would be necessary to consider all three factors. Breeders generally may be unaware of the internal adjustments to density changes made by hybrids, and the need for a deeper evaluation of a hybrid's capacity for compensation.

Chandler (1969) discussed the grain-straw ratio with particular reference to nitrogen response in rice. Research at the International Rice Research Institute (IRRI) had shown that nitrogen-responsive cultivars put twice as much of the total dry matter produced into grain production as did the low-nitrogen responders.

Syme (1970) found that harvest index and yield showed significant positive correlation.

Singh and Stoskopf (1971) found that harvest index was positively associated with yield, but negatively with vegetative parts. This is another way of viewing breeding for short-statured wheats where reduction in height has a direct relation to the reduction of the vegetative or straw component of biological yield. They found much genotypic variation for harvest index, which is encouraging for the possibilities of genetic improvement of harvest index and grain yield.

Syme (1972) reported on comparisons of single plants in pots, and the same 49 cultivars grown in the Fifth International Spring Wheat Yield Nursery (ISWYN) on 63 sites worldwide. The stability of these mean yields was indicated by an *r* value of .964 found between the yields of 31 cultivars that were grown in two successive years at the 63 sites. However, these were overall correlations; they were low for any particular cultivar. Grain yields were not significantly correlated, but correlations were found between single plant and crop performance for height, days to ear emergence, and 100-grain weight. Of 16 plant characters, the strongest correlation with yield was harvest index. Regression analysis showed that harvest index accounted for 71.7% of the variability in ISWYN yields.

Implications for harvest index when applied to less than the whole plant are contained in Chandhanmutta and Frey's (1973) paper on indirect mass selection in oats. They selected for heavy panicles in a massive germ plasm hybrid composite for two cycles. The selection increased panicle weight as expected, but also raised grain yield by 5.6% per cycle—an indirect response. Along with these changes, taller plants and later maturity were observed. Changes of this nature could be countered by an index, visual discrimination, or by choosing parents showing earliness and short plant type.

Nass (1973) studied 20 variables of 22 spring wheat cultivars for two years. His goal was to relate morphological characters to a role in selecting for yield. Harvest index was significantly correlated to yield (0.62 and 0.75 for 25 cultivars in two years), as were ears per plant and yield per ear. Nass suggested that some combination of these three would be effective in a selection program for yield. A point of additional interest was that the photosynthetic area above the flag leaf node, a whole complex which has been suggested as related to yield in the same manner as harvest index, was less important in its association with yield than was expected.

Hamblin and Donald (1974) reported F_3 plant harvest index and F_5 line harvest index significantly correlated at the 0.1% level for F_3 low nitrogen with F_5 low and high nitrogen, but the same comparisons for F_3 high-nitrogen situations were nonsignificant.

Takeda and Frey (1976) made an important contribution to an understanding of harvest index by relating it to vegetative growth rate. Although much interest has centered on harvest index because of its stable relationship to grain yield, Takeda and Frey showed that when harvest index was linked to vegetative growth rate almost all variation for grain yield was accounted for in the backcross material studied. Growth rate was obtained as straw weight from sowing to heading divided by number of days (this is a reliable indicator of weight to maturity). What the authors have done is to identify two phenomena affecting grain yields, growth rate, and harvest index. They suggested that in some cases, selection for increased growth rates, for example, 0.7–0.9 g/day, would be more effective than selection for increased harvest index.

Harvest index has important negative associations; for example, Takeda and Frey (1979) and Bhatia (1975) found this to be so with protein percentage.

Jalani, Frey, and Bailey (1979) arrived at some interesting hypotheses useful to oat breeding goals for their area (Iowa corn belt). From earlier studies (Takeda and Frey, 1976, 1977) it was known that about 90% of the variation in grain yield from interspecific crosses was due to growth rate (GR) and harvest index (HI), with GR having the greater influence, 1.27 : 1, respectively. The present study, which dealt with lines drawn from EMS-derived populations and a check population, confirmed the high influence of GR and HI and led to separate conclusions that induced mutations were deleterious relative to these two traits. In general, selection for GR or HI showed positive correlations with other traits, but there was one exception: responses were negative with heading date. The authors commented:

> Selection for high GR or HI resulted in early HD. This is an interesting result with respect to oat breeding for the Corn Belt, USA, because selecting for high GR and/or HI would result in the saving of early lines with improved GYD. This is the exact goal of our breeding program and is counter to the retention of late lines that occurs when selection is practiced for GYD only. [GYD = grain yield, HD = heading date.]

There remains a question as to whether this has universal application to all environments. For example, in my area of breeding (New York) and others, perhaps Wisconsin and Michigan, midseason oats clearly were superior in yielding capacity; whether they have longer growth duration is not known.

The second observation by Jalani et al. was based on their knowledge that commercial oat cultivars seemed to have the optimal harvest index for Iowa conditions. Because of this they concluded that the amount of vegetative growth was the limiting factor in grain-yielding capacity. There are two ways to increase vegetative growth, by extending growth duration or by increasing the growth rate. The first is not practical because of the hazards from increasing stress which would accompany longer time in the field before harvest. Therefore, the authors concluded that growth rate is the obvious trait to exploit to obtain greater vegetative growth.

Continued attention of researchers to harvest index as an indicator or predictor of future potential high yield in wheat has led to considerations of character associations and forms of index selection. McVetty and Evans (1980b) in a second follow-up paper in 1980 took into consideration previous research that had elicited these findings:

1. *Bhatt (1976):* Harvest index has high heritability.
2. *Okolo (1977):* Harvest index applied to F_2 plants was not successful as a yield potential estimator; however, productivity or biological yield was.
3. *Donald and Hamblin (1976):* Harvest index alone was insufficient as an estimator. Biological and grain yields also must be brought into the equation.

Using the same three spring wheat crosses from their earlier 1980 study and seven tall and seven short cultivars, they generated spaced-plant and solid-seeded information on biological yield, harvest index, and height. From these correlations they constructed a limited index selection.

A crucial finding was that correlations of traits with yields depended on height differences in this manner:

1. *Tall Plants:* There was a high positive r between spaced-plant biological yield and plot grain yield. There was no correlation between harvest index and grain yield.
2. *Semidwarf Cultivars:* They showed the reverse, high positive r between harvest index and yield, and no correlation between spaced-plant productivity and yield.

Therefore, in choosing selection criteria for a cross, the height values tracing to parental phenotypic and genotypic input first must be considered.

These insights were put to the test in the three crosses. First, in the tall × tall cross under stress-free conditions, only productivity was used as a criterion (high was defined as one standard deviation above the cross mean). At 16% retention, nine of 45 F_2s and two of five high-yielding F_4 bulks were identified. This was five times better than random selection. The same cross under normal conditions, using similar procedures, was 11 times better than random selection.

In the two tall × semidwarf crosses McVetty and Evans stated:

> The selection criteria have to be adapted to the height-yield relationships of the plants in the segregating populations. The tall portion of the population (plants taller than the population mean height) was selected in the same way as the progeny of tall × tall crosses. The short-statured portion of the population (plants shorter than the population mean height), in contrast was selected for high harvest index (above the population mean). The expected portion of the F_2 population retained was 8% for tall plants and 25% for short plants, or 33% overall.

Put to the test in the 'Glenlea' × 'Era' progeny the results were:

Stress-Free Environment: 24/45 F_2 (53%), 6/11 HY F_4 bulks (55%)
Normal Environment: 32/75 F_2 (43%), 5/10 HY F_4 bulks (50%)—four times better than random in both environments.

In the Glenlea × 'NB320' cross the numbers were:

Stress-Free: 24/45 F_2 (53%), 6/7 F_4 (86%)—16 times better than random.
Normal: 28/75 F_2 (37%), 7/13 F_4 (54%)–10 times better than random.

The criteria used resulted in substantially higher selection gains than random selection, and as the authors pointed out the gains were higher than by visual selection. Importantly, the procedures generally could be described as environment-neutral.

McVetty and Evans' paper, although short, nevertheless contains several items that mark it as important. First there is the confirmation of earlier hypotheses, the recognition of a dual role for height differences. Then there is the identification and association of characters into a form of index selection, showing its effectiveness over environments. Finally, there is the innovative construction of a variable selection index.

Attempting to account for the higher relative yield of modern 'Houser', a 1977 wheat introduction, over 'Honor' (1920), as reported by Jensen (1978), Gent and Kiyomoto (1985) found the yield advantage to be due to greater harvest index. Their tests showed that increased photosynthesis did not give Houser the yield advantage inasmuch as the yield difference took place at the end of the grain-filling period, at which time photosynthesis had ended.

Although Honor initially showed a higher rate of canopy net carbon dioxide exchange, the two wheats were essentially equal after stem elongation and during the period when Houser gained its yield advantage. Neither could they demonstrate an advantage due to the awns on Houser. Although the greater productivity of Houser was attributed to harvest index, this by itself does not provide a biological explanation or basis for the increased yield, except for the known association of harvest index with efficiency and height. The explanation could be in disease resistance or type of root system.

If a breeder has a spring wheat cross containing genes for photoperiodism, semidwarfism, and awns, what should be the balance striven for in the selected genotypes? Knott (1986) examined this question for a specific environment, that prevailing in Saskatchewan—long-day and short growing season. The effects on several agronomic characters were determined by partitioning according to the three elements in the original question. The test lines were F_6, developed by a random procedure (single seed descent).

The agronomic characters related to the three test elements were:

1. *Photoperiod Insensitive (vs. Sensitive):* earlier, shorter, and lower-yielding plants;
2. *Semidwarf (vs. Tall):* earlier, higher-yielding plants, with lower protein and smaller kernels;
3. *Awned (vs. Awnless):* earlier, higher-yielding plants (in two cases but lower in another), with larger kernels.

Based on the information disclosed in the study, the ideal wheat ideotype for the given environment would be a photoperiod sensitive, awned semidwarf. Selection for the taller end of the height class would be desirable and the presence of awns would aid kernel size.

REFERENCES

Alessandroni, A. and M. C. Scalfati. 1973. Early-generation selection for grain yield of dwarf and semidwarf progenies of durum wheat crosses. Proc. 4th Int. Wheat Genet. Symp., Missouri Agric. Exp. Stn., Columbia, Mo.

Arny, A. C. and R. J. Garber. 1918. Variation and correlation in wheat with special reference to weight of seeds planted. J. Agric. Res. 14:359–392.

Baker, R. J. and K. G. Briggs. 1982. Effects of plant density on the performance of 10 barley cultivars. Crop Sci. 22:1164–1167.

Baker, R. J. and P. L. Dyck. 1974. Combining ability for yield of synthetic hexaploid wheats. Can. J. Plant Sci. 54:235–239.

Beaven, E. S. 1914. Rep. Brit. Assoc. 83:660.

Beaven, E. S. 1920. J. Farmers' Club, London: Part 6.

Bhatia, C. R. 1975. Criteria for early generation selection in wheat breeding programs for improving protein productivity. Euphytica 24:789–794.

Bhatt, G. M. 1976. Variation of harvest index in several wheat crosses. Euphytica 25:41–50.

Bhatt, G. M. 1977. Response to two-way selection for harvest index in two wheat (*Triticum aestivum* L.) crosses. Aust. J. Agric. Res. 28:29–36.

Bhatt, G. M. and N. F. Derera. 1978. Selection for harvest index among near-homozygous lines of wheat. J. Aust. Inst. Agric. Sci. 44:111–112.

Briggs, K. G. and A. Aytenfisu. 1980. Relationships between morphological characters above the flagleaf node and grain yield in spring wheat. Crop Sci. 20:350–354.

Buzzell, R. I. and B. R. Buttery. 1977. Soybean harvest index in hill plots. Crop Sci. 17:968–970.

Chandhanmutta, P. and K. J. Frey. 1973. Indirect mass selection for grain yield in oat populations. Crop Sci. 13:470–473.

Chandler, R. F., Jr. 1969. Plant morphology and stand geometry in relation to nitrogen. *In* J. D. Eastin, F. A. Haskins, C. Y. Sullivan, and C. H. M. Van Bavel (Eds.), Physiological aspects of crop yield, ASA, Crop Sci. Soc., pp. 265–285.

Donald, C. M. and J. Hamblin. 1976. The biological yield and harvest index of cereals as agronomic and plant breeding criteria. Adv. Agron. 28:361–405.

Engledow, F. L. and S. M. Wadham. 1923. Investigation on yield in the cereals. Part I. J. Agric Sci. 13:390–439.

Engledow, F. L. and S. M. Wadham. 1924. Investigation on yield in the cereals. I. (Part II, cont.). J. Agric. Sci. 14:287–324.

Fischer, R. A. and F. Kertesz. 1976. Harvest index in spaced populations and grain yield in microplots as indicators of yielding ability in spring wheat. Crop Sci. 16:55–59.

Gent, M. P. N. and R. K. Kiyomoto. 1985. Comparison of canopy and flag leaf net carbon dioxide exchange of 1920 and 1977 New York winter wheats. Crop Sci. 25:81–86.

Hamblin, J. and C. M. Donald. 1974. The relationships between plant form, competitive ability and grain yield in a barley cross. Euphytica 23:535–542.

Hanway, J. J. and W. A. Russell. 1969. Dry matter accumulation in corn (Zea mays L.) plants: comparisons among single-cross hybrids. Agron. J. 61:947–951.

Hsu, P. and P. D. Walton. 1970. The inheritance of morphological and agronomic characters in spring wheat. Euphytica 19:54–60.

Hsu, P. and P. D. Walton. 1971. Relationship between yield and its components and structures above the flag leaf node in spring wheat. Crop Sci. 11:190–193.

Jain, H. K. and V. P. Kulshrestha. 1976. Dwarfing genes and breeding for yield in bread wheat. Z. Pflanzenzuecht. 76:102–112.

Jalani, B. S., K. J. Frey, and T. B. Bailey, Jr. 1979. Contribution of growth rate and harvest index to grain yield of oats (*Avena sativa* L.) following selfing and outcrossing of M_1 plants. Euphytica 28:219–225.

Jensen, N. F. 1978. Limits to growth in world food production. Science 201:317–320.

King, A. J. and E. H. Jebe. 1940. An experiment in pre-harvest sampling of wheat fields. Iowa State College Agric. Exp. Stn. Res. Bul. 273:624–649.

Knott, D. R. 1986. Effect of genes for photoperiodism, semidwarfism, and awns on agronomic characters in a wheat cross. Crop Sci. 26:1158–1162.

Love, H. H. 1912. A study of the large and small grain question. Proc. Am. Breeders Assoc. 7:109–118.

McVetty, P. B. E. and L. E. Evans. 1980a. Breeding methodology in wheat. I. Determination of characters measured on F_2 spaced plants for yield selection in spring wheat. Crop Sci. 20:583–586.

McVetty, P. B. E. and L. E. Evans. 1980b. Breeding methodology in wheat. II. Productivity, harvest index, and height measured on F_2 spaced plants for yield selection in spring wheat. Crop Sci. 20:587–589.

Moussouros, B. G. and D. C. Papadopoulos. 1935. Correlating yield and phenological averages to increase efficiency in wheat breeding. J. Am. Soc. Agron. 27:715–723.

Nass, H. G. 1973. Determination of characters for yield selection in spring wheat. Can. J. Plant Sci. 53:755–762.

Nass, H. G. 1978. Comparison of selection efficiency for grain yield in two population densities of four spring wheat crosses. Crop Sci. 18:10–12.

Nelson, O. E. and A. J. Ohlrogge. 1957. Differential responses to population pressures by normal and dwarf lines of maize. Science 125:1200.

Okolo, E. G. 1977. Harvest index of single F_2 plants as a yield potential estimator in common wheat. M. S. Thesis, Univ. of Manitoba, Winnipeg, Manitoba, Canada.

Robinson, R. G. and L. A. Bernat. 1963. Dry weight of panicles as an estimate of yield in grain sorghum. Crop Sci. 3:22–23.

Rosielle, A. A. and K. J. Frey. 1975a. Estimates of selection parameters associated with harvest index in oat lines derived from a bulk population. Euphytica 24:121–131.

Rosielle, A. A. and K. J. Frey. 1975b. Application of restricted selection indices for grain yield improvement in oats. Crop Sci. 15:544–547.

Rosielle, A. A. and K. J. Frey. 1977. Inheritance of harvest index and related traits in oats. Crop Sci. 17:23–28.

Schapaugh, W. T., Jr., and J. R. Wilcox. 1980. Relationships between harvest indices and other plant characteristics in soybeans. Crop Sci. 20:529–533.

Sharma, R. C. and E. L. Smith. 1986. Selection for high and low harvest index in three winter wheat populations. Crop Sci. 26:1147–1150.

Simpson, G. M. 1968. Association between grain weight per plant and photosynthetic area above the flag-leaf node in wheat. Can. J. Plant Sci. 48:253–260.

Singh, I. D. and N. C. Stoskopf. 1971. Harvest index in cereals. Agron. J. 63:224–226.

Smith, H. F. 1936. A discriminant function for plant selection. Ann. Eugen. (London) 7:240–250.

Syme, J. R. 1970. A high-yielding Mexican semi-dwarf wheat and the relationship of yield to harvest index and other varietal characteristics. Aust. J. Exp. Agric. Anim. Husb. 10:350–353.

Syme, J. R. 1972. Single plant characteristics as a measure of field plot performance of wheat cultivars. Aust. J. Agric. Res. 23:753–760.

Takeda, K. and K. J. Frey. 1976. Contributions of vegetative growth rate and harvest index to grain yield of progenies from *Avena sativa* × *A. sterilis* crosses. Crop Sci. 16:817–821.

Takeda, K. and K. J. Frey. 1979. Protein yield and its relationship to other traits in backcross populations from an *Avena sativa* × A. sterilis cross. Crop Sci. 19:623–628.

Thorne, G. N. 1963. Varietal differences in photosynthesis of ears and leaves of barley. Ann. Bot. 27:155–174.

Thorne, G. N. 1965. Photosynthesis of ears and flag leaves of wheat and barley. Ann. Bot. 29:317–329.

Turner, H. N. 1959. Ratios as criteria for selection in animal or plant breeding, with particular reference to efficiency of food conversion in sheep. Aust. J. Agric. Res. 10:565–580.

Voldeng, H. D. and G. M. Simpson. 1967. The relationship between photosynthetic area and grain yield per plant in wheat. Can. J. Plant Sci. 47:359–365.

Walton, P. D. 1969. Inheritance of morphological characters associated with yield in spring wheat. Can. J. Plant Sci. 49:587–596.

Watson, D. J., G. N. Thorne, and S. A. W. French. 1958. Physiological causes of differences in grain yield between varieties of barley. Ann. Bot. 22:321–352.

Watson, D. J., G. N. Thorne, and S. A. W. French. 1963. Analysis of growth and yield of winter and spring wheats. Ann. Bot. 27:1–22.

Welbank, P. J., S. A. W. French, and K. J. Witts. 1966. Dependence of yields of wheat varieties on their leaf area durations. Ann. Bot. 30:291–299.

Whan, B. R., A. J. Rathjen, and R. Knight. 1981. The relation between wheat lines derived from the F_2, F_3, F_4 and F_5 generations for grain yield and harvest index. Euphytica 30:419–429.

Yap, T. C. and B. L. Harvey. 1972. Relations between grain yield and photosynthetic parts above the flag leaf node in barley. Can. J. Plant Sci. 52:241–246.

27

STABILITY

There are three generally recognized measures of stability used by breeders, biometricians, and geneticists. The first, denoted b, was proposed by Finlay and Wilkinson (1963), and represented the regression of a variety's performance on the mean yield of a site. The second, D, also a regression technique, was proposed by Eberhart and Russell (1966); it measures the deviation of observed from predicted yields. Wricke (1962, 1966) proposed his ecovalence W, which is derived from a partitioning of the $G \times E$ sum of squares, removing the parts applicable to each genotype's participation. The terms b and D are usually used together to complement each other, with occasional use also of W as a third measure of stability.

STABILITY: MODELS AND MEANINGS

Plaisted and Peterson (1959) proposed a model to evaluate genotypes for stability over locations. Apparently the model may be used only if the combined analysis of variance (the first step) shows a variety × location mean square that is statistically significant. The other three steps are, successively, computing the combined analyses of variance for all pairs of combinations of genotypes, equating observed and expected mean square to obtain estimates of the standard deviation of the variety × location interaction from each variety pair, and summing and averaging these estimates for all pairs having a common varietal member. In essence this isolates the contribution or response that each variety has made to the total variance pool. When applied to the known adaptation characteristics of well-known "old" varieties of potatoes in New York State, the model produced responses whose interpretation agreed with the recognized farm performance reputation of the potato cultivars.

Plaisted and Peterson's model is different and interesting because it uses genotype × environment (site) variance without in any way characterizing the site environment: it is directed solely to developing the genotype's response to any environment in which it might be placed. In this sense the model may be said to be site-neutral. The resolution of the analysis therefore provides a measure of a genotype's stability: high stability if the sum of its partitioned contributions to genotype × environment (site) interaction is low relative to all other genotypes, or at the other extreme, low stability if the sum is high.

Eberhart and Russell (1966) reviewed the attempts to ameliorate the genotype × environment interaction problem. These procedures included the stratification and clustering of testing environments having similar characteristics. This reduced the interaction to a practical minimum. The authors then turned to the consideration of genotype stability, that is, a damped varietal reaction to different environments. They presented a stability model based in part upon Finlay and Wilkinson's model (which used the site mean yield of all entries, and a regression coefficient), but extended by specifying that more (all) environments be included, and by adding a formula that estimates a stability parameter from deviation mean squares from regression. In their words:

> This model provides a means of partitioning the genotype–environment interaction of each variety into two parts: (1) the variation due to the response of the variety to varying environmental indexes (sums of squares due to regression); and (2) the unexplainable deviation from the regression on the environmental index.

Their model has found wide acceptance.

Hanson (1970) defined a "genotypic stability space" derived from genotype × environment interactions in different environments. Stable genotypes would reside in the center of such a multidimensional space and unstable ones nearer the periphery. Stability was measured by the Euclidean distance of a genotype from the center. Hanson's proposals were intended to integrate information, from the Finlay and Wilkinson (1963) and Eberhart and Russell (1966) environmental indices involving regressions of yield on site means and deviations from regression, by introducing another concept based on the homeostatic properties of the genotypes. [See Brennan and Sheppard (1985) for analogous reasoning.]

Joppa, Lebsock, and Busch (1971) used regression analysis on 10 years of regional spring wheat data from 15–20 locations. Eberhart and Russell's 1966 environmental index was used. The analyses highlighted yielding performance (which was known from mean yields) and, particularly, stability. Stability was indicated by the magnitude of mean square for deviations from regression: a low mean square indicated a stable variety and a high mean square suggested instability. In some cases the causes of instability were known, for example, stem rust attack, but in others there was no apparent explanation, suggesting possible genetic deficiencies *in re* adaptation. It seems clear from this study that both high and low mean square deviations merit the breeder's attention.

Pederson's (1974) paper on varietal performance over years was of particular interest to me because two days before reading it, I had been writing about the plant breeder's need for a better understanding of the year interaction effects on variety performance [cf. Briggs and Shebeski (1971)]. Pederson's article is a statistical treatment of the problem. It is clear from his discussions that he brought together many of the elements that affect the interpretation of variety performance. Significant among these is the relationship between accuracy and the number of years sampled.

Pederson asked the question, apropos the conflict between high average yield and stability, which often cannot be answered: "For example, would a relatively stable variety be preferred to a variety which is relatively unstable but has a higher value?"

The aim of his research was to find a bridging parameter that would represent fairly both the interests of stability and the arithmetic mean, and in doing so would consider the relative frequencies with which different environments, "in particular good and poor years," were experienced.

Measures of stability, such as those of Finlay and Wilkinson (1963) and Eberhart and Russell (1966), provide measures of varietal stability that are unbiased estimates so long as linearity prevails over the range of possible environments. Therein lies the "rub": if response is not linear, or if a variety responds differently, uncertainty prevails in the tails of the environmental interaction curve, and it becomes necessary to know the frequency of extremes—good and bad years.

Recognizing that the evaluation of a single genotype, for example, for release as a variety, usually does not encompass many years, Pederson suggested as an alternative that regional long-term data might be used. However, for this to be valid it appeared that testing at at least four sites a year was necessary in order for a mean trial yield to be consonant with the general regional experience. Pederson gave formulas for cases where long-term data are either known or unknown.

It is no secret that the evaluation of genotypes and populations for their stability status has become increasingly sophisticated and complex. Indeed, stability analysis has become almost a discipline in itself. Where once there was a single authority, then two or three, there now are many. In an effort to bring them all together for consideration at the same time, Lin, Binns, and Lefkovitch (1986) assembled and discussed nine stability statistics and nine similarity coefficients. New concepts of stability were introduced and a nonparametric approach (genotype clustering) discussed. This is laudable and useful in that the formulas for all are found handily in one place, but it is questionable whether the authors' goal, "to clarify the apparent confusion," has been reached.

The authors made a statement while speaking of clustering of genotypes, a sound practice, that might lead one to think we have come full circle relative to testing procedures:

> If a well-known cultivar is included in the test, it can be used as a paradigm for the other genotypes in the same subset. These genotypes may be regarded as having the overall characteristics of this cultivar, and extrapolation for them to a much wider range of environments than those tested may be possible.

I agree that the inclusion of a well-known cultivar having the same general characteristics and responses to environments as a clustered set of genotypes can play a special role, but it is only the clustering that is different—we have always had the check cultivar. Paradigm is defined as "model or pattern," and thus is a check or control variety which plant breeders have used from early days.

ENVIRONMENTAL ASPECTS OF STABILITY

Andrus and Bohn (1967) mass-selected cantaloupes for nine generations (index selection using 18 traits), alternately using high-stress and low-stress environments for the growing cycles. The goal of high-performance stability was not realized. They surmised that their hypothesis that alternating environments might lead to general stability might be correct, but "in alternating selection between two widely different environments, we have repeatedly altered or reversed the direction of approach to genetic stability (homeostasis) and hence have delayed the approach to full stability in environmental response."

Attention is called to Freeman's (1973) review of interactions which included a discussion of stability. The paper contained almost 100 references to previous work on interactions.

Findings with implications for selection strategies in crop plants emerged from a study by Jinks and Connolly (1975) of the fungus *Schizophyllum commune*. The regression of rate of growth on temperature over a range of 15 to 30°C showed that sensitivity response varied with the environment, so that the authors were led to the conclusion that selection for a type of mean performance, high (H) or low (L), in a type of environment, good (G) or poor (P), determines the level of stability acquired. To illustrate, selection for high mean performance in a good environment (H × G) and for L × P produces environmentally sensitive selections, whereas H × P and L × G produce more stable selections.

I infer two things from this: (1) "sensitivity" is "stability," or its inverse, and (2) selection of plants for yield in a favorable environment will produce selections with lowered stability (greater variance) when tested across a wider sample of environments. Although these findings were not extrapolated to crop plants, they seem appropriate for testing there. It is unfortunate that one alternative to low stability is to select for high yield under poor conditions. The authors do offer a tandem, compromise approach to selection.

Bains (1976) used grain yield of wheat to analyze genotype × environment interactions in six crosses among six parental lines that were chosen for their linear sensitivity to additive environmental variation. Evaluation of both linear and nonlinear components of genotype × environment was made on the F_2, F_3, and F_4 generations of the crosses, which were designed as two each of high × high (sensitivity), high × low, and low × low. It was found that all components of the genotype × environment interaction were associated with the corresponding components known to be in the parents. This association is presumed to have a genetic base, proof of which came from the observations that segregation for components was noted in high × low crosses (contrasting sensitivities in the parents) but not in either high × high nor low × low crosses, where there were similar sensitivities in the parents.

The cluster analysis used by Ghaderi, Everson, and Cress (1980) not only was useful in matching genotypes with locations for regional testing, but also arranged genotypes according to broad measures of stability.

Borojevic and Williams (1982) tracked genotype × environment interactions for seven variables in three wheat varieties over a 10-year period. The combined contribution of these variables accounted for 83%, 75%, and 34% of the variation in grain yields of 'Sava', 'Bezostaia', and 'Bankut', respectively. Sava was substantially higher yielding than Bezostaia but was not as stable in its yield production; the authors gave a pragmatic assessment of stability in stating that preference in choice of a cultivar should be given to high yield over high stability because yield is more important to the grower.

In a two-year test, Heinrich, Francis, and Eastin (1983) created 14 environments in which they tested 56 genotypes or populations of grain sorghum. The Finlay–Wilkinson model of regression of genotype yield on site mean yield was used (without the log 10 scale). From the group, three stable genotypes (b values from 0.53 to 0.83) and three nonstable ones (b values from 1.27 to 1.51) were picked for comparison testing. The usual yield components involving plants, heads, tillers, and seeds were measured and recorded.

Nonstable genotypes are generally considered to have higher yield potential than stable genotypes because their specific adaptation is usually directed toward performance in favorable environments. This was not the case in this study, however, where it was found that the stable broadly adapted genotypes showed equivalent yield potential; therefore stability and high yield are not mutually exclusive.

A second important finding was that two of the stable genotypes were high in yield across all environments. This did not seem to be due to compensation among yield components because their levels remained high. In stable genotypes in unfavorable environments it was noted that seed weights positively influenced higher yields.

A follow-up paper by Heinrich, Francis, Eastin, and Mohammad Saled (1985), using the materials within the framework of the 1983 study, examined possible causes for differences in yield stability. A clue was found in the average 100-seed weights over 14 environments—they were markedly higher in stable genotypes, 2.47–2.79 g versus 2.10–2.20 g, respectively. The heavier seed weight of stable genotypes seemed to stabilize performance over diverse and, particularly, unfavorable environments, suggesting larger, heavier seed weights as a primary breeding objective. However, tests would have to be conducted to ensure that genotypes retained their heavy seed advantage in low-yield environments.

This appeared to be a case where a reason for greater stability was found. This is suggestive of other explanations of differences in stability. For instance, differences in winter hardiness might show a positive correlation with stability.

RELATIONSHIPS TO CROSSES AND POPULATION STRUCTURE

Huskins (1931) investigated blindness or blast in oats and concluded that genetic resistance to blindness existed. More importantly he noted that a literature review showed that it could be caused by a variety of causes, and there were significant genotypic differences. From a breeding and selection viewpoint it would seem that an expression of this undesirable character might signal a lack of stability in a genotype's response to stress.

Allard (1961) wished to test the validity of associations with stability in the self-pollinated crops. To do this he constructed 10 lima bean populations organized from the following three groups: three pure line cultivars, four synthetic mixtures including the three possible cultivar two-line combinations and one three-line combination, and three hybrid unselected bulk populations (F_7 and F_9) from the three possible crosses of the three cultivars. Testing of the 10 populations covered a four-year period.

Two kinds of comparisons were made to show the relationship between diversity and stability, namely, rank order consistency and relative size of variances. The results were as follows:

1. *Rank Order Consistency:* Consistency in rank is an indication of stability. The order of stability, high to low, was bulks, mixtures, and pure lines.
2. *Relative Size of Variances:* Smaller mean squares and components of variance indicate greater stability. The order of yield stability was

bulks and mixtures approximately equal and both more stable than pure lines.

Allard concluded that stability would be more readily attained in multiline varieties than it would be in combinations of genotypes yielding more than the best pure lines.

In these studies with lima beans it seems clear that a heterogeneous population is more stable than a homogeneous one, but highest yields may dictate that the heterogeneity be derived from the pervasive buildup of the population by the segregational expansion from a hybrid rather than from a mixing of pure lines. This brings to mind the paper by Rosen (1949), who mentioned mixed populations of a cross to be used as a cultivar with relaxed concerns about crop uniformity—I believe the 'Traveler' oat cultivar was such a variety. Perhaps cereal breeders could take another look at this kind of variety. In my experience it was infrequent but not by any means rare to find unselected bulk populations from a cross that visually were remarkably uniform. Such crosses might have been evaluated as possible cultivars, but in fact they tended to be discriminated against because their existence pointed to different hypotheses of origin—all unfavorable—such as selfing, linkage, close parent relationship, and the like. Allard's observations suggest that the origin and creation of these heterogeneous hybrid populations be considered as a breeding challenge.

An important paper in that it delineated the outline, scope, and components of the genotype × environment interaction complex was published by Allard and Bradshaw (1964). They characterized environmental variation as "predictable" and "unpredictable," for example, owing to soil characteristics and year-to-year fluctuations, respectively. They suggested "well-buffered" as a general term in place of the more controversial "homeostatic." Populations were defined as exhibiting "individual buffering" or "populational buffering."

Allard and Bradshaw sought to relate these determinations to stability, and refined the meaning of stability to the practical economic aspects of the phenotype, that is, especially yield and quality, rather than to overall constancy. The authors concluded that the path to populational stability under different environmental regimens was through genetic diversity, via heterozygosity, heterogeneity, or both, and including "mixtures of different genotypes," which would include multline applications.

Allard and Bradshaw suggested several practical plant breeding inferences that might be drawn from genotype × environment interactions. Examples were as follows:

1. Large variety × location interactions indicate a region with different environments.
2. Large variety × treatment interactions indicate a situation where the treatment creates a special environment.

3. Either of the above suggest a program to develop more than one variety, each adapted to the important environmental variation observed.

Eberhart, Russell, and Penny (1964) reported higher variety by year interactions, therefore less stability, in single crosses compared with three-way crosses of corn.

Pfahler (1964) studied seven populations of oats (six cultivars and an equal-seed-number composite of the six) for fitness and its variability over enviornments. Significant differences in fitness were found among the cultivars, but the fitness of the composite was higher than the mean of its components. In particular, the variability in fitness of the composite was "much lower than any of the six component varieties," an indication of population stability.

Smith, Byth, Caldwell, and Weber (1967) made estimates of phenotypic stability in soybean populations using both the Finlay and Wilkinson and the Eberhart and Russell models; that is, they accepted the (ideal) stable genotype to rest on zero on the regression coefficient and deviations from regression scales, and at the same time they dealt with a finding of average regression coefficients waffling around 1.0. Thus closer to zero represented greater stability whereas deviations above 1.0 represented less stability. There was a tendency for heterogeneous homozygous lines to be more stable than homogeneous homozygous lines. I received the impression that testing in better environments (higher yield) was positively associated with stability.

Qualset (1968) compared stability relationships in three wheat population structures, namely, 7 parents, 21 F_2s, and 21 1 : 1 mixtures of parents. He used Finlay and Wilkinson's b, Eberhart and Russell's D, and Wricke's W models, respectively. There was a clear superiority, greater stability, of mixtures over F_2s; the results for the parents were inconclusive and ambiguous. Qualset was puzzled by the mixture–F_2 relationship, believing, as I interpreted it, that a two-line homozygote mixture should not have prevailed for stability over the broad heterozygosity of an F_2 population. Heterogeneity aside, Qualset opined that dominance, epistasis, or linkage might influence stability. These would decrease with advancing generations.

Qualset concluded that commercial use of mixtures likely would be cost-ineffective. His final recommendation is important: the suggested use of advanced generations of intervarietal crosses (analogous to extension of this experiment's F_2s) because of higher yields and the possibility of selection among populations for greater stability.

Byth and Weber (1968) examined the performance of several agronomic and chemical traits over three environments in two differently structured hybrid soybean populations, namely, heterogeneous (F_2-derived) and homogeneous (F_5-derived). The F_2-derived lines were advanced sequentially by bulk procedures, and the F_5-derived lines were random samples drawn in that generation. All lines were advanced to the F_7 for testing.

Lines, regardless of source, performed similarly across environments except for two traits, height and maturity. However, for all traits in all three environments, F_2-derived lines showed lower variances than F_5-derived lines, indicating the action of homeostatic forces. This greater phenotypic stability of the F_2-derived lines was linked to genetic heterogeneity as contrasted to genetic uniformity of the more homozygous F_5-derived lines. The inference may be drawn that heterogeneous lines, because of greater stability, may require less regional testing.

The article on feedback mechanisms by Schutz, Brim, and Usanis (1968) raises an interesting point, which they did not discuss, about the duration of feedback. It would seem that there might be two aspects of feedback, one dealing with the sequence of adjustments made during the growing season and the other concerned with the consequences to the following progenies. The current recommendation to growers of multiline varieties, for example, is that growers should return annually for new seed, just as in the case of hybrid varieties. In such situations only the first aspect of feedback would apply and would have no consequences to following progeny. If the intergenotypic effects resulted in favorable overcompensation, higher fitness, or whatever, then that aspect, subject to environmental variance, is repeatable when the grower returns for the same new seed to plant the next year. Only if harvested seeds are used to continue the population year after year are there consequences to the following progenies.

Rasmusson (1968) compared two sets of three kinds of population structure in barley. The populations were pure lines, mixtures of these pure lines, and mixtures of hybrids of these pure lines; the pure lines were different in the two sets. The mean of component lines provided a good estimate of yield of their mixtures; the same held for mean of parental lines and yield of their hybrid mixtures. The mixtures of hybrids of pure lines showed greater stability than either pure lines or their mixtures (showing the effect of greater variance of genetic heterogeneity).

Kaltsikes and Larter (1970) placed more emphasis on the characteristics and interactions of the breeding materials than on the methods used:

> If selection for stability is the main purpose of a breeding programme any of the three methods used in this study can be used. The breeding material, however, should be evaluated both for sensitivity to environment and relative mean performance.

Superior stability of multiline mixtures of soybeans over their pure line components was shown by Schutz and Brim (1971). Their materials were four commercial cultivars and six 2-line and four 3-line mixtures tested over a four-year period in North Carolina. Mixtures that yielded about as well as the mean of their components were designated as complementary; half of the six 2-component mixtures yielded significantly above this measure and were termed overcompensatory. These responses were attributed to specific intergenotypic reactions. Schutz and Brim pointed out that stability is a relative term, the key to which may be consistency. Strictly speaking, none of the mixtures was stable, but there were differences in their level of stability. Mixtures were found to be more stable generally than pure lines. The authors postulated an important distinction between complementary and overcompensatory mixtures in that the first seem to handle the routine smaller interactions about the same as their components, but the overcompensatory mixtures meet and rise above additional environmental obstacles, for example, different locations. It would be interesting to know if this interpretation could be extrapolated to cover a wider range of environmental pressures, so that it might be seen as a hallmark of mixture excellence. The authors discussed the two types of mixtures, complementary and overcompensatory, and believed both are required for stability but one, overcompensatory, adds to productivity whereas the other does not. Overall, three-line mixtures appeared no better than two-line mixtures.

A finding by Feaster and Turcotte (1973) that has application to multiline varieties was that a composite of three doubled haploids of cotton showed improved stability compared to the parent cultivars and the separate doubled haploids.

Marshall and Brown's (1973) paper, discussed in the section on multilines, should be referred to for its heavy emphasis on stability.

Pfahler and Linskens (1979) reported on the relationship between yield stability in experimental oat multilines formulated from two groups of four pure lines and all their possible combinations. These were tested over five years. The Group 1 lines were the highest yielding among 89 lines tested for six years; Group 2 lines exhibited a range of environmental variability typical of the 89 lines. The responses of the various pure lines and multilines (15 populations per group) were measured using Eberhart and Russell's 1966 suggested procedure wherein mean yield, regression coefficient, and deviation mean square are the determinants. In this procedure, the "best" population has the highest yield, a regression coefficient of 1, and a deviation mean square of 0. The authors found that the combination of pure lines produced substantial differences in response; in fact, "some multilines containing a mixture of diverse pure lines approached this ideal more closely than their component pure lines in the environments tested."

Although their studies showed that multilines reduced genotype × environment interaction, this favorable assessment was countered by the anticipated necessity of testing potential multilines over a range of environments in order to select a properly buffered arrangement. In an earlier paper, Pfahler (1972) had found yield of pure lines to be positively correlated with yield stability; therefore, they suggested the problem could be eased by selecting only high-yielding pure lines for inclusion.

Langer, Frey, and Bailey (1979) examined three measures of variety production stability (r^2, S_d^2, and W). All were found to be highly correlated and any one would have satisfactorily measured variety stability.

Walker and Fehr (1978) working with soybeans found no significant consistency in stability response between pure lines and mixtures, although the trend favored mixtures. It was not possible to select an optimal number of pure lines for a mixture.

AUTHOR'S COMMENTS ON STABILITY

I confess to ambiguous feelings about the meaning and desirability of stability. In the multiline it was given high priority and the multiline is a conservative breeding approach. But one has to see a pure line cultivar from the viewpoint of the variety, the environment, and the goal. Plant breeders, I believe, project their cultivar objectives to anticipated use by the better growers, and it is logical that better growers are attracted to better environments. It is rare to find cultivars developed for poor environments; nevertheless it does happen for a region where there are pressing economic and social reasons. Also, we must recognize that some important crop economic areas are frequently subjected to severe environmental stress, for example, drought-prone areas, where strategy may dictate a stable genotype that, teleologically, strives under all conditions to do the best it can. Such a strategy might deprive better growers in better environments of top performance unless, as Langer, Frey, and Bailey, as well as others, have found, production and stability responses are independent phenomena. For a region having distinct differences in environment there is merit in the strategy of breeding a cultivar for maximum response and aiming it at appropriate clustered environments. In this scenario, stability response would have lower priority, but might be of value in characterizing the environments so that they could be properly clustered.

It seems that a measure of the importance of stability lies in the nature of the environments into which the breeder intends to launch his new cultivars. As the homogeneity of the environments increases, either through natural affinity or planned clustering, the meaning and importance of stability changes and the focus for cultivar development for such situations becomes clearer. The identification of better environments then leaves the breeder with the responsibility of assessing other environments in order to note whether cultivars adapted to them are needed. It is not certain that stability need be sacrificed because it has been shown that it may be inherited independently of production responses.

Frankel (1958) stated the extreme alternative choices breeders face, namely, breeding for narrow or wide range of environments. Finlay and Wilkinson in 1963 embellished further the description of these situations, particularly the wide range and fluctuation of climatic conditions found in southern Australia—conditions that dictate choosing cultivars with general adaptability. They provided two indices, one that measured a variety's average yield over all environments and one that provided a regression coefficient as a measure of cultivar stability. There is some criticism that the regression coefficient (comparison base of 1.0 as average stability) is not the best measure, but whether this is valid or not, their study was a landmark statement to plant breeders. Coupled with the site mean yield cultivar index, their model allows grouping of both varieties and sites. It is only a short step beyond this to clustering to get the best match of similar varieties and locations.

Finlay and Wilkinson's discussion of adaptability and stability is instructive. Their ideal variety with general adaptability is characterized as having maximum yield potential in the most favorable environment with maximum phenotypic stability. Although they found, in the cultivars studied, varieties with general adaptability (having the highest mean yields in all environments), all had above-average phenotypic stability. Their stability levels were still far below the ideal. They offered the hope in breeding for compromises that would raise high average performance.

REFERENCES

Allard, R. W. 1961. Relationship between genetic diversity and consistency of performance in different environments. Crop Sci. 1:127–133.

Allard, R. W. and A. D. Bradshaw. 1964. Implications of genotype–environmental interactions in applied plant breeding. Crop Sci. 4:503–508.

Andrus, C. F. and G. W. Bohn. 1967. Cantaloup breeding shifts in population means and variability under mass selection. Proc. Am. Soc. Hort. Sci. 90:209–222.

Bains, K. S. 1976. Parent dependent genotype × environment interaction in crosses of spring wheat. Heredity 36:163–171.

Borojevic, S. and W. A. Williams. 1982. Genotype × environment interactions for leaf area parameters and yield components and their effects on wheat yields. Crop Sci. 22:1020–1025.

Brennan, P. S. and J. A. Sheppard. 1985. Retrospective assessment of environments in the determination of an objective strategy for the evaluation of the relative yield of wheat cultivars. Euphytica 34:397–408.

Briggs, K. G. and L. H. Shebeski. 1971. Early generation selection for yield and breadmaking quality of hard red spring wheat. Euphytica 20:453–463.

Byth, D. E. and C. R. Weber. 1968. Effects of genetic heterogeneity within two soybean populations. I. Variability within environments and stability across environments. Crop Sci. 8:44–47.

Eberhart, S. A. and W. A. Russell. 1966. Stability parameters for comparing varieties. Crop Sci. 6:36–40.

Eberhart, S. A., W. A. Russell, and L. H. Penny. 1964. Double cross hybird prediction in maize when epistasis is present. Crop Sci. 4:363–367.

Feaster, C. V. and E. L. Turcotte. 1973. Yield stability in doubled haploids of American Pima cotton. Crop Sci. 13:232–233.

Finlay, K. W. and G. N. Wilkinson. 1963. The analysis of adaptation in a plant breeding programme. Aust. J. Agric. Res. 14:742–754.

Frankel, O. H. 1958. J. Aust. Inst. Agric. Sci. 24:112–123.

Freeman, G. H. 1973. Statistical methods for the analysis of genotype–environment interactions. Heredity 31:339–354.

Ghaderi, A., E. H. Everson, and C. E. Cress. 1980. Classification of environments and genotypes in wheat. Crop Sci. 20:707–710.

Hanson, W. D. 1970. Genotypic stability. Theor. Appl. Genet. 40:226–231.

Heinrich, G. M., C. A. Francis, and J. D. Eastin. 1983. Stability of grain sorghum yield components across diverse environments. Crop Sci. 23:209–212.

Heinrich, G. M., C. A. Francis, J. D. Eastin, and Mohammad Saled. 1985. Mechanisms of yield stability in sorghum. Crop Sci. 25:1109–1113.

Huskins, C. L. 1931. Blindness or blast of oats. Sci. Agric. 12:191–199.

Jinks, J. L. and V. Connolly. 1975. Determination of the environmental sensitivity of selection lines by the selection environment. Heredity 34:401–406.

Joppa, L. R., K. L. Lebsock, and R. H. Busch. 1971. Yield stability of selected spring wheat cultivars (*Triticum aestivum* L. em. Thell.) in the Uniform Nurseries, 1959 to 1968. Crop Sci. 11:238–241.

Kaltsikes, P. J. and E. N. Larter. 1970. The interaction of genotype and environment in durum wheat. Euphytica 19:236–242.

Langer, I., K. J. Frey, and T. Bailey. 1979. Associations among productivity, production response, and stability indexes in oat varieties. Euphytica 28:17–24.

Lin, C. S., M. R. Binns, and L. P. Lefkovitch. 1986. Stability analysis: where do we stand? Crop Sci. 26:894–900.

Marshall, D. R. and A. H. D. Brown. 1973. Stability of performance of mixtures and multilines. Euphytica 22:405–412.

Pederson, D. G. 1974. The stability of varietal performance over years. 2. Analysing variety trials. Heredity 33:217–228.

Pfahler, P. L. 1964. Fitness and variability in fitness in the cultivated species of *Avena*. Crop Sci. 4:29–31.

Pfahler, P. L. 1972. Relationship between grain yield and environmental variability in oats (*Avena* sp.). Crop Sci. 12:254–255.

Pfahler, P. L. and H. F. Linskens. 1979. Yield stability and population diversity in oats (*Avena sp.*). Theor. Appl. Genet. 54:1–5.

Plaisted, R. L. and L. C. Peterson. 1959. A technique for evaluating the ability of selections to yield consistently in different locations or seasons. Am. Potato J. 36:381–385.

Qualset, C. O. 1968. Population structure and performance in wheat. Third Int. Wheat Genet. Symp., Aust. Acad. Sci., Canberra, pp. 397–402.

Rasmusson, D. C. 1968. Yield and stability of yield of barley populations. Crop Sci. 8:600–602.

Rosen, H. R. 1949. Oat parentage and procedures for combining resistance to crown rust, including Race 45, and Helminthosporium blight. Phytopathology 39:20.

Schutz, W. M. and C. A. Brim. 1971. Intergenotypic competition in soybeans. III. An evaluation of stability in multiline mixtures. Crop Sci. 11:684–689.

Schutz, W. M., C. A. Brim, and S. A. Usanis. 1968. Intergenotypic competition in plant population. I. Feedback systems with stable equilibria in populations of autogamous homozygous lines. Crop Sci. 8:61–66.

Smith, R. R., D. E. Byth, B. E. Caldwell, and C. R. Weber. 1967. Phenotypic stability in soybean populations. Crop Sci. 7:590–592.

Walker, A. K. and W. R. Fehr. 1978. Yield stability of soybean mixtures and multiple pure stands. Crop Sci. 18:719–723.

Wricke, G. 1962. Uber eine Methode zur Erfassung der Okologischen Streubreite in Feldversuchen. Z. Pflanzenzuecht. 47:92–96.

Wricke, G. 1966. Uber eine biometrische Methode zur Erfassung der Okologischen Anpassung. Acta. Agric. Scand., Suppl. 16:98–101.

SECTION IV

GERM PLASM AND CROSSING CONSIDERATIONS

28

GERM PLASM: CHOOSING CROSS QUALITY LEVEL

All hybridizing operations start with considerations of several related elements such as germ plasm, genotypic relationships, and diversity. From this process come decisions that are translated into beginning plant breeding operations, namely, choice of parents and choice of possible crossing patterns, which taken as a whole are the expressions of a breeder's tactical approach to the improvement objectives. In other chapters some of these elements are treated in great detail. Here we discuss the role of germ plasm and how it influences the quality level of the crosses chosen by the breeder.

GERM PLASM CONSIDERATIONS

Breeders and germ plasm are linked in a way almost implied by a definition of plant breeding. The various concerns expressed in this chapter are only a small part of the whole: most breeders have taken part in cooperative germ plasm projects in addition to their own program activities. For example, I have published a dozen articles that represent cooperative participation in preparation and release of germ plasm; two are mentioned in this chapter.

Jensen (1962) proposed a world germ plasm bank for cereals to be founded on breeder contributions of surplus F_2-embryo seeds of each crop, which would be mixed annually with the previous stockpile. Suitable maintenance procedures would keep the composite available for distribution to any interested breeder. This bank was begun but I do not know whether it is currently in operation. An inherent problem with an enterprise of this kind is that it requires constant publicity and pleading for contributions of the particular hybrid seeds that would make the bank increasingly useful.

A wheat composite hybrid population, derived from 235 crosses, was created by Suneson, Pope, Jensen, Poehlman, and Smith (1963). This composite complements Barley Composite Cross XXI; both of these are available to breeders.

Bennett (1965) presented a thoughtful overview of world germ plasm matters including a section entitled, "Genetic conservation and the future of plant breeding." One is impressed with her general view of a narrowing genetic base, particularly with respect to certain well-known crops.

Information on nine barley composite crosses was filed in germ plasm registration statements by Suneson (1969); all are of California origin. These composites have been widely disseminated and generally are available for world distribution.

A good piece of detective analysis by Qualset (1975) illuminates both the history of plant exploration and the value of germ plasm collections. His analysis focused on barley collections from Ethiopia. The barleys of Ethiopia, once considered a center of origin of our cultivated barleys, are unique. The great range of geographic and climatic environments has allowed a comparable range of plant types. What particularly makes Ethiopian barleys unique is the discovery that their group alone in the world harbors resistance to an important worldwide barley disease, barley yellow dwarf virus (BYDV). Approximately one-fifth of collected Ethiopian barleys are resistant.

Fortunately, organized expeditions from the early days of the great plant explorers have preserved some of Ethiopia's barleys, but one also must agree with those, particularly Erna Bennett, who believe that the job cannot be termed "done" administratively and that collection must continue.

Dr. Wiebe once recounted his recollections from a long-ago trip to Ethiopia: from many places on the well-traveled main roads he could look out to the horizon and see the telltale golden patches that were ripe barley fields. He yearned to visit these fields but it was not feasible—there were only two ways of reaching off-road points, by organized horse and mule pack train or by helicopter. The first was too time consuming and the second neither safe nor allowed, inasmuch as helicopters were routinely fired upon in the back country.

Incidentally, the source or cause of the resistance genes for BYDV in Ethiopian barleys that Qualset surveyed almost surely was mutation.

Hawkes (1977) provided a general overview of the place of wild germ plasm in plant breeding. He gave reasons why the use of wild species should be last in a breeder's armory of weapons (one may expect this priority to change in the near future as biotechnology moves to the forefront of research). However, Hawkes listed a variety of ways in which wild germ plasm is and may be used. Examples from a wide range of crops were given, and it should be noted that the majority of uses involved searches for disease resistance.

In 1978 Smith, Kannenberg, and Hunter published two articles on maize gene pools maintained at high plant densities. The first of these (1978a) dealt with effects on population structure, and the second (1978b) with effects on grain yield. Although these papers concerned a cross-pollinated crop, I have extracted certain information that bears also on self-pollinated crops, for example, on intermating, stability, and application to male sterile breeding methods.

Population Structure: The parameters of the gene pool structure encompassed open pollinations for several generations under high plant densities; however, certain pools had received (earlier) additional generations of cross-pollination. There were three gene pools of different origin, namely, relating to degree of diversity, adaptation, and number of previous cycles of recombination. An important point, not emphasized, is that each year the gene pools were subjected to heavy selection pressure. Any (apparently all) "standing, bordered, disease-free plant" was harvested if it had a well-filled ear, and these seeds were planted for the next generation. Selection ranged from 0.5 to 3% of the total population.

Important findings from the first paper were as follows:

1. Genetic variability decreased with generations (almost certainly primarily due to the selection imposed, even though it may not have been intended as directional).
2. Adaptive changes, for example, height and maturity, shifted to apparent coincidence with the environments encountered.
3. These major changes occurred in the first or second generation.
4. Genetic variability was highest in the composite that incorporated exotic components as compared to one that incorporated adaptive material.

Grain Yield: In the second paper, Smith et al. considered the effect of open pollination for several generations, including the upgrading selection referred to above, on grain yield and stability. They compared different generations within each gene pool with the gene pool itself; four single-cross hybrids were used as checks. The nurseries were grown at five locations.

Important findings from this paper were the following:

1. Significant interaction was found for the check hybrids × location but not for the gene pools × location. This may be interpreted as greater stability for gene pools (diverse genotypes and heterogeneous) as against the single crosses (one genotype per hybrid and homogeneous).
2. More diverse germ plasm in a gene pool increased its tolerance to high density, thus showing adaptability.
3. Greater changes were seen in gene pools that had experienced fewer

opportunities for recombination before this study was started in Ontario, Canada.

4. Grain yields varied in the gene pools; "however, the proportions of genotypes with greater potential at better environments increased in all gene pools."

In barley and wheat there are opportunities for similar gene pools with the use of male steriles. The intermating and cross pollination is not so thorough as in a maize population, where every plant is a potential participant. Nevertheless, one might expect somewhat the same things to happen in the small grain cereals treated this way. The important thing is to use this knowledge with emphasis on directed selection for traits and objectives of interest. One can counter the expected reduction in genetic variance that will accompany heavy selection pressure by creating selection subsets, that is, by not trying to contain everything in one population. In my MSFRS (male sterile facilitated recurrent selection) breeding program, a great part of the interest and enthusiasm was generated from the spin-off populations, for example, for different formulations of malting quality, disease resistance, and height.

Fowler and Gusta (1979), assessing progress in winter wheat breeding for winter hardiness, found that limited genetic variability existed for this trait. The Crimean (USSR) wheats have long been the source of genes for winter hardiness, continuing to the present with the use of cultivars such as 'Alabaskaja' and 'Ulianovkia'. They observed also that progress is usually made in small increments, and improved wheats may be only marginally superior. This is understandable inasmuch as a breeder will choose among the consistent survivors for the best performers. Unavoidably, other traits will be involved. Thus breeders use a form of index selection that is in reality a compromise that does not permit always choosing the highest winter hardiness response. Another reason for progress in small increments is that improvement in winter hardiness is pressing against a limiting barrier, cold.

Fowler and Gusta's paper is important for development of the field survival index (FSI) as a way of measuring relative survival. The index takes into account the various factors and interactions that confuse and mask differences in winter hardiness, for example, tests with genotypes showing 0% and 100% survival. Only plots with differential winter kill, between 5% and 95% survival, were used. A genotype's rating would be the difference between its average survival stand and that of the other genotype of interest. Adjustments were made as sites with different stress levels were added, and bridges were built to extend the range. The authors liken this to procedures used in making linkage maps; I see it also as analogous to matching and extending tree ring series to determine ages. The inclusion of a five-year table of FSI's showed a range of 6 to 229, and the ranking of cultivars seemed to me to reflect practical breeding experiences.

Rines, Stuthman, Briggle, Youngs, Jedlinski, Smith, Webster, and Rothman (1980) examined an important oat germ plasm collection of 2200 samples of *Avena fatua*, our common wild oat, made in the northern and western United States. Preliminary screening was done for resistance or tolerance to barley yellow dwarf virus (BYDV), cereal leaf beetle, and stem rust, as well as for protein and amino acid concentrations. The samples examined were submitted to the World Oat Collection (SEA–USDA), where they will be available to scientists.

The excellent review and research article on performance trends and genetic gains in winter wheat since 1970, by Kuhr, Johnson, Peterson, and Mattern (1985), carries a message about germ plasm: the international nurseries are showplaces for many of the world's best wheats in a roster that changes annually. A wheat breeder well may consider that no better source of exotic and proven germ plasm exists. The addition of even one genotype a year can have far-reaching consequences in upgrading a breeding program.

The development of superior dwarf oat lines has lagged behind the well documented, even dramatic, progress made in reducing height in other cereal crops, notably wheat and rice. According to Meyers, Simmons, and Stuthman (1985), the use of dwarf stature germ plasm for oats has been retarded by the need to capture the economic yield from straw, a critically important factor in a marginally profitable crop. Their objective in this study was to compare yields and other characteristics of dwarf (having an average 30% reduction in height) and normal-height oats.

The results predominantly showed areas of similarity between the two groups. Points of interest from similarities and differences were:

1. The yield of dwarfs ranged from inferior to equal.
2. Harvest indices were similar.
3. The exsertion of panicles from the flag leaf sheath in dwarfs was reduced. This negative trait must be corrected.
4. Lodging resistance was greater in dwarfs, which was expected.
5. Straw yields of dwarfs were consistently lower than those of normal-height varieties.

All in all, the situation may be described as one calling for greater attention by breeders, particularly with reference to broadening the germ plasm base for dwarf × normal height combinations.

THE GENEOLOGY OF GERM PLASM

Cox, Kiang, Gorman, and Rodgers (1985) compared coefficients of parentage ("The coefficient of parentage between two individuals is defined as the probability that a random allele at a random locus in one individual is identical by descent to a random allele at the same locus in the other individual") to similarity indices in four groups of soybeans using enzyme profiles. An extreme case cited involved two time-separated ancestral introductions (presumably from China) having the same name. Were they the same genotype or, as sometimes happens, entirely different genotypes identified by a collection or site name? Unanswered questions such as these could cause problems and increase the complexity and uncertainty of breeding plans.

In this case their studies showed that the two cultivars were identical at all loci compared. Although their study was a statistical exercise to elicit information and resolve disagreements in various indices (their conclusion recommended a composite index), the import of such studies to breeders is well expressed in their final statement;

> If, among all potential parents with suitable performance, the breeder crosses the pairs which are least closely related, he or she can attempt to maximize the number of segregating loci in each cross and thereby satisfy the second criterion, a large genetic variance. Indices such as *r* and *s* may be useful in choosing diverse parents.

Their bibliography contains references to similar studies in the cereals.

If internal genetic diversity in genotypes could be measured, breeders would have a powerful aid in the matter of choosing parents for crosses. In the research of Kleese and Frey (1964) to link serological patterns in oat genotypes to the degree of diversity often hidden in such genotypes, the results were inconclusive although encouraging to further study.

Bassiri (1976) presented a useful background paper on the use of electrophoretic procedures for cultivar identification. Crop cultivars may be identified by their unique "fingerprint" induced by gel electrophoresis. Bushuk and Zillman (1978) described the method and apparatus used for wheat in Canada.

In a paper with important implications for the broadening of germ plasm and choice of parents for crossing, Langer, Frey, and Bailey (1978) compared the relative yield performance of 66 oat cultivars, which had been introduced in seven chronological eras beginning with 1932. Each cultivar met three criteria that focused on length of testing, varietal release, and adaptation to Iowa conditions. Two older varieties, 'Richland' and 'Cherokee', were included as long-term checks and their mean was used as a basis for comparison. One, Richland, was considered to be representative of older oat varieties developed between 1916 and 1936.

The 66 cultivars were grouped according to the years in which they were developed; there were seven such eras. The results showed that there was an overall mean productivity advancement of 9% from 1932 to 1973, as measured against the mean of the two checks, which had been present in all of the nurseries. The startling finding was that this level was attained by the varieties developed in the earliest era, 1932–1942. No further significant jump in productivity was found in the remaining three decades of the development periods. The authors postulated a small pool of genetic variability for yield in the early *Avena sativa* L. introductions from northern Europe. This variability was soon captured and the amount of introduced exotic germ plasm since has been relatively meager. In some cases productivity gains readily could be traced to specific backcross programs. A recent introduction of germ plasm from *A. sterilis* L. has shown promise of providing gains of as much as 20–30% above that of the other parents used in the crosses. The paper by Langer et al. is unusually significant in its documentation of the importance of germ plasm to gains in plant breeding.

A screening procedure for parental adaptation is now evolving in sorghum breeding and may have application to other crops, for example, for hardiness, winter versus spring types, and so forth. Thomas and Miller (1979) had reported that environmental origins of tropic-adapted and temperate-adapted types could be distinguished by their base germination temperatures. Mann, Gbur, and Miller (1985) refined the technique to a laboratory procedure and mathematical model whereby lines might be selected for adaptation for either line or parent use.

CHOOSING CROSS QUALITY LEVEL

Somewhere among the related subjects of prediction and choice of parents, crosses, and lines lies another degree of freedom for the breeder: the choice of cross quality level. This exists first as an actuality—there are distinct differences in the quality of crosses, let's say a range from horrible to outstanding. Breeders are intrigued by the causes of cross quality differences and attribute them to the diversity of parents, that is, to wide crosses and the use of exotics, to ways in which crosses are made, for example, single crosses, three-way crosses, double crosses, backcrosses, recurrent selection, and so forth. There is a recognition of the strong influence of isolation of germ plasm, which is important in itself because it keeps genes from entering the main recombination streams through the breeding process. Isolation occurs because of the common practice whereby a breeder adheres closely to germ plasm associated with his crop breeding goals. Examples are many: spring wheat breeders not using winter wheat germ plasm, hard red winter wheat breeders not using any other market class germ plasm, oat breeders working within one maturity group, and so forth. All this has led to a form of restricted breeding of "like to like," sometimes to the lure of the "high × low" syndrome.

THE ARRANGEMENT OF PARENTS IN CROSSES

The problem facing corn breeders in selecting inbreds for desirable hybrid combinations was illustrated by Eckhardt and Bryan (1940): "Thus with only 40 inbreds, 780 single crosses and 274,170 double crosses are possible."

They investigated two ways of making double crosses, which differed according to whether the single crosses each represented two inbreds from the same variety, that is, (A × B) × (Y × Z), or whether the inbreds from the same variety came in separate crosses, that is, (A × Y) × (B × Z). The results were decidedly in favor of bringing similar inbreds together in single crosses. The double crosses from this arrangement, (A × B) × (Y × Z), outyielded their opponents 11 times out of 12 over three years of testing. The fact that the highest yields and the greatest hybrid vigor in double crosses came from combining single crosses of different parentage emphasizes the importance of diversity.

This research reminded me of a practice in my wheat, oat, and barley programs in New York. A reserve stock of hybrids that might be used in future crossing seasons was maintained in my office. These hybrids were useful in making three-way and four-way crosses and saved time toward meeting various objectives. However, many of these crosses were what I referred to as "pseudo-parents." They brought together different but similar parents; for example, they might be two varieties having many differing characters, but in a different sense, they were similar because they fit into the same quality standards group. Examples in wheat crosses were 'Yorkstar' × 'Arrow' and 'Houser' × 'Frederick'; an example in oats was 'Orbit' × 'Astro'. Usually the genetic relatedness was high, and the intent was to use a hybrid in a cross exactly as though it were a homozygous parent. This kind of hybrid used as a parent introduces additional genetic variability without disturbing the sharp edge of an objective. If the two parents differed greatly in relatedness, they were usually chosen for use in three-way and four-way combinations.

Thorne and Fehr (1970) showed that three-way crosses are an effective way to introduce exotic germ plasm, such as plant introductions, into soybean breeding programs, and were superior to two-way crosses of adapted × exotic. The second cross (third way) is a category backcross to adaptive; that is, any parent used was adaptive. The three-way population yield means (F_5 and F_6) exceeded those of the two-way crosses. Strangely but fortunately, genetic variability was not reduced in the three-way populations. The value of the exotic germ plasm introduced was shown by testing the five highest-yielding lines (upper 10%) in each set-maturity group. At one location 47 of the 60 superior lines came from three-way crosses, and at the other location 18 of 30 came from three-ways.

An intriguing discussion of how and in what ways specific cross forumlations affect population stability and yield performance in rye was presented by Becker, Geiger, and Morgenstern (1982). Years ago I recall that "hybrid" rye prospects were not encouraging, principally due to inbreeding depression of a magnitude that threatened inbred viability and survival. Today there are available cytoplasmic male sterility and restorer genes. I am including this report of research by Becker et al. because it provides insights into crossing methodology: they compared 141 hybrids in crosses formulated as follows, nine single crosses, 33 each of three-way, double cross, and top cross, and two-component blends of three-way crosses. An added experiment 2 included 24 hybrids representing all of the groups except the three-way cross blends. Testing was at five locations (experiment 1 at three, experiment 2 at two) for two years, making 10 environments.

Comparing hybrid types for yield, it was found that single crosses and three-way crosses were alike. Double crosses yielded less than three-ways, and in experiment 1, blends of three-way crosses yielded more than both. None of the yield differences was significant, however.

Top cross yields were decidedly higher than other formulations. This was considered to be due to the pollinators used; an attempt was made to achieve genetic balance among the other crosses by using three "old" pollinator lines, but in the top crosses three contemporary high-yielding commercial cultivars had been used.

As for stability, measured by experimental variance and deviation mean squares, there was a general increase with increasing heterogeneity; that is, top crosses were more stable than double crosses, which were more stable than three-way crosses. An exception was that single crosses showed high stability, greater than expected. There was also much variability within a cross type so that it should be possible, for example, to select among three-way crosses for one that was more stable than double crosses. All in all, Becker et al. chose three-way crosses, or single crosses, over double or top crosses as the more favorable format for the plant breeder, but stressed that at present single crosses were not feasible. A satisfactory program requires single crosses as seed parents, and the pollinator should be heterozygous or heterogeneous in order to extend pollen-shedding time, thus indicating that the male should be a single cross or two-line synthetic.

In further studies on the value of interspecific hybridization between *A. sativa* and *A. sterilis*, Murphy and Frey (1984) compared six hybrid populations, four of which were interspecific and two intraspecific.

The interspecific hybrid populations, "W." were designed to project the effect of 12.5% *A. sterilis* germ plasm. An average effect was attained by crossing one *A. sativa* cultivar ('Noble') with eight F_1s of three-way crosses, where the *A. sterilis* accessions occupied a 25% parental position; this procedure was used in each of the four W populations. The two intraspecific populations, crosses within *A. sativa* germ plasm, were bulked samples of the three 3-way matings of the three *A. sativa* cultivars (C_1), and the nine 3-way matings of each of their single crosses with three other modern cultivars (C_2).

More than 500 F_3-derived lines were extracted from each population for testing purposes in F_4 and F_5. Comparisons with parents were made on six measured traits. All characters except harvest index showed wide parental differences comparing species versus species.

Bypassing the detailed results, the authors summed the comparisons in this way:

> Our study showed that the development of a population from a broad sample of *A. sativa* cultivars (C_2) was a more efficient short-term strategy for grain yield and vegetative growth rate improvement than was the introgression of limited samples of germplasm from randomly chosen *A. sterilis* collections into the cultivated gene pool.

They decried a random approach to selection of *A. sterilis* parents and suggested recurrent cycles of crossing and a clustering of similar accessions to aid breeders in a choice.

My interpretation of this research focuses on the limited sample of introgression entry (of *sterilis* into *sativa* germ plasm) at the 12.5% level only. Even so, the four W populations varied considerably in percent transgressive segregates. Staying within the cross format used, it is possible that 25% or 50% *sterilis* input might provide a more satisfactory production of transgressive segregates, offering possibilities directly for yield or to provide a better base for parents for recurrent selection. The cross format used, in interspecific terms, is analogous to a BC_2 even though three *sativa* parents take part, and emphasizes *sativa* more than it does the different cultivars. If a breeder contemplates using either of two elite cultivars as parents, it may be a sound choice to substitute their F_1 for both. Thus different cross arrangements of the same four parents may provide distinctly different progeny, for example:

		% PI
W_1 =	Otter × PI 317789 2× Grundy 3× Noble	12.5
Examples:	Otter × Grundy 2× PI 317789 3× Noble	25.0
	Otter × PI 317789 2× Grundy × Noble	25.0
	Otter × PI 317789 2× Grundy or Noble	25.0
	Otter × Grundy 2× Noble 3× PI 317789	50.0

Considering only the W_1 cross, the dilution to 12.5% PI may be too great. There is also a sampling problem in the final cross; that is, a large number of W_1 crosses would have to be made to sample fully the potential recombinations presented when the second cross (Otter × PI 317789 2× Grundy) is used as a parent. Only the average of several W_1 crosses could express the value of 12.5% PI input. This may have accounted for the variation shown in percent transgressive segregates in Murphy and Frey's study.

On the other hand, the 50% PI level may not be dilution enough. This suggestion is based on a general experience of dealing with wide crosses where success usually requires at least one backcross or outcross to a non-PI genotype. Therefore, this leaves the 25% PI input level as the most likely desirable strategy, to be followed by early-generation testing, selection, and recurrent selection.

Aware of the restricted genetic base of U.S. soybean cultivars, Khalif,

Brossman, and Wilcox (1984) turned their attention to the type of cross combinations used as a means of increasing the diversity of germ plasm used by breeders. They evaluated the efficiency and potential of three kinds of crosses, namely, two-, three-, and four-parent combinations. They examined the ability of these kinds of crosses to generate genetic variability and promising progenies from adapted and exotic parents. They found the three-parent cross (three-way cross) to be superior to either the two- or four-parent crosses (single and double crosses, respectively).

This paper was of especial interest to me because I confess to an inordinate fascination with breeding methodology. In my cereal programs I used all of these combination procedures and some others (Jensen 1978). The three-way cross was my favorite combination, palpably superior to either the single or double cross. The single cross does well, particularly when both parents are from approximately similar germ plasm quality sources, that is, both adapted. However, when adapted and exotic are mated it seems that a minimum requirement is that a third (adapted) parent is needed to channel the expressed variability into a high progeny potential range. This third parent can be the recurrent adapted parent (backcross) or another selected adapted genotype.

The double cross can be faulted for "getting out of hand" and lacking the controls that the three-way offers the breeder; nevertheless it can be harnessed easily to serve the breeder's needs. Examples of modifications are (1) use of three adapted and one exotic; (2) use of a three-way backcross × a single cross; (3) if equally desirable adapted genotypes are available, subtitution of their single cross for a pure line parent in a cross; and so forth. To illustrate an expansion, one could construct the five-parent cross (single cross × three-way) in such a way that it would have significant increased genetic variability that by design would be channeled to the breeder's desired objectives.

Research in 1986 with recurrent selection procedures for three cycles among parent lines derived from interspecific peanut crosses by Guok, Wynne, and Stalker (1986) showed that substantial gains in yield, both per cycle and compared to commercial cultivars, could be obtained.

THE HIGH–LOW ASPECT AFFECTING CROSS QUALITY LEVEL

In the early days of corn breeding, and possibly to some extent today, a lively interest was shown in the choosing and matching of inbreds for hybrid combinations. Johnson and Hayes (1940) compared the yields of single crosses between low × low, low × high, and high × high yielding inbreds. They found that there was little difference in yield between low × high and high × high, but low × low combinations gave single crosses having substantially lower yields and might well be avoided in planning.

That the linkage between inbred line vigor and cross performance is generally positive is a accepted "given" among corn breeders. Even so the association and apparent combining ability must be checked further, reaching eventually into the specific combinations of the single and double crosses. Murphy (1942) developed important information about the relation of the parent quality level of recovered inbred lines of corn to the yields of their single crosses. On the premise that combining ability and vigor were separate but related heritable traits, Murphy undertook the challenge of upgrading lines of proven combining ability in crosses by improving their vigor. The upgraded lines were referred to as "recovered lines." Backcrossing, with selection, to each parent separately in the respective single crosses was the method used to upgrade lines. The best recovered lines were then crossed to the recurrent and nonrecurrent parents and the crosses tested for yield. From these results were chosen a number of recovered lines showing variations for combining ability when crossed with the nonrecurrent parental inbred line.

A frequency distribution of yields from these crosses is revealing. Yields were expressed in terms of the yield of the original F_1 cross from original inbreds. The distribution of yields of 17 crosses, over the range of −3 to +4 times the standard error of a difference, showed that only high × high and high × intermediate crosses produced yields in the 0 and plus range: 6 of 11 crosses yielded 0 or better. The remaining six crosses, high × low, intermediate × intermediate, intermediate × low, and low × low, with one exception at 0, produced yields only in the minus range.

On the basis of this, Murphy suggested a three-stage evaluation of recovered lines, namely, (1) testing in single crosses to the nonrecurrent parent, (2) discarding those that produced low-yielding crosses, and (3) testing the remainder in all combinations to find the best single cross.

It should come as no surprise that much of our understanding of parent diversity and potential, as well as performance predictions of crosses, arose from early research on corn. The creation of valuable inbreds, single crosses, and double crosses required a degree of credible linkage of prediction and performance; otherwise breeders would be inundated with the great amount of evaluation required to check out every possible combination. Cowan (1943) gave a very readable review of the pertinent corn literature on these subjects. Cowan had two objectives, namely, (1) the correlation between the yields of inbreds in top crosses and their resulting single- and predicted double-cross yields, and (2) the influence of inbred genetic diversity on single-cross yields and predicted yields of double crosses.

Cowan made two groups, A and B, of inbreds representing two origins. Group B was actually more centralized in origin, made up from different plants of the same variety, 'Bailey'. Each inbred was top crossed with the same open-pollinated variety, 'Harrow Golden Glow'. In the next step, Cowan intercrossed group A inbreds (class I single crosses), group B inbreds (class II), and group A × B inbreds (class III). Yield data from these single crosses was used to predict the yields of 118 double crosses. Cowan then characterized and assigned each single cross to one of three groups, high × high, high × low, and low × low, according to the combining ability in topcrosses of their inbreds.

The analysis of these data led to these results and conclusions:

1. High × high and high × low yields were similar, and distinctly higher than low × low single-cross yields.
2. Topcross yields of unrelated inbreds (group A) were positively and significantly related to single-cross yields and predicted double-cross yields. Closely related inbreds (group B) showed no such correlation.
3. Single-cross and predicted double-cross yields of unrelated inbreds were higher than those of related inbreds.

The plant breeding implications are applicable also to self-pollinated crops, namely, diversity is positive, parent performance per se indicates cross and selection potential, low × low crosses are unprofitable, and if made, discardable, and prediction is worthwhile where procedurally feasible.

Green (1948) concluded that the combining ability of corn inbreds was an inherited trait. Using four inbred lines, two high-yielding and two low-yielding, he compared the three combinations, high × high, high × low, and low × low. The yields of their respective single crosses were 80.0, 72.4, and 58.4 bu/acre. The low × low category might be discarded without qualms; however, the high × low category was characterized by a meaningful proportion of desirable F_2 plants (valued according to number of topcrosses which exceeded the average of all topcrosses by more than one standard error of a difference). High × high produced three times as many in this category, so the combining ability of parents is a good prepotency indicator.

Lonnquist (1953) wrote in 1953 concerning the relation between heterosis and yield in corn, pointing out that explanations for the cause of heterosis would affect markedly the practices corn breeders use. The two contrasting causes considered were (1) the action of dominant–partially dominant favorable growth factors and (2) heterozygote advantage (overdominance). Lonnquist devised a series of single crosses using high- and low-combining inbreds in such a way that separate estimates could be obtained for testing both hypotheses.

What he found was a little of both—evidence that could support both theories of gene action—however, peripheral information from genotype × environment (season) interaction and sources of high-yielding lines convinced him that favorable growth factors were more important than overdominance in forming yield. Most interesting were his findings from the experimental crossing format used, namely, high × high combining lines, high × low, and low × low. The test cross group mean yields for all generations (crosses after each S_1 to S_5 inbreeding generations) were 72.6, 66.6, and 56.8 bu/acre, respectively. When the number of single-cross yields significantly exceeding the overall mean from the diallel group of crosses from four low- and five high-combining lines was examined, it was found that, compared to the overall mean (which I interpret to be the "grand mean" of 80.6 bu), the greatest number of highest-yielding lines (six) were found in the high × high group, followed by high × low (four), and low × low (zero).

Lonnquist's findings agree with most published results regardless of crop, and the plant breeding guidance provided may be summarized as follows:

1. Discard low × low crosses with confidence.
2. Crosses of high × high parents are most promising.
3. High × low must be treated with care for several reasons:
   a. They may represent crosses involving introgression of genes from exotic and important sources.
   b. There may be greater variance; for example, Lonnquist found a greater genotype × season interaction in high × low crosses, and
   c. They produce smaller but important numbers of high-yielding combinations, single crosses, or eventual homozygous individuals in the self-fertilizing crops.

Clayton, Morris, and Robertson (1957) crossed fruit fly lines from a high selection intensity group (for abdominal bristles) in F_{12}, making high × high, high × low, low × low, and crosses of all lines back to the base population. The F_1 results, scoring only the females, showed the cross rank order in descending number of bristles to be high × high, high × base, high × low, low × base, and low × low.

Langham (1961) described the high–low method, which he had used successfully to breed commercial sesame cultivars in Venezuela. The system recognized that the conventional restricted germ plasm base in effect imprisons desirable potent genes in a buffered condition. His method attempted to break out of this mold through "crossing the best parent with the worst parent for the character under study (HIGH × LOW)." Selection was at both ends of the curve, seeking individuals showing transgressive segregation for the traits under consideration. Further cycling crosses were made between selected individuals.

I have some sympathy with the philosophy and theory behind Langham's proposals, perhaps less with the details. Breeders do need to break out of isolation patterns, diversify, and open up their procedures. Whether the best crosses with the worst is the best approach to the initial attempt, I cannot say. The idea of selecting off the tails of the curve is intriguing, as is the idea of a second round of intermating in this group.

The influence of germ plasm, and perhaps the choice of breeding strategy, on yield potential is evident in some work of Sprague, Russell, Penny, Horner, and Hanson (1962) on epistasis in corn. In two years of tests involving the comparison of predicted and observed yields of single crosses (components) and their three-way crosses, 5 out of 20 cases showed statistically significant advantages of single-cross means over three-way cross means. A startling observation was that one inbred line, 0s420, was a parent in all five but *had the lowest mean yield in the diallel single-cross series*; the inbred line, B14, least involved in these five cases, had the *highest mean yield in the same diallel series*! Does this suggest something relative to high × low combinations?

Lonnquist and Lindsey (1964) made two diallel series of crosses of the three highest- and three lowest-yielding inbred lines chosen as parents in these two different ways: (1) on the basis of the inbred line test, and (2) on the basis of an unrelated topcross test. On the first basis there was a linear yield trend, namely, high × high, high × low, and low × low. When parents were chosen on the basis of topcrosses however, the order became high × low, high × high and low × low. The interpretation given was that selection of inbreds as parents in the first case represented selection for additive effects, whereas selection on the basis of topcross performance represented a sorting for heterotic effects.

Apropos of germ plasm and hybridizing, Frey (1964) characterized the four oat crosses he used in a study as coming from parents having high and poor adaptation; the results showed that "high × high" produced 50% of the final selected lines, "high × poor" (two crosses), 25% each, and "poor × poor," none.

Nettevich (1968) concluded from a three-year study of 48 spring wheat hybrids that crossing high-yielding × high-yielding parents gave far better results, expressed as heterosis relative to best parent and standard (apparently the standard commercial variety), than were obtained with high × low crosses (next best), and low × low crosses. The importance of diversity was evident when high × high crosses of different ecotypes (from regions of the Soviet Union) were found more productive than high × high crosses of the same ecotype.

A consideration in making hybrids involves the relative status of the parents in their expression of yield or other quantitatively controlled character. An excellent summary of high × high, high × low, and low × low experiences of several researchers was given by Busch, Janke, and Frohberg (1974). They conducted research in these three categories, assessed the predictive value of bulk and midparental yields vis-à-vis random line performance, and evaluated the combining ability of parents.

The results of crosses in the three categories above were interesting: "The greatest range in yield, the highest single line yield, and the greatest genetic variability were found in the high × low group of crosses."

However, one cross in the high × low group, CB115/ND58, had an unusually high range and introduced tremendous variability. When the data from this one cross were removed, the high × high group contained the highest single-line yield and highest genetic variance. The low × low lines were uniformly mediocre. The authors' findings may be summarized thus: (1) some high × low crosses produce lines that outperform high × high cross lines, and (2) these lines are more difficult to identify because they occur with lowered frequency; that is, high × high crosses produced 37.5%, and high × low 8.1%, of lines two or more standard deviation units above the mean.

Midparent values were successful in selecting only high × high crosses, which were shown, for whatever reason, to have the lowest percentage of high-yielding lines among selected crosses.

The authors found testing of F_4 and F_5 bulks was successful in predicting crosses having the highest proportion of high-yielding lines. Practically, a plant breeder would have to assess the efficiency and value of prediction at this late generation or look for modifications that view the bulk method as having wider application beyond serving as a prediction vehicle.

Bains (1976) used four high-yielding and two low-yielding wheat parents in six crosses as follows: two high × high, three high × low, and one low × low (Bains' research was concerned primarily with high and low sensitivity to linear and nonlinear regressions related to genotype × environment interactions and stability; see Chapter 27 on stability). The following information has been developed from Bains's data.

When the known mean grain yields in grams of the parents are inserted as inputs [of parents in a cross, e.g., high × high cross: parent *a* (17.29) × parent *b* (16.40)], the mean (16.84) can be compared with the progeny mean yields (F_2, F_3, and F_4). When this is done for the three combinations of crosses, the results are these:

	Mean Parent Yield Input	Mean Progeny Yield
High × high (two crosses)	16.46	15.05
High × low (three crosses)	13.27	11.48
Low × low (one cross)	10.83	10.12

From this it can be seen that the order of progeny yields was in agreement with the magnitude of parental input, the yields of parents had a predictive component, and no advantage accrued to wide crosses based on parental yield differences.

In a paper in which the combinations of high- and low-yielding oat parents was incidental to the main study, Rosielle and Frey's (1977) data nevertheless show that the greatest positive deviations of grain yield from midparental means was in the order high × low, low × low, and high × high. This ranking was surprising, and may be explained by possible close genetic relatedness of cultivars in the high × high group.

Takeda and Frey (1980) uncovered an example of an outstanding result that could not have been predicted because the oat parents used represented a cross of "low × low" tertiary seed set. This was the case of outstandingly high tertiary seed set found in 'C.I. 8044', out of the cross 'Clintland' × 'Garry 5'. It would be difficult to find two parents less suited to a desired result. Both parents were excellent genoytpes, Clintland a successful cultivar and Garry 5 a good midseason parent for crossing. However, neither was known for high tertiary seed set, and in fact Garry 5, a Cornell selection, was very sensitive to heat and short-day stress (we lost two southern winter increase crops of this genotype because of low fertility).

Wilcox and Schapaugh (1980) compared crosses of three superior group II soybean cultivars with themselves and with nine plant introductions (PIs) from three countries. The PIs were also of group II maturity, but "were distinctly lower yielding than the cultivars." In every case the yields of the cultivar × cultivar bulks were superior to those of the cultivar × PI bulks. These results lend support to the "best × best," and are unfavorable to the "high × low" crossing hypothesis. This is a case where the high × low crosses would have benefited from an additional cross before being introduced into the selection program, by making either a backcross to the superior group parent or a three-way outcross to another superior adapted genotype.

In studies by Bailey, Qualset, and Cox (1980), an interesting version of high × low hybridization occurred: the only wheat single cross (among six) showing heterobeltiosis in two years was 'Ramona 50' by 'Pitic 62'. In 1970 the only three-way cross (among 12) showing high-parent heterosis was the above F_1 × 'INIA 66'. However, every case of *negative* heterosis also had Ramona 50 as one parent!

Bitzer, Patterson, and Nyquist (1982) examined a diallel cross of eight different soft red winter wheats chosen on the basis of low and high yields. Grain yield and three components of yield were measured in hill plots. Considering grain yields only, average heterosis (midparent base) was 30%, 25%, and 19%, respectively, for low × low, low × high, and high × high crosses. Despite this ranking, the authors concluded that successful hybrid wheats would be most likely found in high × high crosses because in them would be found the greatest accumulation of general combining ability effects.

Assuming that the performance of hybrids is related to parental prepotency and cross excellence, implying future sources of superior selections, it is interesting to look at this study beyond its hybrid wheat significance. There is indeed a good correlation between yield of a parent and all the cross combinations in which it participated; that is, a parent may be characterized by its cross yields. For illustration the yields in grams of the eight parents are given here, followed by the mean yield of the seven crosses in which each participated:

Parent	Yield (g)	Mean Yield (g)
1	15.4	29.4
2	19.4	32.2
3	26.0	36.2
4	30.4	36.9
6	32.5	35.6
6	35.1	39.6
7	35.3	38.8
8	41.1	43.9

Further, if we arbitrarily take the 10 highest yields, this picture emerges:

Yield (g)	Cross Parents	Type cross
51.5	7 × 8	H × H
48.9	4 × 8	L × H
47.4	3 × 8	L × H
46.1	6 × 8	H × H
43.7	4 × 6	L × H
43.6	5 × 8	H × H
43.2	5 × 6	H × H
40.4	4 × 7	L × H
39.6	3 × 6	L × H
38.6	6 × 7	H × H

We see that there were five high × high, five low × high, and zero low × low crosses in the top 10. High × high mean yields were 44.6 g, low × high 44.0 g. Thus it would appear that low × low crosses can be eliminated, high × high have the best odds for success, and low × high combinations cannot be ignored.

Lamkey and Hallauer (1986) explored the problem as to what extent maize inbred lines selected for yield transmitted their traits into single crosses formulated according to high × high (HH), high × low (HL), and low × low (LL). The inbred lines were random selections from a large group of lines, classified either as high or low yielding. There were 24 high yielders and 24 low yielders, each representing an approximate 10% selection from the overall group of lines. These were used to produce 48 HH, 96 HL, and 48 LL single crosses. These were compared in nursery trials at six environments in Iowa.

The average group yields fit the pattern of almost all studies of this kind, that is, significant yield differences in the descending order of HH, HL, and LL. The authors found that the original grouping of lines, whether high or low, could be translated into a similar grouping of high or low combiners; however, beyond this no prediction was possible. To single out the lines having the highest yield potential it was necessary to cross to a series of testers. The import of this is that large numbers of inbred lines may be screened on yield performance per se, and a second screening by test crosses would be necessary only on the reduced-number best group of lines.

THE GENETIC DIVERSITY OF THE PARENTS

McKenzie and Lambert (1961) found significant intergeneration yield correlations, F_3 and F_6, in barley crosses, and concluded that success in selection efficiency was related to a high yield range among lines in the F_3 test. By extrapolation this suggested that parents, in order to generate such a range, should be genetically diverse.

Walton (1969) reported that the highest-yielding spring wheat F_1s came from genetically diverse parents (CIMMYT vs. U.S. and Canadian cultivars).

Steady improvement in genetic gains for yield and other characters in malting barley was the verdict of Wych and Rasmusson (1983) from an evaluation of six cultivars that had played important roles in malting barley production in Minnesota and the two Dakotas over a period of more than 60 years, the approximate period between 1920's 'Manchuria' and 1978's 'Morex'. Overall the rate of gain approximated 1%/year, but during the past 40 years the rate had doubled to 2%. This was considered satisfactory progress, particularly in view of the authors' belief that the restrictions of a relatively narrow genetic base, exacerbated by strict quality constraints on breeding, were influential over the lone period.

The dominating influence of malting and brewing quality parameters no doubt has had an effect on breeding procedures in malting barley, placing a premium on conservative parental choices and closely controlled pedigree crosses, teamed with significant laboratory cooperation ('Kindred', a farmer's field selection, remains an anomaly).

It may be, however, that the historical restraints on expanding the germ plasm base for malting barley breeding are less absolute than perceived. I had some indication of this in the 1960–1970s, when for a time the malting and brewing industry thought that the Northeast might serve as a supplementary malting barley production area. During this time my project developed a promising winter malting selection from our NY6005 Series, a cross of spring 'Traill' × winter 'Hudson'. I was surprised that this selection was obtained so readily. One may assume a general consensus among barley breeders that the genetic diversity in this simple cross of a malting-quality spring barley and a nonmalting-quality winter barley was such that the selection of a high-quality winter malting candivar from the bulk population would be precluded, or at least would be considered a prospect highly unlikely to occur.

When the regional interest in malting barley production declined, this selection later found use as the cultivar 'Wintermalt', in a different part of the United States, through a joint release by the Oklahoma and Cornell University Agricultural Experiment Stations. By this time my entire barley program had been converted to the MSFRS system (male sterile-driven bulk populations). Although the dream of northeastern malting barley production had disappeared, I felt the MSFRS methodology would be ideal for breeding malting barleys, especially when coupled with the excellent industry, regional, and national laboratories' support capabilities.

Johnson and Hayes (1940) concluded from their study of corn inbred lines and their single-cross yields that "diversity in genetic origin is an important factor in obtaining the maximum expression of hybrid vigor."

Harrington (1944) demonstrated differences in cultivar heterogeneity in a series of intravarietal hybrids of spring wheat. He found persistent heterosis in crosses made within the variety 'Reliance', which had never been reselected since its initial release about 35 years before, slight heterosis within 'Apex' crosses (Apex had been reselected at F_7), and no heterosis in 'Marquis' × Marquis crosses—Marquis had been twice reselected in advanced generations. The results conformed to expectations based on the decreasing internal diversity of the cultivars, respectively, but would seem to have little practical application.

An important moderately early paper on wide crosses ("radical wheat breeding") was provided by McFadden and Sears (1947). They discussed the technology and opportunities available for the transfer of valuable traits to the hexaploid wheats from rye, the agropyrons, and the diploid and tetraploid wheats. The article contains an extensive citation of previous literature.

Escuro, Sentz, and Myers (1963) examined the genetic gains recoverable from oat crosses differing in the degree of relatedness found in their parents. Two sets of crosses were formed: set I consisted of three crosses of three closely related oat lines; set II involved crosses of the three closely related lines with 'Garry', which was ancestrally unrelated to these lines. Six plant traits were evaluated. It was discovered that genotypic variances were greater in set II crosses for four of the six characters. Furthermore, set II crosses showed greater genetic advance for all six traits. The results stress the importance of genetic diversity in planning crosses; choosing unrelated or genetically diverse parents is a practical and positive response to this knowledge.

In a study that produced favorable heterosis patterns for yield in winter wheat (10 of 21 F_1 hybrids exhibited heterobeltiosis and the average yield of all hybrids was 24% greater than the better parent), Gyawali, Qualset, and Yamazaki (1968) were not able to detect any influence of the parental diversity that might have been expected among the soft red, hard red, and soft white wheat parents used. Low heterosis values were observed generally in soft red × soft white hybrids, however. Relatively, the level of diversity in the parents used would not be expected to be high—perhaps on a par or below that in spring × winter crosses. The range of heterosis found, the different but adapted parents, and the knowledge that very widely diverse parents can show significant heterosis suggest that specific combining ability may play a large role and must be sought for in crosses.

Browning, Simons, Frey, and Murphy (1969) produced an important article that invites reading, first, for the thoughtful chronological treatment of the host–pathogen problem that evolved in concert with our modern scientific and crop production practices, and second, for the innovative scheme to create interregional diversity through regional control and deployment of resgenes.

The authors convincingly showed that the great 2500-mile south-to-north, Mexico-to-Canada "Puccinia Path," which once harbored residual pathogens, now, with the near eradication of the barberry (stem rust) and the buckthorn (crown rust) alternate hosts, has become a single continental region whose chief distinguishing characteristic relative to the rusts is the airborne habit of the pathogens' spores. The authors proposed "the logical sequel to barberry and buckthorn eradication; namely, interregional diversification by deployment of different resgenes north and south to break the unity of the Puccinia Path."

They considered that a possible three-zone delineation would be necessary with different resgenes used in each zone. This would be especially important for the north and south zones. If, as they suggested, more resgenes were to be used per area, multiline cultivars would be ideal vehicles to put them in place. The deployment of oat crown rust resgenes proposed by Browning et al. requires a certain amount of cooperation among cereal breeders. Of course, a prerequisite is that a sufficient number of resgenes be available—they chose crown rust over stem rust for that reason. Only uncommitted resgenes could be deployed; therefore the zone genetic slates would have to be wiped clean, and agreement be reached among breeders on the strict use of the resgenes within the designated zones. The problem at present lies here: breeders are reluctant to trade their independence of action for what might be the common good. I have attended national meetings where the deployment proposal was introduced and discussed, and it seemed obvious that breeders' and state's rights—"turf" questions—were of deep concern. Regional deployment of resgenes may be a case of a good idea broached before its time. Unfortunately, the "proper time" would be a situation when the present procedures have proved inadequate.

Knott (1971) independently (he acknowledged the similarity to the 1969 paper of Browning et al., which he had not seen previously) proposed a North American rust control scheme based on assigning resgenes to three suggested specific regions for seasonal control of wheat stem rust, plus a recommended use of general resistance in the southern overwintering areas.

Assuming a positive association between genetic diversity and heterotic expression, Grant and McKenzie (1970) measured the yield heterosis in three crosses of spring with winter wheat under density conditions similar to farm practice. The results showed that under dryland conditions the hybrids outyielded the spring parents by 35%, 40%, and 2% (the last not significant). Under irrigation all three hybrids significantly outyielded the spring wheat parents by 36%, 25%, and 19%. Accurate comparisons with the winter parent were not possible. Significant changes in maturity, toward lateness, over the spring wheat parent were seen in the hybrids.

Advances in the maintenance of broad-based genetic germ plasm stocks were indicated and described in the release of three sorghum random-mated stocks by Nordcross, Webster, Gardner, and Ross (1973).

McKenzie and Grant (1974) examined the spring wheat yield potential of lines from spring × winter crosses, a strategy of bringing genetically diverse germ plasm together. Four crosses produced F_5 and F_7 lines for the test. Additionally the two spring wheat parent lines used in the crosses were used as controls, and were themselves hybridized. Lines from the spring × winter wheat crosses did indeed show high yield potential: their mean yields exceeded those of the controls by 112% (87%), 114% (46%), 120% (95%), and 121% (64%) (figures in parentheses are the percent of strains whose yields exceeded the highest control).

The Iowa program of incorporating *Avena sterilis* germ plasm into *A. sativa* breeding stocks via introgression received encouragement from the results of a study by Lawrence and Frey (1975). Naturally it was desired to move genes for perhaps complexly inherited quantitative characters from *A. sterilis*. It was not clear that backcrossing could accomplish this. Lawrence and Frey took two productive oat genotypes representing recurrent *A. sativa* germ plasm and crossed each with four *A. sterilis* lines, following which the eight backcross programs were each conducted through BC_5. Extensive testing of populations and lines from the backcross generations showed substantial favorable introgression into the recurrent genomes. Especially, there was increased genetic variance for grain yield. The study highlighted the value of hitherto unexploited exotic germ plasm.

Fedak and Fejer (1975) found that more than half of the spring × winter barley F_1s outyielded their spring parent. This was attributed to genetic diversity stemming from the spring and winter parent differences, and the interspecific origins of the winter parent lines. No inferences were drawn as to the nature of the eventual homozygous lines from these crosses.

An interesting interspecific transfer of germ plasm was accomplished in tobacco by Wernsman, Matzinger, and Mann (1976). Progenitor species ($2n$) germ plasm had been transferred to tetraploid ($4n$) cultivated tobacco on a number of occasions, but always had involved only a single or a small number of genes. The triploid hybrid is sterile. The goal in this study was to accomplish a more integrated transfer that would include and represent the influence of genes for quantitative traits. Cultivated *N. tabacum* was treated with colchicine and the chromosome number doubled ($8n$); when crossed with the progenitor species ($2n$) *N. sylvestris*, a partially fertile pentaploid was produced. This was immediately selfed and backcrossed to the domestic tetraploid. The seed produced became Syn 0 and through a series of selfs, testcrosses, selections, and intercrosses, a number of lines were produced for comparison tests. Results showed that lines were produced with attributes superior to the domestic cultivar used as the parent; for example, the two synthetics formed yielded 7.7% and 12.9% more cured leaf than the cultivar. Breeders interested in incorporating progenitor or exotic germ plasm may see in these testcross and zigzag procedures described by Wernsman et al. suggestions for other ways of going about it.

The extensive work at the Iowa station to evaluate and introgress *Avena sterilis* germ plasm into *A. sativa* germ plasm has been notably successful, as evidenced by concrete advances in protein, yield, and disease resistance. Particularly welcome is new information on physiologically based characters such as growth rate, the subject of Takeda and Frey's (1977) paper. In an earlier paper (1976) they had shown that growth rate and harvest index accounted for 90% of the variation in grain yields in backcross progenies, and growth rate made the larger contribution to this. The ramifications of this association to our understanding and approach to selection for grain yield are great. A simple way of looking at this is that an oat breeder now knows of three major traits, growth rate, harvest index, and grain yield, that are different but interrelated, and may be used separately or in complementary ways in selection, even in a selection index.

In the Takeda and Frey 1977 paper the focus was on *A. sterilis* as the source of the growth rate genes. Takeda and Frey found heritability values of about 0.4 for growth rate, a substantial figure, buttressed by the finding of many segregates showing transgressive (positive) growth rates. One reason for also looking at harvest index was the high association of growth rates with straw yields. However, a positive point was that association was low with plant height; therefore a visual curb could be placed on excessive plant height during selection with little penalty to growth rate, lodging resistance, or grain yield. In fact, Takeda and Frey stated that high growth rate could be combined with an optimual harvest index to produce high grain yielding cultivars.

Holcomb, Tolbert, and Jain (1977) published a diversity analysis (index) for the *Oryza sativa indica* and *japonica* groups in 1977, examining data from about 500 and 1400 accessions to collections as catalogued by the International Rice Research Institute. Insofar as possible an attempt was made to include entries representing original collections in the several countries of origin, and any that traced to known irradiation or to plant breeding and selection procedures were excluded. It was impossible, of course, to discriminate rigidly because the amount of gene exchange over time could not be known with certainty.

Diversity was indicated by summarizing data and creating a diversity index, H', based on the metric data for quantitative traits, and on counts for 27 qualitative traits. References were made to previous analyses of other crops, for example, safflower, durum wheat, and barley. The study showed *indica* collections to be more variable than *japonica*, and in a general way gives some sense of the distribution of traits by countries.

An interesting regional germ plasm diversity maintenance scheme for corn was proposed by Lonnquist, Compton, Geadelmann, Loeffel, Shank, and Troyer (1979). The area improvement scheme involved the cooperation of several regional maize breeders. Each contributed seeds from visually selected desirable plants that were intermated to form the base population. These intermated hybrid seeds were composited, and allowed to intermate for an additional generation during the winter in Hawaii. After establishing suitable reserves, aliquot samples were distributed to the cooperators, who then practiced mass selection for several desirable characters at a selection intensity of about 5%. The seed composite from selected plants from each cooperator was returned to a central location for blending with those from the other locations. This area improvement scheme was conducted through four cycles beyond the intermated base point, following which performance was assessed at each of the participating locations. One point should be noted: to prevent a buildup of specific adaptation, the re-composited seed returned each year to each cooperator did not include any seed from the previous harvest which had been furnished by that cooperator to the central location. The system was described briefly as a form of convergent–divergent selection, and was viewed as a stabilizing selection procedure.

The performance results from two years showed that two hybrid controls significantly outyielded the area improvement cyclic bulks, and also had lower harvest grain moisture. However, within the cycled populations there was a steady improvement in yield with cycles, and a concomitant decrease in grain moisture. There was an increase in stability over cycles.

This scheme is applicable to crops other than corn, and for the same reasons—to maintain broad germ plasm diversification that might otherwise be diminished or lost when stocks are restricted to one location. A somewhat similar proposal was made for a self-pollinating crop, winter oats, in the 1960s. The winter survival of winter oats was so marginal that there was yearly danger of complete loss of source populations. I do not recall that the proposal was formally implemented, and of course no intermating was envisaged, but the scheme did include heterozygous populations.

The focus of McProud's (1979) paper on repetitive cycling and recurrent selection procedures was the narrow band of germ plasm used in three major world barley breeding programs used to illustrate his arguments.

Moderation in the use of exotic germ plasm is indicated by the results obtained in soybeans by Schoener and Fehr (1979). They used crosses between five plant introductions (PIs) and four cultivars, and compared these with crosses between cultivars only. The study incorporated several unusual features: the PIs were preselected for high yield, and the PI proportion in each cross was arranged to be 100% (AP_1), 75% (AP_2), 50% (AP_3), 25% (AP_4), and 0% (AP_5). The 0% category represented crosses involving only cultivars. Four and five intermatings (within the crosses representing each group) were carried out before selfing began and the selection of F_4 lines was completed.

The yields showed a linear increase with decrease in the input proportion of PIs, except that 75% PI was lower yielding than 100% PI. Yields were obtained from 480 lines subdivided into 12 blocks of 40 lines each. When the top 10% from each block of four lines was selected—a total of 48 lines—it was found that 23 came from the 0% PI cross group AP_5, followed by 13 from AP_4, 8 from AP_3, 1 from AP_2, and 3 from AP_1. No conclusions were drawn as to the influence of intermating, and one assumes that the intermating was an effort to increase the genetic variance of all populations for comparison purposes.

Schoener and Fehr distinguished two uses of exotic germ plasm, namely, for either short-term or long-term selection. For the short term, the use of exotic germ plasm might be difficult to justify, which may be another way of saying that sufficient variability for parents at present exists in soybean cultivars and advanced lines. For the long term, they recognized the need for outside germ plasm input and suggested a reasonable balance might be at the 50% level, where the greatest genetic variability was found.

The authors gave two possible reasons for the lower yield of AP_2. A third seems possible: AP_2 and AP_4 achieved 75% and 25% PI germ plasm, respectively, via the backcross. In AP_2, the four PIs each were crossed with a different cultivar, then each hybrid was backcrossed to its parent PI. The same 75% input also could have been achieved through having the first crosses each followed by an outcross to a different PI. It is not clear to me whether this could account for the lower yield of AP_2; the same argument could be made for a pull in the opposite direction in AP_4, which received a strict backcross to cultivars.

Only serendipity can account for a valuable germ plasm discovery by Carrigan and Frey (1980). Studies on different patterns of root development in two species of oats identified an *Avena sterilis* line whose root volume was more than three times larger than the next best line. Significantly higher dry weights of panicles, straw, and roots were also recorded.

General guidelines for introgression of exotic germ plasm into adapted populations were derived from computer simulation models by Dudley (1982). The stated problem is one that continually confronts all breeders of all crops—a recognition that one must go "outside" for germ plasm, accompanied by a realization that introgression of useful genes will be difficult and slow. The guidelines in brief were:

1. When one parent has more favorable alleles than the other (the typical case), backcross at least once to the favorable (adapted) parent.
2. Early selection (or a discard procedure) will be necessary to concentrate useful genotypes and raise population means to wanted levels. This also implies the need for recurrent selection.

3. The more diverse the parents, the more necessary are cycles of backcrossing.

The significance of Dudley's remarks is that wide crosses are the beginning, not the end of introgression efforts, and they must be monitored, worked with, and fostered.

The paper by Vello, Fehr, and Bahrenfus (1984) is an expansion of the soybean germ plasm probe conducted in 1979 by Schoener and Fehr. In the present research the parent pool was expanded to 40 PIs and 40 domestic cultivars; both sets were preselected with heavy emphasis on high relative yield within each group.

Five populations, AP_{10}–AP_{14}, were created with descending PI germ plasm percentages in the order of 100, 75, 50, 25, and 0, respectively. The objective of the study was to determine whether the earlier results found, from only four PI and four cultivars, could be considered to be of a general nature. Therefore, in addition to the increase in numbers of parents, a complicated series of two intermatings involving single crosses, double crosses, and backcrosses effectively mixed the germ plasms to create the designated percentages of PI and domestic germ plasm. In 1980 and 1981, yields, genotypic variance estimates, height, and lodging data were obtained on S_4 lines (S_2-derived).

In 1980 when random lines from the five populations were examined, very few differences were found among AP_{10} (100% PI) to AP_{13} (25% PI), either in yield or genotypic variance. AP_{14}, however, had the highest yield with only about one-half the genotypic variance. It also had the highest percentage of selected superior lines, with the percentage decreasing in each population as PI percentage increased.

When the best selected lines were tested in 1981, the mean yield of lines, by population, increased linearly with decreasing percentage of PI, AP_{14} having the highest yield. Height increased in the same way, AP_{14} showing the greatest mean height, but lodging score decreased, being lowest (best) in AP_{14}. Throughout, the number of best selected lines was positively associated with percentage of *domestic* germ plasm.

The conclusions drawn support a general effect of PI germ plasm use. For short-term breeding use, the introduction of PI germ plasm may be expected to produce populations with lower mean yields, higher genetic variance, and a lower frequency of high-yielding individuals. The ultimate value of exotic germ plasm can only be speculated upon at the present.

YIELD CONSIDERATIONS AFFECTING CROSS QUALITY LEVEL

When Shebeski (1967) rigidly controlled the environmental variance and locked it in by frequent wheat check varieties (each hybrid plot was adjacent to a control plot), the F_3 and subsequent derived-F_5 line yields were highly correlated. With data of such good agreement from generation to generation, the high predictive value for yield of the labor-intensive, control-heavy F_3 is readily seen: the order of ranking correlation for the two generations was 0.847, in practical terms near perfection. Thus high F_3 lines remained high in F_5, and low remained low. With a calculated cutoff point, the breeder would have a reliable method to make selections and discards on the basis of the F_3 yields. There exists, therefore, a sound basis for evaluating the balance between the extra F_3 effort, and the lesser F_4 and subsequent generation costs that would ensue from the early discard based on predictions.

In a thought-provoking article, O'Brien, Baker, and Evans (1978) examined intergeneration correlations and selection response in four wheat crosses having the common female parent 'Glenlea'. Seventy-eight F_2 plants were selected from each cross. These, with Glenlea, were grown in F_3 replicated yield trials, and 10 high-yielding and 10 low-yielding lines were selected from each cross. From these selections there were developed F_4 and F_5 bulks and a family of F_5 lines. These formed the basis for replicated plot intergeneration comparisons.

Although all crosses had the common female parent, Glenlea, there were significant differences in yield, among crosses III, IV, II, and I, in the order of high to low cross mean yield. A sobering finding was that crosses I and II had a wider range in yield (difference between high and low groups selected) than III and IV, but produced fewer F_3 lines better than Glenlea. In this case the higher cross means of III and IV were indicative of a cross's potential for yielding superior lines.

It would be expected that bulks from high-yielding lines would prove superior to bulks from low-yielding lines, and this was found for three of four crosses in each generation of bulks (F_4 and F_5) and in the F_5 family means. The presence of crosses I and II in all these groups was consistent with their greater variability and range in yield, leading to greater selection efficiency. These two crosses also showed greater heritability. Predicted response to selection was greater for I and II than for III and IV. Predicted response was also greater than observed response, the latter aberration being attributed to assumptions having been based on heritability estimates drawn from an inadequate sampling of environments.

The study highlighted opposing movements affecting selection: populations having greater phenotypic and genetic variance may show greater response to selection, but the highest-yielding lines may come from crosses with smaller variance but higher mean yields. The challenge is to combine these; otherwise cross mean yield would seem a superior indicator of potential.

With one exception the yield ranges of generation groups agreed with the indications of their F_3 group, implying positive predictive qualities for the F_3. It was found that the highest-yielding F_5 lines came from the cross with the highest F_3 population average yield; the authors considered this a promising selection indicator.

O'Brien, Baker, and Evan's concluding statement, that early-generation selection for yield will be most effective in those populations showing the greatest amount of genetic variation, must be taken with caution inasmuch as they did not couple this with the earlier corollary that the highest-yielding lines from selection in their study came not from populations with greater variance but from crosses with high initial cross mean yields (and lower variances).

Coffman and Stevens (1937) commented on the low expression of heterosis in a study of the prepotency of oat parents, and attributed this to the known close relationship of the cultivars used as parents.

In a study of heterosis in crosses of diverse winter wheat cultivars by Brown, Weibel, and Seif (1966), substantial heterosis was shown, averaging 113% of the high parent and 126% of the midparent. One variety, 'Racine', contributed a disproportionate share of the positive general combining ability effects.

REFERENCES

Bailey, T. B., Jr., C. O. Qualset, and D. F. Cox. 1980. Predicting heterosis in wheat. Crop Sci. 20:339–342.

Bains, K. S. 1976. Parent dependent genotype × environment interaction in crosses of spring wheat. Heredity 36:163–171.

Bassiri, A. 1976. Barley cultivar identification by use of isozyme electrophoretic patterns. Can. J. Plant Sci. 56:1–6.

Becker, H. C., H. H. Geiger, and K. Morgenstern. 1982. Performance and phenotypic stability of different hybrid types in winter rye. Crop Sci. 22:340–344.

Bennett, E. 1965. Plant introduction and genetic conservation: genecological aspects of an urgent world problem. Scottish Plant Breeding Stn. Rec.:27–113.

Bitzer, M. J., F. L. Patterson, and W. E. Nyquist. 1982. Hybrid vigor and combining ability in a high-low yielding, eight-parent diallel cross of soft red winter wheat. Crop Sci. 22:1126–1129.

Brown, C. M., R. O. Weibel, and R. D. Seif. 1966. Heterosis and combining ability in common winter wheat. Crop Sci. 6:382–383.

Browning, J. A., M. D. Simons, K. J. Frey, and H. C. Murphy. 1969. Regional deployment for conservation of oat crown-rust resistance genes. *In* J. A. Browning (Ed.), symposium on "Disease consequences of intensive and extensive culture of field crops." Iowa State Univ. Exp. Stn. Special Rep. 64:49–56.

Busch, R. H., J. C. Janke, and R. C. Frohberg. 1974. Evaluation of crosses among high and low yielding parents of spring wheat (*Triticum aestivum* L.) and bulk prediction of line performance. Crop Sci. 14:47–50.

Bushuk, W. and R. R. Zillman. 1978. Wheat cultivar identification by gliadin electrophoregrams. I. Apparatus, method and nomenclature. Can. J. Plant Sci. 58:505–515.

Carrigan, L. and K. J. Frey. 1980. Root volumes of *Avena* species. Crop Sci. 20:407–408.

Clayton, G. A., J. A. Morris, and A. Robertson. 1957. An experimental check on quantitative genetical theory. I. Short-term response to selection. J. Genet. 55:131–151.

Coffman, F. A. and H. Stevens. 1937. Influence of certain oat varieties on their F_1 progeny. J. Am. Soc. Agron. 29:314–323.

Cowan, J. R. 1943. The value of double cross hybrids involving inbreds of similar and diverse genetic origin. Sci. Agric. 23:287–296.

Cox, T. S., Y. T. Kiang, M. B. Gorman, and D. M. Rodgers. 1985. Relationship between coefficient of parentage and genetic similarity indices in the soybean. Crop Sci. 25:529–532.

Dudley, J. W. 1982. Theory for transfer of alleles. Crop Sci. 22:631–637.

Eckhardt, R. C. and A. A. Bryan. 1940. Effect of method of combining the four inbred lines of a double cross of maize upon the yield and variability of the resulting hybrid. J. Am. Soc. Agron. 32:347–353.

Escuro, P. B., J. C. Sentz, and W. M. Myers. 1963. Effectiveness of selection in F_2 of crosses among related and unrelated lines of oats. Crop Sci. 3:319–323.

Fedak, G. and S. O. Fejer. 1975. Yield advantage in F_1 hybrids between spring and winter barley. Can. J. Plant Sci. 55:547–553.

Fowler, D. B. and L. V. Gusta. 1979. Selection for winterhardiness in wheat. I. Identification of genotypic variability. Crop Sci. 19:769–772.

Frey, K. J. 1964. Adaptation reaction of oat strains under stress and non-stress environmental conditions. Crop Sci. 4:55–58.

Grant, M. N. and H. McKenzie. 1970. Heterosis in F_1 hybrids between spring and winter wheats. Can. J. Plant Sci. 50:137–140.

Green, J. M. 1948. Inheritance of combining ability in maize hybrids. J. Am. Soc. Agron. 40:58–63.

Guok, H. P., J. C. Wynne, and H. T. Stalker. 1986. Recurrent selection within a population from an interspecific peanut cross. Crop Sci. 26:249–253.

Gyawali, K. K., C. O. Qualset, and W. T. Yamazaki. 1968. Estimates of heterosis and combining ability in winter wheat. Crop Sci. 8:322–324.

Harrington, J. B. 1944. Intra-varietal crosses in wheat. J. Am. Soc. Agron. 36:990–991.

Hawkes, J. G. 1977. The importance of wild germplasm in plant breeding. Euphytica 26:615–621.

Holcomb, J., D. M. Tolbert, and S. K. Jain. 1977. A diversity analysis of genetic resources in rice. Euphytica 26:441–450.

Jensen, N. F. 1962. A world germplasm bank for cereals. Crop Sci. 2:361–363.

Johnson, I. J. and H. K. Hayes. 1940. The value in hybrid combinations of inbred lines of corn selected from single crosses by the pedigree method of breeding. J. Am. Soc. Agron. 32:479–485.

Khalif, A. G. M., G. D. Brossman, and J. R. Wilcox. 1984. Use of diverse populations in soybean breeding. Crop Sci. 24:358–360.

Kleese, R. A. and K. J. Frey. 1964. Serological predictions of genetic relationship among oat varieties and corn inbreds. Crop Sci. 4:379–383.

Knott, D. R. 1971. Can losses from wheat stem rust be eliminated in North America? Crop Sci. 11:97–99.

Kuhr, S. L., V. A. Johnson, C. J. Peterson, and P. J. Mattern. 1985. Trends in winter wheat performance as measured in international trials. Crop Sci. 25:1045–1049.

Lamkey, K. R. and A. R. Hallauer. 1986. Performance of high × high, high × low, and low × low crosses of lines from the BSSS maize synthetic. Crop Sci. 26:1114–1118.

Langer, I., K. J. Frey, and T. B. Bailey. 1978. Production response and stability characteristics of oat cultivars developed in different eras. Crop Sci. 18:938–942.

Langham, D. G. 1961. The high-low method of crop improvement. Crop Sci. 1:376–378.

Lawrence, P. K. and K. J. Frey. 1975. Backcross variability for grain yield in oat species crosses (*Avena sativa* L. × *A. sterilis* L.). Euphytica 24:77–86.

Lonnquist, J. H. 1953. Heterosis and yield of grain in maize. Agron. J. 45:539–542.

Lonnquist, J. H. and M. F. Lindsey. 1964. Topcross versus S1 line performance in corn (Zea mays L.). Crop Sci. 4:580–584.

Lonnquist, J. H., W. A. Compton, J. L. Geadelmann, F. A. Loeffel, B. Shank, and A. F. Troyer. 1979. Convergent–divergent selection for area improvement in maize. Crop Sci. 19:602–604.

McFadden, E. S. and E. R. Sears. 1947. The genome approach in radical wheat breeding. J. Am. Soc. Agron. 39:1011–1026.

McKenzie, H. and M. N. Grant. 1974. Evaluation for yield potential of spring wheat strains from four spring × winter crosses. Can. J. Plant Sci. 54:45–46.

McKenzie, R. I. H. and J. W. Lambert. 1961. A comparison of F_3 lines and their related F6 lines in two barley crosses. Crop Sci. 1:246–249.

McProud, W. L. 1979. Repetitive cycling and simple recurrent selection in traditional barley breeding programs. Euphytica 28:473–480.

Mann, J. A., E. E. Gbur, and F. R. Miller. 1985. A screening index for adaptation in sorghum cultivars. Crop Sci. 25:593–598.

Meyers, K. B., S. R. Simmons, and D. D. Stuthman. 1985. Agronomic comparison of dwarf and conventional height oat genotypes. Crop Sci. 25:964–966.

Murphy, J. P. and K. J. Frey. 1984. Comparisons of oat populations developed by intraspecific and interspecific hybridization. Crop Sci. 24:531–536.

Murphy, R. P. 1942. Convergent improvement with four inbred lines of corn. J. Am. Soc. Agron. 34:138–150.

Nettevich, E. D. 1968. The problem of using heterosis of wheat (*Triticum aestivum*). Euphytica 17:54–62.

Nordcross, P. T., O. J. Webster, C. O. Gardner, and W. M. Ross. 1973. Registration of three sorghum germplasm random-mating populations. Crop Sci. 13:132.

O'Brien, L., R. J. Baker, and L. E. Evans. 1978. Response to selection for yield in F_3 of four wheat crosses. Crop Sci. 18:1029–1033.

Qualset, C. O. 1975. Sampling germplasm in a center of diversity: an example of disease resistance in Ethiopian barley. Crop Genetic Resources for Today and Tomorrow, Inter. Biol. Programme, Cambridge University Press. Vol. 2:81–96.

Rines, H. W., D. D. Stuthman, L. W. Briggle, V. L. Youngs, H. Jedlinski, D. H. Smith, J. A. Webster, and P. G. Rothman. 1980. Collection and evaluation of *Avena fatua* for use in oat improvement. Crop Sci. 20:63–68.

Rosielle, A. A. and K. J. Frey. 1977. Inheritance of harvest index and related traits in oats. Crop Sci. 17:23–28.

Schoener, C. S. and W. R. Fehr. 1979. Utilization of plant introductions in soybean breeding populations. Crop Sci. 19:185–188.

Shebeski, L. H. 1967. Wheat and wheat breeding, *In* K. F. Nielson, (Ed.), Proc. Can. Centennial Wheat Symp. Modern Press, Saskatoon, pp. 249–272.

Smith, C. S., L. W. Kannenberg, and R. B. Hunter. 1978. Development of maize gene pools at high plant densities. I. Effect on population structure. Can. J. Plant Sci. 58:95-99.

Smith, C. S., L. W. Kannenberg, and R. B. Hunter. 1978. Development of maize gene pools at high plant densities. II. Effect on grain yield. Can. J. Plant Sci. 58:101–105.

Sprague, G. F., W. A. Russell, L. H. Penny, T. W. Horner, and W. D. Hanson. 1962. Effect of epistasis on grain yield in maize. Crop Sci. 2:205–208.

Suneson, C. A. 1969. Registration of barley composite crosses. Crop Sci. 9:395–396.

Suneson, C. A., W. K. Pope, N. F. Jensen, J. M. Poehlman, and G. S. Smith. 1963. Wheat Composite Cross I. Created for breeders everywhere. Crop Sci. 3:101–102.

Takeda, K. and K. J. Frey. 1977. Growth rate inheritance and associations with other traits in backcross populations of *Avena sativa* × *A. sterilis*. Euphytica 26:309–317.

Takeda, K. and K. J. Frey. 1980. Tertiary seed set in oat cultivars. Crop Sci. 20:771–774.

Thomas, G. L. and F. R. Miller. 1979. Base temperature for germination for temperate and tropically adapted sorghums. *In* Proc. 11th Biennial Grain Sorghum Res. Util. Conf., Lubbock, Texas, p. 24.

Thorne, J. C. and W. R. Fehr. 1970. Exotic germplasm for yield improvement in 2-way and 3-way soybean crosses. Crop Sci. 10:677–678.

Vello, N. A., W. R. Fehr, and J. B. Bahrenfus. 1984. Genetic variability and agronomic performance of soybean populations developed from plant introductions. Crop Sci. 24:511–514.

Walton, P. D. 1969. Inheritance of morphological characters associated with yield in spring wheat. Can. J. Plant Sci. 49:587–596.

Wernsman, E. A., D. F. Matzinger, and T. J. Mann. 1976. Use of progenitor species germ-plasm for the improvement of a cultivated allotetraploid. Crop Sci. 16:800–803.

Wilcox, J. R. and W. T. Schapaugh, Jr. 1980. Effectiveness of single plant selection during successive generations of inbreeding in soybeans. Crop Sci. 20:809–811.

Wych, R. D. and D. C. Rasmusson. 1983. Genetic improvement in malting barley cultivars since 1920. Crop Sci. 23:1037–1040.

29

PREDICTING AND CHOOSING PARENTS

Conventional plant breeding programs are founded on hybridization. Annual or more frequent crossing programs are usually modest enterprises, tedious perhaps but not large in comparison with the huge populations that they spawn. It is exactly because of this linkage of number of hybrids to size of program, and its consequences in size and quality, that the crossing task assumes an excessive importance. The quality of the first factor determines the quality of the second. In fact, Qualset (1979) said that the most often asked question among plant breeders is how to choose parents.

In this chapter we review the experiences of plant breeders who have given thought and time to this vexing problem. Of course, choosing parents for crosses and choosing among crosses, the subject of the next chapter, are intimately related, and it is inevitable in what follows that there is occasional straying across the mutual chapter boundaries.

EARLY LANDMARK CASES

Harlan, Martini, and Steven's (1940) paper on methods in barley breeding is a gold mine of information on evaluating parents and crosses, as relevant today as when the research was done a half-century ago. The authors had specific limited objectives at the beginning of the research, namely, to make a comparison of pedigree and composite bulk methods in barley breeding. They correctly called this area "experimental plant breeding," but the study also yielded information across a wide front of breeding interest.

The research study was a massive one. The base work involved hand-making all of the possible 378 crosses among 28 selected parents (by coincidence, genetic male sterility in barley was reported about the time this study was completed). Each cross was kept separate and grown for seven generations without selection in yield nurseries. The composite was formed by mixing an equal amount of seed from each of the F_2 cross rows, and also was grown for seven generations.

Selections were first made in the eighth generation from space-planted pedigree crosses and from the composite, the two together occupying a single large field area. All plants were pulled and examined and the best ones saved, apparently by visual determination. The numbers saved per cross were based on a prior grouping of the crosses according to their yield in the years before F_8. From each cross 15, 10, 8, 6, and 5 selections were taken from the highest-yielding group (#1) to the lowest-yielding group (#5), respectively. This was on the assumption that high-yielding combinations should produce more and better selections than low-yielding combinations. The grand total of selections from all the pedigree crosses was 2921. An equal number of selections was made from the one-half acre of composite material. All selections were sown in rows with the 'Trebi' cultivar as a check every 20 rows. At maturity all rows were harvested and selections made on the basis of both yield and quality, not on yield alone.

The information developed about parents can be viewed from different vantage points. The relative yields of the parent varieties were not measured directly (they had been selected on the basis of use, diversity, geographic origin, etc.); instead, the average yield of all selections made in the 27 crosses in which each parent participated was given. There is a complication here in that the yield of the pedigree crosses before selection was not given but was indicated by the number of selections made (in yield groups which stratified crosses from high to low). Therefore, there are several intertwined pieces of data relating to each of the 28 parents:

1. The number of selections made per parent in 27 crosses,
2. The mean yield of this number,
3. The number of selections saved from those made,
4. The number of crosses with *no* superior selections,
5. The number of superior selections,
6. A category of number of outstanding selections.

The data for the top and bottom cultivars among the 28, ranked according to the mean yield of all their selections, were as follows:

	(2)	(1)	(3)	% (3/1)	(4)	(5)	(6)
Atlas	511.4	260	130	50.0	5	78	34
Glabron	389.6	185	52	28.1	20	10	3

where the numbers in parentheses refer to categories in the preceding list. For the varieties in between there was good agreement among columns as one proceeded down the rankings.

What we have here then is a definitive sorting of worth values of both parents and crosses. Looking first at potential parent values, one might proceed with confidence down the list of varieties each ranked by the mean yield of all selections from the crosses in which it participated, and set one's own cutoff point. Harlan et al. tout the best parents as among the top 10–12. However, exceptions suggest a note of caution. For example, 'Han River' made the top 11 in last place, but its favorable percent-saved rating was outstanding. A "natural" cutoff point might be found in the "outstanding selections" category, whose numbers took a significant drop after the top 11 ranked varieties (but eliminated Han River). The authors highlighted Trebi as being unique; it was different in origin, surprising in its areas of adaptation, a good performer, and a parent that readily transmitted its favorable qualities to its progeny.

When we turn to the predictive value among crosses, the study is less revealing. It seems clear that one might concentrate on crosses containing the best parents, for example, 'Atlas'; however, the yield data available were from its (Atlas's) matings with all the other parents, including 'Glabron', the lowest-ranking cultivar. This confounding masks some information that would be valuable to have, for example, the performance of a variety in crosses with other varieties stratified according to high × high, high × low, and low × low. If we glance at Glabron's performance we are reassured because in 27 combinations it produced only three outstanding selections; therefore the loss in eliminating Glabron as a future parent would be relatively minor. Nevertheless, one would like to know what crosses produced selections of candivar status and what ones did not.

Corn breeders have wrestled mightily with the problem of yield prepotency. Successful hybrids trace back to inbred lines, which have been identified by their ability to generate high average progeny from top crosses. This general indication of combining ability is also considered to be correlated with prepotency for yield in combinations of inbreds.

Jenkins (1940) was intrigued with the question of how early in the inbreeding process did inbreds show prepotency, and found that it occurred very early. This raises interesting methodology questions similar to the question, "how many backcrosses should one make?" The situation suggested early-as-possible testing of inbreds in top crosses, but left unanswered the question of what happened with further inbreeding of the prepotent lines. From Jenkins' standpoint the earliest possible identification of prepotent lines would result in greatest breeding efficiency, because effort could immediately be concentrated on only the best lines. The key to this problem hinged on the number and importance of the dominant alleles present. Prepotency yield values might provide a measuring scale.

What Jenkins found in his research was that inbred lines did indeed become stable for yield prepotency very early. Highest potential, of course, was indicated in the first inbreeding generation, but even so the amount and the segregation within these lines for yield prepotency were so limited as to be impracticable for a breeding program. However, when he compared mean squares and variances of yield found in comparable open-pollinated plants (of the 'Krug' variety) with those found among sibling first-generation inbreds, the advantage was strikingly in favor of yield variability in the open-pollinated group. Jenkins concluded:

> These data further emphasize the greater chances of obtaining lines of outstanding performance in hybrids through selection *among* large numbers of inbred lines rather than *within* lines. They also add to the accumulating evidence indicating that the yield prepotency of lines in hybrids, as measured by their top crosses, may be determined very early in the inbreeding period.

If we apply this to self-fertilized crop breeding, it highlights the F_2 and the F_2 family line as important points for selection pressure. Perhaps yield prepotency can be identified from heterosis of the F_1, but the meaningful segregation is contained in the F_2 and its family progeny.

Jenkins' experiences led him to suggest the creation of synthetic varieties, to be used, so to speak, on the fringes of hybrid corn production areas, taking the place of less productive open-pollinated varieties. These would be formed by a single generation of inbreeding, choosing the best lines from topcross tests and intercrossing these lines to form the synthetic.

Thirty generations after Harlan formed Composite Cross II in barley from a bulk of 378 hybrids from 28 parents, Bal, Suneson, and Ramage (1959) examined changes in 21 characters in three F_{30} versions of the composite. The three populations were no selection, selection for large seeds, and selection for small seeds. The authors evaluated parental contributions in different ways and ranked parents according to what we might call potential worth. They found satisfactory agreement among the three rating criteria, namely, (1) yield rank of parents alone, (2) rank of number of superior recombinations, and (3) rank in estimated contributions to the composite at F_{30}.

A barley breeder today could do worse than choosing as parents, for example, the nine top-ranked cultivars in their first two columns, where there was 100% agreement. A closer analysis of the source of superior recombinants, not possible from this paper, would also be worthwhile in identifying unusual contributions of cultivars not highly ranked.

Thus we have several examples of early searches for indicators that would identify genotypic values useful for selecting parents.

THE USE OF CROSS-POLLINATING TECHNIQUES

Suneson and Riddle (1944) used a crossing technique somewhat akin to topcrossing in a comparison of barley F_1s with adapted parents. All crosses were made on male sterile females. Heterosis generally was high, with yields varying in reasonable accord with the yield relationships of their parents. There is promise here that yield prepotency of varieties might be estimated through tests of their F_1 hybrids.

F. D. Richey (1946) had a most interesting short article on multiple convergence crossing in corn. He varied the parental input at the various convergence stages in a series such that he had derived lines for test from these categories:

1. Derived by backcrossing to the recurrent parent
2. Derived from crosses between derived lines from the same recurrent parent and the same nonrecurrent parent
3. Derived from crosses between derived lines from the same recurrent parent and different nonrecurrent parents.

Taking the recurrent parent lines as the 100% yield level, the yields of these groups were 130%, 182%, and 191%, respectively, a dramatic increase coincident with the added diversity.

Some breeders of self-pollinated crops use similar zigzag methods. For instance, Mac Key (1963) gave illustrations of multiple convergent crossing using the occasional backcross. Jensen's (1970) diallel selective mating system (DSM) showed opportunities and illustrations of multiple convergent crossing, and in fact, in a follow-up article (Jensen, 1978) that showed the integration of the DSM with the bulk breeding method in use at Cornell, the procedures were likened to corn breeding techniques.

Lonnquist (1949) demonstrated the value of topcrosses to the parent open-pollinated variety to measure the combining ability of S_1 lines (for a synthetic). Eight high-yielding S_1 maize lines derived from topcrosses (+1 SD units above mean yield of all topcrosses) were used to form a high-yield synthetic. An analogous low-yield synthetic was formed from seven derived S_1 lines (−1 SD units). When these two were compared in 1947, and the results expressed in percent of yield of the unselected open-pollinated variety, the yields were 142% for high, and 85% for low. The implications for other crops regarding testing parents for prepotency, that is, determining the link between parent and offspring performance, is clear.

DIALLEL EVALUATION TESTS

Allard (1956) presented an in-depth look at the use of the F_1 diallel cross to estimate parent prepotency for seed size in a lima bean cross. The parents were nine surviving family lines selected in the F_8 from the cross of 'Hopi 2000' × 'Wilbur'. The 36 F_1 hybrids and their parents were grown and seed size indices prepared from the number and weight of seed of each. The diallel cross analysis methods of Jinks, and of Hayman, were used: these required that six assumptions be recognized. Also seed size is known to be relatively stable in lima beans and it was assumed that phenotypic values were not unlike genotypic ones.

From the array of mean F_1 seed size values in Allard's Table 1, it is evident that the involvement of two of the nine parents was different, based on the observed larger (heavier) seed sizes of their progenies. The identification of these two parents was critical to the ensuing analysis; for example, this knowledge cleared up an aberration observed in the slope of the regression lines: when calculated with these two F_1 parents excluded, the lines approached normality. Thus the basic assumptions of the analysis held for seven of the nine parents, but not for numbers 4 and 9. In these two, the higher "heterosis" observed was deemed due to interallelic interaction, possibly fixable.

Allard discussed, and rejected, the usefulness of the F_2 in diallel analysis. Using the F_2 would eliminate the problem of obtaining sufficient F_1 seed numbers, but it also introduces negative factors such as lowered efficiency in estimating dominance, the requirement of a large number of individuals to establish F_2 family means, and less validity to the basic assumptions.

Allard's paper is a valuable tour de force on the use of the diallel cross in parent prediction. Initially, I was disturbed by the close relationship of the nine parents—all from the same cross, homozygous, and highly selected to the criteria of good agronomic type—however, Allard clearly was aware that the case was not typical. Furthermore, I am reassured through the author's interpretations that the process was meaningful, and should be more so when greater differences and more parents are involved. The question remains as to the capability of many plant breeding programs to absorb the added work load.

Whitehouse, Thompson, and Do Valle Ribeiro (1958) examined diallel cross analyses for their usefulness in making yield predictions. They focused on grain yield per plant and the four components of yield: weight per grain, grains per spikelet, spikelets per head, and heads per plant (sometimes these are compressed into three components, namely, kernel weight, kernels per head, and heads per plant). These are generally considered independent and additive in inheritance and are multiplied together to produce yield per plant, which again must be adjusted for yield per area.

The authors observed and measured 19 wheat cultivars, chose four that exhibited gross values of one or more of the components, and made a diallel set of crosses. They described two ways of choosing parents: the first, a direct method, is through a careful observation of possible genotypes, followed by a subjective pairing or grouping to form fruitful alliances; the second, an indirect method, involves the making of many crosses followed by extensive elimination based on early generation observations, preferably in the F_1. From the performance of the hybrids the potential worth of the parent is deduced.

The problem of increasing components of yield is complicated by compensation, which may involve adjustments in one or more components. Because of this likely but unknown shift, Whitehouse et al. recognized that the parent pairs likely to combine in an advantageous way could not be identified easily without making the actual crosses.

The highly detailed analyses of the parents, F_1, and F_2 hybrids from the diallel were based on the methods independently developed by B. I. Hayman and by J. L. Jinks, in the mid-1950s. The peculiar feature of these methods is that one variety and all its crosses can be isolated and examined in its "array." Variances, covariances, and other statistics are generated from the various tables and graphs, and from these, estimates of dominance, effective factors, and heterosis are produced. All of this provides information as to the location (variety) and best combination of parents to use to advance particular components of yield.

I have given this brief but inadequate description of the study because, on the one hand, it is an elegant presentation of the power of analysis possible from diallels, and on the other hand it indicates the inadequacies of diallel analysis as a staple addition to a plant breeding program. It often is too sophisticated and for proper interpretation requires specialized knowledge or experience, which are sometimes not available to plant breeding programs. The nature of the results, both in quantity (genotypes analyzed) and quality (certainty and reliability), and the prospects held out to breeders, seem meager and unattractive relative to the effort expended. The study nevertheless is an impressive one from which important genetic and statistical information can be extracted.

In a noteworthy paper on choice of parents and cross prediction, Lupton (1965) discussed two general approaches to assessing a genotype's potential, namely, physiological and genetic. His paper was concerned with the second, the genetic analysis of varieties to evaluate their relative potential and worth as parents to improve yield. The initial problem was confounded by a need to reduce two factors to simpler terms: first, the large germ plasm pool of available parents, and second, the procedures needed to make the assessment. The solution to the problem of germ plasm numbers was solved by restricting parent numbers to those meeting reasonable geographic, market, and program targets, and balancing these against the need to increase genetic variability for the improvement of yield.

Lupton solved the second part of the problem by electing to use an incomplete diallel set of crosses in which the cultivars were crossed against five tester varieties. The tester varieties were selected on the basis of their relevance to commercial goals. Also they were assigned a degree of permanence, the same ones being used over a period of several years. The similarity to top crossing in maize is evident.

In the interests of efficiency, no tests were made on the F_1s, and the F_2 and F_3 were grown as unselected bulks. A digression is in order here for plant breeding implications that might not be obvious immediately to beginning or less experienced breeders. Lupton's routine descriptions are good illustrations of some of the ways in which breeders bring excess numbers of crosses down to manageable size: although measurements were not taken on the F_1 it nevertheless was carefully observed and crosses that showed hybrid necrosis, extreme susceptibility to disease, and weak straw were eliminated.

Lupton acknowledged that the precision of results obtained from incomplete diallel sets would be less than from a complete diallel analysis; however, it would be sufficient for his goal of evaluating parents. For details of the analyses one should read Lupton's paper in the original, but a synopsis of his general results and views from evaluating three methods of cross prediction can be given here:

1. *Analysis of the Relation of Dominance to Yielding Capacity:* useful results were absent due to large amounts of variability in dominance relationships over years and locations. Lupton gave an example of two adjacent field trials of the same genetic material where in one case yield inheritance was expressed as a dominant character while in the other as a recessive. The explanation turned on differences in density, rate of seeding, which changed the dynamics of compensation among the yield components.

2. *Analysis of General and Specific Combining Abilities of Parents and Combinations:*

   a. General combining ability: in a perverse way, GCA provided much useful information that otherwise could not have been predicted without the use of hybrid populations and the analysis—perverse because GCA is expected to relate directly to the yielding potential of parents if significant dominance expression is absent, but this was not the case; there were significant differences in dominance.

   b. Specific combining ability: SCA was almost completely absent, suggesting that an overall consistency of performance as parents might be expected of the varieties. Lupton was careful to acknowledge that this conclusion rests on the tester varieties being truly representative of all relevant varieties in the breeding program.

3. *Analysis via mean yield and yield variance of combinations:* this testing was done on a reduced number of combinations (even so, more than 2000 plots were required). These analyses were considered fruitful and the cross predictions made were in general agreement with results of the other methods.

Not mentioned in the above are several instances of useful information uncovered by the research. For example, certain parents and certain combinations were highlighted as especially prepotent. There were also indications of a negative nature concerning some genotypes and combinations.

Lupton cannot be faulted for ignoring the practical impact of studies such as these on the ability to conduct a breeding program. He chose a compromise between two views that bracketed the problem:

On the one hand, it may be argued that if a complex series of trials is necessary for the identification of promising parental combinations, then no expense should be spared in their conduct. At the other extreme it may be claimed that trials of this sort are an unnecessary extravagance, and that the plant breeder should be able to select his crosses by detailed consideration of the parental material available to him without recourse to experimental hybridizations.

It is clear that he had reservations about some of the management procedures. A positive note that I see is that the experimental procedures used left useful selection material in the form of F_3 unselected bulks. Also, I should have liked to know more about the universality of these results as they characterize variety prepotency and combining ability: can the results of one experiment correctly characterize a genotype? If so, it would have to be evaluated only once, and this would open the door to cooperative sharing of results among breeders within a region, provided there was agreement on the same group of tester varieties.

Smith and Lambert (1968) reported on tests with barley crosses for combining abilities and predictive values. The material used for the combining ability studies included 10 cultivars and their diallel crosses. Six parents were used in two crossing systems for the predictive values studies, namely, each crossed to three male sterile tester lines to produce 18 topcrosses, and the six-parent diallel producing 15 crosses for a study of the F_2–F_5 generations.

General combining ability was high and statistically significant for the four characters studied, whereas the opposite was found for specific combining ability. A preponderance of additive genetic effects was postulated.

Parental prepotency as measured by the F_1 topcrosses was not reliable. Performance was based on the relationship to the number of superior F_5 lines produced from a cross. There were 24 superior lines produced by the 15 crosses and all of these came from eight crosses. One parent of the six used was involved in no cross that later produced a superior line—the topcross tester system placed this parent high in ranking and it was this anomaly that contributed to the low regard placed on top crossing. The 24 superior F_5 lines, each with two parents, created 48 parent attachments: the distribution of these among the six parents, ranked by yield, were 13, 13, 11, 8, 3, and 0, respectively, thus showing a clear relationship of parent participation. Overall parent performance (including their yields) was a good indicator of their prepotency, based on number of superior lines produced.

Examining the bulk performance relationship, Smith and Lambert found that the F_2 identified two of the three most prepotent parents (as correlated with parent yield and number of superior F_5 lines) and correctly ranked the poorest parent. The F_3, F_4, and F_5 bulk yields were in even closer agreement with parental performance.

The evidence was favorable for the reliability of predictive values shown by the criteria, midparent, F_3, F_4, and F_5 performance values; the F_2 was deemed less reliable.

ANCESTRAL RECORDS AND TESTS OF RELATEDNESS

Young's (1961) paper on how sire and dam records are used in animal selection has application to the plant breeder's problem of selecting parents. The breeding value of a plant parent can be extracted from different sources, for example, genetic relationships, progeny performance, and so forth, all coming together in a reinforcing index selection manner. Young, for example, dealt with ways of combining ancestor records with contemporary records, something that is now being attempted with crop parents.

Searle (1963) addressed the question of worth of ancestor records in animal selection. In effect, he compared selection indices based on ancestor records for efficiency (the derived expression was the correlation between the index and "true additive genetic merit"). The results showed these relative efficiencies: parent records accounted for 80% of the maximum attainable, and if grandparents are included the figure went to 90%. Searle pointed out, however, that even small progeny tests or performance tests are more efficient than ancestor records.

St. Martin (1982) defined the coefficient of parentage (see Kempthorne 1969) as the probability that alleles found in two individuals are identical by descent. To illustrate, the probability in sibs from a pure line ought to be 100%, or $r = 1.0$, and arbitrarily, 0 in the Asian soybean PIs which were the germ plasm foundation for U.S. breeding programs, and were the base used by the author for comparisons. They examined the coefficients of parentage among 27 modern cultivars of varying maturities that had been released during a five-year period.

A total of only 20 ancestral PIs appeared in the pedigrees, which represented generally three or four breeding cycles, and the average coefficient of parentage was estimated at 0.25%, a figure interpreted by the authors as "the mean inbreeding coefficient of the plants that would result from a diallel cross of the cultivars." Furthermore, a figure of 0.25% translates to a small number of lines recombined, for example, 11 if cycle 3 is examined. This number of lines is relatively low in comparison with corn, for instance. The conclusion was that great progress has been made in the past, paradoxically without a particularly efficient use of the available germ plasm over the historical period.

The methodological problem rests on a balance where too much introduction of exotic germ plasm is often counterproductive. Compromises have to be made in the balanced use of short-range and long-range programs, where the long-range projects deal with the introgression and recurrent cycling of new germ plasm, periodically feeding the genotypes into the adapted short-range programs.

St. Martin's statement, "Implicit in the calculation of coefficients of parentage by this method is the assumption that cultivars derived from bi-parental crosses have obtained half of their genetic material from each parent," was of interest to me because, long before the development of present sophisticated procedures, I used a crude method to try to locate the important parental sources in pedigrees of successful genotypes of wheat, oats, and barley by charting parental contributions in this manner, that is, A × B = 50:50, (A × B) ×C = (25:25):50, and so forth. Summations of strings of pedigrees gave useful information on prepotent parents.

The genetic relatedness of potential parents for a cross is of great consequence to the later observed worth of a hybrid, affecting not only its genetic variance but the eventual genotypes synthesized from the genetic components. I think it fair to say that guidelines to help the breeder are confused and emit ambiguous signals, leading to support for both high × low and high × high crosses. These terms are extended and described further by other appellations such as "exotic," "broad germ plasm," "best by best," and so forth. In the extreme it would be quite possible, in ignorance of parentage, to cross two promising genotypes or similar cultivars and create a hybrid population with little genetic variance and limited potential for genetic advance.

It is from this background that Cox, Lookhart, Walker, Harrell, Albers, and Rodgers (1985) studied the pairwise relationships of 43 hard red winter wheat cultivars. They compared the results of two procedures to estimate degree of relatedness, namely, gliadin electrophoresis to produce similarity coefficients, *s*, as described by Lookhart, Jones, Walker, Hall, and Cooper (1983), and coefficients of parentage (*r*) (see Kempthorne, 1969). The first procedure is based on matching electrophoretic patterns, the second on matching pedigrees. In both procedures there is a positive association between score and relatedness; that is, a score closer to 1.00 (from 0) indicates closer genetic relatedness.

The coefficients of parentage procedures involved the tracing of pedigrees of the 43 cultivars back to "ancestor" genotypes, defined as the point in time where no further geneological relationship could be discovered. Of 35 ancestors found, 7 were considered major contributors and 28 minor (appearing in pedigrees but once). The gliadins and their electrophoregrams were obtained from the 43 cultivars (and 16 of the ancestors) and *s* calculated. There were 903 pair combinations involved for *r* and *s*.

Any discussion of results must begin with the realization that the two measures did not agree well, although there was a small but significant correlation between *r* and *s* between cultivar pairs (but not between ancestor–cultivar ones). One observation that seemed important was that if one took only the relatively few cultivar pairs with high *r*'s (above .50), the correlations between *r* and *s* markedly improved; in other words one could place more confidence in the accuracy of relatedness estimates for a portion of the total comparisons—this observation was based on 52 out of the 903 pairs, or about 6%.

The authors remarked that "both *r* and *s* are imperfect estimates of the relationship between two cultivars." A number of variables substantiating this were discussed. The complexities involved were well illustrated by results of the clustering analysis, which was founded on *r* distributions for all 43 cultivars (the *r* clustering results seemed informative in themselves). These showed that all of the cultivars in one large cluster at the top of the dendograph had one cultivar, 'Scout', in their pedigrees. But this apparent closeness was not corroborated by *s* clustering analysis; in fact, a calculated distortion coefficient between *r* and *s* groups was inordinately high. Another example was 'Turkey', shown by *r* to be the largest contributor to modern HRW wheats: its mean *s* with modern cultivars was only average for all ancestor–cultivar pair values. Furthermore, different sources of accession of the same ancestor genotype often failed to show the same *s* values, suggesting genetic drift and effects of natural or artificial selection.

Cox et al. believed that genetic similarity could be more easily ascertained if the electrophoretic data could be increased by including data from additional loci. They suggested selecting parents for crosses on the basis of high rank in different clusters.

The exciting progress made in understanding parent relationships of genotypes is nowhere better illustrated than in the article by Murphy, Cox, and Rodgers (1986), which dealt with 110 red wheats evenly divided, 55 to 55, between soft red (SRW) and hard red (HRW) cultivars. Using cluster and principal coordinate analyses [coefficients of parentage earlier had been calculated by Cox, Murphy, and Rodgers (1985) for 400 wheats, of which this group is a subsample], these authors convinced me of the power of the analyses when the first cut of the dendograph neatly formed two groups, one containing 49 SRW and the other 55 HRW plus the remaining six SRW wheats. The second cut, on the second principal component axis, placed the 13 clusters (six SRW and seven HRW) into clear geographic origin positions (some cultivars failed to fall into any cluster). Most clusters showed a dominant or prominent ancestral figure to which germplasm and homogeneity could be traced. Naturally, the relationships were not precise, there was overlapping and shared germ plasm, and there were elements of uncertainty. Nevertheless, the authors stated that "a clearer understanding of relationships within the red winter wheat gene pool can help in the selection of parents for crosses."

In the ongoing research to evaluate the parental prepotency of soybean genotypes, St. Martin and Aslam (1986) compared parental influences between and within two broad categories of parental origin and development, adapted and exotic. They used eight genotypes of each, with the restrictions that each be among the high seed yielders of its group and meet certain maturity parameters. Two of the adapted and two of the PI (plant introduction) genotypes were used as males, the others as females. The requisite crosses provided 48 bulk hybrid populations which, with check cultivars, were evaluated in F_3. In addition to recording data on several characters the authors were interested in male and female effects: were there interactions, and if so, were they associated predominantly with adapted or exotic germ plasm?

Female × male interactions were found, but always in a bulk with at least one PI parent. PI males produced bulks yielding an average 10% less than adapted male bulks; PI females produced bulks yielding 4% less than adapted female bulks. Similar results held for characters other than yield.

When the data from all the crosses were examined, regardless of germ plasm source, no male × female interaction was found. This disappearance of the known germ plasm effect simply must be due to the expansion of numbers in this analysis, which overwhelmed the germ plasm source influence. A similar response would be expected in the adapted × adapted crosses.

The search for a male–female interaction might suggest a concern for cytoplasm contribution. The fact that the authors did find yield differences between PI male-derived bulks and PI female-derived bulks is intriguing. In practice, breeders often find an association of lower yield from any cross of a PI parent with an adapted parent, but in this study St. Martin and Aslam have further partitioned the effects to show male and female differences.

The authors concluded that a testcross procedure for identifying yield-prepotent parents could confidently rely on results from any reasonable choice for tester. However, they questioned whether a testcross procedure was necessary because parental yields, that is, of the PIs themselves, may be an adequate indicator of yield potential.

It is not my intention to cover thoroughly the subject of cultivar identification through the new electrophoretic techniques. An up-to-date review of research in barley was given by Gebre, Khan, and Foster (1986). It is apparent that identification techniques make it possible to group cultivars in clusters of genotypes with similar performance patterns. There are also genetic similarities that are subject to tests. The authors stated, "Cultivars that cannot be distinguished by PAGE may have parental relationships." From the plant breeding viewpoint the knowledge of the existence of such relationships is very important because it aids the breeder in his choice of parents. In wheat the coefficient of parentage has been used to provide a second measure of relatedness (see Cox et al. discussed above).

THE PARENT AS INDICATOR OF PREPOTENCY

Qualset (1979) stated, "How to choose parents for crossing is probably the most often asked question among plant breeders." Most breeders get their information for decisions from experience, prediction, diallel cross studies, and the known correlation between bulk population performance and subsequent line performance. Qualset added an example of a wheat protein and lysine content screening method (for full details see Qualset, Zscheile, Ruckman, and Hille Ris Lambers, 1972) for use once a breeder has narrowed and focused his breeding objectives. He chose lysine as a percent of grain protein content as the screening target. Approximately 100 high-lysine sources were chosen from the world collection screening list made available from an earlier USDA/Nebraska project and these were crossed to low-lysine 'INIA 66'. The F_1s, the world collection parents, and INIA 66 were grown under conditions where both spring and winter types thrived, and the lysine figures were obtained. These were then arranged in a scatter diagram two-dimensional frequency distribution that had built into it certain points of reference, specifically, the recurring parent INIA 66 statistics, the ideal correlation line for F_1, and the nonrecurring world collection parents, vertical and horizontal standard error of mean lines for INIA 66, and an 'Opaque-2' reference point. Qualset described the arrangement in this way:

> By this method it is a simple matter to select crosses and/or parents for further evaluation. This single-parent array is simply one array of a diallel table and, if desired, can be utilized to obtain information about genetic variances and heritability.

Qualset's remark about the difficulty of choosing parents is well taken. His proposed screening method, which probably has general use, particularly interests me because it addresses the problem of numbers. Here we see a way in which a wheat breeder can access the 40,000 or so genotypes in the world collection, first by a limiting or focusing procedure such as using the USDA–Nebraska screening list, and second by evaluating a select group (of 100) as possible parents. The opportunities are not narrow—the cereal world collection has already been partitioned into many subsets according to unique characters such as disease resistance.

The results of Allard and Harding (1963) raise a question as to the reliance on early-generation analysis as a prediction tool. Instead of an indicated single gene pair, further research brought them to believe that at least four gene pairs were involved.

Knott (1965) found a correlation of .79* between the yield of wheat F_1s and their parents, implying the likelihood that high-yielding hybrids would come from high-yielding parents.

Quinones (1969) found that parental performance (defined as that performance that led to the choice as parent in crosses) in the year crosses were made was a reliable indication of the value of future selections from these crosses. In 1966 Quinones yield-tested 160 F_8 lines selected by the pedigree method from numerous crosses made over the years. These lines traced to 22 crosses made in 1959. The field performances of these parents in 1959, covering five traits including yield, were correlated with the same traits appearing in the F_8-derived lines. Statistically significant interannual positive correlations were found for yield, maturity, growth habit, and bacterial blight reaction; seed size between the parents and progeny was not significantly correlated.

Parental performance in the year that crosses were made would seem to have only a coincidental relationship to tests made years later. This suggests that the phenomenon Quinones observed occurs either sporadically or is generally related to the later performance of progeny from the crosses. A general relationship would indicate that parental performance need not be specifically tied to the year crosses were made, which would mean that incidental and average records of parental performance could be used to predict progeny performance.

Stuthman and Stucker (1975) made a study in which they concluded that prediction of superior crosses from only parental information was ineffective. If one looks at their study as a source of information on choosing *parents* rather than *crosses*, the results seem more hopeful and useful. For example, the information would induce me to choose for future crossing parents their numbers 10, 5, 2, 12, 1, and 8, the first four because they ranked highest in yield and in number of "best" lines exceeding best parent by 5%, and the last two because, although they ranked next to last for yield, their progeny means and "best" lines numbers were comparable to the first four.

Pederson's (1981) paper on a least squares method for choosing parents and their proportions in crosses is important in several ways: first, in its presentation of methods of choosing parents; second, through drawing together in one discussion other methods such as Grafius' 1963 vector analysis, Whitehouse's 1970 canonical variate analysis, and Bhatt's (1973) multivariate analysis; and third, by an implicit inclusion of selection index thinking—all framed in contexts that plant breeders understand.

Pederson's least squares method begins with assumptions familiar to plant breeders. He assumed a number of parents were available, with known data on many variables (characters); this would be true of most breeding programs. Next he denoted an ideal value and an acceptable deviation from the ideal. Using an illustration with spring wheat, an ideal value for yield of 4.0 tons/ha was derived from actual yields of three cultivars ranging from approximately 3.0 to 3.9 tons/ha, and a preferred maximum deviation of 0.2 was set—when these were applied to several traits, the similarities to a selection index become apparent. Pederson modified a linear additive model of Wright's (see Grafius, 1963) to implement the least squares method and applied this to 25 wheat cultivars. Solving the equations gives the optimal input-to-a-cross proportions of each parent for the type of cross the breeder had in mind (single and three-ways are illustrated, and greater parent number participation, up to five or seven, is implied).

Pederson commented that meeting precise optimal proportions of cultivars in a cross might involve tedious and extensive cycles, and backcrosses, and in this sense a note of impracticality is introduced. I think the very existence of knowledge of the best proportions of parent choice and input is in itself helpful to a breeder, and might be all that is needed in some cases to choose the parents for crosses without resorting to precise parent-input formulation. In any event, Pederson proceeded to search for a more acceptable and practical method.

I shall not attempt to give a detailed review of Pederson's alternative "enumerative" method except to say that it begins from the base of the least squares solution, partitions and examines all possible crosses of the cultivars in the set, and solves for minimum deviations from the ideal, finally identifying the most practical solution.

This is a stimulating idea paper on the subject of choosing parents and needs to be read in its entirety.

ALTERNATIVE CONCEPTS OF MEASURING PREPOTENCY

Grafius (1964) presented a concept of plant breeding that was based on a geometric model (cf. earlier paper by Grafius, 1956). Pointing out that the complex trait, yield, is the product of three (and only three) components, namely, number of heads per given area, average number of kernels per head, and average kernel weight, he likened these to the three dimensions of boxes. Any one of the edges might change in length but this would be accompanied by compensatory changes in one or both of the other edges, so that the volume (read: yield) of the box would remain constant.

Grafius also proposed a vector method for the selection of parents to be used in crossing. This method also was based on a geometric construct. Essentially a vector is the calculated entity that describes the variety in mathematical terms.

Grafius' views are important in that they contribute to a better understanding of problems in plant breeding. Whether they are practical for inserting into a breeding program, I cannot say. On the one hand, the vector method of choosing parents garners a great amount of information about a genotype and would permit the dual choice of parents for a hybrid that would almost guarantee genetic diversity and a large genetic variance in the ensuing population. On the other hand, much additional work would be placed on the project. In particular, I would call attention to the fact that the vector method, in essence, is built around a weighted index procedure. If a breeder has had difficulties coming to terms with indices, this might pose a problem.

Jordaan and Laubscher (1968) compared the F_1 to the F_2 to F_5 progeny bulks sequence, and found the latter to be superior in consistency of results. They recommended F_2 progeny bulks to evaluate the breeding value of parents.

Bhatt (1970) proposed a multivariate analysis scheme, not involving hybridization, as one approach to identifying parent potential in self-pollinating crops. The analysis rested on the assumption that maximum genetic divergence is synonymous with superior parents, and involved the clustering of like-performing cultivars. Using six characters (yield and its comonents), Bhatt formed 12 clusters from a selected group of 40 wheat genotypes. The clustering turned out not to be closely correlated with the wheats' geographic distribution.

Bhatt's proposal is aimed at putting a quantitative measure to genetic diversity which often is thought of as related to geographic diversity (and sometimes is). The multivariate analysis approach appears to accomplish this through grouping and then measuring the statistical distance between groups—the larger the distance, the greater the diversity. The smaller the distance, even to within the same cluster, the less the likelihood that parents within the restricted distance group would yield desirable segregates if crossed. Bhatt suggested ranking the possible cluster combinations in descending order of magnitude of distance between, and using the mean as a guide to a minimum distance that might separate two clusters that, for example, were being considered as sources of parents for a cross. Practical considerations of traits and performance could then be used to choose a particular genotype from each cluster. In a sense, the analysis acts as a rough index scheme where all involved factors contribute and shape the clustering decisions. Bhatt emphasized the theoretical nature of the proposal and suggested ways of testing it.

In further studies on choosing parents for potentially successful hybrids, Bhatt (1973) examined four methods of choosing parents, namely, conventional, random, multivariate analysis, and ecogeographic. The measurement criteria were F_2 hybrid bulk yield, F_5 lines yield means and variances, and number of F_5 transgressive yield segregates. Bhatt found the following order of efficiency in selecting best parents: multivariate analysis, ecogeographic diversity, and the conventional and random methods. The study thus affirmed the earlier (1970) theoretical presentation for the multivariate analysis.

Bhatt recognized the limitations of his study by acknowledging that the crosses might overlap the boundaries of his categories. Moreover, category descriptions themselves might be considered arbitrary and artificial, for example, the choice of only two crosses for "random," when all crosses, whether conventional or ecogeographic, were drawn from the same set of populations and might each be considered "random." The research framework sought both diversity and a means of predicting the (unknown) best parental combinations. A priori, one might predict the canonical analysis—the only method with the power to predict and assess—to be the most efficient, and the ecogeographic, which involved elements of origin and isolation, to be next. Conventional, as described, suggests close relationship, and random rests on chance, so these two are a toss-up, except that no breeder would opt for the random method other than as a phase step in an ongoing directed program such as backcross or recurrent selection.

GENETIC DIVERSITY, WIDE CROSSES, AND INTROGRESSION

I am reminded of a successful oat cross from my project which was conventional, not random. Today I am at a loss to give a logical account for choosing the 'Alamo' parent, except to state that in all the small grain projects there was a conscious effort to introduce promising exotic material into each crossing program, and Alamo, an outsider, did have interesting crown and stem rust resistance traits. The cross was N.Y. 5279: Alamo/4/ Garry Sel. 5 C.I.6589/3/Goldwin//Victoria/Rainbow, a complicated sequence in which the last hybrid made was to Alamo. This cross was outstanding and impressive in the F_2, and from it were obtained two cultivars, 'Orbit' and 'Astro', which I considered my two best oat cultivars.

Lawrence and Frey's (1975) research was done for other purposes; nevertheless, the results from analyzing backcross-induced introgression of *A. sterilis* germ plasm into *A. sativa* germ plasm, measured by grain yield, clearly showed significant evidences of differences in parent potency in the five generations of backcross progeny derived from two oat recurrent *A. sativa* parents crossed to the same four *A. sterilis* donor parents. The evidence of the dramatically higher percentage of transgressive lines in each generation of each cross involving the recurrent 'Clintford' parent is convincing.

Bhatt (1976), in a paper on the application of multivariate analysis to quality characters in wheat (his previous research in this area had been directed toward yield), found that clustering on the basis of six quality characters measured on 12 well-known genotypes grown in four environments did group the wheats having recognizably similar quality traits together. Bhatt extrapolated from this that the method, if applied to breeding lines of unknown quality, would similarly sort them into a series of groups having distinct quality profiles, which could be interpreted to mean that the groups represented genetic divergence for quality.

Aside from showing that the analysis worked for both yield and quality, it would seem that the application to quality would be less useful than its application to yield. Also the reason for clustering must be examined. The analysis might be useful in selecting groups for discard—that is, individuals falling within a group whose profile was undesirable—but because such a group, by definition, shows wide genetic divergence, its discard would run counter to the use of "distance" to measure (crossing) desirability. If selection of parents for crossing were not a major objective, the act of testing, for example, for the six quality traits, would seem sufficient to group the wheats without the use of multivariate analysis. In breeding practice any serious divergence of an individual's quality trait is measured against the known performance of a control genotype.

In an effort aimed at broadening the genetic base for peanut improvement, Isleib and Wynne (1983) collected 27 extics, clustered them into five different morphological groups, and made testcrosses to an adapted genotype. Later, parents, F_1s, F_2s, and F_4 bulk populations were evaluated to assess their value in predicting later generation progeny performance. The traits measured were pod length, pod size, seed size, meat content, and seed yield (meat content × pod yield). The results in brief were as follows:

1. The sorting into morphological groups was deemed a useful procedure—variation for characters among bulks was related to differences among these groups.
2. Yield differences in bulk were associated with subspecies exotic groups but not with morphological groups.
3. The best indicators of F_4 bulk performance were the performance of parents themselves, and of F_2s.
4. F_1 performance (heterosis), epistatic effects, and inbreeding depression were poor indicators, relatively unrelated, of F_4 bulk performance.

In sum, a satisfactory means of selecting parents to advance yield would be simple performance testing of the exotics, retaining the top few. The authors discussed short-term and long-term strategy, which is of special importance when dealing with exotics as parents, and they stressed the need for recurrent selection.

The Iowa station has been a leader in research on recombination and introgression in the two species, *A. sativa* and *A. sterilis*. In continuing research to evaluate the value of recombining these two germ plasms, Cox and Frey (1984) made all possible crosses between six modern *A. sativa* cultivars and 10 *A. sterilis* accessions from eight Mediterranean countries. About 40 nonshattering F_2 plants per cross were harvested to establish F_2-derived lines for testing with parents in F_3 hill plot nurseries. Five traits were measured to provide the three comparison criteria of grain yield, biomass, and vegetative growth index. The objectives of the authors were to elicit information on general combining ability (GCA), specific combining ability (SCA), parental performance, and other parameters to indicate the breeding value of parents. Breeding value was defined as being based on GCA and percentages of significantly high transgressive segregates which exceeded the high parent by one LSD at the 5% level.

A number of interesting observations and conclusions were made, among them the following:

1. SCA effects were greater than GCAs. Interestingly, high SCA matings generally were predictable from any combination of high × low GCA parents, and were more likely to occur than from matings of high × high GCA parents. This is somewhat in opposition to other findings and suggests the necessity of testing progenies for productivity.

2. There was a significant correlation between performance of parents and mean progeny performance of their crosses. However, this relationship did not hold well if one looked for prediction capability for producing extreme progeny types. Here it was found that there was an imbalance in that more extreme segregates were found for biomass and vegetative growth than for grain yield. If extreme progeny types were treated as a group, Cox

and Frey found that the mean genotypic variance of a parent, derived from all crosses, was a better predictor of extreme progeny production than was the parental performance itself.

There seems little doubt as to the value of these interspecific crosses. The general impressions I received were two, namely, a greater understanding of the range of happenings in a large series of interspecific crosses, culminating in conclusions drawn from the accumulation of similar responses, and beyond this, the importance of looking for specific indicators of potential. The authors provided several illustrations of high performance of specific parents and combinations. Such information of this specific nature furnishes a breeder with knowledge that is applicable immediately to the breeding program.

COMMENTS

The breeder is left with no clear-cut solution to the dilemma of choosing parents for the right germ plasm input to the program. One can incorporate sizable prebreeding and preselection inputs into the overall program. These usually involve a two-year delay, which can be accommodated, however, for once established and the initial delay absorbed, the program thereafter will feed a full quota input of crosses into the system annually. Alternatively, one may continue within the broad range of the conventional method, endeavoring to include useful diversity through the known reservoirs of germ plasm diversity, that is, with considerations of parent origin, isolation, and similar desiderata.

Selection programs that attempt to sort out a large number of crosses and identify those most likely to produce the most productive lines are a necessary part of plant breeding projects. They make it possible to concentrate time and effort on the most promising crosses. However, there is an aspect of this, important to parent building and germ plasm, that has not been considered. The selection of crosses to work with means that other crosses are discarded. A number of studies have shown that the poorer (rejected) crosses sometimes produce very superior lines, although in general the numbers are much fewer. A matter of judgment on cutoff points for discarding crosses is required in order not to lose completely these infrequent but superior lines.

One aspect of "discarding" deserves consideration. Selecting the best among a series of crosses, on the basis of quantity and quality of superior lines produced, in effect establishes two streams, a "best" and a "worst" stream. In examining these two categories breeders often see additional things that confirm their assignment of a cross to the "worst" stream. Examples might be "wide" germ plasm differences, germ plasm that has been isolated, poor parent history in combinations, and the like. If we grant that there are factors of genetic isolation and differences between the two streams, and that there are low expectations of superior lines in the "worst" stream, what are the implications if a truly superior line *is* found in a least promising cross? This line must possess unusual genetic recombination, and it may contain unusual formulations of different genes. It may be, in fact, the very germ plasm source needed as a parent for recombining with cultivars and lines out of the "best" stream in order to obtain further yield advance. Except in research studies, the "worst" stream of crosses is discarded and we have little opportunity to see the few superior lines that might have come from them. There is no practical alternative to the discard option; however, the result is a kind of backcross or recurrent selection aspect to a whole breeding program, where the best continually associate with the best—not a bad situation, but certainly one that is limiting to germ plasm breadth.

In summary, I think it fair to say that choosing parents for hybridizing combines the art and science aspects of plant breeding. In this chapter we have dealt with the latter aspect, improving, I hope, the odds for more productive crosses. More information on choosing parents is found in the next chapter on choosing among crosses. Also, for detailed illustrations on how I chose parents for the annual crossing procedures with the small grain cereals, see Chapter 34, "The Four Stages of the Plant Breeding Process."

REFERENCES

Allard, R. W. 1956. Estimation of prepotency from lima bean diallel cross data. Agron. J. 48:537–543.

Allard, R. W. and J. Harding. 1963. Early generation analysis and prediction of gain under selection in derivatives of a wheat hybrid. Crop Sci. 3:454–456.

Bal, B. S., C. A. Suneson, and R. T. Ramage. 1959. Genetic shift during 30 generations of natural selection in barley. Agron. J. 51:555–557.

Bhatt, G. M. 1970. Multivariate analysis approach to selection of parents for hybridization aimed at yield improvement in self-pollinated crops. Aust. J. Agric. Res. 21:1–7.

Bhatt, G. M. 1973. Comparison of various methods of selecting parents for hybridization in common bread wheat (*Triticum aestivum* L.). Aust. J. Agric. Res. 24:457–464.

Bhatt, G. M. 1976. An application of multivariate analysis to selection for quality characters in wheat. Aust. J. Agric. Res. 27:11–18.

Cox, D. J. and K. J. Frey. 1984. Combining ability and the selection of parents for interspecific oat matings. Crop Sci. 24:963–967.

Cox, T. S., J. P. Murphy, and D. M. Rodgers. 1985. Coefficients of parentage for 400 winter wheat cultivars. Kansas State Univ., Agron. Dept. Rep., Manhattan, KS.

Cox, T. S., G. L. Lookhart, D. E. Walker, L. G. Harrell, L. D. Albers, and D. M.

Rodgers. 1985. Genetic relationships among hard red winter wheat cultivars as evaluated by pedigree analysis and gliadin polyacrylamide gel electrophoretic patterns. Crop Sci. 25:1058–1063.

Gebre, H., K. Khan, and A. E. Foster. 1986. Barley cultivar identification by polyacrylamide gel electrophoresis of hordein proteins: catalog of cultivars. Crop Sci. 26:454–460.

Grafius, J. E. 1956. Components of yield in oats: a geometrical interpretation. Agron. J. 48:419–423.

Grafius, J. E. 1964. A geometry for plant breeding. Crop Sci. 4:241–246.

Harlan, H. V., M. L. Martini, and H. Stevens. 1940. A study of methods in barley breeding. USDA Tech. Bull. 720. 26 pp.

Isleib, T. G. and J. C. Wynne. 1983. F_4 bulk testing in testcrosses of 27 exotic peanut cultivars. Crop Sci. 23:841–846.

Jenkins, M. T. 1940. The segregation of genes affecting yield of grain in maize. J. Am. Soc. Agron. 32:55–63.

Jensen, N. F. 1970. A diallel selective mating system for cereal breeding. Crop Science 10:629–635.

Jensen, N. F. 1978. Composite breeding methods and the DSM system in cereals. Crop Sci. 18:622–626.

Jordaan, J. P. and F. X. Laubscher. 1968. The repeatability of breeding values for eleven wheat varieties estimated over generations. Third Int. Wheat Genet. Symp., Canberra, Australia. Aug. 5–9, pp. 416–420.

Kempthorne, O. 1969. An introduction to genetic statistics. Iowa State University Press, Ames.

Knott, D. R. 1965. Heterosis in seven wheat hybrids. Can. J. Plant Sci. 45:499–501.

Lawrence, P. K. and K. J. Frey. 1975. Backcross variability for grain yield in oat species crosses (*Avena sativa* L. × *A. sterilis* L.). Euphytica 24:77–86.

Lonnquist, J. H. 1949. The development and performance of synthetic varieties of corn. Agron. J. 41:153–156.

Lookhart, G. L., B. L. Jones, D. E. Walker, S. B. Hall, and D. B. Cooper. 1983. Computer-assisted method for identifying wheat cultivars from their gliadin electrophoregrams. Cereal Chem. 60:111–115.

Lupton, F. G. H. 1965. Studies in the breeding of self pollinating cereals. 5. Use of the incomplete diallel in wheat breeding. Euphytica 14:331–352.

Mac Key, J. 1963. Autogamous plant breeding based on already highbred material. Akerberg and Hagberg: Recent plant breeding research, Svalof, 1946–1961, Stockholm.

Murphy, J. P., T. S. Cox, and D. M. Rodgers. 1986. Cluster analysis of red winter wheat cultivars based upon coefficients of parentage. Crop Sci. 26:672–676.

Pederson, D. G. 1981. A least-squares method for choosing the best relative proportions when intercrossing cultivars. Euphytica 30:153–160.

Qualset, C. O. 1979. Mendelian genetics of quantitative characters with reference to adaptation and breeding in wheat. Proc. Fifth Int. Wheat Genetics Symp., Indian Soc. Genet. Plant Breeding, New Delhi. 2:577–590.

Qualset, C. O., F. P. Zscheile, J. E. Ruckman, and D. Hille Ris Lambers. 1972.

Prospects for breeding for improved lysine content in wheat. Am. Soc. Agron. Abstr. p. 27.

Quinones, F. A. 1969. Relationship between parents and selections in crosses of dry beans *Phaseolus vulgaris* L. Crop Sci. 9:673–675.

Richey, F. D. 1946. Multiple convergence as a means of augmenting the vigor and yield of inbred lines of corn. J. Am. Soc. Agron. 38:936–941.

Searle, S. R. 1963. The efficiency of ancestor records in animal selection. Heredity 18:351–360.

Smith, E. L. and J. W. Lambert. 1968. Evaluation of early generation testing in spring barley. Crop Sci. 8:490–492.

St. Martin, S. K. 1982. Effective population size for the soybean improvement program in maturity groups 00 to IV. Crop Sci. 22:151–152.

St. Martin. S. K. and M. Aslam. 1986. Performance of progeny of adapted and plant introduction soybean lines. Crop Sci. 26:753–756.

Stuthman, D. D. and R. E. Stucker. 1975. Combining ability analysis of near homozygous lines derived from a 12 parent diallel cross in oats. Crop Sci. 15:800–803.

Suneson, C. A. and O. C. Riddle. 1944. Hybrid vigor in barley. J. Am. Soc. Agron. 36:57–61.

Whitehouse, R. N. H., J. B. Thompson, and M. A. M. Do Valle Ribeiro. 1958. Studies on the breeding of self-pollinating cereals. 2. The use of a diallel cross analysis in yield prediction. Euphytica 7:147–169.

Young, S. S. Y. 1961. The use of sire's and dam's records in animal selection. Heredity 16:91–102.

30

PREDICTING AND CHOOSING CROSSES AND LINES

Major plant breeding concerns center around prediction and potential. Great opportunities are won or lost in decisions made when choosing among parents for crosses and when choosing among crosses for exploitation. Intimately involved are questions of genetic relatedness of cultivars. A knowledge of the degree of relatedness between any two potential parents would aid the breeder in planning future crosses because, other considerations being equal, dissimilar genotypes favor the creation of a hybrid pool having greater genetic variance. Relationship estimates are generally considered more reliable for autogamous species such as the cereals because of closed descent integrity and the existence of voluminous pedigree and performance records.

EARLY STUDIES OF DETERMINING RELATIVE VALUES OF CROSSES

Harrington (1932) reported on the use of F_2 analysis in predicting the potential value of a spring wheat cross. The cross in question was rust-resistant 'Marquillo' × high-quality 'Marquis'. A total of about 40,000 F_2 plants was grown. Harrington judged the experiment a success although the results from this particular cross were disappointing, basically because there were too few plants having the desired combinations and level of quality as expressed by rust resistance and by milling and baking quality. The analysis predicted seven superior plants in the population of 36,800; actually obtained were six promising lines, each of which contained some defects. He recommended that comprehensive F_2 studies of crosses be made. This judgment is based partly, I am sure, on the several years of processing that would be required without prediction based on the F_2.

Harlan, Martini, and Stevens' (1940) huge research study comparing pedigree and composite breeding and selection systems involved 378 crosses among 28 barley parents, and contained much information about predicting parent and cross potential worth (see the preceding chapter on predicting and choosing parents).

Harrington (1940) provided a rational background for a belief that early-generation performance could predict the relative potential value of crosses and that the verification of predicted value would be the performance of lines selected from a cross. He wrote,

> Since yield depends on many genetic factors, it is reasonable to assume that the probability of obtaining high yielding segregates would be greater in a cross between parents with a larger total of yield genes than in one between parents possessing relatively few such genes.

It follows, he continued, that heterosis should be a true measure of the yield value of a cross. To measure and verify this hypothesis it would be necessary to study parents, F_2, F_3, and selected lines from the crosses. Strangely enough, Harrington excluded F_1s because of the difficulty of obtaining the large numbers needed and the unreliability of single-plant yields. Two sets of wheat crosses were studied, one set of six crosses "of compatible varieties differing in yielding ability," and another set containing four crosses ranging from close to wide parent relationships.

In the first set, the bulked F_2 yields of each cross were significantly higher than the parents' average and in four crosses were higher than the better parent. What seemed important here was that relative differences singled out two crosses—those containing 'Quality'—as lower yielding. This became apparent in the bulk F_3 testing when these two crosses were the only ones yielding below the parent averages. Furthermore, when F_6, F_7, and F_8 lines were tested for yield, superior lines were obtained from all crosses except the two containing the Quality parent. Therefore it could be concluded that F_2 results, buttressed by F_3 results, made a reasonable separation of the six crosses, allowing for the discard of one-third of the populations.

In Harrington's second set involving related and less-related parents, the F_2 yield of the cross of two related parents was inferior to one parent and equal to the other parent; the F_2 yield from the "geographically wide" parent cross was dramatically greater, by more than 20%, than either parent. No general inference is possible, however, from such a small sample.

The impact of this on efficient methodology bears on the use of this information. Much of the value to the breeder would be lost if all crosses had to be checked through the line testing stage; that is, the value of the potential early discard would not be felt. Greatest efficiency would be obtained if the procedure were viewed as a prescreening effort, a challenge that all family lines, bulked F_2 and F_3, must meet in order to receive the full attention in the regular nurseries. Discard of whole crosses could begin in the F_2 on the basis of visual judgment and yield. The small screening effort would produce three F_2 categories, namely, elite, doubtful, and discard. Only the doubtful crosses would require F_3 verification. These verification plots would be small, as was the case with Harrington's, yet there would be ample seeds from reserve stocks available for the continuation of the selected crosses.

I think there is an implied suggestion here for a speedup in the making of crosses, so that the number of selected crosses that survive the screening process would approximately balance the known handling capacity of the breeding project. Such a prescreening system would introduce a one-year gap in the program; however, if it were phased in partially and slowly, while cross numbers were being increased, there would be no gap. If Harrington's view is correct that heterosis of the F_1 is an important predictor, equivalent effort could be assigned to the F_1 in prescreening. The net result ought to be a rise in the overall quality and efficiency of a program.

Immer (1941) provided one of the early insights into the yield and yield component character of early-generation hybrid populations vis-à-vis their parents and advancing generations. Heterosis was defined as increase over midparent values. F_1 data were from spaced plants of six parents and six crosses; F_2 data were from drilled rows at crop density. In the F_1 all crosses averaged heterosis figures of 8.3% (number heads/plant), 11.1% (number seeds/head), 4.9% (weight/seed), and 27.3% (yield/plant). In two subsequent years, the average yields of all crosses were as follows (parent average = 100):

	F_2	F_3	F_4
1939	134%	109%	
1940	119%	114%	105%

The author expected, theoretically, that as the bulk progenies approached 100% homozygosity, their yields would approximate the mean yields of the parents. The expected F_4 figure of 107.5% was compatible with the obtained 105%.

Immer made a special comparison of yields of the six parents in drilled rows (600 seeds) and in spaced-seed (5 in. apart) rows. With heavy tillering observed, the spaced-seed rows outyielded the drilled rows by 6.4%. From this the author concluded that (because limited seed supplies dictated space planting) the F_1 generation yields would not be reliable indicators of future generation yields. However, F_2 or F_3 bulk crosses showed promise as reliable indicators of future population yields and should provide a basis for among-cross decisions on the premise that the high-yielding crosses ought to have the highest proportion of high-yielding individuals.

Atkins and Murphy (1949) reported on yield and test weight characteristics of 50 oat segregates drawn from each of 10 crosses established as high- or low-yielding on the basis of five years of bulk testing (F_2 through F_6). They found that of the highest-yielding segregates, based on twice the standard error above the mean of two check cultivars, seven were from the high-yielding and 10 from the low-yielding group. Also, among the 10 crosses studied, two produced the highest proportion of superior yielding lines: these two crosses had been classified in the potentially poor yielding group. Atkins and Murphy concluded, "These results suggest that considerable high-yielding germplasm may be lost if bulk crosses are discarded on the basis of early generation yield performance."

Atkins and Murphy's findings strike a note of caution concerning prediction among crosses and discard decisions. Beyond this, their paper leaves unanswered a perplexing question: how many years or generations of bulk testing would be required to establish a yield base? In this, a research study, five generation years were used, but a breeder would like to spend minimum time, preferably the F_2 only. The question is sharpened by their observation, "Correlations between successive generations for yield of bulk hybrid populations were consistently low."

Grafius, Nelson, and Dirks (1952) explored the usefulness of early-generation bulked progenies as between-cross predictors of sources or producers of high-yielding segregates. The results were disappointing and are explainable on a logical sequence of assumptions:

1. The genetic variance of heterozygous populations is composed of additive and nonadditive genetic input, plus random (environmental) variation.
2. Additive effects can be fixed, and if all variation were additive the progeny means would be the same as the parental means.
3. The nonadditive dominance and epistatic effects disappear, the first through selfing toward homozygosis and the second (general case) owing to nonrecognition. This, supported by certain other cited research, led the authors to conclude that the mean yield of potential parents would be a better indicator of cross value than the early-generation bulk progeny test.

The paper by Grafius et al. left me with some uncertainty as to the use of parental means as an indicator of cross potency. Its validity is not questioned, but clarification as to how the parental yields were obtained, and the high, unselected, and low bulks were created, is crucial. It is clear in this study that these bulks were made up on the basis of yields of spaced plants. the uncertainty arises because of several reports that spaced-plant yields are often negatively correlated with crop density yields (especially, see Hamblin and Donald, 1974). Thus midparent yields for estimating progeny variance should come from crop density trials, that is, the typical nursery trial of cereal breeders. The relationship of spaced-plant-obtained parental yields to crop density-obtained parental yields ought to be explored, for even though spaced-plant data may be artificial, it is possible that its prediction value could be high.

Whitehouse (1953) suggested that selection among crosses to detect those most likely to contain high-yielding segregates might be done in early generations and based on the amount of heterosis observed, adjusting for the particular generation observed. The amount of yield excess over the midparents might be an indication of the number of yield genes involved.

Seemingly, the F_1 would be the best generation, if testing problems of single or few plants could be managed, but two generations would be better in order to get some measure of nonadditive effects specific to a cross.

Using an arbitrary division-of-lines procedure based on best parent and self-performance, Frey (1954) found that the yield, test weight, heading date, and plant height of F_2-derived barley lines were good predictors of F_3 selection performance. Prediction reliability was in the order of heading date, height, test weight, and yield. In one of the two crosses, 'Kindred' × 'Bay', the relationship did not hold for yield. Examination of the methods used showed that F_2 progenies were tested for stem rust reaction and only resistant F_3 progenies were selected; it is possible that this selection pressure isolated a group of closely related lines clustering around a reduced variance for yield.

Frey viewed the bulk hybrid method as a means of choosing between (or among) a series of random crosses, but concluded that selection for yield and other quantitative characters "would be more successful within planned crosses than between random crosses." The distinction between a random and a planned cross does not seem clear, but one can interpret it to mean that a planned cross has thought behind it, whereas this need not be a part of a randomly made cross. His proposal that selection be only between F_2-derived lines represents a sound appreciation of both theory and efficiency and of the important distinction separating selection between and selection within F_2 lines.

Lupton and Whitehouse (1955) considered the twin problems of selecting crosses for further selection and recognizing the low-yielding crosses, which nevertheless might contain high-yielding segregants. Their solution was to use a combination of visual selection and small F_2 plant progeny tests.

THE USE OF DIALLEL ANALYSIS IN EVALUATING CROSSES

In an extensive test of the suitability of bulk hybrid tests to predict future pure line performance under Kansas conditions, Fowler and Heyne (1955) encountered uniformly negative results with yield, and positive results with plant height, maturity, and test weight (the authors felt the latter traits might be selected more efficiently than by the analysis of bulk crosses).

Diallel crosses of 10 winter wheat cultivars were made, which, with the parents, constituted 55 kinds of populations. The bulk populations were brought to the F_5, after which they, with the 10 parents, were grown for three years. The generations were repeatedly grown as follows: F_3—three years, F_4—two years, and F_5—one year. Two years were devoted to selecting random lines from both parents and every cross; some selection may have been applied the second year.

The results for yield, combined over the three generation years, showed that no significant differences existed among the 55 kinds. However, there were highly significant differences between kinds in all generations and in all years. These differences lacked consistency, however, as was shown when yields were ranked and considered by years and by generations.

Fowler and Heyne were able to arrange comparisons in a telling manner, for example:

1. By eliminating generation-to-generation variance, that is, F_3 (1946) versus F_3 (1947)
2. By eliminating year-to-year variation, that is, F_3 (1947) versus F_4 (1947)
3. By involving successive generation-years, that is, F_3 (1946) versus F_4 (1947).

In only one of the 11 comparisons was significance found: yield correlations between F_3 and F_4 when grown in the same year. However, when repeated the following year that correlation was not significant.

A low but significant r between early-generation hybrid kind (crosses) and selections from each kind was found; however, this disappeared when averaged over 55 kinds and three successive generation-years and therefore was not deemed meaningful. Positive and predictive relationships were found among kinds for plant height, maturity, and test weight.

In their discussion, the authors suggested two possible reasons for the absence of significant yield differences:

1. The genetic factors for yield may have been uniformly distributed in the cross populations. This seems credible inasmuch as all of the parents were either regionally bred or adapted, implying common breeding goals, ancestors, and adaptive selection.
2. An extremely variable climate existed, which created wide yearly swings in performance and which confounded and masked possible yield differences.

Gilbert's (1958) "no nonsense" approach to the diallel cross is novel and might be said to have been written from a plant breeder's viewpoint; that is, it rhetorically raises the practical question, "Do diallel crosses help the breeder decide what to do next?" Gilbert was critical of certain assumptions and procedures associated with the diallel cross. He believed the performances of parental varieties to have substantial predictive value for the future behavior of their cross progenies, but conceded that diallel procedures added further information. Overall, he was concerned about the massive work load inherent in a breeding project that uses a given number of parents; a fractional diallel cross was suggested as a partial solution to reduce the number of possible crosses.

Johnson and Aksel (1959) reported a massive genetic study of the inheritance of yielding capacity, including attention to components of yield, in a 15-parent diallel cross of barley. The results are too complex to be treated here in detail, but the authors expressed three plant breeding implications:

1. The highest-yielding arrays and crosses were identified.
2. The number of kernels per head was identified as the most important yield component.
3. Scaling tests on individual crosses were able to detect significant nonallelic interactions. The presence of such interactions associated with a high-yielding cross should indicate crosses with the greatest potential for highest-yielding lines.

Overall, a wealth of information was presented that can be mined for relevance to parents, head type, components of yield, and location × year interactions. The authors commented that the results could be used to eliminate perhaps 95% of the material under consideration, principally on F_2 data. A study such as this, once done, leaves an important residue of information on parents that is analogous to a bank where future deposits and withdrawals are possible.

Lupton (1961) continued studies on cross prediction in winter wheat using a diallel series of crosses among six well-known cultivars. The value of parental combinations was assessed through the first four generations. In some cases plant density was a variable. Estimates of the components of yield were made in the F_1 and F_2. The data were presorted by arranging an array for each character; an array was one parent and all crosses made to it. Graphs were constructed to show the regression of the (covariance) nonrecurrent parents against the variance of the array. The F_3 and F_4 were graphed according to percent of the cultivar 'Cappelle'.

The detailed analyses may be summarized in a brief synopsis:

1. Predictions made in F_1 and F_2 were in good agreement with those made from the F_3, less so with the F_4.
2. F_1 and F_2 trials may be used to eliminate poorer crosses.
3. The procedures are valid.

Overall, Lupton conceded that analysis of cross or component yields alone would have given somewhat similar results, and that the time and labor involved are large. The use of an incomplete diallel series was suggested to reduce the number of crosses required.

Leffel and Hanson (1961) used the diallel cross technique to assess the usefulness of parent and early-generation progeny performance in order to identify the most promising crosses to work with in soybean breeding. Using 10 cultivars and their 45 diallel crosses, they were able to look at performances of the F_1 and F_2 in both spaced and bulked format, and of the F_3 bulked and in line generations. They found that parents made both general and specific contributions. Excepting the F_1, all generations studied showed ability to predict the value of crosses ("value" referred to useful lines obtained from the F_3). Performance of both soybean parents and bulk hybrid populations were considered reliable predictors. F_2 and F_3 bulk performances were highly correlated. In general, F_2 spaced tests gave indications of cross performance similar to F_2 or F_3 bulk generations.

OTHER APPROACHES TO PREDICTION

Sampson (1972) found that yield data from spaced F_1 plants were essentially valueless as a predictor of F_2 and F_3 performance. To a lesser extent, this was true also of midparent yield means. He provided a logical basis for choosing breeding strategy: in discussing early-generation testing he pointed out that testing costs are the same to measure "the agronomic worth of 21 lines from 1 cross as to measure 21 different bulk crosses." Thus, although acknowledging that this ignores variation within crosses (an important point), he advocated choosing among crosses.

St.-Pierre and Jensen (1972) practiced bidirectional selection for yield of large and small seed per F_2 plant in a 15-cross-population diallel of six barley parents. Five statistics were derived from the subsequent data and organized into a key to chart automatically decisions for cross selection during early generations.

Meredith and Bridge (1973) found early-generation correlations between F_2 and F_3 progenies of cotton, visually chosen for high yield from diallel cross of five cultivars or lines, to be nonsignificant and thus not helpful in choosing among crosses. However, the progeny tests gave significant correlations for the traits of lint percent, seed index, and fiber length and strength.

Performance of the 240 progenies was followed to the F_7, at which time there remained only one selection superior to 'Deltapine 16'. This selection came from a cross of Deltapine 16 and 'Coker 4104', which not only were the two highest-yielding parents but also produced the highest-yielding F_2 hybrid.

Keydel (1973) found that a progressive set of screening procedures involving winter wheat parents, F_1, an F_2 drilled trial, and up to 12 component properties of yield, could "effect a preliminary selection of the promising hybrid combinations with a satisfactory degree of reliability." Of perhaps more immediate importance, the mean yield of two parents was significantly correlated with F_1 yield (at 0.25*), and the correlation, not significant, between the F_1 and F_2 yields was 0.36.

In an exhaustive study of an eight-parent diallel set of spring wheat crosses, Bhatt (1973) analyzed the F_1 and F_2 data for cross prediction using three methods. There was good agreement among the methods in classifying the 28 crosses for potential yielding ability. Bhatt was able to identify 10 crosses as high potential, 9 medium, and 9 low. His suggestions for procedure were (1) discard the low potential group, (2) use a form of mass selection shown to be associated with high yield, for example, large seed size, with the hybrid bulks, and (3) select single plants at a later date.

All prediction proposals must be subjected to the exigencies of an operating breeding program. All require additional effort, much of which is time-intensive and labor-intensive. Bhatt's study occupied two years and was of a predictive nature, inasmuch as the projections can be verified only some years later when the merits of selections become apparent. The results were a discard of approximately one-third (nine) of the crosses examined, and a gain in knowledge about eight genotypes, valuable for future use. All breeders would recognize the interest and worthwhile nature of this as a research study, but the practical question is whether it is in a form that can be embraced readily as an integral annual part of a breeding program that may consider 20–50 potential parents a season. Bhatt recognized this concern and suggested ameliorative modifications.

Thorne (1974) proposed an early-generation testing scheme that embraced 31 soybean crosses, established equal-size populations of each, visually selected varying numbers from every cross, and finally, from a yield nursery selected the highest yielding regardless of parentage or cross. The procedure is interesting because it employs visual and testing selection for yield, is cross-neutral (that is, the initial treatment disregarded parentage), and in the end, objective testing criteria prevailed. The result is a scheme that shows a discriminant function at two levels, first in visual selection and second in yield testing, that identifies the better crosses (on the basis of producing most productive lines). From this the better parents can be identified by deduction.

There is a question, illustrated in this study, that is critical to the successful use of a selection system: is the system attractive to the breeder? A breeder wants to see ongoing productivity throughout the program, and would be attracted to a system that is producing a harvest of selected genotypes while it evaluates itself, in contrast to one that does only the latter. In this case 3100 lines from 31 crosses were reduced to 153 partially tested selections. In the process all of the 31 crosses were sampled and the 153 selections showed a range of potential differences among the crosses; for example, six of the crosses produced 10 or more superior lines, whereas 13 produced less than five, and four produced none. The breeder might, on the basis of these small samples, return to the better crosses for further exploitation. An analysis of parent contribution could also be made. In addition to these options advanced lines would already be in hand; in Thorne's study there were 153.

Thorne also analyzed different progeny correlations and found F_3-derived and F_5-derived lines to be positively and significantly correlated for yield in 13 out of 16 cases.

Stuthman and Stucker (1975) made a diallel cross of 12 superior oat parents, obtaining 64 of the 66 combinations. A completely random selection procedure, involving single seed descent, advanced 10 lines from each cross to the F_4 where the plants were advanced to F_5 rows for seed increase. The 640 lines were studied only for yield in two years of testing. The authors attempted to rate crosses according to predictions based on parental information followed by analysis of superior lines obtained. In brief, none of the criteria (parent yield, progeny means, rank order) did well in predicting the later-determined value of the crosses. There was a small correlation between parent rank for yield and progeny mean ranks—the top and bottom ranks were the same, but there were many reversals in between. Furthermore, the source of the best and most superior lines could not have been predicted from parental information. For example, the two parents ranking 10–11 (a tie) and next to last for yield ranked 3–4 and 5–6 for producing number of superior lines, based on yields 5% above the best parent. The authors concluded that prediction of superior crosses from parental information above was ineffective.

Using dry beans, Hamblin and Evans (1976) assessed the value of bulk early-generation yields and midparental yields as predictors of cross yields in later generations. It was found that density, whether spaced or crop density, was crucial. In spaced-plant situations, in both midparental and bulk early-generation tests, rare significant correlations between predicted and observed values were obtained; however, in crop density situations almost all correlations were significant in the same kinds of tests. Their conclusions were that at crop densities midparental yields were good predictors of cross yields, and early-generation cross bulk yields were good predictors of later cross potential.

Jinks and Pooni (1976), using the single seed descent method to advance generations, generated homozygous lines from a cross of two pure lines in a study concerned with the probability that inbreds would fall above or below the height level established by P_1, the tallest parent, or P_2 (lowest) or, if heterosis were involved, exceed the F_1 in height. The lines provided the necessary genetic components, family means, and variances, which allowed prediction of inbred line distribution in the general hybrid population. Perusal of these data would provide the breeder with a picture of the merits of a cross and would indicate whether it would be profitable to exploit the hybrid population beyond the F_2.

The procedures, insofar as they concern methodology, involved the parents, F_1s and F_2s, and represented a screening or ranking scheme to choose among many crosses those most likely to produce the greatest final number or proportion of superior genotypes. In the author's words,

> In general, however, a breeder is more likely to be concerned with the proportion of inbred lines that will prove to be superior to the initial pair of parents, or to their F_1 if it displays heterosis, than with the mean and variance of all possible inbreds. It would be more relevant, therefore, to predict the probability of producing inbreds which fall outside of the parental range or exceed the F_1.

Formulas and procedures were given. The authors believed that if suggested procedures were followed, "there is no reason why we need ever go byond the F_2 of an inbreeding programme without a fairly clear idea of the final outcome."

Grafius, Thomas, and Barnard (1976) chose eight spring barley genotypes that showed extreme manifestations of the components of yield, which are heads per area (X), kernels per head (Y), and weight (mg.) per kernel (Z), and which together represent yield (W). Five pairs of crosses featuring contrasting values of these traits were made; each pair set included backcrosses to both parents. Thus the total of crosses was 15. In the F_4 and BS_3, 20 random selections were taken from each cross population and increased two generation. For testing, 10 plots of each of the parents were added to each pair set of crosses. The components of yield were measured in this yield test.

The results showed that progeny mean yield for a cross could be predicted by heads per area × kernels per head; kernel weight seemed to be neutral as a predictor. Kernels per head carried more prediction weight than heads per area.

Progeny mean yields were predicted in three ways: (1) average yield of parents, (2) $X \times Y \times Z$, and (3) $X \times Y$. The coefficients of determination between observed and predicted W were 0.52, 0.45, and 0.68, respectively; thus the parents provided better information than all the three components, but not as much information as total seeds per area.

A number of outstanding lines were identified, and remember, these were from unselected random draws. The highest yielding were from the backcross populations (single backcross). It was concluded also that the highest-yielding segregants would come from populations with the highest mean yields.

A study by Habgood (1977) is of interest to breeders concerned with the problem of selecting parents and crosses. His procedures pit genotype against genotype in such a manner that the pairwise correlations for deviations from site mean yield may be displayed in a diallel table. One then looks for negative correlations, which indicate that the two genotypes did not respond similarly and thus would have a high probability also of being genetically different. Because the correlations are independent of the mean yields of the genotypes one must be careful not to introduce low-yielding parents into crosses. It does seem logical to work with high-yielding genotypes, especially if they also may be genetically different.

Cregan and Busch (1977) examined cross prediction for selection among crosses prior to line selection from the standpoint of the predictive value of bulks of crosses. They chose yield improvement as the criterion of response, and selected parents for crosses that had relatively high performances for yield. The early-generation bulk tests were supplemented for comparison purposes by midparent and F_1 yield measurements. A diallel of eight hard red spring wheats produced 28 crosses. From these, with their parents, were derived the F_1s, F_2–F_5 bulks, and the series of F_5 lines needed for direct comparisons.

Statistically significant positive correlations were found in all comparisons among the F_1, and F_2–F_5 bulks. All comparisons were made in the same year, thereby eliminating sequential year interaction effects. The best match for prediction purposes to choose best crosses involved the F_2 bulk; however, close behind were the mean of the combined F_2 and F_3 and the average of all F_5 lines. The result of these procedures was the identification of seven of the 28 crosses as the most likely to produce superior lines (which were directly measured and known). The prediction procedure worked best at the 25% selection intensity because beyond that, prediction differences became blurred. F_1 and midparent values were less efficient in predicting best crosses. It was clear from the authors' results that bulk hybrid yield tests can identify superior crosses. Against this must be balanced the risk, at 25% intensity, of loss from the 75% of crosses discarded, and the cost of the labor involved. This raises the question as to whether there ought not to be some form of selection applied to the unselected crosses before discard. In one sense, the answers to these questions depend on the size and scope of a breeding program—some programs are large enough in cross numbers that heavy discard can be absorbed without due concern.

Bassett and Woods (1978) outlined dual concurrent procedures to obtain inheritance information during the first stages of a breeding program in a study that addresses one of the important time and efficiency problems in plant breeding. The situation posed for the study concerned the difficulty of transferring a desirable quantitative character from a background of unknown genetic nature to accomplish an improvement in a desirable cultivar; the number and value of the quantitative genes are not known. The breeder has the choice of two options: proceed with an inheritance study, or proceed with a breeding program; both options are reasonable and are defensible. The authors offered a third option: do both simultaneously.

The method involved estimating the number of genes using the means of F_3 progenies of F_2 plants measured for the character in question (in this case, pod length of the common bean). The successful use of comparisons requires a good correlation between random F_2 and their F_3 progeny means, in this case, r = .82. The two parental cultivars chosen previously had given indications that more than two genes were involved in pod length inheritance.

In the first step the parents, F_1, F_2 and both backcrosses were grown, and pods were measured on each plant. The following year F_3 progenies of the 39 longest-pod F_2 plants, 41 random F_2 plants, and the long-pod parent, 'Sprite', were grown in a replicated trial.

The results of the experiment's first stage were ambiguous and inconclusive, the data not fitting any of the models used, although predominant additive genetic variation was suggested.

The second experiment was useful in that the authors found the best heritability indicator to be an adjusted correlation coefficient between F_2 and F_3; the adjustment corrected for the additional inbreeding of the F_2 to F_3, reducing an otherwise upward bias. This coefficient of selection was found to be a direct and reliable indicator of the value of any F_2 phenotype's F_3 progeny.

The use of the backcross generation data, which were required for some of the models, seemed to play a minor part in the analyses. Nevertheless, initially it seemed to offer help toward the solution of the problem. It was deduced, however, that the long-pod trait was controlled by four or five genes, a situation that becomes more difficult for the backcross to handle. Moreover, the breeder wished not to use the backcross toward the long-pod parent because he wished to avoid that genetic background.

Procedures that produce information about the potential pitfalls and estimates of success are more acceptable to breeders if they are obtained while the practical breeding objectives of the cross also are being pursued. The most stimulating part of the paper was the estimation of the number of genes involved. It was found that pod length equal to Sprite was not recovered in any of the 425 F_2 progenies. The authors proposed identifying the F_3 line having the highest mean pod length, and selecting within it for the maximum pod length individual(s). These individuals could then become the parents for the first backcross to the desirable recurrent parent. In this way, the inheritance study blends with the completion of the first cycle of the breeding program.

The F_2/F_3 analysis and procedures just mentioned seem equally valuable for the beginning of a recurrent selection program. To keep a broad population variance, the intermating elements could be expanded to include not only the longest-pod F_3s but also backcrosses in both directions, crosses to F_1s, and so forth.

Nass (1979) published results of a study of early-generation evaluation of spring wheat crosses using the midparents, F_1, and F_2 yields. The methods were applied concurrently to two sets of about a dozen crosses between entirely different cultivars. First, large numbers of crossed seeds were obtained,enough to permit a yield test of three replications. This test included all hybrids and parents. The two highest- and the two lowest-yielding crosses in F_1 were selected from each set. Second, the F_2 of these four crosses from each set were planted in 5000-plant populations. In set 1, 190 random F_2 head samples were drawn from each cross. In set 2, all plants were pulled and "plant head weight" measured, and the highest-weight 10% were saved from each cross. Set 1 and set 2 plants were increased for seed for one year and tested for yield the following two years (nonreplicated). Finally, the parents and bulk F_2 populations of all crosses in both sets were planted in a yield test of four replications.

Within the same year, the only direct comparison was between the F_1 and midparent yields: in set 1 a statistically significant correlation was found, but not in set 2. The correlation between the F_1s and their midparent yields in different years was not significant in either set. Highly significant correlations were found between the bulk F_2 and midparent yields in both sets taken in the same year; there were nonsignificant correlations in both sets when taken in different years. The correlation between F_1 and F_2 yields in different years was significant in both sets. Finally, the correlation between parents in different years was statistically significant in set 1, but not in set 2, where disease severity was responsible for ranking changes.

Turning to the two high-yielding and two low-yielding crosses in each set, selected on the basis of F_1 yield and tested in F_4 for two years, the mean yields of the high crosses exceeded that of the low crosses, 611–541 g/plot (set 1) and 572–494 g/plot (set 2), and 592–516 g/plot combined, respectively.

The final assessment in the study was measuring and counting the number of F_4 lines yielding in the top 10% in the four crosses (two high, two low) in each set (set 1: random draw each cross; set 2: yield-measured selection each cross). The results were as follows:

1. The number of high-yielding lines was substantially greater in the "high" classes than in the "low," overall 146 to 40.
2. The percentages were remarkably similar in both sets: 77.6% "high," 22.4% "low" in set 1, and 79.0% "high," 20.8% "low" in set 2.

General observations about this study highlight the confounding effects of years (not always avoidable) on data variance. Extra effort was required to obtain larger stocks of F_1 seeds. The time–cost–effort balance of these procedures vis-à-vis the whole program must be considered. From this study, however, Nass recommended a progressive sequence of screening tests for superior crosses that involved midparental yield followed by F_1 and F_2 yield tests. A second suggestion was to use midparent yield followed by F_2 bulk yield tests. The two suggestions are quite different. If confidence can be placed in midparent value (and this may depend on the conditions of the test, suggesting that sometimes special attention should be given to reduce genotype × environment effects) then, of course, it would not be necessary even to make many of the crosses to obtain the high seed numbers needed for the F_1 test. Perhaps a combination of the two suggestions is feasible. Midparent tests could be used in advance to screen and identify the crosses to be made (a continuing annual program), followed the next season by same-year midparent and F_1 yield tests of the selected crosses for further elimination. An optional procedure would be to eliminate the population or, if it is continued, use F_2 yield tests. An important point made by Nass was to use only larger, well-filled F_1 seeds because of the known effect of seed size on productivity. This paper is valuable because it integrates the separate problems of choosing parents and choosing crosses.

I was concerned about one important aspect of plant breeding relations that was not addressed in this study. A fundamental distinction between set 1 and set 2 populations from the two highest- and the two lowest-yielding of the 13 crosses was that head selections in set 1 were randomly chosen whereas in set 2 the selections were the highest 10% chosen on the basis of individual plant head weight. There were, first of all, fewer random than yield-selected lines, 190/5000/cross verus 10%/5000/cross, respectively. Nass did not comment on the outcome of these two types, random and test-selected, in which the results, both same year and different years, showed essentially no average differences in yield. There are two diametrically opposed implications, however, depending on the way in which random selection of the 190 heads per cross was obtained. If the selection were truly random, for example, either plants unseen or every *n*th individual chosen rigidly—and this is the choice that the reader must accept lacking further elaboration—then the implications are devastating: random selection was just as good as a measured yield procedure! On the other hand, if there were any relaxation of the random selection, for example, by visually choosing the plants or heads, then the implications of equal merit are refined to a comparison of visual versus measured yield discrimination, a result that may be seen as mutually favorable or at least nonderogatory to either.

I think it fair to say that computer-simulated results lack the cachet of experimental results, principally because they are not real and are based on assumptions. But, playing the devil's advocate, one could argue that real-time experimental results, as from a yield nursery, sample only one combination of environmental factors per test, and such results are not considered reliable until several have been collected and means and variances established. If we then argue that assumptions are derivatives of real-test parameters, who is to challenge the validity of computer-assisted results? I use this circular reasoning to introduce two papers whose author has attempted to bridge the gap between theory and practical application. Rosielle (1983, 1984) is an experienced breeder who used biometry and computer simulation in these studies to evaluate breeding and selection procedures. He asked the general question of how to identify, among a large number of bulk hybrid populations, which ones and how many should be continued. The nature of his 1983 simulation programs and results may be summarized as follows:

1. The problem of cross prediction is very real, particularly for breeders with an active hybrid-generating program. It is easier to generate hybrid populations than it is to field-evaluate them. (Rosielle addressed this potential bottleneck in field procedures.)
2. Rosielle used the criterion of maximum potential gain, which he defined as "that which would be achieved if the particular bulk which maximizes genetic advance were chosen for selection." The body of the paper dealt with the measurement of this attribute. Illustrations were based on the application to a breeding program unit of 100 bulk populations, generation not stated, but implied preferably to be later than F_2.
3. The computer-generated data identified a selection intensity of 5% among bulks as optimal.

Rosielle's practical experience as a breeder surfaces in a short section near the end of the paper, easily overlooked, which could have been developed more fully with profit. He wrote of options, and it is from this that one deduces that the research in the paper is pointed to generations later than the F_2, more likely F_4. Rosielle refers to logical options that would be exercised before selection for yield (thus in F_2 and F_3): single-plant selection against obvious defects (or for obvious positive traits with continued bulking or rebulking implied).

Drawing upon my own experiences in handling bulk populations, for example, of wheat, approximately 2000 annual crosses would be reduced to 400 first-year F_2 bulk populations, that is, bulked if seed amounts were large enough to justify machine sowing. These bulk populations were reduced, by discard or combining, at about the following rate: F_3 200 bulks, F_4 100 bulks, and F_5 50 bulks. The 50 F_5 populations were sown in the special twin-row mechanically space-planted field arrangement described elsewhere for plant selection (actually head removal for head rows). Rosielle's selection intensity of 5% as optimal seems overly severe applied to early truncation. Early visual screening and discrimination must play a part in reducing the F_2 and F_3, but even at the F_4, practically, it would be difficult to settle for the best five statistically determined bulk population selections out of a hundred. An acceptable mix might be 10 (including the five best) plus an additional 10 chosen by the breeder for other reasons.

Rosielle not only evaluated the bulk populations but also samples of lines from them, in order to gauge the output of superior segregants. This has led me to think of a procedure, which I did not think of using in my programs, but which might have merit in any bulk population program similar to mine. I recall that even in F_5, when lines were being withdrawn for head rows, there were bulk populations which, when evaluated by a series of spaced individuals—and thus seen in a new light—were discarded without any selections being made. These undesirable bulk populations could have been detected earlier and labor saved through their immediate discard if there were in place a special procedure of each year adding small sample row lengths of every F_3 bulk population to the standard F_5 twin-row spaced nursery. These would be for visual examination only, and of course would not affect the standard F_3 bulk nursery. If the breeder did not like the segregation pattern seen in the spaced plants, that population could be immediately marked for discard on the F_3 bulk population field plan, with a time saving of two years.

Rosielle's 1984 paper, also a computer simulation study, explored the efficiencies of bulk versus direct-line testing. "Direct-line testing" can be understood from a description of the two methods. The bulk testing procedure evaluated 100 populations, selected five, and in the following year evaluated 20 lines from each. In direct-line testing, two versions were tested: (a) one line from each of the same 100 populations were drawn and evaluated, and (b) 20 random lines each were drawn from a random draw of five of the 100 populations. Whether or not the direct-line procedure is relevant to breeding and selection methodology is not clear—suffice it to say that it was inferior to bulk testing in almost all comparisons, except when the bulks were very similar.

DISCUSSION OF PREDICTING AND CHOOSING AMONG CROSSES

Calculations that show the rarity of a single plant, containing all the favorable alleles, being found in hybrid populations of hundreds of thousands of individuals are sometimes used as arguments for large F_2 and F_3 populations, and for intensified selection procedures to identify such plants. Although these efforts are laudable and worthwhile, breeders should not lose sight of the general experience that the perfect plant is seldom if ever found. Therefore the quest for this elusive grail need not encumber the breeding program unnecessarily. The task of working with F_2 populations might equally be thought of as searching among crosses for the better ones, with a view to concentration on these elite crosses even to the extent of returning to the sowing of reserve seeds, and as discarding to make room for more crosses. Seen this way, searching among crosses is in direct conflict with the general use of large F_2 populations which inevitably are bound to influence and set the subsequent scale of the breeding project's operations. This introduces a "have your cake and eat it too" thought: holding in reserve the large F_2-embryo seed stocks from all crosses while small pilot F_2-sample nurseries are run for assessment.

Without questioning Harrington's 1932 judgment that one season of comprehensive F_2 analysis is preferable to several seasons of processing a hybrid population, there still remain questions in my mind about the suitability of this particular analysis in a breeding program. There is the question of breeding programs that process many cross populations—could they absorb this procedure? There is the matter of the detailed work inherent in the analysis itself. There is also the matter of what use can be made of the information derived from Harrington's study. Early on it became apparent that his numbers projected a dismal prospect; for example, only seven promising lines were predicted from 36,800 plants. The fact that six promising lines were obtained proved the power of the prediction procedure. Of course, Harrington had to complete his study because of the research objectives; however, the practical use of such predictive knowledge would lead to discarding the population. Because most breeders have options of choice among many hybrid populations, a predicting mechanism must be fast and easy and yet make the odds more favorable when the plant breeder makes a decision as to what level of interest and effort to devote to a population. I do not even rule out visual analysis for experienced breeders, but F_2 analysis does not seem to offer a solution that fits the problem.

Indeed, it is important to bring visual judgment early into the among-cross analysis of value. Moving too quickly into sophisticated procedures on the basis, for example, of theoretical considerations of cross value can be counterproductive, and in many cases inefficient, for the practical reason that the true value of crosses is often beyond rigid prediction, because a high proportion of crosses fail to exhibit the segregational products expected. Therefore it is only prudent to evaluate a cross visually before committing major time, effort and money to it.

Hamblin and Evans in 1976 suggested that "F_2 crosses should be tested in bulk replicated yield trials, together with controls, which could be the parental genotypes, at crop densities and preferably at more than one site. Any low yielding crosses should then be discarded, so allowing the maximum of effort to be concentrated in succeeding generations on the more promising material."

Of course, if midparental yield is a valid predictor of cross potential it would be good strategy for a plant breeder to search comparable past nursery yield records of cultivars as a guide in planning crosses.

Harvest index as a selection tool is attractive to breeders because it offers the hope of bridging the gap between the frequent inverse relationship between performance (yield) in heterogeneous or heterozygous populations, often spaced, and that found in single-line populations, the latter usually at crop densities. Selection for yield in early generations cannot be relied on as a predictor of future performance. A number of studies have shown that harvest index does have the measure of stability to bridge the early generation–pure line gap. Odds are improved when it is used in conjunction with grain yield itself, and with measures of biomass growth rate and yield.

A breeding aspect of density that relates to selection method is this: whether the great heterotic effects of early-generation hybrid populations, which are known to be a factor along with spacing in the dilemma of selecting under spaced conditions for eventual performance under crop stands, can be damped or held in check by a restrictive level of high density in the early generations. I had assumed they could and operated my composite breeding populations at twice the normal seeding rate. I wanted to "force" the populations to pass through early generations in the crowded environment that was to be their future, should any become cultivars. I envisioned this heterotic force or vigor as still imprisoned in the plant which, however, because of crowding, must now find outlet in a changed structure, perhaps a Donald-like ideotype. The results might be seen as a favorable form of competition, an accommodation between heterosis and density influences. These procedures worked for me, but today I would be less sure of the competition reason. The more likely reason was that my use of the bulk hybrid population breeding and selection methods was based on processed, uniform-sized seeds and early removal of lines at F_4 or F_5, a generation point calculated to avoid any crippling effects of natural selection. This thought is developed at greater length in the earlier sections on the distinctions between the bulk population and the evolutionary breeding methods.

I believe plant breeders would be helped in their evaluation of crosses if we could create a new type of early-generation bulk population field plot. It would require a seed drill having a mechanism that operated from a logarithmic base. A standard plot area would begin with a seeding rate to produce a designated thin, widely spaced density. This rate would automatically and gradually increase to a designated thick, densely packed population at the other end of the plot. The breeder could see both the "exploded" individual aspects and the crowded aspects of the same population. If desired, selected individual F_2 or F_3 plants could be extracted for family-derived line tests. The overall yield of such plots would represent a compromise of density and competition interactions.

One answer to the question of how to rate and choose the elite crosses among all crosses hinges on some form of advance look at the populations, perhaps through the use of small samples with the remaining seeds kept in reserve. This seems to imply accepting, at some point in the program, a one-time program delay of one or two years; however, once put in place the entire program would proceed as before with one major exception: only the elite crosses would advance byond this point. The early sample nursery would have a positive aspect also in that there would be available in the selected crosses small pilot populations, or individuals, already one generation advanced, which would be useful for testing and monitoring the continued worth of the cross.

PREDICTING AND SELECTING LINES

It seems logical that the subject of selection of lines follows the choosing of parents and the selection of crosses with which to work. Line selection usually is thought of as deliberate choices of attractive individual plants that are approaching or have reached homozygosis. A "working homozygosis" level usually begins about F_4. Lines can be withdrawn at any F level beyond that. Nevertheless, line selection in a broader sense begins with the F_2 where individuals and family lines are started.

The following options are among those the breeder has available as he moves toward line selection from his crosses:

1. *F_2 Selection.* Selection procedures are usually active and positive; that is, specific desired plants are chosen. Selection also has a passive and positive component that manifests itself in the breeder's awareness of the importance of *discard*. Discard is passive when no further action is required once the selections of the desired individuals have been made and they have been removed from the populations. Discarding is positive because it brings gains in efficiency.

2. *F_3 Selection.* Basically the breeder deals with the consequences of decisions at the F_2. Examples of F_2 activities may include:

a. F_2-derived family lines: yield or other testing may be involved.

b. Pedigree (individual plant selection): this involves continued selection and discard.

c. Bulk population: options include growing the unselected F_2, now F_3 population, and growing the bulk of selected F_2 plants, which may be further divided into maturity, type, or disease reaction groups.

3. *F_4, F_5 and Generations Beyond.* This is considered to be the final line selection. All options end here once the choice of line withdrawal generation has been made.

Table 30.1 Cornell Oat Project, 1955 Cohort

		Number of lines		
Year	Nursery	Grown	Selected	Comments
1955	Head row	6440	2256	No yields taken
1956	Rod rows	2256	505	Yields taken
1957	Rod rows	505	233	Replicated, yields taken
1958	Rod rows	233	97	Replicated, yields taken
1959	Rod rows and plots	97	62	Replicated, yields taken
1960	Rod rows and plots	62	27	Replicated, yields taken
1961	(projected)	27	3*	

*Two varieties, Tioga and Niagara, came from this 1955 cohort of many crosses. It is to be understood that the project produced a new cohort, from different crosses, for each of the years listed; that is, there was a 1956 cohort, a 1957 cohort, and so forth.

Line selection identifies specific genotypes. All lines enter nurseries in the following season. In the cereal grains, these are usually head rows about 4 or 5 ft long that are observed by the breeder, who may make notes and collect data about them.

It might be helpful for those not actively engaged in plant breeding, and who therefore have not had the experience of viewing truncation or attrition in action, to see an example from the Cornell oat project in the 1950–1960s (see Jensen, 1961). In this illustration, the numbers began with the head row nursery of 1955, where each head row represented an individual plant selected the year before from a cross population. In Table 30.1, of the 1955 cohort, only the totals are given although many crosses were involved.

REFERENCES

Atkins, R. E. and H. C. Murphy. 1949. Evaluation of yield potentialities of oat crosses from bulk hybrid tests. Agron. J. 41:41–45.

Bassett, M. J. and F. E. Woods. 1978. A procedure for combining a quantitative inheritance study with the first cycle of a breeding program. Euphytica 27:295–303.

Bhatt, G. M. 1973. Diallel analysis and cross prediction in common bread wheats. Aust. J. Agric. Res. 24:169–178.

Cregan, P. B. and R. H. Busch. 1977. Early generation bulk hybrid yield testing of adapted hard red spring wheat crosses. Crop Sci. 17:887–891.

Fowler, W. L. and E. G. Heyne. 1955. Evaluation of bulk hybrid tests for predicting performance of pure line selection in hard red winter wheat. Agron. J. 47:430–434.

Frey, K. J. 1954. The use of F_2 lines in predicting the performance of F_3 selections in two barley crosses. Agron. J. 46:541–544.

Gilbert, N. E. G. 1958. Diallel cross in plant breeding. Heredity 12:477–492.

Grafius, J. E., W. L. Nelson, and V. A. Dirks. 1952. The heritability of yield in barley as measured by early generation bulked progenies. Agron. J. 44:253–257.

Grafius, J. E., R. L. Thomas, and J. Barnard. 1976. Effect of parental component complementation on yield and components of yield in barley. Crop Sci. 16:673–677.

Habgood, R. M. 1977. Estimation of genetic diversity of self-fertilizing cereal cultivars based on genotype–environment interactions. Euphytica 26:485–489.

Hamblin, J. and C. M. Donald. 1974. The relationship between plant form, competitive ability and grain yield in a barley cross. Euphytica 23:535–542.

Hamblin, J. and A. M. Evans. 1976. The estimation of cross yield using early generation and parental yields in dry beans (Phaseolus vulgaris L.). Euphytica 25:515–520.

Harlan, H. V., M. L. Martini, and H. Stevens. 1940. A study of methods in barley breeding. USDA Tech. Bull. 720, 26 pp.

Harrington, J. B. 1932. Predicting the value of a cross from an F_2 analysis. Can. J. Res. 6:21–37.

Harrington, J. B. 1940. Yielding capacity of wheat crosses as indicated by bulk hybrid tests. Can. J. Res. 18:578–584.

Immer, F. R. 1941. Relation between yielding ability and homozygosis in barley crosses. J. Am. Soc. Agron. 33:200–206.

Jensen, N. F. 1961. A case history of the plant breeding process. 1960 Oat Newsletter 11:61.

Jinks, J. L. and H. S. Pooni. 1976. Predicting the properties of recombinant inbred lines derived by single seed descent. Heredity 36:253–266.

Johnson, L. V. P. and R. Aksel. 1959. Inheritance of yielding capacity in a fifteen-parent diallel cross of barley. Can. J. Genet. Cytol. 1:208–265.

Keydel, F. 1973. Investigations on the yielding properties of winter wheat. Z. Pflanzenszuecht. 69:239–255.

Leffel, R. C. and W. D. Hanson. 1961. Early generation testing of diallel crosses of soybeans. Crop Sci. 1:169–174.

Lupton, F. G. H. 1961. Studies in the breeding of self-pollinating cereals. 3. Further studies in cross prediction. Euphytica 10:209–224.

Lupton, F. G. H. and R. N. H. Whitehouse. 1955. Selection methods in breeding of high yielding varieties. Heredity 9:150–151 (abstr.).

Meredith, W. R., Jr., and R. R. Bridge. 1973. The relationship between F_2 and selected F_3 progenies in cotton (*Gossypium hirsutum* L.). Crop Sci. 13:354–356.

Nass, H. G. 1979. Selecting superior spring wheat crosses in early generations. Euphytica 28:161–167.

Rosielle, A. A. 1983. The effect of variation of genetic variance and correlation between mean and variance on efficiency of bulk yield testing. Euphytica 32:49–56.

Rosielle, A. A. 1984. A comparison of bulk testing and direct-line testing in self-fertilizing crops. Euphytica 33:153–159.

Sampson, D. R. 1972. Evaluation of nine oat varieties as parents in breeding for short stout straw with high grain yield using F_1, F_2 and F_3 bulked progenies. Can. J. Plant Sci. 52:21–28.

St. -Pierre, C. A. and N. F. Jensen. 1972. Evaluating the selection potential of crosses of barley. Can. J. Plant Sci. 52:1029–1035.

Stuthman, D. D. and R. E. Stucker. 1975. Combining ability analysis of near homozygous lines derived from a 12 parent diallel cross in oats. Crop Sci. 15:800–803.

Thorne, J. C. 1974. Early generation testing and selection in soybeans: association of yields in F_3- and F_5-derived lines. Crop Sci. 14:898–900.

Whitehouse, R. N. H. 1953. Breeding for yield in the cereals. Heredity 7:146–147.

31

HYBRIDIZING OR CROSSING TECHNIQUES

STERILITY AND THE NATURE OF NATURAL CROSSING

Among cereal crops oats seems to experience more damage from sterility and lower seed set than other small grains. Oats apparently is more vulnerable to interactions with diseases and environmental changes, such as wetting or drying, affecting pollen viability or stigma receptivity. Elliott (1922) studied this problem, suspecting halo blight, but concluded instead that much of the sterility observed in oat fields was the result of too much moisture about the panicles.

A most exhaustive treatment of natural hybridization of wheat was written by Clyde Leighty and John Taylor (1927). They considered that the phenomenon was environment-sensitive, occurring more often in certain years, and was associated more with some varieties or species than others. From a breeding or genetic study standpoint, precautions had to be taken to minimize the consequences of natural hybridization.

The proclivity of certain genotypes to natural crossing in the field is nowhere better illustrated than in the case of 'Chinese Spring' wheat. Leighty and Sando (1928) documented the behavior of this peculiar wheat. In a preliminary study of crossing between eight wheat and two rye varieties grown in alternate wheat–rye rows, they planted the seeds from individually harvested wheat plants the following year and not a single wheat–rye hybrid was found among the approximately 20,000 plants. This would seem to settle the question of natural crossing; however, a year later it was observed that a single row of the Chinese wheat (C.I. 6223) showed 18% wheat–rye hybrids—the preceding year it had been grown contiguous to a large block of different ryes. This discovery revitalized the research on these hybrids. A repeat experiment resulted in 19% wheat–rye hybrids. Additionally, wheat was manually pollinated with rye pollen and a seed set of more than 90% obtained. The uniqueness of this figure may be gauged by wheat–rye manual crossing experience using other wheat genotypes: there the expected seed set had been 1–1.5% seed set. Manual crossing of Chinese Spring by another wheat yielded 89% seed set; hence the Chinese wheat crossed as readily with rye as with wheat.

This research is quite familiar to me inasmuch as wheat–rye research at Cornell University was substantial for decades, extending into the 1950s. I met both Clyde Leighty, who would stop to visit Dr. Love at Cornell, and W. J. Sando in Beltsville.

POLLEN CHARACTERISTICS AND PRODUCTION

Pope (1939) showed that barley spikes containing ovules and spikes with anthers and pollen could be removed from plants at appropriate stages of maturity and stored for substantial periods of time. This made it possible to hybridize genotypes that flowered at different times in the field. The maximum safe storage of pollen was found to be 26 days at 36°F; the maximum elapsed time before pollination for meaningful production of seed on stored emasculated spikes, placed in water in test tubes, was about the same, 42 days at 36°F.

Coffman and Stevens (1951) found that oat plants regulated their peak pollen-shedding periods to avoid extremely high temperatures. Adjusting crossing procedures on very hot days to this observation, they were able to increase seed set from 5.4% in the afternoon to 24.8% in the early evening.

The quantity of pollen production by a plant is important in the self-pollinating cereals, but is generally of little consequence as long as the cultivar yields consistently well. For hybrid wheat production, sufficient pollen is crucial, but there are certain self-fertilizing situations where it is also important. One case is in seed increase, where a potential new cultivar may be taken hundreds or thousands of miles into a different environment in order to obtain a certain advantage, for example, an over-winter increase of a spring-sown cereal, or a disease-free environment. We encountered this situation in the Cornell oat program when, in the early 1950s, we attempted Arizona winter increases of two candivars, one of which became 'Tioga', and the other of which was a Victoria blight-resistant selection from 'Garry'. Both attempts were near disasters, especially the Garry Sel. 5, which grew 9 ft high and barely returned the planting seed.

A second case is in the breeding of cereals, where breeders learn from experience that different genotypes or cultivars are "good" male parents because of obvious high pollen production, and that others with poor pollen production should always be used as female parents. To illustrate from my experience with oats, 'Rodney' was a heavy pollen producer, a good male, and Tioga a poor pollen producer, a poor male and always to be used as a female parent.

De Vries (1974) showed variety differences in anther length and number of pollen grains per anther, and listed citations to research by other scientists.

EXTENT OF NATURAL CROSSING

H. K. Hayes (1918) determined that 11 of 880, or 1.37%, of common wheat plants examined in 1917 were natural hybrids, and it was concluded that, with doubling, 2–3% of natural crossing had occurred.

Garber and Quisenberry (1923) made a progeny analysis of more than 3000 head selections from "nineteen impure varieties of winter wheat grown in rod rows." Unfortunately the basis for head selection was not given, so we do not know whether the draw was random or subjective. The final results showed less than 1% natural crossing.

Oats is one of the most closed of self-pollinating species. Stanton and Coffman (1924) measured the amount of crossing among black- and white-kerneled cultivars in adjacent rows and found an average of 0.36%, which logically is doubled to 0.72% on the reasonable assumption that within-row sibbing would be occurring at a comparable rate. The most susceptible variety had about 2% natural crossing.

Self-pollinated flax might be considered vulnerable to cross-pollination because of the openness of its flowers. Henry and Tu (1928) found from progeny tests of seeds harvested from a white-flowered type grown in rows 1 ft from a blue-flowered kind that there was 1.25% natural crossing (or 2.50% when factoring in the assumed equivalent crossing between white × white). the following year the count was 1.71% (or 3.42% doubled). A linear decrease in observed natural crossing was found when distances between rows were increased. The authors observed and considered thrips a possible agent of natural crossing.

Using hull color differences (white and black) Stevenson (1928) measured the amount of natural crossing between eight cultivars of barley over a period of three years. From an examination of more than 10,000 plants it was found that natural crossing in Minnesota was very low, ranging up to only 0.15% (or doubled, 0.30%) when it occurred. It was closely associated with cultivar, some varieties showing no natural crossing at all.

Harrington's (1932) paper contained a good review of previous knowledge about natural crossing in the small grain cereals. His studies spanned a period of five years with wheat and oat tests and three years with barley, and are noteworthy in that progeny tests were required of all suspected individuals. This procedural requirement made it possible to distinguish between hybrids and mere admixtures. The scale of the project was massive:

in round numbers of plants examined there were 77,000 wheat, 27,000 oats, and 30,000 barley.

The procedures centered on growing alternate rows of selected pairs of varieties in which hybrids that occurred could be identified by segregation products. Seeds of each variety were harvested in bulk from center parts of the block and space-planted the following year, and any suspected hybrids removed and progeny tested the next year.

The general results at the end of five years were as follows (the observed results were doubled to include the expected within-row sibbing, which was not detectable):

	Mean (%)	Range (%)
wheat	0.88	0.00–2.16
hulled oats	0.07	0.00–0.20
hulled barley	0.07	0.00–0.17

The observed exceptions were fully as important. Species and morphological differences were many; for example, hulless oats and barley showed higher natural crossing. Varietal differences were great; for example, hull-less 'Liberty' oats showed 9.82% natural crossing one year, whereas 'Gold Rain' oats showed none in three years. Natural crossing was higher in wet seasons for wheat and barley, but oats showed no unusual response. Some hybrids found were outcrosses to some other parent outside the test block.

The effect of distance on natural crossing was found to be linear—fewer natural crosses with increased distance. Harrington found the vast majority of hybrids could be traced to adjacent row crossing but there were always a few where the putative parent was from 7 to 30 ft distant. In the production of pure seed the isolation distance between fields of the same crop to prevent mixtures and hybrids is of great practical importance. Harrington advised a distance of at least 10 rods.

The phenomenon of natural crossing occasionally can be turned to good advantage. It is important for an oat breeder to know of varieties, such as Gold Rain, that seldom outcross. At the other extreme are genotypes that invite outcrossing and are good hybrid producers. The promiscuous proclivities of Chinese Spring wheat are well known. In my Cornell project I had certain semidwarf wheat lines, of 'Gaines' × other percentage, which could not be kept pure in normal nursery growing conditions. The outcrossing was so marked that these lines were continued in larger plots from which head and plant selections were extracted annually and bulked for growing and later pure line selection.

CROSSING TECHNIQUES AND SEED SET

The difficult problem of obtaining hybrid seeds in oat crossing was eased through the use of the "approach" method, first reported by Rosenquist (1927). The technique involved positioning adjoining pots so that the heads of the plant to be used as male were in a slightly superior position to the emasculated florets on the female heads. A large glassine bag was placed over the panicles of both parents. With a reasonable match of maturity, gravity pollen shed fertilized the female florets.

I have discovered the source of the "twirling" method of hand pollination, which was shown me long ago by Gus Wiebe: it was probably derived from Rosenquist's paper described above, and from Pope's (1933) article, which described how emasculated and receptive female spikes were pollinated by tapping them with a chosen excised male head, producing a pollen shower. It is a small step from tapping to twirling.

The well-known difficulties encountered in hybridizing oats were described by Coffman (1937). He presented the sum of his 12 years of experience crossing oats, with details on techniques.

Pope's (1944) article is a primer on barley growing and hybridizing techniques. Although written after the discovery of the genetic male sterile genes, the article deals only with hand emasculation and pollination of normal genotypes.

Hadden (1952) achieved unusually high seed set in crossing oats. Although many breeders emasculate one day and pollinate the next, Hadden found that in the South Carolina climate this produced, for example, only 2.7% seed set. If, however, he followed morning emasculations with pollinations about mid-afternoon of the same day, the observed natural pollen-shedding time, the hybrid seed set rose to 69.5%. Oat breeders will also note that more than half of the hybrid seeds obtained came from pollinating secondary florets.

Coffman (1956) found that bagging panicles immediately after field pollination resulted in a decrease in percent seed set. An increase in temperature under the bags is a possible cause of the reduced seed set.

Brown and Shands (1956), using standard hand crossing techniques, evaluated several variables that might affect the seed set of oat hybrids. The following observations were recorded:

1. Experience in crossing ranked high as influencing seed set; experienced workers had consistently higher seed sets.
2. The choice of which of the two parental varieties should be used as female in a cross played an important role; for example, 'Rodney' had higher seed sets than 'Vicland' (Rodney coincidentally is known to be outstanding in pollen production for use as a male parent).
3. Temperature was found to have an inverse relationship to seed set.

Maximum temperatures lowered seed set, whereas low field temperatures aided seed set.

4. Little difference was found between morning and afternoon pollinations.
5. A two-day interval between emasculation and pollination gave highest percent seed set, with one-day and three-day intervals close behind.
6. Short intervals between emasculation and pollination favored seed set on top florets whereas longer intervals favored the bottom florets.
7. Experience with the use of stored pollen was promising but further study would be required.

Two observations on Brown and Shand's report are that the position-of-floret effect is probably related to the age differences of panicle florets, which generally bloom from the top downward, and that although the variety used as male was not explored, my own experience has been that there are very great differences in seed set percentage depending on which way a cross is made—so great that I would never use certain varieties as males. The differences seemed to be related to the pollen production capabilities of different genotypes.

Wells and Caffey (1956) showed a labor-saving technique for crossing: scissor emasculation, which eliminated the tedious removal of separate anthers in wheat and barley.

Recognizing the difficulties in obtaining seeds in oat hybridization, Coffman (1961) detailed a field preplant germination procedure that increased the number of surviving F_1 plants by 10–25% over direct field planting. The difficulties referred to, experienced by most oat breeders, are serious enough so that the additional effort involved in this technique could be absorbed readily.

Marshall (1962) countered the debilitating effects of high temperatures and low humidity on hybrid seed set in oats by using a combination of shading and wet pollinating bags. Glassine or other bags were immersed in water just prior to being placed over the pollinated panicle. He was able to show increased seed set from these procedures.

A discussion of several methods of hybridizing was given by Wells (1962). The techniques, which were intended as mass methods, included cone mass pollination, male sterility, and scissor emasculation.

The approach method has been used extensively by Hamilton (1953) with barley, Curtis and Croy (1958) with wheat, and McDaniel, Kim, and Hathcock (1967) with oats. The last-named authors showed the tremendous seed sets in oats obtainable through the approach method: the seed set in more than 10,000 florets subjected to pollination by this method was 63%. They showed that field-grown plants could participate in matings by cutting and placing a (male) panicle in a test tube filled with water and attaching this at the proper height to a stake by the female plant. Multiple female plants surrounding a single male are also possible.

HANDLING PEDIGREES AND POPULATION NUMBERS

Purdy, Loegering, Konzak, Peterson, and Allan (1968) outlined a method for presenting pedigrees in the small grain cereals that has since become the standard.

A warning against using too small populations was voiced by Bliss and Gates (1968) in remarks principally applicable to mass selection:

> Considerable efficiency can be gained in a breeding program if selection is practiced in populations of an optimum size. When populations are extremely small genetic gain may be restricted by random loss of favorable alleles, while very large populations within families may preclude the use of more families. Allard (1960) stated that no rules can be drawn as to the number of F_2 individuals and F_3 families that should be grown, but that few breeders use less than fifty F_3 families and usually grow many more. The ratio of F_2 individuals to F_3 families ordinarily varies from about 10 : 1 to 100 : 1.

The maize breeding program at Iowa State University reproduces and advances heterogeneous breeding stocks by hand random mating in a 500-plant sample. Pollination of every plant is made but since "pollen from one plant was used on only one or two other plants," the female representation was broader than that of the male. Omolo and Russell (1971) wished a more scientific basis for number of plants to be crossed. They formed random mating samples of 500, 200, 80, 32, and 13 plants per generation and cycled them through five generations, following which the populations were examined for genetic changes. The only character change that could be related to sample size was yield, which showed significant decrease with decrease in sample size. It was concluded that a practical compromise could be fewer than 500 plants per stock, and the figure of 200 was recommended.

A survey of Canadian wheat breeders by Hurd (1975) gave a picture of wheat breeding programs in quantitative terms. Nineteen breeders made from 2 to 10 crosses a year, five others made 25–200. F_2 plant numbers in a program ranged from one-third under 2000 to one-third between 10,000 and 20,000. One-third of the programs used the single seed descent method and one-third practiced early-generation yield testing. The mass pedigree method was used less.

Yonezawa and Yamagata (1978) used a statistical model to explore the question, given the restriction of total population size in a breeding program, what is the optimal relation between number and size of crosses? Their general conclusion was that "the number of crosses rather than the size of a cross should be increased if circumstances permit."

A general impression received from all population-size studies I have seen is that the numbers of individuals in populations are smaller than I would have anticipated. Although this probably means breeders work with larger populations than necessary, a conservative reaction would be to look at published and recommended numbers as minimum figures, subject to a cautionary expansion.

REFERENCES

Allard, R. W. 1960. Principles of plant breeding. John Wiley, New York.

Bliss, F. A. and C. E. Gates. 1968. Directional selection in simulated populations of self-pollinated plants. Aust. J. Biol. Sci. 21:705–719.

Brown, C. M. and H. L. Shands. 1956. Factors influencing seed set of oat crosses. Agron. J. 48:173–177.

Coffman, F. A. 1937. Factors influencing seed set in oat crossing. J. Heredity 28:297–303.

Coffman, F. A. 1956. Bagging possibly reduces success in oat crossing. Agron. J. 48:191.

Coffman, F. A. 1961. A method of obtaining more F_1 plants from hybrid oat seed. Crop Sci. 1:378.

Coffman, F. A. and H. Stevens. 1951. Relation of temperature and time of day of pollination to seed set in oat crossing. Agron. J. 43:498–499.

Curtis, B. C. and L. I. Croy. 1958. The approach method of making crosses in small grains. Agron. J. 50:59–51.

Elliott, C. 1922. Sterility of oats. USDA Bull. 1058. 8pp.

Garber, R. J. and K. S. Quisenberry. 1923. Natural crossing in winter wheat. J. Am. Soc. Agron. 15:508–512.

Hadden, S. J. 1952. Unusual success in crossing oats. Agron. J. 44:452–453.

Hamilton, D. G. 1953. The approach method of barley hybridization. Can. J. Agric. Sci. 33:98–100.

Harrington, J. B. 1932. Natural crossing in wheat, oats, and barley at Saskatoon, Saskatchewan. Sci. Agric. 13:470–483.

Hayes, H. K. 1918. Natural crossing in wheat. J. Hered. 9:326–330.

Henry, A. W. and C. Tu. 1928. Natural crossing in flax. J. Am. Soc. Agron. 20:1183–1194.

Hurd, E. A. 1977. Trends in wheat breeding methods. Can. J. Plant Sci. 57:313 (abstr.).

Leighty, C. E. and W. J. Sando. 1928. Natural and artificial hybrids of a Chinese wheat and rye. J. Hered. 19:23–27.

Leighty, C. E. and J. W. Taylor. 1927. Studies in natural hybridization of wheat. J. Agric. Res. 35:865–887.

McDaniel, M. E., H. B. Kim and B. R. Hathcock. 1967. Approach crossing of oats (*Avena* spp.). Crop Sci. 7:538–540.

Marshall, H. G. 1962. Effect of wetting and shading bags on seed set of oat crosses. Crop Sci. 2:365–366.

Omolo, E. and W. A. Russell. 1971. Genetic effects of population size in the reproduction of two heterogeneous maize populations. Iowa State J. Sci. 45:499–512.

Pope, M. N. 1933. A rapid method for making small grain hybrids. J. Am. Soc. Agron. 25:771–772.

Pope, M. N. 1939. Viability of pollen and ovules of barley after cold storage. J. Agric. Res. 59:453–463.

Pope, M. N. 1944. Some notes on technique in barley breeding. J. Hered. 35:99–111.

Purdy, L. H., W. Q. Loegering, C. F. Konzak, C. J. Peterson, and R. E. Allan. 1968. A proposed standard method for illustrating pedigrees in small grain varieties. Crop Sci. 8:405–406.

Rosenquist, C. E. 1927. An improved method of producing F_1 hybrid seeds of wheat and barley. J. Am. Soc. Agron. 19:968–971.

Stanton, T. R. and F. A. Coffman. 1924. Natural crossing in oats at Akron, Colorado. J. Am. Soc. Agron 16:646–659.

Stevenson, F. J. 1928. Natural crossing in barley. J. Am. Soc. Agron. 20:1193–1196.

Vries, A. Ph. de. 1974. Some aspects of cross-pollination in wheat (*Triticum aestivum* L.). 3. Anther length and number of pollen grains per anther. Euphytica 23:11–19.

Wells, D. G. 1962. Notes on the hybridization of wheat and barley. Crop Sci. 2:177–178.

Wells, D. G. and H. R. Caffey. 1956. Scissor emasculation of wheat and barley. Agron. J. 48:496–499.

Yonezawa, K. and H. Yamagata. 1978. On the number and size of cross combinations in a breeding programme of self-fertilizing crops. Euphytica 27:113–116.

SECTION V

PLANT TRAITS OF SPECIAL RELEVANCE

The overwhelming importance of seed characteristics to plant breeding, selection, and methodology can scarcely be doubted. For that reason Chapter 32 is devoted solely to this subject. Chapter 33 deals with a few other traits whose characteristics also have a high association with methodology.

32

SEED CHARACTERISTICS

The attention seed size merits today in relation to crop improvement is admittedly minor, yet it is interesting to note how completely it occupied the attention of early researchers. In the very first issue, Volume 1, 1907–1909, of the Proceedings of the American Society of Agronomy, no less a person than W. M. Jardine of the Bureau of Plant Industry, U.S. Department of Agriculture, took as his subject, "Methods of studying the relative yielding power of kernels of different sizes," and the article preceding his in the proceedings was by C. A. Zavitz (1908). Zavitz reported that, without exception, in six to eight years of testing oats, barley, field peas, and spring and winter wheat, large seeds produced higher average yields than medium or small seeds, specifically in 37 out of 40 tests. It should be noted, however, that tests were on equal numbers of seeds per standard area. Other and later tests have shown that with equal weight of seeds sown at crop densities there is often little difference in yield.

Probably no other aspect of the cereal grains has been studied as long and as thoroughly as the nature of seeds and seedlings, and one is well entitled to want to let the matter rest. Nevertheless, there are aspects of seeds that continue to be very important under modern conditions and are of special importance to the plant breeder. Examples that come to mind are the influence and interaction of the seed drill on genetically or environmentally conditioned seed characteristics, the effect on the crop from planting seeds of different origins, the nature and magnitude of yield compensations occurring in plants grown from seeds of different characteristics, and the relation between equal weights and equal numbers sown.

For convenience, the subject of seed attributes has been divided into four parts, beginning with a section on seed size characteristics.

INFLUENCE OF SIZE

In six years of testing heavy, light, and nonprocessed wheat seeds, Georgeson, Burtis, and Otis (1896) found heavy and control seeds to give essentially equal yields, slightly superior to yields produced by light seeds. In a separate experiment with oats, using the same three categories of seed, they reported (1897) on an eight-year series of experiments in which heavy seeds were found to outyield unscreened seeds by 1 bu/acre and light seeds by 3 bu/acre.

Montgomery (1910) analyzed the results of 30 experimenters who had published on the seed value of light and heavy kernels in cereals. As a result of his evaluation he urged the general adoption of a standard method for making such tests, and it may be assumed that this proposal arose from the great variation shown in the results. To the extent that conclusions were possible, Montgomery stated,

> It appears that where equal numbers of large and small hand-selected seed were sown the advantage was in favor of the larger or heavier seed but no check with the original seed was included in these cases. Where machine separation was practiced, no marked variation in results has been secured.

Lill (1910) tested the germination of wheat seeds graded according to size and density. Grading according to size showed little correlation with germination, but grading according to density showed a direct correlation. It is interesting that density grading by projection was used before 1910. Lill stated, concerning the method he used,

> The separation of the wheat according to the density of the kernels was accomplished with a wind-blast grader—the theory being, and later by actual tests was proven to be a fact, that kernels of greater density would fall first, while kernels of lesser density would be carried farther in proportion to their density, so that by catching the falling kernels in divisions or boxes placed at certain distances from the blast they would be graded according to density, with the first grade nearest the blast.

Also in 1910, Waldron (1910), using correlation and regression in oats, concluded that plants grown from large grains produced higher yields owing to increased vigor and amount of food supply. It did not follow necessarily that they came from the best genotype.

Hutchinson (1913), in a spaced-plant study with oats, obtained positive correlations between weight of seed used and plant total weight and yield and number of kernels at maturity.

Williams and Welton (1913) conducted a five-year study of yields from large and small oat seeds when planted at equal weights and numbers per acre. Large seeds showed a substantial yield increase over small in both situations. In another four-year experiment large seeds were not higher yielding at either rate than unprocessed seeds, but small seeds gave yields equal to large and unprocessed seeds at equal weights sown, but yielded less when sown at equal numbers. This suggested that compensation occurs with equal weights sown (approximating farm drill dispensing of seeds).

Zavitz (1915) reported on extensive trials that explored the seed size question in wheat, barley, and oats. As a general conclusion, yield advantage favored large plump seeds over medium plump and small plump seeds. The yield differences were substantial.

The unpredictable nature of seed size studies is exemplified by the experience of Williams (1916), who found no yield advantage of larger seeds in an eight-year trial of wheat seeds of different sizes, yet found almost a 50% yield advantage of large over small in another experiment conducted for six years with selected large and small wheat seeds.

Kiesselbach and Helm (1917) found that small wheat kernels, two-thirds the weight of large kernels, yielded 11% less grain than large kernels, but when small and large seeds were alternated in rows, the yield from the small seeds was 24% less than from the heavy seeds.

Comparing unselected, large, and small seeds, where the seeds were planted in equal numbers at a normal rate for large seeds, they found in two winter wheat varieties that yields from large seeds were 2.3% better than unselected, and small seeds were 3.1% inferior to unselected. In two spring wheat varieties, large seeds outyielded unselected seeds by 11.8%, and small seeds were 7.7% below unselected.

When small and large seeds of oats and spring wheat were space-planted to permit maximum plant development, the small seeds produced only 72% as much grain yield as the large seeds.

Kiesselbach and Helm used a fanning mill for 12 years for continuous grading of two winter wheat varieties. The result of this bidirectional selection of one-fourth heaviest and one-fourth lightest was a spread in yield of approximately 1 bu/acre, with the heavier seed producing yields averaging 0.4 bu/acre over the unselected. The situation seemed to be different with oats. 'American Banner' produced a striking reversal after eight years of similar fanning mill screening—the unselected seed yielded 1.6 *fewer* bu/acre and the heavy seed 3.67 *fewer* bu/acre than light seed.

Arny and Garber (1918) reported on a four-year experiment with precisely weighed seeds of 'Marquis' wheat. The number of plants per year appeared to be between 200 and 300. The emphasis of the study centered on correlations between plant characters. They found little correlation between weight of seed sown and plant characters.

In an extensive review of the influence of planting seed size on yield of small grain crops, Kiesselbach (1924) found from published literature and Nebraska tests that large seeds do indeed produce substantially higher yields than small seeds. However, the differences in yield became smaller and even negligible when sowing was by volume in an ordinary grain drill of the times; that is, smaller seeds made up in quantity for any deficiencies in quality. This necessarily would not hold true today because of the use of precision-sowing drills; with such drills the advantages of larger seeds might be exploited.

The importance of seed size in plant experiments was shown by Trelease and Trelease (1924) in water culture experiments over 41 days using seeds of the 'Marquis' wheat cultivar. They found that the mean top and root yields, that is, whole plant yields at harvest prior to grain formation, were directly related to original seed weights. Of special interest to plant breeders was the finding that the narrower the weight range of selected seeds used, the narrower was the range of variation in dry weight produced. For example, the means of medium and unselected seeds were practically identical (31.0 versus 29.8, respectively) but the medium group was selected within a 4 mg range (32.5–36.5 original weight) whereas the unselected group had an 18–46 mg range. The coefficient of variability for the medium (selected) group was $10.7 \pm 0.6\%$, but it was $21.2 \pm 1.2\%$ for the unselected group. The significance of this for precise experimental results with plants is obvious. Trelease and Trelease calculated that to obtain the same degree of accuracy from 80 plants of the medium (selected) seeds, it would require 316 plants of the unselected group, despite the fact that the two groups had similar means.

Plump, heavy seeds of wheat, as evidenced by 1000-kernel weight, were highly correlated (.69) with yield in studies conducted by Waldron (1926).

Hayes, Aamodt, and Stevenson (1927) found that yield and plumpness of grain were highly correlated (.51 and .62, respectively) in both spring and winter wheat. They suggested selecting for grain plumpness in the early generations.

Taylor (1928) found that yields of large and small seeds, separated from a wheat cultivar and continually screened and planted in line of descent for the same seed size, differed with the advantage to the large seeds. Furthermore, the advantage of large seeds became greater with time, there being almost a 7 bu/acre advantage by the fifth and sixth years. Seeding rates were adjusted to sow approximately equal numbers per area. It was not suggested that any genetic change had occurred. An inverse relationship between seed size and loose smut incidence was also demonstrated.

The importance of grain morphological quality was shown convincingly in correlation studies in oats with yield and several other traits reported by Immer and Stevenson (1928). Yield and plumpness (defined only as measured in percent, presumably visually scored on a 0–100% scale) showed the highest correlation, .78, among the primary traits, which also included date of heading, height, crown rust, and lodging. A multiple coefficient of .82 was found when all were considered with yield.

The slight gain from index selection would seem not to be justified, particularly because most breeders have efficient ingrained opinions as to acceptable ranges of the last four traits mentioned.

Bridgford and Hayes (1931) correlated 11 characters with yield in hard red spring wheat. Yield was positively correlated with plumpness of grain (.42), and 1000-kernel weight (.38).

Bonnett and Woodworth (1931) felt that "the yield characteristics of a variety can best be determined by a study of single plants." Yield per plant in barley was found to be correlated with spikes per plant and with grain weight. They expressed the breeder's difficulties associated with testing at different spacings and with seeds of different sizes this way: "The analysis showed that if seeded at the same rate (pounds per acre) a small-seeded variety may outyield a large-seeded variety on account of the larger number of plants per unit area rather than because of superior plant–yield characters."

Bartel and Martin (1938) reported that seed size of corn, sorghum, and proso was directly related to early stalk growth. The logarithm of weight per seed planted was directly proportional to the logarithm of the weight per stalk after 10–12 days.

According to Waldron (1941), Mallach (1929) obtained 16% greater yield per plant in wheat from heaviest versus medium-weight kernels. There was little difference between the yields of plants sown with lightweight or medium-weight kernels, nor were there consistent differences in the weights of the seeds produced from these two classes of seeds.

Waldron conducted two experiments with spring wheat, comparing light to very heavy seeding rates, and comparing seeds of different sizes and origins. In the first experiment two classes of seed weights (26.6 mg/kernel and 40 mg/kernel) were chosen from one hybrid-origin seed lot. Rates of seeding were adjusted according to a varying number of kernels in equivalent-size areas receiving the same total seed weights, and to a varying weight of kernels with numbers of kernels held equal per equivalent-size sowing areas.

In the experiment where weights of seeds sown were held constant, in eight categories of increased weight of seeds planted (17.2–292 lb/acre), there was an average advantage of 1.52 bu/acre or 10% from the heavy-weight seeds. In the two lowest rate categories, 17.2 and 25.5 lb/acre, the yield advantage was to the lightweight seeds, suggesting a density-dependent relationship with overcrowding at the higher seeding rates (weights). Weights of kernels produced were also in favor of heavy kernels planted (statistically significant).

Results when equal numbers of kernels were planted per unit area were that yields from heavyweight kernels exceeded those from lightweight kernels in each of nine seeding rates, averaging a statistically significant 1.76 bu/acre, or 12%, more.

In the experiment comparing seed origin (greenhouse vs. field grown) there was no significant overall difference in yield (43.2 vs. 42.1 bu/acre); however, 12 of the 16 greenhouse families produced seed heavier than their field-grown counterparts. When these 12 were considered together the yield advantage from the heavier greenhouse-produced seeds was a statistically significant 1.97 bu/acre.

Middleton and Hebert (1950) reported on a trait, purple straw color (an anthocyanin), that was associated with heavier kernels that were significantly heavier by a mean 1.2 g/100 kernels in all three testing years. The yield per bushel effect was inconclusive: in one year there was a highly significant advantage of purple lines, but in the other two years the comparisons were nonsignificant. It is not known whether or not the color advantage is specific to the 'Purplestraw' cultivar.

Katz, Farrell, and Milner (1954) described a grain projection apparatus that separated wheat into density classes that correlated with test weight. In a companion paper, Milner, Farrell, and Katz (1954) analyzed 14 samples of hard red winter wheat subjected to a seed projection device for separation according to density. They found again that density was correlated with test weight. Protein percent was negatively correlated with density. This is a well-known phenomenon in drought and disease situations, where lightweight grain and high protein content are associated. Whatever the reason, seed projection did stratify for protein level.

Kneebone and Cremer (1955), using hand-screened caryopses, found a direct relationship between seed size and seedling vigor in five species of native grasses.

McFadden (1958) showed that large seeds resulted in greater tillering and yields than small seeds. Another paper by Kaufmann (1958) in the same year showed that large seeds produced seedlings with greater vigor than small seeds.

Kaufmann and McFadden (1960) compared equal numbers of large and small seeds in greenhouse and field tests in interplant, interrow, and control situations. In all tests plants and plots sown to large seeds outyielded those from small seeds. The advantage accruing to large seeds increased in every case of competition, a higher number of heads per plant being found on plants grown from larger seeds.

Kaufmann and McFadden had an interesting suggestion, reminiscent of Gardner's grid (1961):

> Results of tests involving inter-plant competition have a bearing on seed preparation for testing in early generations, since there is a wide range of seed size from a single head or plant of six-rowed barley. To virtually eliminate this source of non-genetic variation in the F_2, seed should be separated into different size groups prior to seeding. Selections could then be made from within each size group. Conditions for single plant selection from within lines in the F_3 and subsequent generations could be improved by using uniform seed for any line.

They were concerned that apparent productivity differences in lines or cultivars caused by differences in size of seed sown might mask or dominate genetic differences.

An interesting association of seed size and another trait, loose smut incidence in barley, was shown by McFadden, Kaufmann, Russell, and Tyner (1960). It was shown conclusively that small and medium-sized seeds produced plants with a higher incidence of loose smut, *Ustilago nuda* (Jens.) Rostr., than large seeds. This verified earlier reports by Taylor (1928) of the same phenomenon in wheat.

Murphy and Frey (1962) studied the nature and inheritance of groat weight components (density, length, and width) in oats. Density of healthy kernels was found to be essentially a constant. Length showed high heritability, with its development occurring in a short period of time after anthesis. Groat width had the greatest influence on groat weight.

Frey (1962), noting that seed weight is a component of grain yield, determined that the regression of grain yields on 100-seed weights was positively and significantly correlated in four of six oat crosses studied. The degree of correlation was so small, nevertheless, that practically there was little advantage in using both in preference to selection for yield alone. Even so, the desirability of having high values for both in the same genotype is an argument in favor of tandem selection.

It is the reserve material of the endosperm, not the embryo meristematic tissue, that is responsible for seed size differences influencing seedling growth, according to Bremner, Eckersall, and Scott (1963). In their research with the wheat cultivar 'Cappelle' they used two excised embryo sizes, two endosperm weights, and two depths of planting. They found seed size and plant size directly related to the amount of reserve material in the seed (until it was exhausted and prior to the impact of photosynthesis).

Kaufman and McFadden (1963) found in barley that large seed produced plants with greater vigor, more tillering, and greater yield than small, medium, or bulk seeds. In some cases cultivar yield rankings were changed depending on seed size used. They also noticed that plants from large seeds headed and ripened earlier.

Bradshaw (1964) said:

> It is a conspicuous characteristic of all wild plant species that while numbers of seeds may vary widely, seed size is held extremely constant. But this is not necessarily true in cultivated species. . . . Breeding for higher yield seems to have caused loss of stability in this character.

Sharma and Knott (1964) observed the inheritance of seed weight in crosses of two parents that exhibited extreme size differences: 'Selkirk', a heavy seed weight Canadian commercial spring wheat, and 'Chagot', an Indian spring wheat whose kernels were about one-half the weight of Selkirk's. They found that seed weight was highly heritable, was controlled by few genes, and did not exhibit transgressive segregation. Conclusions were that it should be relatively easy to improve seed size of wheat through breeding.

Tandon and Gulati (1966) determined the effect of seed size differences on seven barley characters in a study showing two unusual approaches. First, seed size differences of approximately 2 : 1 by weight were obtained from the natural differences inherent in spike location on the six-row barleys, central or lateral, respectively (central kernels are the larger). Second, the genetic influence on variability was eliminated by (1) using genetically similar F_1s, (2) using F_2s of two crosses (segregating, kept separate by family line), and (3) using homozygous pure lines.

Larger seeds produced progenies superior to those of smaller seeds in all categories tested except germination percentage. Yield and number of tillers per plant showed the greatest response to larger size. The authors discussed two consequences of this information; that is, a drastic bias could be introduced into theoretical studies in quantitative genetics, affecting estimates of heritability and the like, if seed size differences were not controlled, and seed size control applied to commercial crop production could increase crop yields (25% greater in this study).

Using two cultivars of spring barley, Kaufmann and Guitard (1967) studied the effect of large and small seeds, the former being approximately twice as heavy, on seedling growth in greenhouse soil and vermiculate pots. Growth was measured by length and width of first and second leaves at days after emergence.

Large seeds without exception produced the longest and broadest leaves. There was some indication that the relationship with seed size carried into maturity where heaviest spikes and greatest number of florets and culms came from plants grown from largest seeds. However, it was noted that as the plants developed the differences became smaller and less distinct and cultivar influences could be seen.

Kaufmann and Guitard found a better correlation between yield and first leaf length than yield and flag leaf length, which was not enough for ready use as a selection indicator, but did suggest further study.

Fehr and Weber (1968), using mass selection, found that the selection combination of large seed size and high specific gravity resulted in the most progress toward high protein and low oil, whereas small seeds and low specific gravity gave maximum progress for low protein and high oil.

Test weight in the eastern (U.S.) soft white winter cultivars is a recognized varietal trait (the soft whites as a group have lower test weights than the soft reds). Yamazaki and Briggle (1969) examined the components of test weight in seven cultivars. Surprisingly, it was discovered that varieties did not differ significantly in kernel density. Great differences were found, however, in packing efficiency. Cultivars with low packing efficiencies showed some deviation from the typical high-efficiency shape, which was a cylindrical kernel with oval ends. The authors believed that breeders could raise test weight by selecting for appropriate kernel conformation.

I know of at least one exception to the generally low test weights found in the soft white winter wheats, and that is 'Fredrick', a fine Canadian cultivar. I believe some of the newer Michigan whites are also excellent in test weight. Fredrick's test weight was generally higher than all of the Cornell cultivars including 'Avon' and 'Arrow'. It would be interesting to compare its packing efficiency with the varieties used in Yamazaki and Briggle's study.

Frey and Huang (1969) found a significant correlation between 100-seed weight and yield of grain, and proposed a tandem selection procedure.

Ali, Atkins, Rooney, and Porter (1969) showed that kernel size varied according to the portion of the wheat spike in which it grew. In general, the grain from the central third of the spike seemed to exemplify the most desirable characteristics, that is, the highest 1000-kernel and test weight, and the longest and widest kernels with average protein content and milling yield. These findings were in general agreement with those of McNeal and Davis (1954).

The effect of initial seed weight on mature crop characters was studied by Austenson and Walton (1970) using 30 heads each (approximately 900 seeds) of three spring wheat cultivars. The tests were conducted under dry field and spaced-plant conditions. In general, initial seed size in all three cultivars was significantly correlated with total yield, grain and straw yields, heads per plant, and seeds per plant. The increased yields from large seeds were less than 5%, large enough, however, for the authors to recommend sieving the seeds for planting to obtain this "free" advantage in yield.

Detailed information was taken on seven mature plant characters (those mentioned above plus seeds per head and 1000-kernel weight). For reasons of the study of the detailed raw data were not included, but it suggested that a wealth of information exists that would be useful in other analyses such as harvest index.

Incidentally, Austenson and Walton clearly showed the origin of different seed size; within a single head large seeds could be as much as three times the weight of the smallest. In sieving the seeds no genetic advantage occurs, merely the elimination of those least likely to compete successfully because of small size.

Background for the following study included the knowledge that 'Selkirk' spring wheat normally produced seeds 15–20% heavier than 'Thatcher', and earlier research had shown the weight genes were relatively simply inherited [also, a paper by Geiszler and Hoag (1967) showed that large, heavy seeds of Selkirk produced higher yields than small, light seeds]. Knott and Talukdar (1971) reported the transfer by backcross of the heavy seed character to Thatcher. In yield tests of 39 lines, 20 were significantly higher than Thatcher and only one was significantly lower; the average yield increase was about 6%.

Knott and Talukdar were interested in the effect of greater seed size on the components of yield. They wished to discover where the compensation that created the higher yield occurred. The higher yield has been attributed to a combining of the heavier seeds of Selkirk with the larger number of seeds per plot of the recurrent parent, Thatcher. It is a possibility that the higher yield may simply have been a direct response of larger or heavier seed size, which has been shown often to be related to seedling vigor, growth, and yield. In some cases, large versus small seeds from the same genotype have given yield differences of the magnitude described in this paper.

Hunter and Kannenberg (1972) reported a study in corn that showed seed size to have no effect on rate or extent of emergence regardless of hybrid used, planting depth, or temperature imposed. Grain yield was slightly higher for large seed but apparently not significantly different.

In a mass selection experiment in wheat, Derera and Bhatt (1972) reported that selection for large seed size resulted in higher grain yields per plot. In a subsequent paper, Bhatt and Derera (1973) found that a series of quality attributes (test weight, milling extract, wheat and flour protein, kernel hardness) were size neutral. Aside from a positive association between seed size and height, they concluded that selection for seed size posed little negative selection risk.

When Roy (1973) used the single seed descent method of closely spaced plants growing in a favorable environment, that is, stressed by density but not by nutrients and moisture, he found that the two grades of seed used, large and small, which were planted in various density and competition formats, produced plants with uniformly similar weight and size seeds; that is, whether the seed sown averaged 47 or 32 mg, the seed weights produced were in the range of 37–33 mg.

The author conceded, however, that if growing conditions were not optimal the competitive disadvantages that plants from small seeds suffer early in the growth cycle, but overcome under favorable conditions, might be magnified. The consequences of competition in heterogeneous or segregating populations, separate from seed size, were not explored fully. The conclusions were limited to dense but otherwise favorable conditions, where it was found that seed size differences would not be expected to persist into the next generation.

Hamblin (1975), working with bean cultivars having substantial seed size differences, found that seed size played an important role in reproductive, and hence competitive, fitness. He suggested screening the F_2 to select the consumer-preferred size.

In a study of the heritability of quality parameters in wheat, Bhatt and Derera (1975) determined test weight to be an independently inherited trait, not correlated with any of the nine traits examined.

Hamblin (1975) suggested a preliminary seed screening of the F_2 to stratify the population's seed size and minimize competitive interactions that stem from variable seed sizes. This raises the question whether the principles of clustering and the Gardner grid might be applied in a general way to several other characters that have high influence on fitness in early generations, and a consequent impact upon choice of breeding system. If, for example, the F_2 or F_3, were mechanically space planted, and the plants selected according to subpopulation groups, bulk breeding with its greater efficiency might proceed at or near crop density conditions. The subpopulation groups, for example, might be constructed around height, erect leaves, tiller numbers, or disease resistance. Extra labor would be required during this generation, but the positive gains might be substantial; for example, an important immediate benefit would be the discard of all unwanted plants.

Plant breeders have a legitimate interest in knowing whether cultivar performance under test conditions might be different under production field conditions because of an interaction between seed characteristics and farming equipment and practice. Carver (1977) explored this question, focusing on sowings with both constant seed weights and constant seed numbers. Seed lots were prepared from one winter wheat variety where small, medium, and large sizes characterized the different 1000-grain weights of 26.6, 47.3, and 56.7 g, respectively. For the constant weight studies, each size class was sown at four seeding rates with the same weight per plot within each rate: the weights per plot were 90, 134, 179, and 202 kg/ha (the last three figures cover the range of commercial seed rates).

For the constant seed number experiments, each size lot of seed was sown at the four rates, but the number of seeds per plot at each rate level was constant, and matched the numbers drilled at the 26.6-g (smallest) 1000-grain sample.

Carver found yield differences of 12% between seed lots of the same variety, owing to differences in 1000-grain weight of the planted seeds when grown under the constant weight system. Small seeds had an advantage over medium and large. The explanation may rest on the fact that heavyweight seeds, thus fewer per plot, require adequate tillering to enhance yield, whereas the more numerous small seeds can rely more on their single main culm. In other words, there is an interaction of density and ideotype a la Colin Donald.

When seed numbers were held constant over each rate, the yield differences related to size were much smaller. Larger seeds tended to show an advantage at lower seeding rates.

The conclusions drawn from the study were two:

1. The constant seed number option is preferable to a constant weight system for testing (cereals) because performance more closely matched on-farm performance.
2. Uniformity in seed size is important. It must match, at least, the level achieved by seed growers in their common practice of sieving and discarding small seeds.

My personal experience with oats supports the need to test for yield using seed lots with kernel size and quality equivalent to that of seed being sold to farmers. The New York seed producers took pride in selling a heavyweight seed oat, and I found it necessary to match that quality when preparing seed for nursery tests. This involved buffing and polishing the field-harvested seeds, using a commercial paint shaker–mixer (Jensen 1964), followed by a screening out of the smaller kernels.

In controlled hardiness tests of two hardy winter wheat cultivars, 'Kharkov 22MC' and 'Winalta', Freyman (1978) found increased hardiness in plants from large kernels, and suggested that in winter kill risk areas the seeding be done with large kernels.

If seed weight is important, what is a reasonable sample size for its determination? Havey and Frey (1978) found the optimal sample size to be 100 kernels taken from two replicates, one sample per plot.

The possibility that increasing the seed set in tertiary florets could increase oat production sparked the research of Takeda and Frey (1980). Most oat grain production is based on two kernels in an oat spikelet, the primary and secondary. The tertiary kernel, sometimes called a "pin oat," is not uncommon, but more frequently is ignored and removed as part of processing, especially if the crop is to be used as seed. Nevertheless, if the caryopsis attains reasonable size its proportionate contribution to yield cannot be ignored.

Takeda and Frey explored the range of genotypic expression in 35 oats, and the effect of environment (two seeding dates) on the development of the tertiary seeds. The unexpected finding was that three of the oat genotypes were outstanding for high tertiary seed set. These three were related, all tracing to C.I. 8044, an issue of the cross of 'Clintland' × 'Garry 5'. Heritability values were high and it seems plausible that selection for higher tertiary seed set is feasible.

A second interesting finding was that late sowing depressed tertiary seed set. This was attributed to higher temperatures and greater stress.

In soybeans small-seeded types (12–19 g/100 seeds) are generally higher yielding than large-seeded ones (weighing more than 20 g/100); nevertheless there is a special demand for the latter for human consumption. Breeding efforts for the two types have tended to choose parents having seeds of like size, and one might speculate that this may be the result of historical experience. Bravo, Fehr, and de Cianzio (1981) examined the seed size segregation results from a series of two-way and three-way crosses. The two-way crosses were between small- and large-seeded parents and the three-way crosses involved these F_1s crossed with parents having intermediate-seeded sizes. All of the parents were adapted agronomic types of equivalent maturity.

The F_2 population means of the two-way crosses were related closely to the mean of the parents, suggesting predominant additive gene action. In the three-way crosses the F_2 population means followed the seed size of the third parent; that is, if the third parent had a seed size larger than the mean of the two-parent cross, then the mean of the three-way cross would be higher. Estimates of genetic variability and range of seed sizes were essentially similar for three-way crosses. Transgressive segregation was almost nonexistent. Conclusions were that improvement of large-seeded types would be difficult if approached via the small × large cross route. The authors suggested approaches that might be tried, all based on keeping the large-seed component high during crossing:

1. Use as parents the largest-seeded cultivars within the "small" category.
2. Backcross to the large-seeded parent.
3. Outcross (three-way) to another large-seeded parent.
4. Select for large seed in F_2 and remate with large-seeded parent.

Evidence that large heavy seeds contributed to yield stability in low-yield environments was found by Heinrich, Francis, Eastin, and Mohammad Saled (1985) in grain sorghum. Both stable and nonstable genotypes had the ability to increase seed weights in favorable environments, as when sink capacity was deliberately reduced through lowering seed numbers. The inherent larger-seed genotypes (stable) also had higher numbers of seeds in low-yield environments, thereby showing they could utilize a broader range of sink capacity.

Lafond and Baker (1986) found seed size to exert a greater influence on seedling shoot dry weight than on median emergence time or Haun stage of development. It seemed to be the largest factor affecting spring wheat cultivar differences in seedling vigor.

DISCUSSION OF SEED SIZE EFFECT

Ali, Atkins, Rooney, and Porter (1969) showed that a breeder can screen out smaller kernels with confidence: in the process there are differential reduction and changes in gene frequencies, but probably not a complete elimination of genotypes.

Waldron (1941) believed that advancing maturity and favorable environmental conditions brought about a narrowing of differences in yield results from different grades of seeds.

Kiesselbach and Helm (1917) concluded from an exhaustive historical survey of 19 investigators' findings that yield is favored from larger seed size and weight when seeds are space planted or planted in equal numbers (at a rate favoring large seeds), but when planted in equal weights all grades of seed yield equally.

These reiterations are only a few of researchers' beliefs. In my cereal breeding projects with wheat, oats, and barley at Cornell University I considered seed attributes to be very important. After harvest I reserved to myself the processing of all hybrid population seeds in the wheat, oats, and barley projects. I used fanning mills, sieves, gravity graders, seed projection, paint shakers, and air blasts—all the available tools. Practicality played a part because the harvested seed increases from plots were so large, many pounds, whereas only a small fraction was needed to plant the next generation plot. It was desirable to reduce the amount for storage in consideration of the large amount of space needed for all reserve seed stocks.

The time between harvest and fall planting of winter wheat in New York was approximately one month, and it became increasingly difficult to incorporate the laborious seed processing with all the other details of analyzing data and preparing new plans, not to mention the pressing barley and oat harvests that coincided and followed the wheat harvest. Finally, a system was developed whereby we processed as many seed lots as possible and readied them for the plans and fall planting. The seed lots that could not be got ready were held back and processed during the winter, the next year's field plans started, and both set aside as the nucleus for the following fall's planting when the cycle would be repeated. This meant, in effect, that the size of the wheat project overall was doubled on paper, with one crop in the field and a second one waiting in reserve.

Another problem in the wheat project concerned kernel color. Our commercial wheat was soft white winter. White kernel color is a recessive trait that breeds true. Crosses of white wheats posed no problem, but F_2 progenies of white × red crosses were handpicked for white before sowing. This represented weeks of day and night eye-straining work for the breeder. Again, what could be got ready in time was planted and the remainder sorted over winter and made ready for the next year.

Seed quality was carefully observed. Seeds of all crops were visually scored on a simple 1–10 scale with appropriate description notes appended. Even today, I find somewhat incredible and difficult to convey to others the fact that the Cornell cereal project had no quality laboratory, no milling and baking facilities, no Pelshenke tests, no hardness tests, no quality tests whatsoever, yet the project produced world-class wheats with superb milling and baking qualities, for example, 'Yorkwin', 'Cornell 595', 'Genesee', and 'Yorkstar', and oat cultivars such as 'Orbit', which for years in regional and national tests showed the largest kernel. Of course before becoming a cultivar, a Cornell wheat candivar would have passed through the sophisticated milling and baking facilities maintained by the U.S. Department of Agriculture at Wooster, Ohio, but only a small select group made this invaluable annual trek. The point I want to make to beginning plant breeders is this: personal, hands-on familiarity with cereal seeds becomes with experience an effective gauge of quality, and I am sure this applies to any crop species. After the seed underwent threshing and some processing at our nurseries it was the practice in my projects to prepare and send to my office laboratory a small package of a few grams of seed from each genotype. No Cornell seed lot or line, whether from a yield-testing nursery or an early-generation hybrid nursery, escaped being poured into my hand, being examined, and receiving a visual quality score: 5 was average, anything below was subject to discard, and anything above merited saving.

Scores of 7 and higher represented elite material (a 9 was rare and no 10 was ever assigned). This system was not exclusive to wheat, but was applied to all the crops in the breeding program.

Today I would expand this genotype screening procedure in one respect: it would be applied to bulk seed lots of crosses beginning in the F_2 and used as a guide to choose among crosses—an indicator for discard of whole populations.

With spring-planted oats, processing could be done over the winter. The first operation was a general cleaning out of straw, weeds, chaff, and dust. The oat seeds were then placed in a paint shaker for a given time (Jensen, 1964). I was always conscious of the special pains that good seed growers took to market a bright, plump, uniform product, and we found that the paint shaker buffing exposed the seeds' quality, or lack of it, more than any other treatment. Following this procedure, our attention turned to further grooming of the seed lots, beginning with a second screening for discard of small kernels.

With wheat, I initially experimented with each hybrid seed lot. Different sieve combinations were tried in order to appreciate the seed lot's general profile of characteristics. We never removed the larger kernel fractions, but concentrated our efforts on removing sufficient smaller fractions to reduce a large volume to a small volume. Frequent dramatic size differences appeared in different hybrid lots—a set of sieves that were suitable for lot after lot would suddenly be totally wrong for a seed lot that happened to be large and plump with little cleanout, and sieves would have to be changed to accommodate the particular seed quality attributes of this particular cross.

One suggestion to cereal breeders that I think would be helpful would be the preparation of a seed profile for each new cultivar. This would be obtained from a series of measurements of both bulk grain and typical heads, and data from cleanout after treatments. The profile would show at a glance the characteristics of a new cultivar as compared with the profiles of past cultivars.

I think that the procedures used in my project to screen all seed lots for uniformity had much to do with the success of the bulk population breeding method. The use of uniform size seeds removed an important source of variability in the competition environment of the bulk population.

This section has highlighted the importance of seed size differences in breeding, selection, and nursery yield testing. If, in addition to the above discussion, a single recommendation is desired, it would be this: a seed processing arrangement to screen seed lots for size uniformity should be implemented throughout a breeding program.

SEED ORIGIN: PRODUCTION SOURCE EFFECT

Relatively little research has been done on a subject of importance to plant breeders—the effect of seed source on the performance of the following crop. To what extent are yield differences in regional trials influenced by the conditions under which the seed was produced? The breeder wants a true representation of the genetic potential and interactions of the entries in a nursery. If seed origin is shown to be an added source of variation in regional trials, the solution, which is a requirement that all entries be grown at a given location, would entail a year's delay for most entries. However, this delay might be minimized or eliminated by the breeder through a more perceptive anticipation of his new releases, and perhaps by sending more entries to the seed increase location.

Waldron (1944) suggested that local seeds produced crops with higher competitive values.

Sarquis, Fischer, Parsons, and Miller (1961) tested the stocks from three locations at Coalinga, California, over a three-year period using equivalent-quality seed stocks of the same variety, 'California Mariout', which had been grown for two or more generations at the same location. There were no statistical differences in yield performance. They concluded that the "geographic source of California grown barley planting seed of the same variety has no significant effect on yield, if the quality is otherwise equal."

There are, however, special reasons for choosing planting seed produced in certain areas, for example, for assuring disease-free seeds.

The influence of seed origin has long been known with respect to certain plant pathogens, where the smut mycelia and seedborne molds had invaded the kernel during the previous growing and storage season (Holton and Heald, 1936; Tervet, 1944; Christensen, 1955). In a slightly different case, Fezer (1962) found that the incidence of field loose smut was inversely related to seed size, and suggested screening infected seed lots to obtain the largest plump kernels. Also, Gregory and Purvis (1936, 1938) and Weibel (1958) have shown that differences in vernalization can be caused by climatic changes during the seed-growing season.

Suneson and Peltier (1936) chronicled large variations in cold resistance among the same winter wheat cultivars grown from seed samples produced at different geographic locations. The study emphasized the importance of a common seed source for comparison or performance nurseries.

McFadden (1963) recounted that he had observed in 1959 that barley varieties whose seed came from different locations showed significant yield differences that were attributable to their seed source. To investigate this phenomenon he had four barley varieties grown for seed increase at five widely separated locations in Canada. When the seeds from the different locations were brought together after two years, striking visual differences in growth vigor due to seed source were found. Yield differences due to seed source were significant at the 1% level for all four 1960 location tests and three of six 1961 tests, and at the 5% level in one 1961 test. Depending upon which seed source results were chosen, a number of different varietal yield rankings could be shown. McFadden suggested as a solution that a number of seed increase centers should be established. These would raise and contribute uniform seed stocks for use in uniform nurseries.

According to McFadden, Findlay (1956) reported that seed of a variety brought in from an early-ripening area would produce a crop that would ripen a week earlier than the same variety grown from home-grown seed.

McNeal, Berg, Dubbs, Krall, Baldridge, and Hartmann (1960) planted seed from the same lot of 'Thatcher' spring wheat at 11 locations in Montana. From these they chose from four locations samples that exhibited extremes of test weight and protein and showed high germination and variations in kernel weight. These were sown at six seeding rates at four locations, with testing at one site, Moccasin, being done under both irrigated and nonirrigated conditions.

Significant yield differences were obtained only from rates of seeding; the effects of source, protein content, and test weight were negligible. The test involving seed origin would have benefited from a test of all 11 seed sources.

Kinbacher (1962) reported that cold resistance of 'Dubois' winter oats was markedly affected by different areas of seed production in four states.

Further evidence of the influence of seed origin was provided by Quinby, Reitz, and Laude (1962), who took seeds of two hard red wheat cultivars, 'Pawnee' and 'Wichita', grown at eight locations in five states, and tested them for yield in common nurseries in three states over a three-year period. Statistically significant yield differences were found within each cultivar that were associated with the seed lots produced at different locations.

Walsh, Reinbergs, and Kasha (1973) pointed out that seed for yield testing of doubled haploid (DH) lines, which are homozygous and immediately ready for testing, can be obtained in two ways: at once, from production of tillers on the DH plant, or from the seed increase of a single head of the doubled haploid. However, the seeds from the two options are different: the original seeds are smaller and less plump. A study of progenies of seeds from these two sources revealed that seed source was important. On the one hand, if all the seeds of lines being tested were from original seeds, there was a smaller problem because the source × genotype interaction tended to be uniform or common to all. In the more likely and desired situation where DH lines were being compared with check cultivars and lines from conventional hybridizing, the differences in seed quality, such as reduced emergence and higher mortality with consequent changes in density and differences in vigor, were very important and worked to the disadvantage of original seeds. The authors concluded that, for all testing that would involve non-DH lines, the DH original seeds should undergo one generation of seed increase to obtain "normal" seed on a par with that produced by ordinary conventional lines. It was noted also that this was important even when only DH lines were being tested from original seeds: the reduced density meant that the breeder would be vulnerable to selecting for the compensating high tillering associated with more open stands, which is often counterproductive to high yield in later dense stands.

Evidence on the importance of seed source was presented by Ries and Everson (1973) in a paper that is also a good review of research linking protein content of seeds to seedling vigor in several crops. The authors' studies showed that, although large seeds generally produce vigorous seedlings, the seedling size and vigor regardless of genotype are positively related to the protein content of the seed. These findings relate to cereal breeding in that seed source and history can have profound effects on the relative performance of entries in nurseries, and as the authors suggested, foliar applications of nitrogen might aid seed growers in producing a higher-quality product.

Sterling, Johnston, and Munro (1977) at Charlottetown, Prince Edward Island, Canada, made a detailed study of barley emergence, yield, and kernel weight of two cultivars using planting seed tracing back to the same lot of Foundation Seed of each. The crops were grown for three successive years at 10 locations. Each year the seed of the two cultivars grown at the 10 locations was sent to Charlottetown where processing for uniform seed size was applied and the seed lots returned to their location source to be planted for the next crop. Samples from each yearly crop of each cultivar from all locations were provided to the Charlottetown station, which then ran a year-later test of all samples, continuing for the stated three years.

The focus of these experiments was on pathogen contamination of the seed, related to source. Therefore, although yields in grams were given, no attention was given to genotype yield variability related to source of seed. There were two treatments applied, fungicide and no fungicide. Taking the yield data from the treated plots (a standard good farming practice) and converting to bushels per acre, the yields from harvested seed from 10 locations, when planted and grown at one location and for the three years, ranged as follows:

'Volla':	50–66, 89–101, and 66—73 bu/acre
'Keystone':	44–50, 76–88, and 57–61 bu/acre

The important statistic is the great range in yields, when expectation would be for seeds of the same cultivar to yield in a close range. Remembering that in each series uniformly processed seed from the same Foundation Seed Stock was furnished the growing station each year, and that the above results came from replicated trials at one location, I think the evidence is strong that seed source powerfully influenced performance. In effect, the above trials could be considered uniformity trials, but the results showed yield range comparable to some variety trials.

GERMINATION AND SEEDLING VIGOR

Kiesselbach and Helm (1917) used the term *sprout value* to denote the inherent life-force bound up in seeds. It is a measure of the reserve food material of a seed. Sprout value was defined as "the moisture-free weight of the maximum plant growth derived from the seed when planted and grown in a nonnutritive quartz medium and in absolute darkness."

All seeds tested over a two-year period gave an average recovery figure of 50.2% of the seed substance in the sprout. A close relationship was found between size of seed and sprout value, with a distinct disadvantage for small or shrunken seeds planted at greater depths.

Cutler (1940) made a study of the effect of clipping or rubbing oat grains of six varieties on test weight and viability. In three years of testing he found that test weights were increased by an average 8.8 lb/bu (29.4%). Furthermore, clipped grain stored up to three years showed no loss of viability.

Fryxell (1954) called attention to European research on the use of high osmotic pressure as a possible selection tool in plant breeding. The idea is intriguing, although it is difficult to assess the validity of the positive claims made. Seeds that germinated and grew vigorously in salt or sugar solutions would be expected to have higher osmotic values than those that did not—the ability to "pull' water from the solution being the key. If correlations with useful plant traits were found associated with high osmotic readings, the technique would lend itself to large-scale treatments of hybrid seed lots.

Rogler (1954) found high correlations between seed size (weight) and seedling vigor, as indicated by emergence from various depths of planting. Among cultivars and strains of crested wheatgrass, large-seeded types showed greater seedling vigor.

Pointing out that the rate of seedling growth was an expression of genetic factors for vigor, Christiansen (1962) devised a rag doll method of measuring early seedling development using cotton and soybean seeds.

White (1962) described experiments with cloning wheat. He had no difficulty in obtaining four to five clonal units per plant. The author suggested early seed increase applications and use in recurrent selection by slowing growth on one clone while sister clones were being evaluated for parents. Cloning is of interest as a research tool: Grafius, Nelson, and Dirks (1952) suggested its use to reduce estimates of standard error, but it would seem to have limited value in a breeding program, largely because of restrictions on numbers that could be prepared and handled.

Maguire (1962) offered a refinement to the daily counting method of germination (cf. Throneberry and Smith, 1955). This modification incorporated the rate of germination, which may be quite different in two seed lots having the same germination percentage, for example, in Kentucky bluegrass.

Williams (1963) measured the emergence force exerted by forage seedlings, using weighted glass rods. Seed size was found to be significantly correlated with vigor of emergence.

Plant breeders need efficient selection procedures with capacities to screen large populaions of seeds, seedlings, or plants. Williams, Snell, and Ellis (1967) presented three methods for measuring drought tolerance in corn that met these requirements. The relatively simple procedures involved temperature and seedlings, germination in mannitol, and seedling wilting tests.

Jennings and Aquino (1968) found seedling vigor to be synonymous with competitive ability:

> Alternatively, greater tillering, height, dry weight, leaf number and length, and LAI may be considered as a definition of greater early vigor in the development of competitive types. The characters that increase size and early vigor, therefore, are considered to be associated with competitive ability.

Drawing upon data from previous research (Ries, Schweizer, and Chmiel, 1968), Schweizer and Ries (1969) confirmed the positive correlation of seed protein and amino acid content with seedling vigor. Seeds from oat and wheat cultivars that had received supplemental nitrogen showed an accumulation of greater dry weight of total amino acids and of Kjeldahl nitrogen. The crops grown from these seeds in tests that included controls showed dramatic increases in seedling growth and yield.

The authors pointed out the importance of seed protein enhancement or fortification in the production of seed stocks, where the influences on nutrition and yield would carry over to the commercial crops grown from the seeds.

Accepting the premise that seedling vigor contributes to yield, Boyd, Gordon, and LaCroix (1971) studied its relationship to seed size and germination resistance. They used two-row and six-row barleys. For those not familiar with barley, the seeds of two-row barleys are larger, heavier, and more uniform than those of six-row barley. The greater variability of six-row barley seeds is a consequence of the smaller satellite side kernels in the spikelet. Because of this, a bimodal seed size distribution pattern is normal in progenies of two-row × six-row crosses. This distinction is of importance in barley breeding because of its implications for seed characteristics of a cultivar. For example, Boyd et al. showed 1000-kernel weight mean differences of 34.74 g (six-row) and 49.86 g (two-row).

Multiple regression showed that seed size and germination resistance accounted for more than 90% of the variability in seedling vigor among the F_3 lines tested. Germination resistance introduced a new factor, dormancy, into the picture. They pointed out that "greater dormancy at harvest is the same phenomenon as high germination resistance at harvest." In other words, the phenomenon often carries over into germination in the spring. The import of this is that breeding efforts to lower field sprouting (greater dormancy) predispose the following seeded crop to a slower or less vigorous start.

Boyd et al. used the germination resistance test devised by Gordon (1971). This test measures the relative rate of germination, or the phenomenon may be viewed as measuring the resistance to germination. The value of the test rests on its precise preparations based on uniform seed quality, water adjusted to seed weight, and so forth. Readings are taken at 24-hr intervals and a mean germination time is calculated. The test offers a standard for comparisons.

Working with snap bean seeds, Ries (1971) found that both seed size and protein content were correlated with subsequent seedling size (growth) and protein content. Protein content of seed seemed to be more important than seed size. Field studies corroborated the greenhouse studies and showed that protein content, especially, increased yield.

Lowe, Ayers, and Ries (1972) found that seedling vigor of wheat, measured as shoot dry weight after three weeks, was positively correlated with seed protein content. The authors concluded, "When the seed protein level is increased, either by the use of N fertilizer or by selection of large seeds for sowing, the seedling vigor of the next generation can be increased. As a result, an increase in grain yield is possible. . . ."

The findings have relevance not only to commercial productivity but especially to growers of Foundation, Registered, or Certified seeds. The authors suggest a novel explanation for the belief that varieties deteriorate (or 'run out' in the common parlance): gradual protein depletion of seeds grown under low nitrogen conditions.

Extensive tests by Lowe and Ries (1973) showed that the factors for protein content in wheat seeds lie in the endosperm and not in the embryo parts: "The absolute amount of endosperm protein in wheat seed bears a linear relationship to seedling growth. . ."

Dasgupta and Austenson (1973) reported on a two-year study of the association between seed characters and yield of the ensuing crop. The seed traits examined were bushel weight, 1000-kernel weight, cracked pericarp, germination, and variations such as cold germination tests, rate of oxygen uptake, contents of crude protein, fat, fiber, and ash. The yield tests showed that a range of 26–55% of yield variability could be assigned to seed characteristics and stand. Variation was high enough so that the authors said, "There were no simple quality characteristics in these seed samples that could be used consistently to predict the yielding ability of a particular seed lot."

Nevertheless, it seems obvious that the quality aspects of the seed lots were exerting a huge influence on the yields of the sown crop. Dasgupta and Austenson concluded that, considering the variation observed over years and locations, seed size and germination were the most useful indicators of seed quality.

Ellis and Hanson (1974) explored the use of scutellar respiration of hybrid seeds during germination as an indicator of the heterotic potential of F_1 hybrid barley. Their research impressed upon them the profound confounding effect of varying seed quality, resulting in loss of viability and lowered vigor. They concluded that genetic effects on seed performance must be measured in a milieu devoid of environmental effects, that is, a mandated common production of seed lots: same year, same site, same processing.

Noting the several studies that show positive relationships between protein content and seedling vigor, grain yield, and large seed size, Ries, Ayers, Wert, and Everson (1976) confirmed that seeds with higher protein content and large size produced more vigorous seedlings. Protein per seed gave the closest correlation with seedling vigor. Their study also documented the relation of seed position in the head with protein content. Protein quantity per seed was increased by foliar applications of urea after anthesis. According to the authors, Holzman (1972) also found an association between high protein content of seed and high seedling vigor.

DePauw and Clarke (1976) examined and corroborated procedures for accelerating generation advancement in wheat proposed by Mukade, Kamio, and Hosada (1973), who had been able to grow four to six generations in one year, whereas two generations are about all that may be expected even with a spring wheat in northern latitudes. DePauw and Clarke modified the technique, which focused on inducing immature seeds to germinate, through changes in light, hydrogen peroxide applications, and preharvest clipping and drying. As a result, the authors were able to record a shortening of the plant-to-harvest period amounting to 12 days in one cultivar, to a maximum of 23 days in another. They suggested, regarding sprout-resistant wheats that begin to exhibit dormancy about 24 days post-anthesis, that harvesting seeds 21–22 days after anthesis should avoid dormancy problems. Robertson and Curtis (1967) had shown earlier that harvest at 21 days post-anthesis and air-drying seeds provided high germination in winter wheat.

These results are reminiscent of vivipary induced in barley by Pope (1949), who showed that water applied to the developing kernel seven days after pollination caused continuous growth, with the plumule visible on the fourteenth and roots on the fifteenth day after pollination.

The pertinence of these results to plant breeding, as suggested by De Pauw and Clarke, is that they can shorten by years particular breeding projects. The results are adaptable to backcross, single seed descent, mass and recurrent selection, and diallel selective mating projects, and are particularly useful for greenhouse and controlled growth chamber applications.

Holmes and Burrows (1976) reported on a potential seedling screening test for high-protein oats based on the high correlation of an early leaf protein test and the known protein content of the parent seed. Interpolation to a segregating population might indicate application to the F_2, but there is a problem here with numbers because there is a substantial laboratory effort required for the leaf analysis, which is completed by the twelfth day. If the method is found feasible, it may be that greatest efficiency would come at a later generation after discard has been practiced on any number of other important criteria. The seedling test then could be applied to screened large kernels as a predictor of protein among selected individuals.

A succinct literature review of the relation between protein and seed size and seedling vigor preceded the report by Ayers, Wert, and Ries (1976). Protein seed levels of a winter wheat cultivar, 'Logan', were enhanced by foliar applications of nitrogen at anthesis and three weeks after. A portion of their studies showed that both protein content and seed size were positively correlated with seedling vigor, but the correlations of protein with seedling vigor were higher than those with seed size.

Evans and Bhatt (1977a) conducted four greenhouse experiments to study different aspects of seedling vigor. They found seed size positively correlated with protein level and also with seedling vigor. Protein content was positively correlated with seedling vigor. The multiple correlation of the three phenomena was highly significant. Within genotypes (cultivars) larger seeds were positively associated with higher protein contents. Within a series of 12 cultivars significant differences in seedling vigor were found, even when seed weight was held constant and the effect of protein content minimized.

Evans and Bhatt considered that seed size and protein content might be used as a selection index for seedling vigor (SV). They stated further,

> The genotypic differences for SV observed lend encouragement for using SV measurements as a selection criterion in segregating hybrid populations. By far the most important factor influencing SV is the seed size variation. if this factor could be eliminated as a variable, the differences in SV could be attributed largely to genotypes.

Evans and Bhatt (1977b) coincidentally developed a selection method for seedling vigor in wheat by growing seeds in sand and water culture and harvesting at 20 days. They first determined that green shoot weight was correlated with other measurements (roots, caryopsis, whole plant, and all dry weights) and could be used as a reliable estimate of the whole plant dry weight. The selected high-vigor plants could then be resuscitated from the root and crown base. They stated, "The above procedures could be adapted to single seed descent, mass selection or recurrent selection where seedling vigor is used as a selection criterion in the breeding program."

Noting that standard germination tests of sugar beet seeds did not correlate well with observed field emergence, Akeson and Widner (1980) devised a laboratory packed-sand test that was highly correlated with field emergence.

The research of Bulisiani and Warner (1980) provided new insights into the interlocking relationships of seed size and seedling vigor in wheat with protein content and nitrogen availability. If we go beyond the research details, it was seen that field research under high-nitrogen conditions limited or camouflaged the response that might be expected from seed protein content—the response was not really evident in either vigor or yield. Yet the growth chamber experiments showed response both from the endogenous protein content of the seed and from experiments with exogenous applications of nitrogen. In fact, the latter essentially canceled differences in seed protein content which were substantial in the absence of additional nitrogen. Throughout, large seeds exerted more influence on seedling vigor than small seeds—there was some confounding here because large seeds also contained more protein. The authors speculated that under conditions of low soil nitrogen at seeding, the advantages of higher-protein seeds might be more evident, even to producing higher yield, as had been shown in some other research.

The problems of selection of any kind within accession germ plasms, which are seldom genetically pure, has long been a concern in the storage and maintenance of seeds. Roos (1984) of the National Seed Storage Laboratory in Fort Collins, Colorado contributed results of actual and simulated studies on genetic shifts in such populations following seed storage and crop regeneration. He found when seeds of mixed bean populations fell to 50% or less germinability, and this factor was coupled with the small sample sizes used for planting, that some components were gradually eliminated. The simulated determination of number of seeds of each genotype entering each cycle (a cycle population size was 64 seeds, initially made up of equal seed contributions from the eight genotypes) was calculated from a series of equations, and a component was considered eliminated when its seed contribution number fell below one.

There are messages here for scientists conducting population and competition studies, particularly if the studies extend over long time periods. The storage-related condition of planting seeds can easily be responsible for field results that, without this knowledge, might be attributed to competition effects.

Working with nine cultivars of spring wheat, which is often planted under adverse conditions of moisture and temperature that are inimical to early seed establishment, Lafond and Baker (1986a) studied the relationships among seed size, speed of emergence, seedling shoot dry weight, and Haun stage. The description of cereal plant development was systematized by Feekes (cf. Large, 1954); later, Haun (1973) quantified this, making it more applicable to the seedling stage. These traits, collectively, often are referred to as seedling vigor. Lafond and Baker's experiment involved three main effects: the nine cultivars, three seeding dates, and two seed sizes. All of these showed highly significant differences, except for seed size in one year. Significant interactions were seeding date by seed size, both years; seeding date by cultivar, in one year; cultivar × seed size, neither year (although Lafond and Baker's Table 3 seemed to show a significant interaction in 1982).

The significant differences observed for median emergence time suggests that the quicker-emerging types might be used in breeding. There was, however, a correlation between seed size and speed of emergence that could influence this interpretation: smaller seeds tended to emerge sooner than large.

Seed size seemed to play a dominant role in seedling vigor; that is, larger seeds always led to greater shoot dry weight than plants grown from small seeds. Estimates were that seed size accounted for about 50% of the differences between cultivars in their seedling shoot dry weights. Lafond and Baker concluded that selection for seedling vigor might begin with selecting greater seed weight (the easiest trait), shorter median emergence time, and earlier Haun stage of development. Perhaps index selection is indicated for a later stage.

In another paper in 1986, Lafond and Baker (1986b) looked at relative speeds of germination in the dark of nine spring wheat genotypes when two years of tests were conducted under different temperatures and osomtic moisture stress. The defined criterion for germination required the radicle to penetrate the coleorhiza and attain a length of 2–3 mm. Five temperature regimes were established: 5, 8, 12, 20, and 30°C. A further preparation was the sizing of seeds into large, medium, and small groups.

A major finding was that, given time, all samples exceeded 90% germination; therefore final germination percentage could not be used as a screening criterion. No interaction between cultivar and seed size was noted; however, seed size reflected an interesting relationship to speed of germination: small seeds germinated more rapidly, and the larger the seed size, the slower the germination regardless of temperature. The authors recommended that this phenomenon could be avoided by using seeds of similar size from a common source.

Temperature × cultivar interaction of median germination time was significant but was not due to ranking changes over temperature regimes: instead it was traced to greater differences shown by cultivars at the low temperatures than at the high. Lafond and Baker stated that a screening procedure using speed of germination would be but the first step in determining this phenomenon's relationship to other useful plant characteristics.

Differences in seedling vigor of cereals can be impressive—often dramatic—and many breeders have speculated as to its import and whether it might be used in selection. If shown to be important, seedling vigor has a number of attributes recommending it to the breeder as an aid. Some of these are early detection, nondestructive means of detection, and a logical association with positive plant characteristics such as seed size, protein content, seed density, fast germination and establishment, osmotic value, and competitive advantage.

SEED QUALITY: YIELD AND PROTEIN RELATIONSHIPS

There are few more dramatic illustrations of what pure and continued selection can accomplish than the high–low protein and oil experiments conducted at the Illinois Station. Woodworth, Lang, and Jugenheimer (1952) made a progress report at the end of 50 generations of selection. The results *were* dramatic: mean oil percent moved bidirectionally with selection from the 1896 base of 4.70% to 15.36% for high and 1.01% for low. The protein moved from the 10.92% base to 19.45% high and 4.91% low. The selection has resulted in great morphological and other changes, and it should be noted that yields are only about one-half those of adapted hybrids. Remarkably, the variation remaining is still high, especially in high oil and high protein.

'Atlas 66', a wheat cultivar from North Carolina, has long been known for its high protein content and leaf rust resistance (Middleton, Bode, and Bayles, 1954). Protein content and rust resistance are presumed linked. Johnson, Schmidt, Mattern, and Haunold (1963) had noticed in Nebraska tests that Atlas 66 was poorly adapted, barely surviving winters, and lower yielding, yet high in protein. To find out whether high protein levels would persist in adapted segregants they crossed Atlas 66 with high-yielding adapted 'Comanche' and examined the F_2-derived families. F_2 plants were screened for protein, and F_3 lines tested for protein and yield. The evaluation was based on 15 selected lines that were high in both categories, hence adapted.

Three years of testing showed that all 15 families exceeded Comanche, a good variety, in both yield and protein content. The association between high protein and leaf rust resistance was evident from the fact that all families were either resistant or segregating for disease reaction. Increased genetic variation for other agronomic and quality traits reinforced the conclusion that higher protein was a reasonable goal to pursue in hard wheat programs. It may be pointed out that Atlas 66's lack of adaptation to Nebraska conditions is not in any way derogatory—its germ plasm development and performance in North Carolina may be considered equivalent to that of Comanche.

Two 1964 papers dealing with the effectiveness of early-generation selection in spring wheat hybrid populations for various quality traits are those by McNeal, Berg, Bequette, Watson, and Koch (1964) and by Lebsock, Fifield, Gurney, and Greenaway (1964). The varied results indicated that F_3 and F_4 selection was effective for certain characters, for example, when F_3 selection was based on farinograph results.

Rasmusson (1965) reported a study to determine whether F_3 seed analysis could reliably assess four characters important to malting quality in spring barley. He found that selection for diastatic power was effective. Gain from selection for kernel plumpness, extract, and protein was slight, leading to the conclusion that it was not worthwhile. He recommended that prediction tests be delayed until advanced generations. Correlations for the traits were given.

In a presentation to the Third International Wheat Genetics Symposium in Canberra, Australia, Johnson, Whited, Mattern, and Schmidt (1968) reported further on the 'Frondoso'-derived high-protein variety, 'Atlas 66', and in the process gave additional information useful in breeding for high-protein wheats. The Atlas 66 case is unusual because two very desirable characters, high protein and leaf rust resistance, were found to be associated, presumably linked in the coupling phase. This combination counters the erstwhile commonly held belief that increasing the total yield of protein per acre was impossible because grain yield and percent protein acted as counterbalances—this did not happen in the Frondoso-derived lines, where no reductions in yield accompanied increases in protein.

Extensive testing of lines with this combination of rust resistance has been done at the Nebraska station. It is believed that differential nitrogen uptake by the plant is not significant in high grain protein; rather it seems to be due to more efficient and complete translocation within the plant.

The amino acid lysine is an important part of protein. It is of plant breeding interest that the most satisfactory way of expressing amounts found in tests is as a percent of protein—this more accurately expresses between-genotype relative differences.

Briggs, Bushuk, and Shebeski (1969) used control wheat cultivars in a form of uniformity trial to map fields for variations in fertility. This information was then used to evaluate the quality characteristics of lines according to where they were grown in the nursery.

The most obvious finding with plant improvement implications from a study on the relation of wheat seed protein on yield by Ries, Moreno, Meggih, Schweizer, and Ashkar (1970) was that higher seed protein was reflected in higher yields in the next crop. Increase in seed weight under the same conditions did not show the same relationship; that is, there was no correlation between yield and seed size.

Briggs and Shebeski (1971) examined early-generation selection for breadmaking quality and for yield in hard red spring wheat. Using three different hybrid populations, which later were evaluated singly in three successive years, they measured F_3 line yields directly, and also expressed them as percentages of adjacent controls (every third plot was a check). After selection for yielding ability, the F_3 lines were subjected to 13 separate quality (breadmaking) tests. The double truncation selection reduced the retained F_3 lines to relatively few. Because the selection considered 13 quality factors (independently—not on an index basis) the range and variation of quality was great.

The selected F_3 lines were grown for seed increase and further screened, with accompanying discard, for agronomic and other characters. The final groups were then grown in comprehensive F_5 evaluation trials, one cross group per year. The evaluation results are as follows.

Although incidental to the research, it must have pleased the authors that in all three years the average yield of the lines was significantly higher than that of the controls (commercial and bulk). Also pertinent to improvement via selection, the mean of the selected lines was greater than the mean of the ancestor lines from which they were selected. At least three selection regimes were imposed on the lines prior to the F_5 evaluation, so this is clear evidence of the effect of selection.

With three exceptions in one year (1000-kernel weight, blend loaf volume, and farinograph mixing tolerance index), significant differences for the other 10 quality traits were found between means of the F_5 populations in all three years. The F_3–F_5 relationships of the 13 quality and 2 yield criteria were mixed; that is, there were positive, negative, and ambiguous examples, which I believe must be attributable in part to the situation that the authors recognized: "The performance of an unreplicated F_3 selection at 1 location in 1 year was assessed by an F_5 population grown at a nearby but different site in the subsequent year."

Broad-sense heritabilities calculated in the F_5 were encouraging enough so that the authors recommended selection between F_3-derived populations for the breadmaking traits.

McNeal, Berg, McGuire, Stewart, and Baldridge (1972) studied the nitrogen–plant relationships with respect to yield and protein content in eight spring wheat crosses. From each cross high- and low-protein composites were formed from F_3 progeny row seeds. The crops grown from these composites showed that plants, regardless of low- or high-protein category, took up the same amount of nitrogen from the soil and translocated equal amounts to the grain. Nevertheless, the nitrogen percent of the high-protein composites was higher because of lower yield. The inverse yield–protein relationship persisted, but cases are known where the two increase together. The authors suspected that plant growth rather than genetic relationships may hold the key to dual increases.

Dubetz (1972) tested the generally held belief that yield and protein content are inversely related in wheat. With two spring wheat varieties grown under irrigation he found that protein contents generally rose with each added increment of nitrogen. However, first increments were not always effective, suggesting the need of larger amounts of nitrogen to trigger a difference. The difference in yield (large) of the two varieties was essentially canceled by differences in protein content, such that the two varieties only differed in total protein production by 3%. The results showed that nitrogen application under irrigated conditions produced high-protein bread wheats.

The responses of two wheat genotypes, differing in their grain protein percent, to nitrogen applications were studied by Johnson, Dreier, and Grabouski (1973). One of the wheats was a high-protein germ plasm release (2–3% higher protein) and the other the commercial high-yielding 'Lancer' (average protein content). The varieties were pair-tested across five and seven rate ranges of nitrogen application. The results showed (1) there were significant general protein responses to nitrogen; (2) regardless of yield differences and rate of nitrogen application the high-protein genotype maintained a consistent 2% protein advantage over Lancer; and (3) the genotypes responded differently—for example, "less nitrogen fertilizer was required for maximum yield of Lancer." However, from a breeding prospect it was evident that high grain yields could be attained simultaneously with high protein.

Bhatia (1975) considered the relationships of three criteria for early-generation selection to improving the yield of protein in wheat. The three were grain protein percent, per kernel, and per yield per unit area. It is well known that there are some negative correlations among and between these criteria, and yield and other characters affecting yield. Using 21 wheats selected for both high and low protein from 120 cultivars, Bhatia correlated the criteria with grain yield, biological yield, harvest index, 1000-kernel weight, and number of grains per square meter. The important associations found for the three criteria were:

1. *Grain Protein Percent:* It was negatively correlated with grain yield, grain number, and harvest index.
2. *Grain Protein Per Kernel:* Correlations were positive with grain weight, negative with grain number.
3. *Grain Protein Yield Per Unit Area:* Correlations were positive with grain yield, weight, and number; harvest index; and grain protein percent.

Bhatia concluded the best early-generation selection scheme to be the third, grain protein yield per unit area.

The listing of the 21 wheats and their attributes (his Table 1) can be viewed also as an index array, and in fact, Bhatia chose five wheats that ranked high in more than one of the desirable traits measured. Looking for the moment at breeding strategy rather than early generation selection, it would seem that a recurrent selection program beginning with intermating of top-ranked individuals might have promise. Because negative associations are so important, I would expect the program to branch into concurrently moving streams at the end of the first cycle; for example, one population might ignore the negative harvest index relationship, another might involve tandem selection, and so on. Improved selections from any stream could be shared by being fed back into the cyclic streams. The use of genetic male sterility, if available, would reduce the cost of the program over cycles.

Frey's (1977) paper on protein of oats stands as a modern summary of knowledge on this subject.

Takeda and Frey (1979) made the point that grain protein percentage of oats has meaning only when associated with high grain-yielding capacity. In their Table 1 they brought together the correlations between these two important attributes found in 11 research studies in five crops, barley, corn, oats, sorghum, and wheat. All of the correlations were negative and all were statistically significant. Because of this negative relationship, increase in percent protein of grain in oats, which has a favorable heritability of approximately 50%, is not a simple goal and probably is not as important as increased protein yields per unit area.

The Iowa station has pioneered in the use of *A. sterilis* germ plasm a source of genes for high protein. The authors used five successive backcrosses to the domestic *A. sativa* parent to produce backcross generations to study heritability and association relations among 13 traits. In general, the regression of backcross means to the recurrent parent was normal (and without selection in a breeding program would have advanced desired protein levels only by chance). Heritability percentages, of course, varied with each trait, but overall seemed favorable, falling within a total range from 34 to 68%. Groat protein percent showed positive genetic associations with plant height (unfavorable) and protein per groat, and negative ones with grain yield (expected) and harvest index (new).

Using the obtained data, Takeda and Frey ran a simulated study at a 5% selection intensity to see the expected advance for protein yield. Taking the top 5% of lines for grain yield, for groat protein percent, and for protein yield, they found that selection for grain yield alone, and for protein yield alone, both gave equal genetic gains in protein yield; however, selection for groat protein percent was ineffective in advancing protein yield.

To balance this simulated approach, Takeda and Frey detected four lines with significantly higher protein yields than the recurrent *sativa* parent (showing some introgression of the *sterilis* genes for protein yield). In one of the lines the increased protein yield was due entirely to higher grain yield. In another case there was a balanced contribution from grain yield, groat protein percentage, and interactions, suggesting different pathways to improving more than one character.

REFERENCES

Akeson, W. R. and J. N. Widner. 1980. Laboratory packed sand test for measuring vigor of sugarbeet seed. Crop Sci. 20:641–644.

Ali, A., I. M. Atkins, L. W. Rooney, and K. B. Porter. 1969. Kernel dimensions, weight, protein content and milling yield of grain from portions of the wheat spike. Crop Sci. 9:329–330.

Arny, A. C. and R. J. Garber. 1918. Variation and correlation in wheat with special reference to weight of seeds planted. J. Agric. Res. 14:359–392.

Austenson, H. M. and P. D. Walton. 1970. Relationships between initial seed weight and mature plant characters in spring wheat. Can. J. Plant Sci. 50:53–58.

Ayers, G. S., V. F. Wert, and S. K. Ries. 1976. The relationship of protein fractions and individual proteins to seedling vigour in wheat. Ann. Bot. 40:563–570.

Bartel, A. T. and J. H. Martin. 1938. The growth curve of sorghum. J. Agric. Res. 57:843–847.

Bhatia, C. R. 1975. Criteria for early generation selection in wheat breeding programs for improving protein productivity. Euphytica 24:789–794.

Bhatt, G. M. and N. F. Derera. 1973. Associated changes in the attributes of wheat populations mass selected for seed size. Aust. J. Agric. Res. 24:179–186.

Bhatt, G. M. and N. F. Derera. 1975. Genotype × environment interactions for, heritabilities of, and correlations among quality traits in wheat. Euphytica 24: 597–604.

Bonnett, O. T. and C. M. Woodworth. 1931. A yield analysis of three varieties of barley. J. Am. Soc. Agron. 23:311–327.

Boyd, W. J. R., A. G. Gordon, and L. J. LaCroix. 1971. Seed size, germination resistance, and seedling vigor in barley. Can. J. Plant Sci. 51:93–99.

Bradshaw, A. D, 1964. Inter-relationship of genotype and phenotype in a varying environment. Scottish Plant Breeding Stn. Rec. (1964), pp. 117–125.

Bravo, J. A., W. R. Fehr, and S. R. de Cianzio. 1981. Use of small-seeded soybean parents for the improvement of large-seeded cultivars. Crop Sci. 21:430–432.

Bremner, P. M., R. N. Eckersall, and R. K. Scott. 1963. The relative importance of embryo size and endosperm size in causing the effects associated with seed size in wheat. J. Agric. Sci. 61:139–145.

Bridgford, O. R. and H. K. Hayes. 1931. Correlation of factors affecting yield in hard red spring wheat. J. Am. Soc. Agron. 23:106–117.

Briggs, K. G. and L. H. Shebeski. 1971. Early generation selection for yield and breadmaking quality of hard red spring wheat. Euphytica 20:453–463.

Briggs, K. G., W. Bushuk, and L. H. Shebeski. 1969. Variation in breadmaking quality of systematic controls in a wheat breeding nursery and its relation to plant breeding procedures. Can. J. Plant Sci. 49:21–28.

Bulisiani, E. A. and R. L. Warner. 1980. Seed protein and nitrogen effects upon seedling vigor in wheat. Agron. J. 72:657–662.

Carver, M. F. F. 1977. The influence of seed size on the performance of cereals in variety trials. J. Agric. Sci. 89:247–249.

Christensen, C. M. 1955. Grain storage studies. XXI. Viability and moldiness of commercial wheat in relation to the incidence of germ damage. Cereal Chem. 32:507–518.

Christiansen, M. N. 1962. A method of measuring and expressing epigeous seedling growth rate. Crop Sci. 2:487–489.

Cutler, G. H. 1940. Effect of "clipping" or rubbing the oat grain on the weight and variability of the seed. J. Am. Soc. Agron. 32:167–175.

Dasgupta, P. R. and H. M. Austenson. 1973. Relations between estimates of seed vigor and field performance in wheat. Can. J. Plant Sci. 53:43–46.

DePauw, R. M. and J. M. Clarke. 1976. Acceleration of generation advancement in spring wheat. Euphytica 25:415–418.

Derera, N. F. and G. M. Bhatt. 1972. Effectiveness of mechanical mass selection in wheat (*Triticum aestivum* L.). Aust. J. Agric. Res. 23:761–768.

Dubetz, S. 1972. Effects of nitrogen on yield and protein content of Manitou and Pitic wheats grown under irrigation. Can. J. Plant Sci. 52:887–890.

Ellis, J. R. S. and A. D. Hanson. 1974. Tests for cereal yield heterosis based on germinating seeds: a warning. Euphytica 23:71–77.

Evans, L. E. and G. M. Bhatt. 1977a. Influence of seed size, protein content and cultivar on early seedling vigor in wheat. Can. J. Plant Sci. 57:929–935.

Evans, L. E. and G. M. Bhatt. 1977b. A nondestructive technique for measuring seedling vigor in wheat. Can. J. Plant Sci. 57:983–985.

Fehr, W. R. and C. R. Weber. 1968. Mass selection by seed size and specific gravity in soybean populations. Crop Sci. 8:551–554.

Fezer, K. D. 1962. Differential incidence of loose smut among seed-size classes within barley seedlots. Crop Sci. 2:162–164.

Findlay, W. M. 1956. Oats. Their cultivation and use from ancient times to the present day. Oliver and Boyd, Edinburgh and London.

Frey, K. J. 1962. Inheritance of seed weight and its relation to grain yield of oats. Iowa Acad. Sci. Proc. 69:165–169.

Frey, K. J. 1977. Protein of oats. Z. Pflanzenzuecht. 78:185–215.

Frey, K. J. and T. F. Huang. 1969. Relation of seed weight to grain yield in oats, *Avena sativa* L. Euphytica 18:417–424.

Freyman, S. 1978. Influence of duration of growth, seed size, and seeding depth on cold hardiness of two hardy winter wheat cultivars. Can. J. Plant Sci. 58:917–921.

Fryxell, P. A. 1954. A procedure of possible value in plant breeding. Agron. J. 46:433–434.

Gardner, C. O. 1961. An evaluation of effects of mass selection and seed irradiation with thermal neutrons on yield of corn. Crop Sci. 1:241–245.

Geiszler, G. N. and B. K. Hoag. 1967. Wheat seed size influences yield. N. Dak. Farm Rev. 24:12–14.

Georgeson, C. C., F. C. Burtis, and D. H. Otis. 1896. Experiments with wheat. Kansas Agric. Exp. Stn. Bul. 59:89–105.

Georgeson, C. C., F. C. Burtis, and D. H. Otis. 1897. Experiments with oats. Kansas Agric. Exp. Stn. bul. 74:195–211.

Gordon, A. G. 1971. The germination resistance test—a new test for measuring germination quality of cereals. Can. J. Plant Sci. 51:181–183.

Grafius, J. E., W. L. Nelson, and V. A. Dirks. 1952. The heritability of yield in barley as measured by early generation bulked progenies. Agron. J. 44:253–257.

Gregory, F. G. and O. N. Purvis. 1936. Vernalization of winter rye during ripening. Nature 138: 973.

Gregory, F. G. and O. N. Purvis. 1938. Studies in vernalization of cereals. II. The vernalization of excised mature embryos, and of developing ears. Ann. Bot. 2:237–251.

Hamblin, J. 1975. Effect of environment, seed size and competitive ability on yield and survival of *Phaseolus vulgaris* (L.) genotypes in mixtures. Euphytica 24:435–445.

Haun, J. R. 1973. Visual quantification of wheat development. Agron. J. 65:116–119.

Havey, M. J. and K. J. Frey. 1978. Optimum sample size and number per plot and replicate number for seed weight of oats. Cereal Res. Commun. 6:113–122.

Hayes, H. K., O. S. Aamodt, and F. J. Stevenson, 1927. Correlation between yielding ability, reaction to certain diseases and other characters of spring and winter wheat in rod-row trials. J. Am. Soc. Agron. 19:896–910.

Heinrich, G. M., C. A. Francis, J. D. Eastin, and Mohammad Saled. 1985. Mechanisms of yield stability in sorghum. Crop Sci. 25:1109–1113.

Holmes, D. P. and V. D. Burrows. 1976. Development of a seedling screening test for predicting relative grain protein content in oats. Euphytica 25:51–64.

Holton, C. S. and F. D. Heald. 1936. Studies on the control and other aspects of bunt of wheat. Washington Agric. Exp. Bull. 339.

Holzman, M. 1972. Untersuchungen uber den einflur der dungung auf den saatgutwert und dessen lokalisation in getreidekorn. Z. Acker- Pflanzenbau 135:279–309.

Hunter, R. B. and L. W. Kannenberg. 1972. Effects of seed size on emergence, grain yield and plant height in corn. Can. J. Plant Sci. 52:252–256.

Hutchinson, T. B. 1913. Correlation characters in *Avena sativa* with special reference to size of kernels planted. M. S. Thesis, Cornell University, Ithaca, NY.

Immer, F. R. and F. J. Stevenson. 1928. A biometrical study of factors affecting yield in oats. J. Am. Soc. Agron. 20:1108–1119.

Jennings, P. R. and R. C. Aquino. 1968. Studies on competition in rice. III. Mechanisms of competition among phenotypes. Evolution 22:529–542.

Jensen, N. F. 1964. Processing equipment for small grain test weight samples. Crop Sci. 4:438–439.

Johnson, V. A., J. W. Schmidt, P. J. Mattern, and A. Haunold. 1963. Agronomic and quality characteristics of high protein F_2-derived families from a soft red winter–hard red winter wheat cross. Crop Sci. 3:7–10.

Johnson, V. A., D. A. Whited, P. J. Mattern, and J. W. Schmidt. 1968. Nutritional improvement of wheat by breeding. Third Int. Wheat Genet. Symp., Aust. Acad. Sci.; Canberra, pp. 457–461.

Johnson, V. A., A. F. Dreier, and P. H. Grabouski. 1973. Yield and protein responses to nitrogen fertilizer of two winter wheat varieties differing in inherent protein content of their grain. Agron. J. 65:259–263.

Kaufmann, M. L. 1958. Seed size as a problem in genetic studies of barley. Proc. Genet. Soc. Can. 3:30–32.

Kaufmann, M. L. and A. A. Guitard. 1967. The effect of seed size on early plant development in barley. Can. J. Plant Sci. 47:73–78.

Kaufmann, M. L. and A. D. McFadden. 1960. The competitive interaction between barley plants grown from large and small seeds. Can. J. Plant Sci. 40:623–629.

Kaufmann, M. L. and A. D. McFadden. 1963. The influence of seed size on results of barley yield trials. Can. J. Plant Sci. 43:51–58.

Katz, R., E. P. Farrell, and M. Milner. 1954. The separation of grain by projection. Cereal Chem. 31:316–325.

Kiesselbach, T. A. 1924. Relation of seed size to the yield of small grain crops. J. Am. Soc. Agron. 16:670–682.

Kiesselbach, T. A. and C. A. Helm. 1917. Relation of size of seed and sprout value to the yield of small grain crops. Nebr. Agric. Exp. Stn. Res. Bull. 11, 73 pp.

Kinbacher, E. J. 1962. Effect of seed source on the cold resistance of pre-emerged Dubois winter oat seedlings. Crop Sci. 2:91–93.

Kneebone, W. R. and C. L. Cremer. 1955. The relationship of seed size to seedling vigor in some native grass species. Agron. J. 47:472–477.

Knott, D. R. and B. Talukdar. 1971. Increasing seed weight in wheat and its effect on yield, yield components, and quality. Crop Sci. 11:280–283.

Lafond, G. P. and R. J. Baker. 1986a. Effects of genotype and seed size on speed of emergence and seedling vigor in nine spring wheat cultivars. Crop Sci. 26:341–346.

Lafond, G. P. and R. J. Baker. 1986b. Effects of temperature, moisture stress, and seed size on germination of nine spring wheat cultivars. Crop Sci. 26:563–567.

Large, E. C. 1954. Growth stages in cereals. Illustrations of the Feekes' scale. Plant Pathol. 3:128–129.

Lebsock, K. L., C. C. Fifield, G. M. Gurney, and W. T. Greenaway. 1964. Variability and evaluation of mixing tolerance, protein content and sedimentation value in early generations of spring wheat, *Triticum aestivum* L. Crop Sci. 4:171–174.

Lill, J. G. 1910. The relation of size, weight and density of kernel to germination of wheat. Kansas State Agric. Coll. Exp. Stn. Circ. 11:8 pp. Manhattan.

Lowe, L. B., G. S. Ayers, and S. K. Ries. 1972. The relationship of seed protein and amino acid composition to seedling vigour and yield of plant. Agron J. 64:608–611.

Lowe, L. B. and S. K. Ries. 1973. Endosperm proteins of wheat seed as a determinant of seedling growth. Plant Physiol. 51:57–60.

McFadden, A. D. 1958. Breeding for high yield. Proc. Can. Soc. Agron., pp. B7–12.

McFadden, A. D. 1963. Effect of seed source on comparative test results in barley. Can. J. Plant Sci. 43:295–300.

McFadden, A. D., M. L. Kaufmann, R. C. Russell, and L. E. Tyner. 1960. Association between seed size and the incidence of loose smut in barley. Can. J. Plant Sci. 40:611–615.

McNeal, F. H., and D. J. Davis. 1954. Effect of nitrogen fertilization on yield, culm number and protein content of certain spring wheat varieties. Agron. J. 46:375–378.

McNeal, F. H., M. A. Berg, A. L. Dubbs, J. L. Krall, D. E. Baldridge, and G. P. Hartmann. 1960. The evaluation of spring wheat seed from different sources. Agron. J. 52:303–304.

McNeal, F. H., M. A. Berg, R. K. Bequette, C. A. Watson, and E. J. Koch. 1964. Early generation selection for flour absorption and dough mixing properties in a Lemhi × Thatcher wheat cross. Crop Sci. 4:105–108.

McNeal, F. H., M. A. Berg, C. F. McGuire, V. R. Stewart, and D. E. Baldridge. 1972. Grain and plant nitrogen relationships in eight spring wheat crosses, *Triticum aestivum* L. Crop Sci. 12:599–602.

Maguire, J. D. 1962. Speed of germination—aid in selection and evaluation for seedling emergence and vigor. Crop Sci. 2:176–177.

Mallach, J. 1929. Untersuchungen uber die Bedeutung von Korngrosse und Einzelkorngewicht beim Saagut. Wiss. Arch. Landw., Abt. A, Arch. Pflanzenbau 2:219–295.

Middleton, G. K. and T. T. Hebert. 1950. Purple straw color in relation to kernel weight in wheat. Agron. J. 42:520.

Middleton, G. K., C. E. Bode, and B. B. Bayles. 1954. A comparison of the quantity and quality of protein in certain varieties of soft wheat. Agron. J. 46:500–502.

Milner, M., E. P. Farrell, and R. Katz. 1954. The separation of grain by projection. II. Systematic differences in physical properties and composition of wheat fractions. Cereal Chem. 31:326–332.

Montgomery, E. G. 1910. Methods of testing the seed value of light and heavy kernels in cereals. J. Am. Soc. Agron. 2:59–60.

Mukade, K., M. Kamio, and K. Hosada. 1973. The acceleration of generation advancement in breeding rust-resistant wheat. Proc. 4th Int. Wheat Breeding Genet. Symp., Missouri Agric. Exp. Stn., Columbia:439–441.

Murphy, C. F. and K. J. Frey. 1962. Inheritance and heritability of seed weight and its components in oats. Crop Sci. 2:509–512.

Pope, M. N. 1949. Viviparous growth in immature barley kernels. J. Agric. Res. 78:295–309.

Quinby, J. R., L. P. Reitz, and H. H. Laude. 1962. Effect of source of seed on productivity of hard red winter wheat. Crop Sci. 2:201–203.

Rasmusson, D. C. 1965. Effectiveness of early generation selection for four quality characters in barley. Crop Sci. 5:389–391.

Ries, S. K. 1971. The relationship of protein content and size of bean seed with growth and yield. J. Am. Soc. Hort. Sci. 96(5):557–560.

Ries, S. K. and E. H. Everson. 1973. Protein content and seed size relationship with seedling vigour of wheat cultivars. Agron. J. 65:884–886.

Ries, S. K., C. J. Schweizer, and H. Chmiel. 1968. The increase in protein content and yield of Simazine-treated crops in Michigan and Costa Rica. Bioscience 18:205–208.

Ries, S. K., G. Ayers, V. Wert, and E. H. Everson. 1976. Variation in protein, size and seedling vigor with position of seed in heads of winter wheat cultivars. Can. J. Plant Sci. 56:823–827.

Ries, S. K., O. Moreno, W. F. Meggih, C. J. Schweizer, and S. A. Ashkar. 1970. Wheat seed protein: chemical influence on and relationship to subsequent growth and yield in Michigan and Mexico. Agron. J. 62:746–748.

Robertson, L. D. and B. C. Curtis. 1967. Germination of immature kernels of winter wheat. Crop Sci. 7:269–270.

Rogler, G. A. 1954. Seed size and seedling vigor in crested wheatgrass. Agron. J. 46:216–220.

Roos, E. E. 1984. Genetic shifts in mixed bean populations. I. Storage effects. Crop Sci. 24:240–244.

Roy, N. N. 1973. Effect of seed size differences in wheat breeding by single seed descent. J. Aust. Inst. Agric. Sci. 39:70–71.

Sarquis, A. V., B. B. Fischer, F. G. Parsons, and M. D. Miller. 1961. Geographic origin of barley seed produces no effect on yield. Calif. Agric. 15(4):3.

Schweizer, C. J. and S. K. Ries. 1969. Protein content of seed: increase improves growth and yield. Science 165:73–75.

Sharma, D. and D. R. Knott. 1964. The inheritance of seed weight in a wheat cross. Can. J. Genet. Cytol. 6:419–425.

Sterling, J. D. E., H. W. Johnston, and D. C. Munro. 1977. Effects of seed source and seed treatment on barley emergence, yield, and kernel weight. Can. J. Plant Sci. 57:251–256.

Suneson, C. A. and G. L. Peltier. 1936. Effect of source, quality, and condition of seed upon the cold resistance of winter wheats. J. Am. Soc. Agron. 28:687–693.

Takeda, K. and K. J. Frey. 1979. Protein yield and its relationship to other traits in backcross populations from an *Avena sativa* × *A. sterilis* cross. Crop Sci. 19:623–628.

Takeda, K. and K. J. Frey. 1980. Tertiary seed set in oat cultivars. Crop Sci. 20:771–774.

Tandon, J. P. and S. C. Gulati. 1966. Influence of non-genetic variation in seed size on quantitative characters in barley. Indian J. Genet. Plant Breed. 26:162–169.

Taylor, J. W. 1928. Effect of continuous selection of small and large wheat seed on yield, bushel weight, varietal purity and loose smut infection. J. Am. Soc. Agron. 20:856–867.

Tervet, I. W. 1944. The relation of seed quality to the development of smut in oats. Phytopathology 34:106–115.

Throneberry, G. O. and F. G. Smith. 1955. Relation of respiratory and enzymatic activity to corn seed viability. Plant Physiol. 30:337–343.

Trelease, S. F. and H. M. Trelease. 1924. Relation of seed weight to growth and variability of wheat in water cultures. Bot. Gaz. 77:199–211.

Waldron, L. R. 1910. A suggestion regarding heavy and light seed grain. Am. Nat. 44:48–56.

Waldron, L. R. 1926. A partial analysis of yield of certain common and durum wheats. J. Am. Soc. Agron. 21:295–309.

Waldron, L. R. 1941. Analysis of yield of hard red spring wheat grown from seed of different weights and origin. J. Agric.Res. 62:445–460.

Waldron, L. R. 1944. How wheats behave in competition with one another. N. Dak. Exp. Stn. Bi-monthly Bull. 6:7–14.

Walsh, E. J., E. Reinbergs, and K. J. Kasha. 1973. Importance of seed source in preliminary evaluations of doubled haploids in barley. Can J. Plant Sci. 53:257–260.

Weibel, D. E. 1958. Vernalization of immature winter wheat embryos. Agron. J. 50:267–270.

White, W. J. 1962. Clonal propagation of wheat plants for the evaluation of yield. Can. J. Plant Sci. 42:571–581.

Williams, C. G. 1916. Wheat experiments. Ohio Agric. Exp. Stn. Bull. 298:447–484.

Williams, C. G. and F. A. Welton. 1913. Oats. Ohio Agric. Exp. Stn. Bull. 257: 255–283.

Williams, T. V., R. S. Snell, and J. F. Ellis. 1967. Methods of measuring drought tolerance in corn. Crop Sci. 7:179–182.

Williams, W. A. 1963. The emergence force of forage seedlings and their response to temperature. Crop Sci. 3:472–474.

Woodworth, C. M., E. R. Lang, and R. W. Jugenheimer. 1952. Fifty generations of selection for protein and oil in corn. Agron. J. 44:60–65.

Yamazaki, W. T. and L. W. Briggle. 1969. Components of test weight in soft wheat. Crop Sci. 9:457–459.

Zavitz, C. A. 1908. The relation between the size of seeds and the yields of plants of farm crops. Proc. Am. Soc. Agron. 1:98–104.

Zavitz, C. A. 1915. Farm Crops. Results of experiments at the Ontario Agricultural College. Ont. Dept. Agric. Bull. 228. 80 pp.

33

MORPHOLOGICAL AND OTHER TRAITS

This chapter, particularly the section on awns, is admittedly more associated with my career experiences with the cereal small grains than with other crops.

HEIGHT RELATIONSHIPS

Allan, Vogel, and Craddock (1959) published descriptions of the effect of gibberellic acid applications to dwarf, semidwarf, and standard-height wheats.

Allan and Vogel (1964) showed a continued concern about coleoptile length and first leaf growth because of their importance to fall-sown wheat emergence in Washington. Using the well-known semidwarf stock 'Norin 10-Brevor 14' and monosomic analysis, they found positive agreement between coleoptile length and mature plant height. Coleoptile length and first leaf growth appeared to have complex inheritance patterns.

Using F_2 and monosomic analysis, Gale, Marshall, and Worland (1975) identified a gene, Gai_3, for insensitivity to gibberellic acid in the cultivar 'Minister Dwarf'. The trait may be common to dwarf types, and because it has been shown that a similar lack of response is in Norin 10-Brevor 14, the ancestor of many commercial varieties, genic commonality may be assumed. A single dominant gene exerts control; in one cross the authors found F_2 ratios of 79 insensitive : 29 sensitive.

Coleoptile length is an important varietal trait; if it is short, emergence may be impeded. Although not conclusively determined, it appears that Gai_3 depresses other gene action that might increase coleoptile length.

Gale and Marshall (1976) showed in a brief literature review that Gai_1, Gai_2, and Gai_3 have the same chromosomal locations as the height-reducing genes in descending height order, Rht_1, Rht_2, and Rht_3, respectively. Whether linkage or pleiotropism is involved is not fully known. Gale and Marshal give an excellent discussion of the two hypotheses. From a breeding standpoint, linkage would be preferable to pleiotropism; linkage can be broken, but pleiotropism could present an unbreakable association not subject to manipulation.

The paper by O'Brien and Pugsley (1981) confirmed the intricate relationship between dwarfing genes, yield, gibberellic acid, and pleiotropism described and proposed in the earlier series of papers (Allan, Vogel and Craddock, 1959; Allan and Vogel, 1964; Gale et al., 1975; Gale and Marshall 1976). The 'Norin 10' dwarfing genes (*Rht* series) were associated with an apparent pleiotropic yield effect. It was separately noted that the *gai/rht* genotype was sensitive to gibberellic acid application, resulting in visible elongation of internodes, whereas the *Gai/Rht* genotype did not respond and thus could be separated in the seedling stage and later tested for yield. O'Brien and Pugsley involved a number of varieties of conventional height and semidwarf 1 and 2 categories of reduced height, in such a manner that there were F_2 populations segregating in the gibberellic acid insensitivity groups, *Gai/Rht* 1 and *Gai/Rht* 2.

Selection for GA_3 insensitivity (reduced height) by spraying with acid at the F_2 seedling stage resulted in a later correlated response to selection for higher grain yield in the *Gai/Rht* genotypes; thus an indirect selection effect had a positive pleiotropic influence. The homozygous insensitive individuals could be visually identified. The heterozygous plants could not be easily separated from the homozygous sensitive; however, it was found that measurement of the height to the second leaf node was fairly reliable as a guide to dividing this latter group.

AWNS

Grantham (1919) noted that farmers had a general prejudice against awned wheats. However, his studies showed that bearded wheats had outyielded beardless ones by 3.3 bu/acre in 26 tests over 10 years. Grantham stated that the difference in yield was caused in most cases by the poorer quality of the grain from the smooth-awned varieties. It is well-known that one of the positive effects of awns is on test weight of grain. He also noted that the bearded types tillered more and were higher yielding in poor years.

In a detailed study Harlan and Anthony (1920) measured the effect of awn removal from six-row and two-row barleys. They found that awn removal occasioned no injury; the clipped and unclipped kernels developed normally for a time after clipping—a period during which injury, if inflicted, would be evident. Unclipped normal-awned barleys outyielded clipped sister spikes. The difference in ultimate kernel weight was due to quantity of starch deposited. The authors speculated that the awns may act as a "sink," in today's parlance, for deposit of surplus or unneeded minerals. In the absence of awns, as in the clipped spikes, such materials are deposited in the rachis, resulting in brittleness, a characteristic more often observed in awnless and hooded types than in awned ones. It is interesting that this research was done in an era when smooth-awned or barbless types were still in the developmental stage.

Hayes and Wilcox (1922) accepted the premise that awned barleys usually outyield awnless ones. An important early, pre-combine-harvester consideration in favor of awnless barleys rested on the elimination of the "unpleasant features of handling rough-awned varieties." Hayes and Wilcox were unable to find any obvious physiological limitations peculiar to smooth-awned versus rough-awned types from their yield and transpiration studies, implying no impediment to breeding for smooth-awned barleys.

The unpleasant features of rough-awned barleys referred to above have been partly erased by the advent of the combine harvester, which eliminated the handling of grain bundles from the grain binder, the chore of shocking the bundles, and the loading and unloading of bundles during threshing (anyone who lived during this era would have difficulty choosing between the pleasures of growing rough-awned barley and harvesting of Russian thistles in the drought years!). But not all of these features have been eliminated, as farmers and ranchers will attest. Barley straw of a rough-awned variety is shunned both for feed and bedding. The awns cause mouth sores in livestock and become embedded in hair, wool, and rancher's clothing. The breeding of smooth-awned varieties has been a significant advance in plant breeding.

Hayes (1923) compared awned versus awnless lines from the cross 'Marquis' × 'Preston' in three generations, F_3–F_5. The yield advantage in favor of awned wheats amounted to 7%, 17%, and 8%, respectively, over the three years. The yield advantage from awns was relatively greater in unfavorable seasons. Hayes attributed the awned yield advantage to a physiological relationship.

In an extensive inheritance study in crosses between two hard red spring wheats being bred for drought and rust resistance, Clark (1924) found a very slight advantage in yield associated with awnedness; this occurred under drought conditions.

Hayes, Aamodt, and Stevenson (1927) found plumper grain associated with awned wheats rather than with the awnless types.

The accepted classification of wheat awns by Clark, Florell, and Hooker (1928) in a sense recognizes the range and variation of awn types typically found in crosses and represented in cultivars. Five classes were recognized: (1) awnless, (2) apically awnletted, (3) awnletted, (4) short awned, and (5) awned. Awnless is typically dominant. They found the presence of awns to favor higher yields.

Goulden and Neatby (1929) found, in a group of wheat lines from the cross of 'H–44–24' × Marquis, that 34 were awned and 48 were awnless. The average yield advantage of the awned lines was 14% (245.6 g vs. 215.6 g).

Clark and Quisenberry (1929) found awnletted plants to be higher in yield than awned plants in a cross between awnletted Marquis and awned 'Kota' wheats. The study was marred by greater shattering in the awnletted types.

Clark, Quisenberry, and Powers (1933) found no association between awn type and bunt reaction (independent), and also none between awns and yield in the cross of awned, resistant 'Hope' and awnless, moderately susceptible 'Hard Federation' spring wheats.

Aamodt and Torrie (1934) studied 57 F_3 lines from the cross of awnless 'Reward' × awned 'Caesium' and found no statistically significant differences between type of awns (awnletted, segregating, and awned) and yield. A similar nonsignificant relationship was found in comparisons of 62 awnletted and awned lines of 'Marquillo' 2× Marquis × 'Kanred'. There was, however, a slight skew in favor of higher yields of awned types.

Parker (1934) repeated in wheat Harlan and Anthony's 1920 experiments with barley to ascertain the effect of awn removal at different stages of maturity. He concluded that removal of the awns resulted in lowered yields. The earlier the awns were removed the greater was the yield reduction.

Lamb (1937) reported on a study of awns and the productivity of Ohio wheats. He found a very slight advantage, not important, of awns to yield and 1000-kernel weight. He saw a trend toward the increased use of beardless wheats in the soft wheat area.

Our knowledge of the influence of awns on yield has been ambiguous. Bayles and Suneson (1940) made comparisons between homozygous awnletted (i.e., awnless) and awned composites separated in early generations from a cross of western winter wheats and a cross of western spring wheats. Over a four-year period at several western experiment stations, the effect of awns or the absence of awns was measured on kernel weight, test weight, and yield with these results:

1. *Kernel Weight:* In 36 and 42 comparisons the awned composites showed 2.6% and 4.3% higher kernel weights in the winter and spring crosses, respectively, than did the awnless or awnletted bulks (differences statistically significant at the 1% level).

2. *Test Weight:* A decided advantage in heavier grain test weight in favor of awns in both crosses was found: 0.7 and 1.0 lb/bushel average superiority in 31 and 35 comparisons for winter and spring crosses, respectively.

3. *Yield:* The winter and spring crosses responded differently: the winter cross showed no differences in yield (means were equal over 48 comparisons). The awned spring composite, despite showing no difference in means the first two years, gave a decided advantage to the awned group in the final

two years, and for the 50 comparisons as a whole there was a 2.5 bu/acre advantage to awns.

Middleton and Chapman (1941) found that smooth-awned barleys rapidly disappeared from an early generation (F_3) of Harlan and Martini's 28-variety bridging cross population planted in the fall in North Carolina. Further studies showed that there was an association of early growth with smooth awnedness such that these plants were injured by early fall freezes.

In yield trials of segregates bulked by awn type, six-row rough awn surpassed six-row smooth awn 41.83–32.62 bu/acre (nonsignificant); however, yields of F_2-derived F_3 lines in 5-ft rows showed significant yield differences in favor of rough awns. The findings are particularly relevant to winter barley breeding; the impact upon spring barley selection is not known.

Miller, Gauch, and Gries (1944) showed that the deawning of winter wheats reduced transpiration.

Suneson, Bayles, and Fifield (1948) reported on comparisons of original awned 'Baart' with its backcross-derived awnless Baart, and original awnless 'Onas' with its backcross-derived awned Onas. The comparisons were measured by 20 cooperators at 18 locations. The advantage of awns from weighted means from the locations over three years were:

Baart vs awnless Baart:	43.5 : 39.4 bu/acre
Baart vs awnless Baart:	61.2 : 60.1 lb/bu test weight
Awned Onas vs Onas:	49.6 : 47.8 bu/acre
Awned Onas vs Onas:	59.1 : 58.2 lb/bu test weight

The importance of the ear to the development of grain dry matter was shown by Porter, Pal, and Martin (1950). Using a barley cultivar they monitored carbon dioxide absorption for 33 days and found that during 496 daylight hours 622 mg CO_2/ear was taken in, and during 296 night hours 117 mg was lost. The net assimilation gain of CO_2 accounted for 34% of the average grain dry weight of 1.09 g/ear. Awn presence or absence was not indicated.

Suneson, Schaller, and Everson (1952) used the backcross method to transfer genes for degrees of awnedness into genetic backgrounds so that their association with yield could be tested. There was initially some question as to whether the higher yield noticed in the variety 'Lion' was due to genes for the awn type or to those conditioning rachilla hair length; this was resolved later and shown to be due to genes for awn barbing. Composites of semismooth-awned individuals (recessive) outyielded rough-awned composites by 0.6, 32.5, 13.3, and 15.5 bu/acre in four years of testing, respectively. The only visible association with yield was the observation that yields and kernel weight were associated. The barbing effect is especially interesting inasmuch as a presence or absence of awn is not involved. Why barbing should be associated with yield is unknown; it may represent a linkage with a component of yield, likely kernel weight.

Acting on procedures proposed by Atkins and Mangelsdorf (1942), Atkins and Norris (1955) reported on research using 10 pairs of isogenic lines of awned and awnless wheats to measure the influence of awns on yield and four other traits, namely, number of culms per area, number of kernels per head, size of kernels, and test weight.

Over a four-year testing period at Denton, Texas, the awned lines yielded an average 1.1 bu/acre (4.4%) more than the awnless lines. In drought years the advantage of the awned lines was greater. Significantly heavier kernels and higher test weights were also found in the awned lines.

The association of semismooth awns of barley and high yield reported by Suneson, Schaller, and Everson in 1952 was found to be a case of linkage by Everson and Schaller (1955). Crossing over within the Chromosome V segment, and segregation, indicated that different factors for yield were involved. An interesting finding was that the increased yield, which was substantial over that of the recurrent parent, Atlas, occurred with early plantings only. There was no increase with late-planted isolines.

The practical consequences of awns in wheat was interpreted from the standpoint of grain and plant moisture relationships in fields of soft red winter wheats by Pool and Patterson (1958). Using near-isogenic lines differing for awns and awnlessness, they found that the presence of awns increased grain moisture absorption during rains and periods of high humidity, but also facilitated grain drying. They considered the latter more valuable and recommended awn presence as an aid to grain drying.

Using near-isogenic lines, Suneson and Ramage (1962) confirmed earlier studies that awned wheats (they tested three basic cultivars) received a 7% yield advantage over awnless types. It was found that the seeds from awned types were larger, which may relate to the yield advantage; however, in mixtures of awned and awnless wheats no competitive advantage was shown by either. Rough-awned barley showed a small but significant yield advantage over awnless types, and this advantage carried over in competition tests.

Patterson, Compton, Caldwell, and Schafer (1962) corroborated the general finding that the presence of awns is favorable to increased kernel weight and test weight of grain. Tests were made on 12 visually similar near-isogenic wheat lines. Awns increased grain yields also; however, there were few statistically significant increases and they all occurred in the year most favorable to high yields. The parental source of awns seemed to exert little influence.

Vogel, Allan, and Peterson (1963) studied correlations of characters with performance in semidwarf wheats adapted to eastern Washington State. They concluded that awnedness was more important to performance in semidwarf than in standard-height varieties. The best semidwarfs were found to have coarse awns providing yield and test weights at least 10% above otherwise comparable awnless types.

Grundbacher's (1963) excellent review of the anatomy and function of the cereal awn highlights two biological processes, photosynthesis and transpiration, the latter's role not fully understood. More understandable is photosynthesis: the awn is a green photosynthetic organ happily attached to the grain structure. Grundbacher estimated that spike and awn contributed about 30% of the carbohydrates in the mature grain, and may account for more than 10% of total kernel dry weight. Practically, the effect of awns appears in higher yields and higher test weight of cereal grain.

McKenzie (1972) presented a clear case of awns having a negative effect on yield of wheat. When long awns were backcrossed to awnless Thatcher (actually awnletted), yields were reduced. The donor of the awns was the cultivar 'Lee', also involved in the reciprocal crosses and backcrosses. Awnless Lee increased in yielding ability upon the removal of awns.

McKenzie gave explanations for this relation of awns to yield. Based on my experience with breeding the awned winter wheat 'Houser', I favor linkage. I suspect that Houser is awned and high yielding partly because negative linkages were overwhelmed by pyramiding genes for yield, or possibly a repulsion linkage was broken. The Houser case was surprising to me in that the presence of awns did not result in higher test weight. This phenomenon observed in Houser raises a challenge to explore what type or source of awn added to Houser would result in higher test weight. The opposite situation occurred in winter barley in New York where awned types are markedly superior to awnless ones: a coupling linkage of genes for awns with favorable performance genes seems indicated here.

Dyck and Baker (1975) found no effect of awns on yield in a cross of an awned Italian spring wheat with Canadian 'Manitou'. However, an association was noted between the awn condition and date of heading: awnless lines averaged 61.0 days to heading and awned segregates averaged 58.4 days. Thus selecting for awns or earliness resulted in an automatic selection for the other trait in this cross.

DISCUSSION OF AWNS

In the early 1950s I obtained from Dr. Rollo Woodward at the Utah Station a brittle-awned genotype which I inventoried as "Woodward's deciduous awn," or simply WDA. This barley had a normal long awn until near maturity when the kernels on the spike were turning from green to tannish yellow. At that time a thinner section of the awn just at the tip of the kernel became brittle and within a few days all of the awns fell off, leaving a stand of completely awnless barley. This was viewed as an oddity because in the arid areas of the West it was associated with shattering of the spike and consequent loss of grain before it could be harvested. However, in the more humid East shattering is seldom a problem, and this seemed to be a possible case of "having one's cake and eating it too," that is, all of the benefits of an awned variety with none of the disadvantages at harvest. The WDA line was crossed extensively into the Cornell winter barleys and performed as expected. Deciduous awned derivatives were exceptionally high in test weight, as much as 4–6 lb/bu over average figures, but were unsuitable for malting purposes because the weak part of the awn was slightly below the upper end of the kernel and resulted in insufficient protection and uneven germination during malting. At retirement I regretted that I had not completed the developmental research to capture the high yield and high test weight for a deciduous-awned feed barley.

The fact that awns can have a direct and significant influence on the components of economic yield seems reason enough to attract cereal breeders' attention. Awns have not been neglected in breeding, but it is difficult to escape the view that the advantages of awns under certain environments have not been fully exploited. The range of results, some seemingly diametrically opposed, is ground for thought. Some of the findings may be traced to environmental response, for example, dry and wet climates. If I were continuing in plant breeding, the enigma of inconsistent awn influence would be a target for fundamental research. I cannot explain what I observed under New York State conditions in two different crops, winter barley and wheat, where the effect of awns was profoundly different: in barley, awned lines were almost always superior to awnless types—they were the survivors in yield tests—and this was true despite two awnless cultivars, 'Wong' and 'Catskill', produced at Cornell. Wong showed an inherent range of awn expression that could not be stabilized. The opposite was true in winter wheat; in the mid-1940s I took the challenge of breeding Cornell's first awned wheat. In the next few decades we found only one awned line that was competitive and superior to the best awnless types. This was introduced as 'Houser', named for Dr. Harry Houser Love, and released in 1977. Higher test weight of grain did not accompany the presence of awns and high yield of Houser. I suspect strong linkage, set up in different phases in the two crops. Today I would approach this problem with mutagens, nuclear and chemical, intermating, parent diversity—particularly centered on different sources and types of awns—and similar tactics.

The case of Wong, generally considered an awnless type, is interesting. Field populations of Wong showed a range from awnless to awnletted to long awns on the two major kernel rows. One of my first tasks as a graduate assistant at Cornell University in 1940 was the roguing of Wong fields and the selection for a constant type. Pedigree methods were used to attempt to find a pure-breeding line. We were never able to select a Wong type that did not have the variable awn character even though this effort continued for decades.

In the 75 years of wheat breeding experience at Cornell, awns exerted a negative influence on yield. Because of this and the difficulty of breeding and selecting a superior awned cultivar for the Northeast, I think it likely that Houser must contain either superior genes for yield or another trait perhaps not recognized and associated with yield, for example, disease resistance or root characteristics, which compensates for a possible negative drag of awns. Nevertheless, the Houser cross remains the only Cornell wheat cross, among thousands made, in which awned lines were superior to sister awnless lines. The challenge remains to raise the test weight for a high-yielding awned wheat for New York. A simple approach might be a backcross program in the cross of awned Houser × awnless high test weight Fredrick. All vigorous F_2 awned plants would be examined for kernel weight and the best crossed to Houser. Progress toward a high test weight Houser would be evaluated and a new cycle initiated if necessary. Alternatively, the vigorous F_2 awned plants might be crossed to Fredrick—this would produce an awned Fredrick, and the backcross made in this direction would provide additional backcross generations of segregation to enable selection to continue for awnedness, high test weight, and high yield. Another option would be to cross Houser to other sources of awns.

ROOTS

The important role of roots in plant productivity is unquestioned; no plant breeder would dispute this. With rare exceptions, however, roots are ignored in actual breeding and selection practices. The reasons are obvious, and not simply "out of sight, out of mind": measuring root characteristics is labor intensive and costly.

I recall only two instances of doing something of a positive nature concerning roots. Arden Colette (1962), one of my graduate students, did a Master's thesis on root attachment of different oat and winter wheat genotypes. He used a fulcrum and weight device suspended from a tripod to measure the exact point at which plants could be lifted free from their root attachment to the soil. The other instance concerned an unusual genoytpe discovered while roguing wheat head rows many years ago. I had been pulling rogue plants from early-generation line rows by grasping the straw with one hand, pulling and lifting the plant for discard. This plant had no "give"; in fact, the unexpected resistance resulted in a cut finger as my hand grip slid upwards. This line was explored further and the unusual root attachment was confirmed, and I did use it as a parent in crosses.

Engledow and Wadham (1923) insisted that the root system had to be related to yielding ability. Explorations into this limiting aspect of productivity, neglected then by almost all breeders, are still neglected today.

Trelease and Trelease (1924), in a water culture study of growth of Marquis seeds of different weights, found that root yields (and top yields) were directly related to original seed weights; that is, the heavier seeds produced the higher yields.

Lamb (1936) reported on a study of roots of winter wheat cultivars that had the objective of identifying characteristics associated with injury from soil heaving by frost, a type of crop injury in cold regions where soil moisture conditions are high. Varietal differences in heaving were known and it was suspected that resistance was due to some property of the roots. Lamb found that vigorous, well-developed plants in the fall led to maximum heaving resistance in the spring. Varieties most resistant to heaving had a longer fall growing period and of course had high cold resistance. Size, extensibility, and especially the breaking tension of roots were found important to heaving resistance.

In a further study Lamb (1939) reported on the best season for sampling roots, the value of stele measurements for strength, and the influence of crop soil fertility level. He found that late fall was the best season for research. Stele diameter he reported to be "a better measure of root size with which to associate breaking tension than diameter or cross-sectional area of the whole root." Lamb also found that fertility level affects both root size and strength, but the rank of cultivars is not altered over a range of levels.

Aamodt and Johnston (1936) obtained evidence suggesting that the type of root system may play a part in drought resistance in wheat. Two Russian drought-resistant cultivars were shown to have a more highly branched primary root system than the nonresistant varieties Marquis and 'Reward'.

Pavlychenko (1937) reported the results of a study that measured the entire root systems of different cereal crops and weeds. The methods used—excavating blocks of soil to a depth of several feet, followed by washing away the soil—underscore the difficulties of obtaining information on roots (I have personal experience, having worked as a laborer excavating and washing root soil blocks for George Rogler at the Northern Great Plains Experiment Station, Mandan, North Dakota in 1939). Some of the data are astounding; for example, total root length of a spring rye cultivar plant without competition was 3,114,373 in., but under competition of 6-in. rows and 18–20 plants/ft, a single rye plant produced only 38,408 in. of roots.

Following up field observations that straw strength (lodging resistance) and root development were associated in oats, Derick and Hamilton (1942) examined five varieties and found distinct differences in the total amount of root growth and in the number and fineness of branching anchorage roots.

Hamilton's (1951) paper on lodging resistance in oats is an in-depth overview of related plant morphology, specifically, culm diameter, coronal root development, and height. It was significant to me because it gave me a working base for all considerations of lodging resistance: the aboveground plant with a heavy weight (head) at top acting as a fulcrum, attached to a base (roots) by a culm or stem (height and diameter).

The reported advantage in mixtures of 'Atlas' over 'Vaughn' barley was interpreted by Hartmann and Allard (1964) as occurring at moderate moisture stress levels and all levels of increasing fertility. The cause was attributed to the larger root system of Atlas.

Edwards and Allard (1963) attempted to discover whether light intensity might be responsible for the intense competitive effects in mixtures shown by the barley cultivars, Atlas and Vaughn (Atlas dominates in mixtures whereas Vaughn excels in pure culture). Shading treatments from full to one-third full sunlight significantly affected both cultivars, but did not change their relative competitive abilities. They favored the hypothesis that the competition may be for moisture or nutrients, particularly because Lee (1960) had found Atlas to show rapid root growth at about the time competition effects were noticed, whereas Vaughn did not.

Stucker and Frey (1960) determined the root system dry weights of five cultivars, sampling successively deeper 4-in. intervals to a depth of 24 in. Roots of all cultivars extended below this depth. There was differential distribution of dry weight according to cultivar. The Corn Belt cultivars placed about 60% of total root dry weight in the top 4 in. of soil, whereas values for the two nonadapted cultivars were 50% and 35%. The authors found no increase in root dry weight after the boot stage in the areas sampled.

Patterson, Schafer, Caldwell, and Compton (1963) found that "dwarfness, compact panicles, and excellent straw strength of the Scotland Club oat were completely associated in inheritance." Although root system was not examined it seems likely that it, too, is inherited as part of the complex.

Hurd (1964) reported on observations of roots of three wheat cultivars studied in glass-faced growth boxes. He wished to understand the relationship between root growth and structure, and drought and soil cracking. The three cultivars showed different root growth patterns regardless of the environmental regime imposed. One suggestion was that the near-surface root pattern might aid in preventing soil cracking.

There are so many factors involved in the overall relationship between roots and yield of spring wheats that Hurd (1968) acknowledged that it was difficult to know what role root growth played in final yield determination. Earlier research and farmer experience had shown that the variety 'Thatcher' had better than average ability to yield well under varying dry conditions. In this study it was tested with six other varieties for root growth response at two moisture levels and in two soil types. Two findings unique to Thatcher appeared: (1) it produced more roots at the 20–30 cm depth, and (2) it penetrated drier soil more rapidly than other varieties. Tests were conducted in two soil types, a coarser loam and a fine clay. The loam was superior as a testing medium because it was less retentive of water, thus responding more quickly to moisture changes. There were some interactions between variety and moisture and soils, leading to the suggestion that selection and yield testing be done on soils representative of intended use.

Derera, Marshall, and Balaam (1969) examined the roots of 15 wheat varieties in an effort to arrive at a basis for a breeding improvement program to increase drought tolerance. Tolerant genotypes were distinguished primarily by early maturity, a trait inversely related to maximum yield when moisture conditions are favorable. They found varietal differences in seminal root penetration; however, except for high correlation between nodal root number and grain yield, there was no clear guidance for breeding objectives.

Passioura (1972) reported on an experiment with wheat roots. On the premise that soils in drought-prone environments have a finite amount of stored water, he suggested that it would be desirable that this water not be squandered before anthesis to produce heavy vegetative growth that could not be sustained later. His experiments restricted the root system to one seminal root (wheats normally have three) per plant. This decreased the rate of water use during the first weeks of growth, but allowed more water later so that single-rooted plants produced twice the grain yield of normal plants, which had exhausted the available water before maturity. Passioura stated:

> There seems little doubt that the larger yield of the single-rooted plants was due to the larger amount of water available to them during the last few weeks of their lives. Hence the most important feature of the performance of these plants is the means by which they conserved water during early growth.

A much greater branching of the roots compensated for the loss of seminal roots to the extent that dry weights were equal in both classes. Passioura concluded that major resistance to the flow of water through (the xylem of) the single seminal root resulted in water conservation in the soil. Apropos breeding considerations, Passioura stated,

> When Gabo wheat is growing on limiting stored water, the grain yield can be greatly increased by forcing the plants to rely solely on one seminal root. The reason for this increase is that the single-rooted plants use less water before anthesis than do normal plants, and the reason that they use less water seems to be that they have a much larger resistance to vertical flow through the root system. Since this resistance depends on the fourth power of the diameter of the main xylem in the roots, the possibility arises of breeding for high root resistance (i.e., for small vessels).

There is an interesting relationship here to plant ideotype: Donald's uniculm and what might be called the "uniroot."

The sharp efficiency of a good screening system in plant breeding is illustrated in a paper by Harding (1974). Harding screened entries from the more than 10,000 entry USDA World Collection of Small Grains (wheat) by examining the subcrown internodes and assigning them to four classes: clean, slight, moderate, and severe. The first 5500 lines examined produced 112 genotypes promising for resistance, suggesting that the collection might contain a total in excess of 200 promising lines.

Phung and Rathjen (1977), in a study of frequency-dependent advantage in wheat, believed that competition took place principally between the root systems and was influenced by the varying distances between plants in a field situation. They concluded, "The available resources of a certain environment seem to be exploited more efficiently by a mixture of several genotypes than when there is only one."

Although Corrigan and Frey (1980) used only six genotypes (two *A. sativa* and four *A. sterilis*) in a study of root volumes in oats, the results included an exciting discovery: one of the lines, P.I. 324748, an *Avena sterilis*, showed a root development pattern so different from the others as to make it seem foreign to the set. At maturity its root volume was more than 331% of the next best line. Generally, in the other lines, root volume ceased increasing when heading occurred, but this was not the case with P.I. 324748; almost one-third of its gain was added after heading. Its root volume was matched also by higher dry weights of panicles, straw, and roots.

Because no crossing barriers exist between the species, the identification of this genotype may be useful in transferring valuable genes to domestic oats.

Vertical-pull resistance ("kilograms of force required to lift a plant vertically from the soil") was used by Jenison, Shank, and Penny (1981) to evaluate the root characteristics of 44 corn inbreds in four environments. The plant breeding concern was for the relationships between root structure and the economic factors of root rot resistance and stalk and root lodging. Although inbred × environment interaction was noted, its significance was minor compared to the highly significant differences among inbreds, which persisted across environments. Screening of inbreds for positive root characteristics was recommended. A substantial list of references on corn root systems is attached to this paper.

In two papers that appeared in 1981 Richards and Passioura (1981a, 1981b) successively developed a novel hypothetical basis for the influence of wheat seminal root morphology on the water use of the crop, treating first the environmental effects and second the genetic variation to be found in a study of more than a thousand genotypes. They started from the premise that if a plant imposed no barrier to water removal by, for example, a desiccating wind, then the water available to the roots would be sucked away without restraint. They postulated three ways of increasing the flow resistance through the upper levels of the seminal root system: (1) decrease the number of flow axes, (2) decrease the diameter of main xylem vessels, or (3) replace main xylem avenues by a multiplicity of smaller vessels. In the first paper they examined environmental influences on these categories, especially looking at seed history and seedling establishment conditions.

Results from the first experiments singled out vessel diameter as the most promising avenue to increase hydraulic resistance. It was found that vessel diameter was positively correlated with size of seed sown. Other characters shown to have an effect were a period of drought during seed filling (which had the same effect as reducing seed size), and drought and high temperature before seedling emergence. For practical reasons only seed size could be manipulated. This led to a search in the second experiment for genetic variation in vessel diameter.

Richards and Passioura examined more than 1000 different wheat genotypes for seminal root resistance to the longitudinal flow of water and for their number of seminal axes. Little variation in number of axes was found, but variation in vessel diameter was substantial, with several narrow vessel types found. A mean range of 55–76 μm was found among individuals of several F_3 populations, and it was shown that xylem vessel diameter was both under multigenic (quantitative) control and heritable (71%). As a result of these experiments it was decided to initiate a breeding program for narrow xylem vessel wheats for dry Australian conditions.

Can the vertical-pull resistance found in maize inbreds be transmitted and detected in their topcrosses? The answer is "yes," as shown by research reported by Penny (1981). Thirty-three representative inbreds were chosen from the 1981 Jenison, Shank, and Penny study. These were crossed on two inbred testers (W64A and Oh545) that were known to be below average in resistance to vertical pull. The pulls were done in four environments at two stages: pretassel, and one month later at ear early milk. The differences in pull were found to be slightly more pronounced at the latter stage.

The results showed high and highly significant correlations between the vertical-pull resistance of inbreds and their topcrosses. Penny concluded that these specific inbred traits were transmissible and detectable, and could be used to predict the inbreds' performance in hybrids. The relationship to corn rootworm control was discussed.

Murphy, Long, and Nelson (1982) studied changes in cultivars developed during the decades of oat improvement. Of all the characters studied the most obvious trend was the greater root development found in the modern selections. Because root characteristics are seldom considered, much less seen, in breeding and selection, this represents a natural association with performance of a selected line, and the inference might be drawn that root development and deficiencies are a limiting factor in yield gains.

Root-pulling resistance has been used extensively in corn to study root system characteristics (it measures the force in vertical pull required to lift a plant from its attachment to the soil). Peters, Shank, and Nyquist (1982) checked 66 hybrids and found no correlation in either of two years between pulling resistance and grain yield.

The article by Barbour and Murphy (1984) on field selection for oat seedling root length is important for two findings: (1) long roots were associated with greater grain yield, higher seed number, and greater plant height—this means that root length can be improved indirectly through selection for the other characters, either individually or by index selection; and (2) the correlation was observed only in nonstress, that is, favorable environments.

Understandably, root systems have been the last plant morphological frontier. Ekanayake, O'Toole, Garrity, and Masajo (1985) took a hard look at root systems and drought resistance in rice with encouraging results. Overall, much detail needs to be discovered, but breeding and selection for desirable root traits seem promising. Examples of desirable characters may be illustrated by the authors' comparison of two genoytpes: "...MGL-2 has greater maximum root length, root thickness, number of thick roots, root volume, root length density, and root dry weight than IR 20, indicating a much superior root system."

Of all characters that have been considered, roots, being invisible, would benefit most from indirect selection via a marker trait. One possibility is a visual score on recovery from drought, because root length, thickness, number, and volume were significantly correlated with drought recovery. Other associations were discussed. The authors suggested that a breeding system for root improvement might be bulk breeding and pedigree selection in a recurrent selection framework.

Asady, Smucker, and Adams (1985) examined a nondestructive seedling test for root tolerance in compacted soil. They found root penetration ratios to be highly correlated (greater than .90) with both field plant dry weights and yields of dry edible bean cultivars.

Although useful for understanding and research studies, it is unlikely that procedures such as proposed by Asady et al. can be easily adapted to selection programs that traditionally deal with many populations and large numbers of segregates. The plant breeder would like to see good research knowledge of this kind applied to a plant and soil growing arena, either greenhouse or field, where the roots would receive a standard stress, visually apparent in its effect on plant growth, and thus amenable to selection. Further research would be needed to evaluate associated effects of the stress upon other plant characters. For example, what would be the relation of selected plants to the population's genetic variance for yield?

Selection for improvement of roots is complicated by their obvious underground invisibility. An ingenious marker screening system was devised by Robertson, Hall, and Foster (1985). They banded a herbicide at different depths and lateral distances from the rows of 33 cowpea genotypes that were known to be vulnerable to herbicide damage. A deep extensive root system would be indicated from leaf symptoms, showing that a genotype had penetrated to or through a particular herbicide boundary.

Lauer and Simmons (1985) found in a photoassimilate partitioning study in barley, using radioactively labeled CO_2, that early tillers promoted plant establishment and root growth, but as soon as the main shoot stem began efficient dry matter accumulation (about three weeks after emergence) the shift of photoassimilate from tillers to main stem began. This period of about two weeks coincided with the "premature" death of excess tillers, which until then were still developing. Coincidental in time, the senescence was hastened by increased canopy shading.

REFERENCES

Aamodt, O. S. and W. H. Johnston. 1936. Studies on drought resistance in spring wheat. Can. J. Res. 14:122–152.

Aamodt, O. S. and J. H. Torrie. 1934. The relation between awns and yield in spring wheat. Can. J. Res. 11:207–212.

Allan, R. E. and O. A. Vogel. 1964. F_2 monosomic analysis of coleoptile and first-leaf development in two series of wheat crosses. Crop Sci. 4:338–339.

Allan, R. E., O. A. Vogel, and J. C. Craddock. 1959. Comparative response to giberellic acid of dwarf, semi-dwarf and tall winter wheat varieties. Agron. J. 51:737–740.

Asady, G. H., A. J. M. Smucker, and M. W. Adams. 1985. Seedling test for the quantitative measurement of root tolerances to compacted soil. Crop Sci. 25:802–806.

Atkins, I. M. and P. C. Mangelsdorf. 1942. The isolation of isogenic lines as a means of measuring the effects of awns and other characters in small grains. J. Am. Soc. Agron. 34:667–668.

Atkins, I. M. and M. J. Norris. 1955. The influence of awns on yield and certain morphological characters of wheat. Agron. J. 47:218–220.

Barbour, N. W. and C. F. Murphy. 1984. Field evaluation of seedling root length selection in oats. Crop Sci. 24:165–169.

Bayles, B. B. and C. A. Suneson. 1940. Effect of awns on kernel weight, test weight, and yield of wheat. J. Am. Soc. Agron. 32:382–388.

Clark, J. A. 1924. Segregation and correlated inheritance in crosses between Kota and Hard Federation wheats for rust and drought resistance. J. Agric. Res. 29:1–47.

Clark, J. A. and K. S. Quisenberry. 1929. Inheritance of yield and protein content in crosses of Marquis and Kota spring wheats grown in Montana. J. Agric. Res. 38:205–217.

Clark, J. A., V. H. Florell, and J. R. Hooker. 1928. Inheritance of awnedness, yield, and quality in crosses between Bobs, Hard Federation and Propo wheats at Davis, California. U.S. Dept. Agric. Tech. Bull. 39:40 pp.

Clark, J. A., K. S. Quisenberry, and L. Powers. 1933. Inheritance of bunt reaction and other characters in Hope wheat crosses. J. Agric. Res. 46:413–425.

Colette, W. A. 1962. Measuring uprooting forces in the small grains. M. S. thesis, Cornell University Graduate School, Ithaca., NY, 52 pp.

Corrigan, L. and K. J. Frey. 1980. Root volumes of *Avena* species. Crop Sci. 20:407–408.

Derera, N. F., D. R. Marshall, and L. N. Balaam. 1969. Genetic variability in root development in relation to drought tolerance in spring wheats. Expl. Agric. 5:327–337.

Derick, R. A. and D. G. Hamilton. 1942. Root development in oat varieties. Sci. Agric. 22:503–508.

Dyck, P. L. and R. J. Baker. 1975. Variation and covariation of agronomic and quality traits in two spring wheat populations. Crop Sci. 15:161–165.

Edwards, K. J. R. and R. W. Allard. 1963. The influence of light intensity on competitive ability. Am. Nat. 97:243–248.

Ekanayake, I. J., J. C. O'Toole, D. P. Garrity, and T. M. Masajo. 1985. Inheritance of root characters and their relations to drought resistance in rice. Crop Sci. 25:927–933.

Engledow, F. L. and S. M. Wadham. 1923. Investigations on yield in the cereals. Part I. J. Agric. Sci., Cambridge 13:390–439.

Everson, E. H. and C. W. Schaller. 1955. The genetics of yield differences associated with awn barbing in the barley hybrid (Lion × Atlas^{10}) × Atlas. Agron. J. 47:276–280.

Gale, M. D. and G. A. Marshall. 1976. The chromosomal location of *Gai 1* and *Rht 1*, genes for gibberellic insensitivity and semi-dwarfism, in a derivative of Norin 10 wheat. Heredity 37:283–289.

Gale, M. D., G. A. Marshall, and A. J. Worland. 1975. The genetic control of gibberellic acid insensitivity and coleoptile length in a "dwarf" wheat. Heredity 34:393–399.

Grantham, A. E. 1919. Bearded vs smooth wheats for Delaware. Delaware College Agric. Exp. Stn. Bull. No. 6. 11 pp.

Goulden, C. H. and K. W. Neatby. 1929. A study of disease resistance and other varietal characters of wheat—application of the analysis of variance, and correlation. Sci. Agric. 9:575–586.

Grundbacher, F. J. 1963. The physiological function of the cereal awn. Bot. Rev. 29:366–381.

Hamilton, D. G. 1951. Culm, crown and root development in oats as related to lodging. Sci. Agric. 31:286–315.

Harding, H. 1974. Screening wheat lines for resistance to common root rot. Can. J. Plant Sci. 54:823–825.

Harlan, H. V. and S. Anthony. 1920. Development of barley kernels in normal and clipped spikes and the limitations of awnless and hooded varieties. J. Agric. Res. 19:431–472.

Hartmann, R. W. and R. W. Allard. 1964. Effect of nutrient and moisture levels on competitive ability in barley (*Hordeum vulgare* L.). Crop Sci. 4:424–426.

Hayes, H. K. 1923. Inheritance of kernel and spike characters in crosses between varieties of *Triticum vulgare*. Univ. Minn. Studies in Biol. Sci. 4:163–183.

Hayes, H. K. and A. N. Wilcox. 1922. The physiological value of smooth awned barley. J. Am. Soc. Agron. 14:113–118.

Hayes, H. K., O. S. Aamodt, and F. J. Stevenson. 1927. Correlation between yielding ability, reaction to certain diseases and other characters of spring and winter wheat in rod-row trials. J. Am. Soc. Agron. 19:896–910.

Hurd, E. A. 1964. Root study of three wheat varieties and their resistance to drought and damage by soil cracking. Can. J. Plant Sci. 44:240–248.

Hurd, E. A. 1968. Growth of roots of seven varieties of spring wheat at high and low moisture levels. Agron. J. 60:201–205.

Jenison, J. R., D. B. Shank, and L. H. Penny. 1981. Root characteristics of 44 maize inbreds evaluated in four environments. Crop Sci. 21:233–237.

Lamb, C. A. 1936. Tensile strength, extensibility, and other characteristics of wheat roots in relation to winter injury. Ohio Agric. Exp. Stn. Bull. 568. 44 pp. Wooster.

Lamb, C. A. 1937. The relation of awns to the productivity of Ohio wheats. J. Am. Soc. Agron. 29:339–348.

Lamb, C. A. 1939. Further studies on root characteristics of winter wheat in relation to winter injury. J. Agric. Res. 59:667–682.

Lauer, J. G. and S. R. Simmons. 1985. Photoassimilate partitioning of main shoot leaves in field-grown spring barley. Crop Sci. 25:851–855.

Lee, J. A. 1960. A study of plant competition in relation to development. Evolution 14:18–28.

McKenzie, H. 1972. Adverse influence of awns on yield of wheat. Can. J. Plant Sci. 52:81–87.

Middleton, G. K. and W. H. Chapman. 1941. An association of smooth-awnedness and spring growth habit in barley strains. J. Am. Soc. Agron. 33:361–366.

Miller, E. C., H. G. Gauch, and G. A. Gries. 1944. A study of the morphological nature and physiological functions of the awns of winter wheats. Kansas Agric. Exp. Stn. Bull. 57, 82 pp.

Murphy, C. F., R. C. Long, and L. A. Nelson. 1982. Variability of seedling growth characteristics among oat genotypes. Crop Sci. 22:1005–1009.

O'Brien, L. and A. T. Pugsley. 1981. F_3 yield response to F_2 selection for gibberellic acid insensitivity in eight wheat crosses. Crop Sci. 21:217–219.

Parker, J. H. 1934. Relation of the awn to yield in Kansas wheat. Northwestern Miller, April 18, 1934, p. 180.

Passioura, J. B. 1972. The effect of root geometry on the yield of wheat growing on stored water. Aust. J. Agric. Res. 23:745–752.

Patterson, F. L., L. E. Compton, R. M. Caldwell, and J. F. Schafer. 1962. Effects of awns on yield, test weight and kernel weight of soft red winter wheats. Crop Sci. 2:199–200.

Patterson, F. L., J. F. Schafer, R. M. Caldwell, and L. E. Compton. 1963. Inheritance of panicle type, height, and straw strength of derivatives of Scotland Club oats. Crop Sci. 3:555–558.

Pavlychenko, T. K. 1937. Quantitative study of the entire root system of weed and crop plants under field conditions. Ecology 18:62–79.

Penny, L. H. 1981. Vertical pull resistance of maize inbreds and their testcrosses. Crop Sci. 21:237–240.

Peters, D. W., D. B. Shank, and W. E. Nyquist. 1982. Root-pulling resistance and its relationship to grain yield in F_1 hybrids of maize. Crop Sci. 22:1112–1114.

Phung, T. K. and A. J. Rathjen. 1977. Mechanisms of frequency-dependent advantage in wheat. Aust. J. Agric. Res. 28:187–202.

Pool, M. and F. L. Patterson. 1958. Moisture relations in soft red winter wheats. II. Awned vs awnless and waxy versus nonwaxy glumes. Agron. J. 50:158–160.

Porter, H. K., N. Pal, and R. V. Martin. 1950. Physiological studies in plant nutrition. XV. Assimilation of carbon by the ear of barley and its relation to the accumulation of dry matter in the grain. Ann. Bot. (NS) 14:55–68.

Richards, R. A. and J. B. Passioura. 1981a. Seminal root morphology and water use of wheat. I. Environmental effects. Crop Sci. 21:249–252.

Richards, R. A. and J. B. Passioura. 1981b. Seminal morphology and water use of wheat. II. Genetic variation. Crop Sci. 21:253–255.

Robertson, B. M., A. E. Hall, and K. W. Foster. 1985. A field technique for screening for genotypic differences in root growth. Crop Sci. 25:1084–1090.

Stucker, R. and K. J. Frey. 1960. The root-system distribution patterns for five oat varieties. Proc. Iowa Acad. Sci. 67:98–102.

Suneson, C. A. and R. T. Ramage. 1962. Competition between near-isogenic genotypes. Crop Sci. 2:249–250.

Suneson, C. A., B. B. Bayles, and C. C. Fifield. 1948. Effects of awns on yield and market qualities of wheat. USDA Circ. 783, 8 pp.

Suneson, C. A., C. W. Schaller, and E. H. Everson. 1952. An association affecting yield in barley. Agron. J. 44:584–586.

Trelease, S. F. and H. M. Trelease. 1924. Relation of seed weight to growth and variability of wheat in water cultures. Bot. Gaz. 77:199–211.

Vogel, O. A., R. E. Allan, and C. J. Peterson. 1963. Plant and performance characteristics of semidwarf winter wheats producing most efficiently in Eastern Washington. Agron. J. 55:397–398.

SECTION VI

PROJECT MANAGEMENT

The following several chapters will emphasise more of the hands-on procedures so necessary to the conduct of an efficient and successful crop improvement program. We begin with a stylized treatment of the breeding and selection processes.

34

THE FOUR STAGES OF THE PLANT BREEDING PROCESS

The plant breeding process divides itself into four natural stages. I have found these useful for teaching purposes, and for convenience in presenting the problems in the process I briefly list and describe the divisions before treating each in detail:

Stage I	Planning and hybridization (breeding)
Stage II	Early generation (selection)
Stage III	Line evaluation (evaluation)
Stage IV	Variety release (release)

The major unresolved "sticky" problems of plant breeding lie in the first two stages. The concerns in stage I are as follows:

1. The alignment and synchronization of objectives and procedures.
2. The choice of parents.
3. The appropriate use of germ plasm resources.
4. The choice of breeding method(s).
5. The evaluation of the predictive information available on parent prepotency and the population worth of different hybrids.

Stage II concerns begin with the F_2-embryo seed produced by hybridization. The basic objective is how best to move the populations through the segregating generations to an acceptable "working" homozygosis point. The major concerns of stage II, therefore, are the following:

1. Choosing the selection method.
2. Deciding whether to use generation information to predict the value of crosses.
3. Deciding whether to select at all during the segregating generations. Indeed, C. H. Goulden (1939) described the difficulties arising from the breeder's penchant for operating on a moving target, that is, practicing selection for superior lines during the period when individuals are changing from maximum heterozygosity to homozygosity.

Stages III and IV are more straightforward and routine in application and are discussed in their proper order.

STAGE I—PLANNING AND HYBRIDIZATION

Most plant improvement projects have major objectives that are clearly understood. For example, those of my cereal improvement programs at Cornell might be briefly described in this way:

Wheat	Soft white winter milling and baking types
Barley	Barley Spring and winter feed types
Oats	Feed types

These major objectives had not changed in the three-quarters of a century of the Cornell breeding program; however, objectives had been added. For example, there was a hybrid wheat program for a brief period and there was a high protein project. Malting quality had been added to the barley program and high nutrition and energy to the oat. Within each major objective were numerous minor goals, each of which made a contribution to the performance of the eventual variety. To illustrate, a breeder might have research in oats underway on disease resistance (stem and crown rust, barley yellow dwarf virus), insect resistance (cereal leaf beetle), plant morphology (lodging resistance, height, roots), kernel characters (plumpness, high test weight), or nutritional quality (protein, fat, and oil content).

It is taken for granted that any new variety must show a high level of performance—it would be a mistake to think that yield is taken lightly. It is in the forefront of planning and hybridizing considerations and is often the deciding criterion in discard–retain decisions.

Stage I renews and repeats itself, usually on an annual basis. Objectives change slowly and there seldom is a final solution to a breeding problem. It is hoped that the latest variety establishes a new benchmark of performance and a new challenge to the breeder, who then prepares for another round of hybridization. Although final solutions are rare, breeders can often enjoy long periods after the successful achievement of an objective. For example, the incorporation of the *A* and *B–C* genes for stem rust resistance in oats gave protection for a quarter of a century in New York State. The common and dwarf bunt resistance of 'Avon' wheat, released in 1959, has given crop protection since then, although the cultivar is not widely grown at present.

Thus the first concern in planning is to reevaluate objectives and to plan so that the yearly hybridization program is broadly focused on these goals. One need not worry about the apparent repetition that occurs year after year in the crossing plan. Seldom, except in the case of single crosses, will identical crosses be made, given the new parents, new thinking, and the new environment of the next season.

In fact, it is difficult to avoid seeing breeding projects as analogous to massive backcrossing programs where instead of the recurrent parent we have the recurrent objective. The breeder hews to the standards by crossing to liberal interpretations of a recurrent parent, that is, to any one of similar-quality genotypes that are genetically different.

CHOOSING PARENTS FOR BREEDING PROGRAMS

A typical way in which a breeder chooses parents for the annual or oftener hybridizing program would show choices somewhat as follows:

1. The "recurrent" reliables, that is, genotypes that represent past and present advances
2. Gene sources for particular objectives
3. New sources to broaden the program's genetic diversity
4. Single crosses and other hybrids.

The first group of proven performers reappear in each year's crossing program, and the makeup of the group changes very slowly with perhaps one or two additions or deletions a year. The group is composed largely of cultivars, enriched as a group by promising lines from the breeding project. Varieties from other programs, for example, another state, may be used if breeding objectives are similar.

Here is an illustration of how I chose the parents for a typical crossing season of the Cornell wheat project. First, a page for this purpose was set aside in the annual field index book for a list of varieties and lines that were basic to the program. This index book, which I carried about with me, was a personal ledger that contained an abbreviated description of all nurseries with their entries. A new parent list would be started when the old list had been closed for the current season, usually at planting time for the fall-sown crops, which was just before the greenhouse planting season. Thus the new list would be open for approximately a year until the next greenhouse crossing season. During this period the list frequently would be referred to and many names of prospective parents written in. Variety or genotype names would come to my attention through stimuli received from reading (newsletters, publications, nursery results, etc.), conversations, memory recall, nursery experience, or a trip, indeed from sources so varied as to be unpredictable. Final decisions on parents would be made about a month before planting in the greenhouse in order to allow time to prepare seeds and records. The index page notes at this time would have increased, and would have been transferred to a working file for the crossing program.

Cultivars would be chosen because of specific information about them that had come to my attention. A special problem was posed by the large public germ plasm collections. For example, what is a reasonable basis for entering a large collection of genotypes and choosing for a parent any one of the thousands available? In 1974 I wrote a draft proposal (not published) to give a breeder a bias-free selection of outstanding genotypes in a collection. Because this is a problem of concern to all breeders, and basic to stage I considerations, I append a synopsis of this scheme herewith:

A Supplemental Program To Broaden The Parent Germplasm Base

Public germ plasm collections such as the USDA's World Collection of Small Grains, which for wheat numbers more than 30,000 entries, pose a problem of use to the breeder. Concerted efforts are being made annually to describe the entries; nevertheless, aside from specific requests based on knowledge for certain genotypes, the collections remain largely unexploited by breeders.

A simple screening and selection scheme could enable a breeder to identify valuable genotypes, a few of which might seem desirable as parents. This could be done by forming a seed composite composed of equal amounts of collection entries; entries may be as small as 1–5 g per genotype. I have used bulk populations of an entire collection (oats) formed in this way, but groups classified according to disease resistance or geographical source, for example, may be used. If selection is to be made in the first growing season, the composite should be planted in machine space-planted rows sufficiently far apart to permit walking between the rows when making selections. If the population is to be grown for one season before selection, a standard field plot may be used the first year, with the space-planting being used the following year when selections are to be made. The composite may be grown at multiple sites and repeated in subsequent years.

Compositing destroys all genotype identity. The breeder thus makes a series of bias-free selections insofar as knowledge of genotype is concerned. The selection is based on the breeder's concept of desired type and performance in a particular environment. The head row nursery of selections the following season provides a second look, following which the reduced numbers of selections are inserted into early performance trials.

This scheme is effective in selecting environmentally adapted and agronomically useful genotypes to which the breeder otherwise would not have access because collections are too large today for entry-by-entry growing. The screening will uncover many plants with useful characters suitable for use as parents. As such they will be unique—some will never have been used before—and furthermore, they will represent an outreach of germ plasm diversity because it is improbable that they would be duplicated in any other breeding program.

To continue on the subject of choosing parents in stage 1, the single-cross hybrid seeds would be chosen from the reserve bank of several hundred single crosses and crypto hybrids. Many breeders use hybrids as parents in order to broaden the genetic base and the options available within their program. A reserve bank of 400–500 selected wheat hybrids was maintained in my office and augmented annually with new combinations from the crossing program. Their availability provided options for three-way and double crosses and backcrosses, and greatly expanded the potential germ plasm breadth of the project. Genetic diversity and potential solutions to breeding problems were represented in this collection of single crosses by a range of narrow, for example, Yorkstar × Houser, to wide, for example, Houser × 'Hussar', crosses. This reserve bank included hybrids embodying genotypes capable of leading to the solution of an insect or pathogen problem, for example 'Ticonderoga' × 'P.I. 178383' for smut resistance.

The breeder's tentative list of perhaps 30–40 pure lines and a large number of single crosses represents a typical position at the approach of a crossing season. The breeder "fine-tunes," the list, with additions and deletions occurring right up to greenhouse planting time.

A typical planting, therefore, may involve from 20 to 40 parents plus many hybrids. Forty parents a year represents a fairly large number, and experience has shown that it is often difficult to utilize all (the number of crosses made per year in the Cornell wheat program ranged from a few hundred to 2000). Sometimes a few of the parents planted never take part in a cross.

This is approximately the procedure by which breeders choose parents for each year's program. Of course, there is much more to germ plasm resources and utilization, but the choice of parents as described here represents the breeder's leverage for germ plasm use in his programs.

A BEGINNING IS A BEGINNING—SO START

A major portion of this book is concerned with breeding and selection methods. For the beginning breeder, however, there need be little pressure to identify or adhere to any given procedure. With the passage of time a breeder's preferences will become a factor, but early on the important thing is to make a beginning and experience the options that present themselves at every turn. To illustrate, a single cross is a beginning. The first option is whether or not to make a further cross. If so, what should it be, a backcross, a three-way outcross, or a double cross? Regardless of whether or not the single cross is used as a future parent, once planted it will produce a progeny. Should this progeny be handled by pedigree or by bulk method, or marked for immediate discard? These seemingly elementary options constantly reappear in the life of a breeding project. With experience and a honed sense of judgment, a breeder can forge a system of procedures with which she or he is comfortable.

I think the majority of breeders are not monolithic in their thinking and operation of breeding projects; zealots enthusiastically supporting one system over a career lifetime are few (but they do exist). Many breeders favor a "system" that translates into an easily handled set of work rules, but they may embroider this with any variation suggested by particular breeding objectives, and especially by the goals envisaged for any special hybrid. To illustrate, I finally came to rely predominantly upon a bulk hybrid population system of breeding and selection; nevertheless, the system embraced several other approaches; for example, the project that produced the semi-dwarf wheat Yorkstar was conceived and conducted throughout as a backcross program. I am also intrigued by the F_2-derived family approach.

The final point under stage I is the use of prediction as an aid in choosing parents and in choosing among hybrid populations (see Chapters 28–30). This is a highly complex subject with which beginning breeders have difficulty because (1) prediction has a base in prior practical knowledge (experienced breeders have a certain amount of learned experience that certain genotypes are potent parents and others are not), and (2) prediction is still an emerging and evolving area of breeding research.

STAGE II—EARLY GENERATION (SELECTION)

Stage II presents opportunities for decisions that will have profound influence upon the size and efficiency of project operations. Typically, the breeder is presented with all the potential populations from a high-geared crossing program. The populations each year exist only as F_2-embryo seeds. He or she must find a way either to discriminate among populations or to commit them all. This is one of the most vexing problems facing plant breeders. In the past they have relied upon knowledge of the parents, intuition, and visual judgment of the growing populations to help make these decisions.

A basic decision awaiting a new breeder is whether to select at all during the early generations of a hybrid population. It is both possible and feasible to move whole populations efficiently to a level of "working" homozygosity, say, F_4 or F_5, all the while imposing options ranging from doing nothing, to taking yields, to practicing negative selection, which is another expression for discarding genotypes. The reason for considering this bulk hybrid population approach traces to the uncertain reliability of many of the available alternatives, namely, the unreliability of selection among heterozygous plants (or indeed among homozygous plants) and the low correlation between the performance of a selected genotype and its progeny, whether it be the next generation or farther down the time line. There is also the question of whether to employ yield testing as an aid to selection.

The breeder, confronted with so much uncertain and largely negative information, easily could become timid and settle for no selection at all, going directly to homozygous genotypes. On the other hand, the breeder might develop a program that could discriminate among the available hybrid populations. This should begin with visual discrimination. However, today there is the opportunity, made possible by specialized machinery, more precise statistical and field designs, and the help of the computer, to do more efficient sorting during the early generations. The emphasis in stage II is to show selection as one of the phases or stages of the plant breeding process. It is inseparable, however, from the several breeding methods discussed throughout this book.

STAGE III—LINE EVALUATION

The third stage begins with the establishment of the genotype as a homozygous line. This might be as early as F_3 or relatively late at F_6 or F_7. Except fortuitously, few genotypes at this stage are truly homozygous and the breeder must settle for a pragmatic "working" homozygosity. The breeder pays a small penalty for working with the earlier generations of a population—a higher proportion of the lines will show obvious segregation. This penalty is slight, however, for only the planting effort is lost, because the rejected rows need not be harvested. A more serious penalty comes from working with too-early-generation material because of the selection of rows exhibiting the illusory and transitory influence of heterozygosity, for example, for vigor and yield. These excellent-appearing rows must be harvested, processed, and cared for until their true and constant traits become evident with the passing of time (usually resulting in their discard).

In the early phases of line evaluation the breeder is under heavy pressure to rely upon visual discrimination: the numbers are large, seed supplies small, and yields only moderately helpful. In my project with wheat at Cornell University, the records showed that I visually selected F_5 or F_6 head rows at a 0.15 level; that is, from a 10,000-line nursery about 1500 lines would be selected. The following year the selections would be in a one-replicate, guarded, three-row plot yield nursery. Here we used a combination visual and data guideline involving a cautious field discard—plots not harvested—followed by data examination of those selected and harvested. This resulted in a 0.33 selection rate. From that point on, the discard rate slowed and evaluation became more and more based on recorded performance data coming from more rigorous testing involving replications, sites, and so forth. In the later stages of line evaluation, regional and national cooperative testing was added for elite lines.

STAGE IV—VARIETY RELEASE

A decision to release a line as a cultivar signals the end of the formal plant breeding process. The decision may or may not be an easy one. It is easy in the rare cases where an evident need for a new variety exists, or when confidence in the superiority of the new entity is high. The more usual situation prevailing, however, might be characterized as a blend of doubts sustained by hope. The breeder knows the candivar's (candidate variety) weaknesses as well as its strengths. These must be brought to the attention of farmers, the eventual growers of the new cultivar. In the United States, first contacts are usually with a group of responsible elite seed growers, specialized growers who take the new variety through the seed buildup stages necessary before its mass availability to farmers. These contract growers are usually part of a seed growers' association and represent a facilitating process between the breeder, the agricultural experiment station, the private seed companies, and the farmer. Seed growers serve the public interest, but they also have personal interests at stake in the success or failure of a new variety. The result is a careful, scientific, and business-oriented scrutiny of all aspects of a potential new variety.

The breeder's responsibilities lessen but do not end with the introduction of a variety. He or she is intimately involved in the two- to three-year seed buildup and the monitoring of seed and crop purity and of production practices. The breeder serves as consultant to the specialized agencies which by law have responsibility for these practices. For the life of the variety, the breeder remains a resource person to whom people turn for information and appropriate seed stocks.

The four stages of the plant breeding process do not represent a breeding or selection system that plant breeders use, but they do help in describing and understanding this complex scientific discipline that is founded in theory and emerges as a practical basic agricultural "tool."

REFERENCES

Goulden, C. H. 1939. Problems in plant selection. *In* R. C. Burnett (Ed.), Int. Genet. Congr., Proc. 7th (Edinburgh). Cambridge University Press, pp. 132–133.

35

MINIMUM STAFF PLANT BREEDING

For six years following retirement I operated a private plant breeding project. As a career professional plant breeder I had been intrigued with the way procedures were dictated to some extent because an operation fell under the aegis of an institution. Would it be possible to conduct a more efficient set of procedures? What if the operating staff consisted only of the project leader? I decided to test some ideas about methodology and began with the oat project. The following section was originally written (but not published) for my personal guidance during the second year of the project (1980) and is presented with only minor changes.

A MINIMUM STAFF BREEDING PROGRAM

This section might be titled, "How to run a one-person breeding program." The background traces to my early retirement from Cornell University in 1978. Giving thought to what I might do to keep busy and interested in retirement, I decided to try my hand at running a private cereal breeding program. Over the years I had noticed many parts of the plant breeding operations where greater efficiencies might be obtained, but these involved techniques and approaches not suitable to a reasonably staffed and equipped institutional breeding program. Nevertheless, without staff I would have no choice but to rely on simpler procedures.

Initially it was planned to operate separate wheat, barley, and oat programs and indeed these were all started. Later, it was decided to limit activity to just one crop. Oats was chosen, not because it was the most promising from an economic view—that choice would have been either wheat or barley—but because one could foresee fewer eventual problems in the quality and nutritional areas of research. Wheat and barley would require rigorous testing for baking, milling, or malting characteristics. Oats is primarily a feed grain and much less a food grain, so quality would rest more in the visual and mechanically determined areas of size, plumpness, and weight of grain.

The procedures for a one-person program are not suggested for adaptation to standard breeding operations, but are described because the evolution of thought and procedures concerning efficiencies of operations are subjects of general interest to other breeders. The following is a detailed accounting of the genesis and operation of this program, now in its second year.

The first steps were to define the scope of the project, list resources, and lay out a general plan of operations. These appeared to be as follows:

1. *Scope.* Develop commercially acceptable oat varieties for the central United States and similarly adapted areas.

2. *Resources.* These were my professional experience, time to devote to the project, family agricultural connections for testing sites in North Dakota and New Mexico, an operating base in Arizona where climate permits two growing seasons annually, and not least, a reservoir of good will among colleagues nationwide upon which I could draw for needed advice or assistance.

3. *General Plan of Operation.* This involved breeding work in Arizona, rapid advancement of generations, early selection, rigorous discard, early reliance on visual discrimination for good agronomic type, and early use of miniature-size yield nurseries.

The challenge of operating a fully competitive oat breeding program without supporting staff translates itself into operations markedly different from those used at any experiment station. Restrictions confronted me in every direction. To begin with, my options for breeding methods were curtailed and I could no longer use the bulk population breeding system because it was too expensive in machinery and land requirements. Accordingly, I settled on using my favorite cross, the three-way, which by definition includes the backcross. The three-way cross requires a single cross to be used as one parent.

Currently, small plot F_2 composites, thickly sown, are being used; however, this is not firm as I see modifications incorporating SSD and prediction-among-crosses options.

I want to establish lines out of the F_3 and this opens many possibilities ranging from spaced plants to bulks, and from harvesting one seed per plant (SSD) to harvesting the entire plant. My current thinking is to study the field formulations during one year and select on the basis of ease of handling; however, the decision may already have been made in favor of a thickly planted F_3 bulk based on screened seeds. The rationale for this is as follows: a major factor in the acceptance and degree of grower satisfaction with an oat variety is the high quality of seeds, that is, kernels that are plump, of high test weight, and with low hull content. Assuming that yield genes are evenly distributed, one could handpick the F_4-embryo bulk seed lot for visually plump seeds. These "elite" seeds are all plump and usually constitute in numbers less than 10% of the original bulk lot. Of course, it is not known whether the plumpness is due to genetic or environmental causes or an interaction between the two. Nevertheless, the breeder now has a manageable space-planted next generation, in which he or she can make a later judgment on plumpness and also apply rigid selection based on agronomic type. The efficiency of this is appealing. Question: can the procedure be advanced one generation and the line selection of plump seeds be made on the F_3-embryo seed?

In planning our Arizona home we added a small 20 × 30 ft low-walled garden to the backyard landscaping. Half of this space is adequate for the breeding and F_1 plant increases. I plant 30–40 pure line parents and about the same number of single-cross hybrid parents, which taken together generate the three-way crosses. After being used as parents, the F_2s of the most promising of the single crosses go to the New Mexico testing site, as do the increases from all the three-way F_1s. There the F_2-embryo seeds are thickly planted in small areas with the intent of limiting production per plant for the *quid pro quo* of many plants (Grafius, 1965). The seeds harvested from the New Mexico site, after buffing and screening out the smaller ones, are planted in the fall for a winter Arizona increase. The Arizona-harvested lots of seed, after severe buffing and polishing (Jensen, 1964), are screened to remove the smaller kernels, after which they are handpicked for plump kernels. The plump kernels (F_4-embryo) are space-planted close together, for example, 4–6 in. Line selection is made during summer and fall and enough seed harvested per plant to sow three plant rows 1 ft long plus a seed reserve, that is, a total of at least 4 g of seed. The lines are now ready to go to two sites in North Dakota and one in New Mexico.

This system provides a reasonably rapid acceleration of generations. If we take the every-spring production of F_2 seeds from the three-way and single-cross F_1 plants in the garden as a time base, in 16 months there are F_4 line plants ready for the next spring's planting. These lines already have been screened for kernel morphological quality and a range of agronomic type conformations. Rigid discrimination in the plant row nursery will reduce numbers to the point where extra individual attention can be given for yield, disease resistance, and protein. At this writing procedures are by no means locked in. For example, total time required could be shortened by applying a modified SSD procedure to the growing F_2 generation by picking a few kernels from selected plants and planting a spaced F_3 nursery for plant or line selection.

Equipment for a minimum staff breeding operation is also scaled down. My nursery sites are roughly fitted for planting by the farm operator. I own a light pickup truck. The other equipment needed is as follows: a small portable rototiller, a lightweight hand push planter, a head thresher of the Vogel type, a hand scythe, hoe, rake, miscellaneous items and supplies such as crossing tools, bags, bird protective netting, seed screens, balances, and so forth. Other equipment that would be useful includes a paint shaker for buffing seeds to seed-quality level and a gravity grader. A small rock tumbler has substituted for the paint shaker.

Prediction methods to aid in evaluating and choosing among crosses would be useful; however, I have been relying on visual evaluation and working with what appear to be the best cross populations of each year. This is reasonably effective because the pedigree system approach displays the F_2 progeny of each hybrid seed. In the case of single-cross progenies from the hybrid seeds of a cross of pure line parents, if the breeder sees one F_2 progeny of a particular cross he has effectively seen them all, for the segregation patterns and ranges are similar. However, with three-way crosses there can be as many genetically different small populations as there are hybrid seeds from the same three-way cross. This situation is tailor-made for choosing among a group of F_2 populations having the same parentage.

Germ plasm use, specifically the choice of parents, requires stricter control in a minimum staff program. Pragmatically, the breeder must curb any indulgences (no doubt the close linkage between thinking and actual work makes this curb more attractive!). The breeder need not narrow the breadth of germ plasm use, but it must be channeled into more rigidly defined paths. To illustrate, exotic parents are chosen for specific reasons, and care is taken to see that they enter three-way crosses in the proper parent slot—it makes a difference in results whether a given parent is inserted in the single cross as against inserting it into the 50% slot in a three-way cross. If the three-way pure line parent position is considered a recurrent parent slot (which it is in a backcross) where one is trying to give the three-way its best chance of success, it becomes apparent that this slot is *always* important [Cockerham (1961) has provided a theoretical basis for the importance of order of matings in three-way crosses]. Consequently, the number of parents deemed good enough to fill this slot is relatively small, and exotic or not, they must be carefully chosen. Candidates chosen include proven performers in a region—even old varieties that have shown remarkable "staying power," for example, the oat varieties 'Russell' and 'Garry'. Growers' experiences are part of the input also; for example, 'Otana' has been a consistently good oat on North Dakota farms. Material I brought back from an extended stay in Australia in 1979 illustrates carefully chosen exotic material. Varieties such as 'Swan', 'West', and Irwin' represent productive varieties with excellent seed size and agronomic type, possessing the further advantage of having been used little in U.S. breeding programs. I was expecially fortunate in obtaining some surplus hybrid single-cross seeds from

Andrew Barr, whose breeding program in Western Australia I much admired.

Many of my views on visual discrimination came from thoughts about efficiency in plant breeding operations. Selection procedures not only must be firm, they must be adhered to savagely. The outside selection perimeter could well be set at no more than 0.10—and it's difficult indeed to limit oneself to 10 lines from every hundred. Any suggestion of a questionable weakness, such as of stem strength, must be taken as a visual signal for immediate discard. Because I always have used a visual score of 1 to 10 for grain quality, a cutoff level of 6 might be required for retention of a line. Grain quality can also be evaluated using the micro-method of Aamodt and Torrie (1934), who published a method of determining an estimate of grain test weight of wheat based on only 4 g.

A labor-saving procedure based on visual observance is the "no harvest" option. Armed with the knowledge that there is a reserve seed stock of each line in the initial plant row nurseries sown at three sites, one could observe, select, and collate the information on a 1000-line nursery at the three locations and return to base without harvesting anything. After examination of the observation data one could save all lines selected at any station. The remaining lines, perhaps as many as 900 reserve seed lots, could be discarded. For planting the following spring a modest over-winter short-row increase of the retained 100 lines could be made using the reserve seed lots. Furthermore, in cases where yield is not to be taken, hand stripping of the desired amount of seed may be more efficient than hand cutting and threshing. Screening among crosses in order to focus effort upon the best populations has been mentioned.

Although observance of a crop can tell an expert much about probable performance and productivity, there can be no substitute for yield trials. These need not be extensive on newer material. For testing new lines the year following the plant row nursery, it is planned to use linear hill micro-plots which had been found useful earlier (Jensen and Robson, 1969). These are plots 1 ft long requiring a minimum of 1 g of seed. These are sown end-to-end, with between-end gaps of 4 in., in rod row furrows. One rod row (16 ft) contains 10 plots complete with row ends buffered. More replications are required than with larger plots (twice as many are recommended), but the nurseries are so easy to sow and handle that they should be sufficient for the next round of selection. About 10% of the plots would consist of check varieties, and a replicated field design to enhance efficiency and precision would be chosen. When line numbers had been reduced to an elite few, short three-row plots with center row harvested would be used. Final testing of a potential cultivar cannot be done by shortcut methods, however, and I would also hope to get independent assessments by colleagues.

In a small program the problem of nutritional quality, especially protein, can best be handled through control of parents in the crossing program. It is possible, of course, to find three high-protein parents to use in a three-way hybrid, but a cross with two such parents, making certain that one is the pure line for the second cross, should be productive. If only one high-protein parent is used the chances of any promising progeny having high protein is lessened. There is also the backcross.

Depredation from birds is a more serious hazard where plots are small and all crossing and single-plant increase operations are out of doors. Crossing bags are used routinely, but a technique recently adopted involves a "pinch an inch" of transparent tape around each pollinated oat floret. This may seem an incredibly tedious operation, but with a little practice the pollinated florets on a typical panicle can be taped in about a minute. The left hand holds the tape dispenser, cups the panicle, and holds the floret while the right fingers tear about an inch of tape and guide it around the spikelet (usually one or two flowers). This method gives complete protection against birds without a crossing bag. Hybrid oat kernels are so difficult to obtain that one has only to lose a few to birds to appreciate that almost any protective method that works is worthwhile. The choice of site is important in avoiding damage from birds; I recall that at the Cornell University Agricultural Experiment Station we had one location (Mt. Pleasant) that never experienced bird damage. Another effective deterrent to birds is saturation dummy plantings surrounding and in the midst of nurseries. It is helpful in these protective plantings to use mixed seeds of crops and maturities, and to have different planting densities approximating spaced plants and solid sowing. Even a late growth of weeds has been known to provide protection to spaced plants!

The hybrid (F_1) plants from three-way crosses, the progenitor bases for the entire breeding program, also require special protection from birds. A few birds can strip an unprotected plant in an hour. Therefore, I cover all three-way hybrid plants using a variety of lightweight and light-colored bags, each of which is liberally punched with air holes to minimize heat buildup. In addition, my breeding nursery is planted in four-row units covered with two canopies of bird netting. The F_1 plants from three-way and single crosses are planted in two close-together center rows over which a fine nylon net is draped on a ridge pole the length of the row. The two outside rows are pure line parents, which are expendable after being used for crossing. A bird netting with larger openings is draped over the width of the four-row unit. This gives effective bird control for a most vulnerable part of the breeding program.

Protection from birds and insects such as crickets is needed when valuable hybrid seeds are planted. I excavate a portion of the garden to a depth of about 6 in., making certain the bottom is quite level. In this area, small biodegradable pots with one seed to a pot are snugly packed. The spaces between pots are filled with sand and a fine screen laid over the top. Watering and an insect spray each evening give excellent protection.

Hybridizing oats under desert conditions requires one to learn a few adaptive techniques. First, the crossing period must avoid seasonal heat extremes. This means early fall planting (September or October), which can bring the plants into flowering by late January. The following things have been noted:

Late afternoon and evening crosses are more successful in setting seeds.

Adding humidity by wetting the bags that were applied after pollinating aids seed set.

Air holes in the bags help, and using the tape technique on crossed florets, with the wet bags removed later helps even more.

A general note: certain pure line parents are better pollen producers than others and should be used as males.

For breeders wondering whether it is worthwhile to emasculate and pollinate secondary flowers in a spikelet, the answer is "yes." My results show that fully one-third of the hybrid kernels come from the smaller second floret.

DISCUSSION

The preceding text, which was written during the early period of experimenting with procedures for an efficient one-person oat breeding program, has been reproduced virtually unaltered so that the evolution of a management system might be followed. The procedures in use after five years showed many changes and are detailed in Table 35.1 (the steps in the process repeat each year):

Thus in 15 months from the spring harvest of F_1 plants, F_5-embryo lines were harvested. Much was known about these lines; for example, each line was ready for its fourth quality screening for plumpness and shape. Height, maturity, and vigor were known, and usually there had been one or two readings on stem and crown rust. After processing and further discard, seeds of the remaining lines were divided into four parts, three for observation head row sites as shown and one part saved for reserve. The rationale or justification for these procedures follows. First, the highest initial priority was placed upon the visual quality of the grain. I knew from professional experience the difficulties in "selling" an outstanding cereal variety that had only average grain quality. Farmers take pride and satisfaction in harvesting a plump, heavy test weight oat. In retrospect, I should have paid more attention to this in the planning of crosses and early-generation screening in my breeding project at Cornell University—especially with regard to choosing among cross populations. Theoretically, one must assume the random distribution of genes for other desirable characters, such as yield, but the initial goal was to reduce quickly a large hybrid population to a small population made up of plump seeds, recognizing that their plumpness may

TABLE 35.1 Steps in Minimum Staff System

Season	Year	Site	Seed	Embryo	Nursery Type	Material	Operation	Manner
Spring	1	AZ	200 g/plant	F_2	Spaced	F_1 plant	Harvest	Hand
Spring	1	NM	All	F_2	Bulk	F_2 seeds	Plant	Drill
Fall	1	NM	2 kg/row	F_3	Bulk	F_2 seeds	Harvest	Strip
Fall	1	AZ	/row	F_3	Lab	F_3 seeds	Thresh	Vogel
Fall	1	AZ	/row	F_3	Lab	F_3 seeds	Buff	Tumbler
Fall	1	AZ	50 g/cross	F_3	Lab	F_3 seeds	Select	Visual
Fall	1	AZ	50 g/cross	F_3	Spaced	F_3 seeds	Plant	Drill
Spring	2	AZ	50 g/plant	F_4	Spaced	F_3 plants	Select and harvest	Strip
Spring	2	AZ	50 g/plant	F_4	Lab	Seed of F_3 plants	Select	Visual
Spring	2	NM	Save reserve	F_4	2-ft head rows	F_4 lines	Plant	Hand
Fall	2	NM	50 g/row	F_5	2-ft head rows	F_4 lines	Harvest	Strip
Spring	3	NM (1) ND (2)	Save reserve	F_5	2-ft head rows	F_5 lines	Plant	Hand

be due to genetic or environmental causes, or an interaction. If the quality was not present in a cross, the entire population should have been discarded.

The critical screening was done in the handpicking of the F_3-embryo seeds. This resulted in a tremendous down-sizing of the populations. For example, I had a group of about 40 F_3-embryo seed lots that ranged in size from one-half to 4 lb each. After processing, the largest handpicked seed lot weighed only 44 g and contained 934 kernels; thus the continuance of this hybrid population rested on a population of slightly less than 1000. The screening sometimes resulted in the discard of entire hybrid populations and in certain cases this was so despite excellent notes on the appearance of the F_2 population in the field. From past experience I would expect this 1000 plant population to yield perhaps 40–50 outstanding plants or lines.

Handpicking plump seeds from the F_2 increase was time-consuming and tedious but could be done between fall harvest and fall planting. It was well worth the effort, for in no other way could the vast differences and range of kernel quality for the different hybrids be grasped, nor the potential value of a cross so quickly assessed. The selection was done by pouring the seed supply in the center of a table, sitting down, and placing a pan on one's lap under the table (a gold prospecting pan with its wide sloping sides was ideal). A thin layer of kernels was brushed toward the selector, letting the seeds roll and slide as they would. Using tweezers all kernels with plump undersides (palea) were placed to one side. Then a separate glance was made over the group for those kernels with pretty backs (lemma); these were lifted with tweezers and examined for plumpness. Finally the remaining seeds in the lot being examined were brushed over the table edge into the pan and a new wave brought out from the central pile. After the experience of selecting plump kernels from a year's collection of hybrids one could *see* the contribution of particular parents.

In the F_3 single-plant nursery, which was grown from planted plump seeds, selection was primarily for agronomic type (height, maturity, vigor, etc.) and disease reactions. Harvest again was by plant stripping. It was too time-consuming to check for kernel quality in the field—that could be done after harvest. However, with the help of pans and wind we did hand rub-thresh in order to fit the seeds into small envelopes. A major discard was made in the following head row nursery at one location. The survivors were then packaged to plant for observation again in short rows at three locations, with a portion of the seed of each kept in reserve. Harvest at the three locations was not necessary. It was easier merely to take notes, collate the information from the three sites, and plant an over-winter short-row increase nursery of the survivors. At this point in the procedures one was ready to begin to test for yield.

In 1984 we moved our residence from Arizona to Albuquerque, New Mexico. If the project were to be continued the change in climate would require a change in breeding and selection procedures. At this point I decided to terminate the program and subsequently made a gift of the seed stocks inventory and records to Cornell University. I felt this program was a success, proving that unusual and efficient procedures could sustain a program whose size and quality, I believe, would compare favorably with many commercial or institutional programs. Excluding the factors of land availability and preparation, and occasional voluntary help, all of which certainly was appreciated, labor costs to operate the project were zero, although there were substantial other expenses. The program terminated, however, before extensive large-scale yield testing procedures were started; it seems clear that these would be costly additions.

REFERENCES

Aamodt, O. S. and J. H. Torrie. 1934. A simple method of determining the relative weight per bushel of the grain from individual wheat plants. Can. J. Res. 11:589–593.

Cockerham, C. C. 1961. Implications of genetic variances in a hybrid breeding program. Crop Sci. 1:47–52.

Grafius, J. E. 1965. Short cuts in plant breeding. Crop Sci. 5:377.

Jensen, N. F. 1964. Processing equipment for small grain test weight samples. Crop Sci. 4:438–439.

Jensen, N. F. and D. S. Robson. 1969. Miniature plots for cereal testing. Crop Sci. 9:288–289.

36

CHOOSING SITES AND ENVIRONMENTS

We readily accept that a *testing site* can be chosen by the plant breeder, but really, can an *environment* be chosen? Perhaps not, but adaptations can be made to the program to fit the environment given, and in some ways actually modify the environment to fit the program. That is the theme of this chapter, that the apparent natural domination of an environment is less than absolute. We shall explore ways of adjusting to, dealing with, and even changing and manipulating the testing environment.

Environments can be vastly different, so different that testing sites often do not represent the intended production areas for a new cultivar. In fact, this is likely to be the case with the home location of a breeding, testing, and selection program that may have been chosen by reason of its attachment to some institution. Soils illustrate an important element of the environment that is subject to great differences. For example, I have worked as a plant breeder in the homogeneous productive soils of North Dakota's Red River Valley, in the glaciated and varied soils of central New York, where numerous soil classification types may be found within one field, and in the arid Southwest, where irrigation smooths some problems and creates others. It is not possible to offer neat, definitive answers to the environment question. In this chapter, however, are presented the trials and experiences of many scientists who have jousted with this problem, sometimes not successfully, but always leaving a legacy of ideas. The answers are in these ideas. We shall learn how environments are differentiated and described, how they may be grouped or clustered to minimize variance, how genotypes also may be arranged and matched to particular testing sites, and how elements such as soil fertility levels, years of testing, replicates, and end-use environments interact with testing programs.

Burton's (1959) paper on crop management, and especially the section on water efficiency, highlights an environment whose dominating influence is water in all its aspects, including drought. Burton cites a dozen or more ways in which an environment can be modified. Some are cultural practices, but other involve a breeding approach.

Allard and Bradshaw (1964) opined:

> Within a region over which it is likely that a set of varieties will be adapted, it is essential that tests be conducted in a series of locations over a series of years. Great precision in the conduct of any one trial at any one location is unnecessary and may be wasteful.

Finlay and Wilkinson (1963) published an influential paper on adaptability based on the performance over several seasons of 277 barley genotypes from a world collection. Finlay and Wilkinson independently [Yates and Cochran (1938) earlier had used the technique] developed the technique of linear regression of yield of each variety on the mean yield of all cultivars, for each site and each season, and used this to compare performances of genotypes over years and sites. The site mean yield value enables the breeder not only to compare cultivars, but also provides a grading measure to characterize environments, sites, and seasons. This concept of characterizing environments is an important contribution. Their general discussion of adaptability and stability merits reading.

The impact of Finlay and Wilkinson's 1963 paper on adaptation cannot be minimized because of its influence on our concept of environments. They argued that the mean yield of all cultivars at a testing site best expressed the nature of that environment.

A quote from Sandfaer (1970) ties together thoughts of Sakai (1955) and Mather (1961) and says something about the controlling nature of the environment and why genotypes interacting within an environment may act differently in the same environment minus the competing element:

> When competition occurs in a population, each competing plant is an effective part of the others' environment without competition. Selection arising from competition therefore does not necessarily make succeeding generations more fit to meet circumstances other than those of the competition itself (Mather, 1961). Selection arising from competition will favour nothing but the ability to survive that particular type of competition.

The role of the environment was further enlarged when de Wit's (1961) ideas about competition for space were introduced (space may be thought of as an environmental "thing"). Additionally, there are the noncompetitive aspects of the environment, for example, ecological niches permitting different species to live in harmony in what otherwise is generally perceived to be a single environment where they would be in competition.

Knight (1970) used the Finlay–Wilkinson model as a vehicle to sound a note of caution about the assumed linear relation between genotype × environment interactions and a true environment. In this model, yields of genotypes were plotted against a statistic representing the environment, specificially the mean of all genotypes at a site. Knight's point was that site mean yield is an average and fails to disclose what factors are involved in, for example, equally low mean yields. These could have been caused by any one of several factors of the environment, such as drought, disease, or frost. These underlying influences are important to making correct biological interpretations.

Frey (1972) pointed out that a cultivar responsive to improved agricultural practices, that is, high yielding, is a requirement for continued oat production in the northern Corn Belt.

Pedersen, Everson, and Grafius (1978) introduced a gene pool concept as a means of characterizing an environment and to serve as a stable basis for cultivar performance comparisons (in a lesser sense the site mean yields fill the role of a cultivar check). Strict repeatability of identical entry lists over years or sites is not required; one merely accepts the view that the gene pool base is a representative sample of the larger population gene pool with which the breeder works. Therefore, routine changes in entries over time have a smaller effect when seen as changes in the gene pool than when viewed as entry changes, which at times may seem drastic. In execution, the genotype mean yield is regressed on the site mean yield, which represents the gene pool. Graphically, these regressions appear as line slopes that may be compared. To illustrate, a cultivar mean yield line (derived from several sites) might give a line whose slope initially is above the site mean yield, but later the lines cross and continue, widening the angle; the interpretation is that this cultivar does better under low-yielding conditions and does poorer in high-yielding situations. Furthermore, examination of cultivar points (sites) carry deductive information on cultivar–site interaction. The gene pool concept has basic similarities to the linear regression, site mean yield techniques of Finlay and Wilkinson.

Keim and Kronstad (1981) contributed to knowledge of drought response in winter wheat by refining definitions that in a sense deal with a genotype's response to its environment. *Drought escape* was found to be synonymous with earliness. *Drought tolerance* refers to the ability of plants under high internal water stress to maintain grain-bearing tillers until maturity. *Drought avoidance* refers to the capacity of a plant to maintain a high plant water status while under drought stress. The cultivar 'Yamhill' is an example of avoidance, accomplishing this through high osmotic potential.

Fox and Rosielle's (1982) paper is important because it highlights and expands the idea of a target population of environments as propounded by Comstock (1977). The scatter of long-term performances of a chosen set of cultivars would represent a field of environments in which the calculated center would represent the target. Selection trial results could be compared with the target values, and distance from the target center (the closer the more valuable) would give an estimate of a trial's worth relative to the breeding goal. The target population concept is an attempt to provide a running mean through unpredictable years, sites, and other variables.

At first glance, the reference set of genotypes seems synonymous with cultivar checks or controls. Fox and Rosielle pointed out that they are in fact different. The difference resides particularly in numbers; one or a few check cultivars would not monitor sufficiently the complexities of an environment and larger numbers would be needed for statistical reasons. Because large numbers are needed, one wonders why the Finlay–Wilkinson site mean yield might not be used as the reference set—it would require only the acceptance of the assumption that the changing of entries does not materially affect the group's performance, year to year, and acceptance of the requirement that numbers in the group remain large. The relationship of such use to the goal of breeding for stability in the original site mean yield context would be irrelevant here.

SELECTION UNDER DIFFERENT ENVIRONMENTS

The desirability of doing variety testing under the best conditions was implied by Meyers (1912), who found that an increase in soil fertility reduced variability in wheat, and by Roberts (1912), who discovered that a favorable growing season reduced variability.

Immer (1942) compared the distribution of single-spaced plant yields of four varieties of barley, and four F_2 crosses among them and one other parent. He found in the varieties a linear relationship between means and variances, highly significant varietal differences, and different patterns of single-plant yield distributions. For the F_2 crosses there was no linear relationship between means and variances, cross differences were just significant, and the distribution of single-plant yields was similar for all crosses. He concluded that the yield variation of spaced single plants was determined almost completely by environmental factors. Furthermore, he noted,

> If different varieties or crosses are to be compared, it would appear that only the means, and possibly the variances, would need to be calculated. The yields of single plants in F_2, spaced five inches apart, would supply essentially no information on the yield of F_3 progeny rows.

Immer's paper is difficult to interpret in an age of greater statistical sophistication; however, the primacy of means and variances in comparisons is clear—it is the spaced-plant behavior for which one wishes more enlightenment.

There is another part of the paper that merits comment. Immer also studied skewness and kurtosis in the distributions. The information on kurtosis might be considered neutral or ambiguous, but skew was a different matter. In both varieties and crosses the yield distributions were skewed in such a way that more individuals than expected piled up at the low yield end. This is a favorable situation for selection, because this is the discard end of the curve and it is scarcely credible that inherently high-yielding individuals, above a breeder's reasonable cutoff point, would fall into this group and be discarded.

The paper by Kelker and Kelker (1986) shows the profound effect skewness of character distribution can have. Skewness can have several causes or explanations, for example, linkage; however, the act of selection itself, either by natural or human initiation, can induce skewness.

Hammond (1947) reviewed the relationship between type of environment and selection with animals, and concluded that the determining factor should be the type of environment in which the animal was likely to live; that is, if the organism were likely to live in an optimal environment, the selection should be made in a similar environment.

A different way of looking at selection in two different environments was presented by Falconer (1952). At the base of his hypotheses lies genotype × environment interaction such that, in the case of two different environments, the interaction can be seen as a genetic correlation. This line of thinking leads to "unthinkable" views, for example, that yield is not one widely varying character. Instead, if the same genotype were grown and measured in two widely different environments, yield might be seen to behave as two different characters: different in the sense that the yield response may not be expressed in the other environment, and in order to be expressed may require the presence of the environment in which the response was selected. This is opposed to one view that selection in the most favorable environment is always riskfree and that such a selection is thereby always better—"a selection for all seasons" as it were.

Falconer viewed two individuals, initially carrying essentially the same genes, being separated and exposed to two environments. If these individuals were heterozygous there would be a gradual convergence of genotype and phenotype to match the demands of each environment. Finally, one can see that two different yield characters could emerge.

Falconer's logic is backed by formulas and he sums his views that genotype × environment interactions should be viewed as genetic correlations:

> In this way a precise answer can be given to the question whether it is better to carry out selection in the environment in which the improved breed is required eventually to live, or in some other environment more favorable to the expression of the desired character. Performance in the two environments is regarded as two different characters which are genetically correlated. Selection for one character will then bring about a correlated response of the other character. The magnitude of this correlated response may then be compared with that of the direct response to selection for the desired character itself.

This leads one to ask, would it be profitable to intermate high-yielding individuals from two environments, thereby merging different genes affecting yield expression? Although couched in animal breeding terms, Falconer's statement poses the plant breeder's selection problem vis-à-vis selection in a spaced-plant situation for eventual performance under crop density conditions.

Sandison (1959) pointed out that "it is more important to cover the main environmental conditions than to achieve high accuracy in individual trials."

He thought it likely that varietal differences would be greater under high-yield levels; thus he supported the view that testing should be done in plus rather than minus environments. Sandison quoted Engledow and Yule (1930) as stating that it was "no use spending great pains on the endeavour to reduce the effects of one sort of error (within trial) when another is left uncontrolled."

Sandison's data are impressive: using data from 21 oat trials of four varieties, six replications, seven locations, and three years, he found the standard deviations of variety yield expressed as a percentage of the mean plot yield to be 11.5%, 11.3%, 11.2%, 11.1%, 11.2%, and 12.4%, for the decreasing replications 6 to 1, respectively; thus there was little gain from increasing replication number.

Gotoh and Osana (1959a) had reported that concurrent selection in three density levels of the same wheat cross had shown that more superior-yielding selections had come from the lowest density, wide spacing. This may also be interpreted to mean that the lowest density was the most favorable environmental condition for the plant, perhaps even the most favorable fertility situation.

Somewhat later in the same year, Gotoh and Osana (1959b) tested the efficiency of selection in a hybrid of winter wheat cultivars, 'Minturki' × 'Tohoku No. 103', under three fertilizer regimes, namely, half, standard, and double amounts. Though it was not explicitly stated, it may be assumed that the F_2, F_3, and F_4 were space-planted (because plants continued to be selected). Populations of 1000 plants from the F_2 were planted in each of the fertilizer treatments. Selection of desired plants was at 30% intensity in the field, later changed to a random selection of 150 from the 300 of each treatment. In the F_3, the three sets of 150 lines were yield-tested in the three fertilizer treatment plots, the best 10% of lines selected, and the three best lines within each line group were retained. Thus there were 15 line groups and 45 lines for each of the three original subpopulations. In F_4, the line groups and lines were again tested for yield in the three fertilizer treatment plots.

The results were surprising. The total number of superior yielding lines found, when all lines from the three fertilizer-derived selected groups were tested across the three fertilizer environments in the F_4, was greatest from the lines derived from selection in the lowest fertility level group, the half standard application. Superior lines were defined in relation to the higher-yielding parent. The highest mean yields were from this group and they persisted across fertilizer environments, indicating wide adaptation. Also, means of lines from all three fertilizer level-derived groups were superior to parent means, indicating that all selection was effective. The data show an inconsistency in linearity: the standard fertilizer (middle application) group was less productive of high-yielding lines than either the half- or double-amount groups. On the other hand, each of the groups of derived lines showed linearity when tested across the increasing fertilizer amount environments, that is, lowest mean yields in the half-amount application and highest yields in the double amount. Information on the basic soil fertility and a no-fertilizer control would have been helpful in interpreting results. The overall conclusion was that genotype differences were magnified the most, and selection was done best at the lowest fertility level.

Jones, Matzinger, and Collins (1960) reported a study with tobacco to assess the influences of variety, environment, location, and years. They found that adequate evaluation of varieties could be obtained from two years of tests at five locations using three replications.

The close association between problems in plant and animal breeding is shown admirably in James's (1961) paper on the subject of selecting animals in two environments. James reviewed various earlier options such as confining selection to each environment, selecting under most favorable environment, Falconer's (1952) quantitative approach to dealing with genetically correlated characters in two environments, and so forth, James dealt with extensions of solutions where the animal (or plant) breeder has different goals, for instance, genetic gain desired in two environments that, however, do not have the same importance. An example easily recognized by plant breeders is testing for adaptation to two environments using index selection as a guide:

> An individual cannot usually be itself tested in both environments, but such procedures as progeny testing provide a suitable method. For instance, in progeny testing to improve egg production in the domestic fowl, groups of daughters from each tested cockerel may be divided, some being tested on farms with an endemic disease, others on farms where this disease is absent. Then the two 'phenotypes' of a cockerel are the means of his progeny groups.

Of the three procedures considered, namely, selection in one environment, separate selection in two environments of two strains, and index selection combining performances for both environments, James concluded that index selection showed the greatest overall genetic advances, but he cautioned that a breeder would need to balance this against the greater cost of index selection.

Balaam and Hunter (1962) found that lattice designs were superior to randomized block designs in wheat varietal trials, with gains up to 143%. They noted that increases in locations caused a more rapid decline in variance of means than did increases in replication. They concluded that "the optimum allocation is one replication per location and to increase the number of locations indefinitely within the limits of the research programme."

Although the study by Miller, Robinson, and Pope (1962) dealt with cotton, the findings have wider relevance. On the basis of three years' testing of 16 cotton cultivars at the same widely separated 11 sites, they concluded that about that number of sites, 10–20, would be needed to get precise estimates of performance. Increasing testing environments would increase precision more than would increasing replicates. The authors commented

> The relatively small variety by location (particularly in the area east of Texas) and variety by year interactions indicated that neither location nor years, per se, were having consistent effects on differential varietal responses. Thus, it apparently makes little difference how this sample of environments is distributed over years and locations,

This called to mind a paper (whose author and source I now cannot retrieve) in which it was proposed that a judicious selection of environments could substitute for years and thereby save time in the testing program.

Hanson and Brim (1963) described autogamous breeding programs (soybeans) as having three levels of evaluation, namely, the initial selection of plants and lines, a numbers-reduction phase focusing on yield, and regional testing to involve different environments. The authors concentrated on the optimal allocation of test material for a given cost restriction, with emphasis on environments, replications, and genetic variability. Their model involved a two-stage test that involved an initial discard, followed by retesting of selected material. Examples were given of sample allocations for certain situations.

Hanson and Brim's paper is also of interest because their approach to locations and years permits a generalized treatment of environments vis-à-vis genotype × environment interactions. This approach states that locations and years may not be unique elements of the environment, but that environment can be sampled by considering these as random effects (I am extrapolating somewhat here). The importance of this to breeders is that they can consider dispensing with years (thereby saving time) if they can substitute additional samples of the environment within the shorter time span.

Frey (1964) chose two essentially adjacent fields which because of historical use and terrain features could be classed as infertile "stress" and fertile "nonstress" areas. The ratio of past mean oat yields on these areas were of the order of 1 : 2, respectively. The questions asked in the study were, "What environmental regimen is best for early-generation selection?" and "What effect do stress and nonstress environments have upon the adaption traits of the selections made in those environments?" All lines selected over a five-year period experienced only one of the two environments.

Frey found selection for yield from both areas to be equivalent; that is, when the selections were grown in common tests they yielded similarly. It appeared more difficult to choose selections in the stressed environment, as evidenced by twice as many lines remaining after each selection cycle, when compared with the nonstressed group. This seems logical when one considers the restricted response for characters such as height or vigor, under the droughty, infertile stress conditions, and when one bears in mind that selection was largely based on visual discrimination. The nonstress environment was superior for producing lines with good (wide) adaptation response. Frey's general conclusion was that high production conditions are best for selection and testing in an oat breeding program.

Levine and van Valen (1964), working with *Drosophila pseudoobscura*, found that mixed-race populations maintained at a higher temperature environment, 25°C, were able to respond and change their proportions when placed in a 22°C environment. However, the reverse was not found: populations kept at 22°C were not able to change when placed in 25°C temperatures. The authors concluded that "this finding indicates that the effects on a population of a series of environments may depend on the order in which the environments are experienced."

Translated into crop breeding, this suggests that the choices when assessing multisite possibilities for early-generation screening or testing are important; for example, selection from the same population at different environmental sites may produce different types of plants. Perhaps a more analogous example would be selecting in one environment in one year, and observing the selections in a different environment the second year: in this case the second selection is not complementary to the first, but rather proceeds from a base already fixed by the first genotype × environment interaction.

Heterozygote advantage is a phenomenon of value in commercial F_1 hybrids, but a troubling one to breeders in early-generation selection for pure lines in autogamous crops. Allard and Hansche (1964) have connected it to type of environment "Apparently the advantages of heterozygotes over homozygotes is particularly associated with stress environments."

Indeed, one of the advantages hybrid corn is supposed to have is better performance under stressed conditions.

Johnson and Frey (1967) reported on the influence of stress on genotypic and environmental variances of quantitative traits of oats. Stress was defined in terms of limiting plant productivity, and in general it has an inverse relationship to productivity, although there are special cases where the opposite is true. The traits examined were heading date, height, plant weight, grain yield, panicles per plot, spikelets per panicle, and weight per 100 kernels. Environmental stress levels were fashioned through the addition of phosphorus or nitrogen and changes in planting dates. Overall, there was a trend that as stress decreased, and the productivity of traits increased, variances increased. Specifically, this was the case for grain yield, plant weight, and number of panicles per plot. There were, however, negative responses, such as that of 100-seed weight to nitrogen. The authors pointed out that it may be misleading to characterize an environment as having a certain stress level without relating this to the attribute under consideration.

An objective of this study was to find guidance to the question, "Should selection be under stress or nonstress conditions?" The authors found that this question could not be answered conclusively. They found greater genotypic variance under decreasing stress conditions; however, this was accompanied usually by an increase in environmental variance so that heritabilities did not increase in a consistent manner.

St. Pierre, Klinck, and Gauthier (1967) stated that the adaptation of cereal genotypes is usually determined at the regional testing level when the lines are being evaluated as candidates for release, whereas little was known about early-generation plant or progeny responses to selection for wide adaptation. They crossed two presumably pure lines of barley genotypes and grew the 27 F_1 plants in the greenhouse. The F_2-embryo seeds were then divided into two lots and space-planted at 6-in. intervals at the two sites chosen for the research, Macdonald College (M) and La Pocatiere (P). At each location, 300 plants, approximately 26% selection intensity, were saved on the basis of visual appearance and yield per head. Each plant yield was divided, and 600 F_3 progenies, spaced, were again planted at each location. The pedigree selection was repeated through the F_5 with the F_3, F_4, and F_5 intensities at 10%, 10%, and 35%, respectively, with 500 progenies selected at each station in the F_5 generation. The result of this generational splitting and sharing of all selections at both sites resulted in the F_5 of 16 pathways, ranging from MMMM through all combinations to PPPP.

The concern in this experiment was with the effect on the same progeny material of a series of pathway environments. It was not possible to evaluate yield on all the F_5 progenies; however, in F_7 and F_8 the strains tracing back to three of the 27 original F_1s were compared in yield trials. At this time the tested lines were nearing homozygosis, and the lines coming out of each pathway ought to reflect the pressures exerted on them over their history. For example, a line from the always-at-Macdonald site (MMMM) ought to say something about that location.

The results, indeed, did have something to say about the lines' treatments: PMPM was found to be the highest-productivity pathway, with the mean of its lines appearing in the highest-yielding group 18 times out of 24. In contrast, the MMMM pathway produced the lowest yields. The interpretation is that selection in the PMPM pathway produced lines with wide adaptation whereas the MMMM pathway yielded lines poorly adapted to all environments. A seeming paradox, but one that breeders must learn to watch for, was that the PPPP pathway, which was found favorable for adaptability, both in itself and in combinations, was the location with the lower yields and therefore the stressed environment. Nevertheless, it was the more favored environment in which to select for wide adaptation.

Johnson, Shafer, and Schmidt (1968) made an important point when they stated that general adaptation is important not just for large areas, but also at times for even a single location because of year interaction variances. They used Finlay and Wilkinson's site mean yield indicator for type of environment in a study of yield potential and stability of performance of several hard red winter wheat varieties grown in regional nurseries in several states for many years.

There was excellent positive correlation between mean yields of a cultivar and its regression coefficient. The study presented a favorable view of newer varieties that exhibited solid yield advances with predictions of greater stability. Specifically, the newer varieties yielded well in poor environments and relatively better than older varieties when grown in favorable conditions. One regional difference was observed: progress has been relatively better in the central and southern plains than in the northern plains. This is thought to be related to the difficult problem of increasing winter hardiness for this latter area.

The problem addressed by Roy and Murty (1970) was to develop procedures to select wheat genotypes for wide adaptation, yet with the response capability to react positively to stress environments, in their case, drought. To do this they intuitively felt that continued selection in one environment, even if it represented the stress environment, was inadequate.

The plant material used represented a massive wheat composite initially involving 3 million plants from 533 crosses. An important preliminary test involving only 17 crosses was conducted in three of the four created environments, which were (A) high fertility irrigated, (B) moderate fertility irrigated, (C) low fertility with starter irrigation, and (D) low fertility rain-fed. In these aliquot F_2 populations, selection was practiced for developmental and other traits such as seedling vigor, synchronous tiller percent, and yield. On this basis, 84 F_2-derived families from the three different environments were established for F_3 testing in all of the four environments. Two additional factors were measured, survival at harvest and overall superior yield in all environments.

First a comparison was made of the performance (criterion: number of best lines) of the 84 families when they were all grown in a stress environment (D) with these results: 14 families were found to be outstanding. In the F_2, eight of these came from A, four from B, and only two from D. This means that the best selections under stress testing did not derive from a stress environment, but rather from the most favorable and presumably less stressful environments. These were not random F_2 selections, it must be remembered, but were chosen principally for developmantal traits plus plant yield.

The test of the 84 F_3 families under four environments involved measurement of the traits mentioned plus several others, principally components of yield and days to heading. From these measurements, the relation of the environment of selection to progeny performance, to estimates of components of variance, and to estimates of heritability were worked out separately for each environment. One may summarize Roy and Murty's findings to show that the F_2 performance under the favorable irrigated A environment was "consistent with the performance of the F_3 progenies in each of the four test environments (A, B, C, D), while no such relationship is found between the material selected under rain-fed conditions and their progenies tested in the other environments... The data, therefore, are in favour of selection under a favourable environment (high to moderate fertility with irrigation) for all the characters studied."

Based on heritability values, the authors believed that selection for developmental traits in the diverse environments was instrumental in shaping the progeny performances seen later. Synchronous tillering, days to heading, and ear length seemed to be critical traits. Synchronous tillering is important and might be critical where a crop must mature evenly before a known seasonal stress period (I was always impressed with a plant that put up an even canopy of heads from a few tillers). It appeared that much of developmental selection could be based on visual discrimination.

Roy and Murty selected 1200 F_2 plants from the large cross composite, reducing these to 600 after a grain quality test and eventually to 100 after other tests. Using the information from the preliminary experiments, they tested these under A and D environments for two years. The results confirmed the efficacy of the tandem system, producing F_3 progenies showing a range of adaptations from adapted to both environments to adapted to each specific environment. Many of these were superior to the check cultivar.

These experiments led Roy and Murty to propose a selection scheme employing forms of screening and selection, for example, visual, laboratory, and use of two environments. The A environment was used as the central one, with occasional simultaneous yield testing and observation in a D environment.

When I first read this paper I was reminded of two papers on animal breeding, (Falconer, 1952) and (James, 1961). In Roy and Murty's case, however, they used a zigzag form of selection in two environments in a deliberate attempt to select for wide adaptation. One can find several variations of these modifications among breeders' practices, some involving many cooperators participating in growing aliquots of a broad-based hybrid bulk at different locations, with parts of the harvested samples recombined, split, and redistributed from a central location.

Roy and Murty are correct in assuming that the probabilities for wide adaptation are enhanced by their procedures. In fact, with enough locations one can see that the expanded number of environments would act to preserve between-line, thus populational, genetic variance. In this sense the results parallel those of the single seed descent method, whose dynamics also maintain between-line variance. One might note that nursery location is not the only variable among the breeder's options: Roy and Murty also created different environments using soil fertility and water changes.

Vela-Cardenas and Frey (1972), in the continuing Iowa studies of aspects of the environment, looked at the relations of heritability and genetic gain to productivity. The genotypes were 240 near-homozygous lines from six crosses. Ten environments were created through combinations of years, fertility, density, and dates of planting. Heritability values were estimated for five traits, heading date, height, spikelets per five panicles, 1000-kernel weight, and grain yield, and from these in turn were calculated the respective expected genetic gains, all based on foundation analysis of variance calculations (heading date alone was measured in fewer than the 10 environments).

The plant breeder is interested in the optimal environment for genetic gain and heritability, actually two different things, and the outcome of the studies was fortunate in that these two factors were in agreement; that is, what was an optimal environment for one was also optimal for the other. For example, maximum heritability and genetic gain for grain yield were both found under normal density and high density conditions. A second general finding was that the optimal environments for maximum heritability and genetic gain were related to and varied with the trait under consideration. The authors discussed the optimal environment for each trait and when compromise or use of index selection was indicated.

A highlight of the paper was the authors' application of findings to the methodology of plant breeding. The following alternatives were suggested:

1. Use a different environment optimal for each trait.
2. Use the environment exhibiting the greatest average heritability for all traits.
3. Use one environment, first applying (economic) weights to the characters. Choose that environment most favorable to the highest priority trait.

All of these, especially the third, touch in some way the concept of index selection without actually embracing it.

Other alternatives suggest themselves. If seed supplies are sufficient they might be divided for planting at more than one location, thus making selection possible in more than one environment. In using diallel selective mating populaions, I found it necessary to use aliquots from the same F_2 reserve seed lot in the same year and in successive years, thus sampling selection under different location and year environments.

Vela-Cardenas and Frey's review of literature on genotypes and type of environment interaction is unusually good.

Their study was directed, of course, toward choosing the best environments in which selection for different traits could be done, but it is difficult to avoid the related matter of choosing environments for testing lines. I hark back to a concept of "universal environment" (Jensen, 1976) that has always intrigued me. ("Universal environment is the potential of all possible environments in the target area for a projected variety.") A common problem facing all breeders with candivars awaiting release is to decide the validity of performance results obtained from a limited sampling of the infinite possibilities of the universal environment. There are many inputs, but final release decisions rely heavily on averages, especially concerning yield. It behooves the breeder, therefore, to build up these averages in such a way that they are representative, that is, arising from the sampling of sufficient numbers and of sufficiently varied expressions of the environment. Accumulating the required data base is a slow process; this does, however, speed up as a candivar is entered in broader-based trials such as regional and national ones. Breeders are aware that these geographically broader trials also may sample environments somewhat different from the breeder's target environment. If by analogy, we take a clue from Vela-Cardenas and Frey's study, where 10 environments were created, all clearly within their target area, it would seem that a breeder swiftly could accumulate a representative data base at little cost. For example, at Ithaca, New York, we had valley soils, the Cornell station fields, and the Mt. Pleasant area, at elevations of roughly 400, 1000, and 1700 ft, respectively. Add to this the options of date of seeding, rate of seeding, choices of soils, irrigation capabilities, fertilizer rates, and so forth, and we have the ingredients for a confidence-building set of data that could supplement and in some instances substitute for the more extensive regional nurseries and years.

Jinks and Connolly (1973) studied the capacity of basidiomycete isolates arising from selections made in separate temperature regimes to adapt to changed environmental conditions (temperature). The criterion measured was rate of growth. The results were attributed to allelic fixation of genes favorable for growth in each temperature regime. The patterns were consistent; that is, low growth rate selected at low temperature reinforced environmental sensitivity, whereas low growth rate selected at high temperature was less reinforcing and the selection more sensitive to environments, and selections that had been made in alternating environments showed better adaptation and stability in changing environments. The authors concluded that the end use should dictate the selection environments. Paraphrasing in plant breeding terms, breeders should use:

1. Restricted environments for selecting individuals destined for specific environments.
2. A wide range of similar environments (clustering) for typical recommended general uses, as of a cultivar.
3. A wide range of different environments where general and unrestricted use is anticipated.

McNeill and Frey (1974) assessed the selection among 1200 pure line genotypes in six environments (three sites used for two years), trying to relate selection efficiency or value to productivity of an environment. Selection was for the three traits, grain and straw yields, and 100-kernel weight. The statistics involved were heritability percentage, genotypic coefficient of variation, phenotypic correlations, and actual gains from selection. They concluded that the productivity of an environment was not a reliable indicator of its value as a selection site: "Except for heritability of 100-seed weight, none of the trait statistic combinations showed an apparent association between the relative productivities of the six environments and the magnitudes of the selection evaluation statistic."

Illustrative of the results was the case of environment 5, the lowest-productivity environment, which showed the highest genotypic coefficient of variation, suggestive of good selection opportunities. In fact, this environment had the lowest actual gains from selection.

The six environments could be described as representing two environments, a group of five like sites, and a single aberrant location. There is food for thought in the knowledge that environment 5 was the same site as environment 2 (Sutherland, Iowa), differing only in year of use; thus the aberrance could not be said to be location only. These apparent reversals have been seen in other research; for example, Finlay and Wilkinson in 1963 reported a similar observation. This complicates the task of evaluating environments, for example, of a testing region, where the objective is to cluster environments to improve testing efficiency; however, most breeding programs do have recorded data on a site's characteristics over a period of time. In the final analysis, the authors concluded that the best guide for choosing a restricted environment—restricted in the sense that a single environment is being sought for a special purpose, selection—is to choose one that is similar to the environment of expected use.

Findings with implications for selection strategies in crop plants emerged from a study of the fungus *Schizophyllum commune* by Jinks and Connolly (1975). The regression of rate of growth on temperature over a range of 15–30°C showed that sensitivity response varied with the environment. The authors were led to the conclusion:

> Selection for high mean performance in a good environment or for low mean performance in a poor environment leads to selections that are more sensitive to environmental variations than selection for high mean performance in a poor environment or for low mean performance in a good environment.

From this I infer two things: "sensitivity" is similar to "stability" or its inverse, and selection of plants for yield in a favorable environment will produce selections with lowered stability (greater variance) when tested across a wider sample of environments. Although these findings are not extrapolated by Jinks and Connolly, they seem appropriate for testing on crop plants. It is unfortunate that one alternative to low stability is to select for high yield under poor conditions. The authors do offer a tandem, compromise approach to selection.

Boyd, Goodchild, Waterhouse, and Singh (1976) studied climatic environments in Western Australia for genotype × environment clues that might aid regional testing practices. Rainfall within-year distribution was found to be of major importance with a region × season interaction. Nevertheless, the mean regional yields were noticeably alike over long periods, suggesting that regional and seasonal differences canceled out each other. They did find a centrally located region that showed greater balance in wheat production. Their conclusions apropos plant breeding testing were that the central region represented the best initial sampling area and that specific adaptation needs could be approached later:

> Traditional regional testing should continue but with more stringent and specific rules for the inclusion of entries; the initial interpretation of results should be based on the central region with its small within-year variability. The inclusion of a reference set of genotypes for characterizing region × season combinations of environments would serve a valuable role in identifying area subdivisions with specific adaptive requirements.

Hinz, Shorter, Dubose, and Yang (1977) considered the problem of selecting genotypes at several locations. Because of genotype × environment interactions and heritability of traits, the rank order of performance of the genotypes over locations has a high probability of being different. Thus if a given percentage or numbers are saved at each of several locations, the total number saved will exceed the cutoff level. The authors address this problem according to the influence of number of locations and heritability.

In analyzing this problem one must distinguish between two kinds of location tests: the genotype or cultivar yield nursery, often called a regional test, and the selection nursery. In the regional nursery, it is customary at harvest to bring the harvested bundles to the central station for curing and processing (although it is quite possible today to carry the "laboratory" to the site and at proper maturity complete all operations including weighing and statistical analysis on the spot). The other case of the selection nursery is, I interpret, the situation dealt with in the paper by Hinz et al. In practice, breeders usually do initial screening at one central location because of convenience, and because of restrictions imposed by seed supplies and genotype numbers. A secondary screening thus seems to be the kind most likely to be done at more than one location. Also, the breeder may wish to see the genotype × environment interactions in different environments to aid in his selection. Hinz et al. provided a basis for predicting expectations. In actual practice a breeder can handle the problem in different ways. Probably the most efficient when dealing with large numbers would be visual selection alone at each location, with all selections from all locations then harvested at only the central station location. In this case only short rows would be needed. The least efficient, of course, would be longer rows, or more replications, for more precision in measuring yield; cutting all of them; and retaining selections on the basis of collated data from all sites.

Frey (1977) suggested that the commonly accepted negative correlation between grain yield and grain protein of cereals was probably due to a deficiency of nitrogen in the soil. This suggests that advances in both yield and protein might be possible under high-productivity conditions.

Eagles and Frey (1977), and Eagles, Hinz, and Frey (1977), came to the conclusion that oat lines superior at all yield levels could be selected based on their mean yields over many environments. They expected such lines to have low stability variances.

Richards (1978) tested two species of rapeseed for maximum selection response in grain yield under optimal (irrigated) and suboptimal (water restricted after flowering to 25% of control) conditions. He concluded that selection for yield was more efficient under the stressed (drought) conditions, or more generally stated, the most efficient selection method is the one performed in the environment of expected use. Such selections also showed yield advances when tested under optimal conditions, or when using a "drought response index," which was the ratio of yield under drought to that under irrigation.

Interestingly, the two species responded differently under drought, one suffering more than the other; nevertheless broad-sense heritabilities were higher under these conditions. The author attributed this to a greater expression of genetic differences under stress.

Walker and Fehr (1978), in an extensive testing of pure lines and mixtures of soybeans in 12 environments, found that they performed relatively the same regardless of environment.

Allen, Comstock, and Rasmusson's (1978) study was directed toward defining the parameters of the optimal testing environment for selection. They analyzed the results of tests with five different crops species: barley, wheat, oats, soybeans, and flax. Not surprisingly, an important finding was that crops may differ considerably in their response, suggesting specific rather than general phenomena. Data were obtained from the results of cooperative regional nurseries that spanned many years, trials, entries, and locations. The authors used genotype × environment interactions and heritability, in addition to the Finlay–Wilkinson site mean yield as the base index, which characterizes environments as favorable, intermediate, or unfavorable.

The extensive volume and nature of the data must be noted. The characteristics of the three environments were strikingly different for each of the five crops. For example, except for oats, yields in favorable environments were twice those in unfavorable ones (oats may have been influenced because all trials were within one state, a more restricted environment). As has been shown in other studies, error variance increased with yield level at about the same rate, so that in favorable environments it was twice the level of that in unfavorable ones. An interesting crop response was that these differences were pronounced in barley, oats, and flax, but less so in wheat and soybeans.

Genotypic variance was positively correlated with environment also. In soybeans and wheat the differences in favorable and unfavorable environments were especially marked. The authors interpreted this as being due to large genotype × environment interactions, an area of special interest to a breeder trying to find the optimal environment for a new cultivar.

The differences in response to environmental class were small in oats, barley, and flax, leading to the conclusion that no environment was superior for plant selection. In the case of wheat and soybeans, where yields were verh high in the favorable environments, it was also noted that intermediate environments were almost as good. There was little solid evidence that favorable environments were better for testing than intermediate or unfavorable ones. Allen et al. were led to the conclusion that a test environments typical of the target population would be best. I interpret this to favor the intermediate environment, and in practice to involve a certain amount of selection and clustering in order to achieve the maximum among of homogeneity.

When Brennan and Byth (1979) examined all environments they found that the performance of wheat in high-yielding environments was a better predictor of line performance across all environments than was performance in low-yielding environments. The best selection strategy was to express line mean yield at each environment in percent of environment mean yield, average these across environments (sites), and use these composite figures to select for high relative mean yield.

In a breeding methodology study, McVetty and Evans (1980) grew and selected the same hybrid populations of three wheat crosses in a normal environment, and a stress-free environment having adequate irrigation, fertilizer, and weed control. Among 360 F_4 (F_2-derived) bulks, 53 were considered "high yielding," defined as a yield equaling or surpassing that of the check cultivar, 'Glenlea'. The authors were surprised to find that the percentages of high-yielding selections found in the two environments were very similar, 37% and 30%, but were marginally better in the normal environment. Optimal conditions in this case did not favor higher selection efficiency.

The focus of this study was an attempt to link any physiological or morphological traits of individual F_2 plants to their F_4 bulk yields. Twenty components were measured or derived from measurements. The results were as follows:

1. Single F_2 traits identified 17 of 53 high-yielding F_4 bulks (32%).
2. Multiple regression analysis identified 24 of 53 (45%).

3. A combined cross analysis associated significant F_2 plant to F_4 bulk yield correlation with these characters: yield, productivity, kernel number, peduncle length, and productivity per tiller.

When a multiple regression analysis was applied to the combined cross data it was found that productivity and peduncle length were the only significant common parameters. Peduncle length had a negative relation to productivity. Productivity was defined as the total aboveground dry weight per plant, presumably as-is at harvest. Harvest index was shown to be nonsignificantly correlated with productivity.

The authors emphasized the unexpected findings concerning productivity, and believed that this trait, with further refinements, might be satisfactory for selecting F_2 spaced plants with high potential future yield.

In cotton breeding, specific adaptation to environments can be seen in varieties adapted to differences in elevation. Heat tolerance is a requirement for low-elevation production. Experience has shown that varieties developed at low elevations often are found to be superior at high elevations, but the converse is found rarely. In this study, Feaster, Young, and Turcotte (1980) compared artificial and natural selection in a cross of derived high-elevation ('Pima S-3') × low-elevation ('Pima S-4') cultivars at three elevation sites, namely, one low site at Phoenix, Arizona at 350 m, and two high sites at Safford, Arizona (885 m) and E1 Paso, Texas (1200 m). Although Safford is considered high elevation, it is agreeably more like Phoenix than E1 Paso. Although eight traits were measured, relative lint yields were used to describe the environments.

Natural populations were harvested in their entirety and a random bulked sample used for the next planting; artifically selected populations were selectively harvested by plants and a random bulked sample used for planting. The populations were grown from F_2 through F_8, each in its isolated site area. For comparisons, the original F_2 seed (cycle 0), the F_5 (cycle 3), and the F_8 (cycle 6) from each location—the latter two included natural and artifically selected components—plus the two parents, were grown at each location for two years, making a total of 15 populations plus parents at each location.

The breeding and selection implications of the results were these:

1. Artificial selection was superior to natural selection in the order of Phoenix (greatest), Safford, and E1 Paso (least).
2. Selection was effective in the environment where the population was to be utilized.
3. Selection based on phenotypic differences of genotypically productive plants "was most effective at Phoenix, somewhat less effective at Safford, and the least effective at E1 Paso."

4. Phoenix would be the logical site choice if an improvement program were to have a single base.

In summary, the most productive area, Phoenix, was best for genetic advancement. Phenotypic expression was magnified in such an environment. Furthermore, it was found that by utilizing the three traits lint yield, fruiting height, and plant height, it was possible at low elevation (Phoenix) to distinguish plant types predictably adapted to high elevations. Low-elevation types began fruiting relatively low on the plant; high-elevation types were usually taller, and fruiting began relatively higher on the plant. Low-elevation genotypes must also resist heat, a normal part of the Phoenix environment. Although none of the environments was free of stress, the weight of evidence supported choice of the "best," most productive environment for maximum selection efficiency.

Shabana, Bailey, and Frey (1980) tested the productivity responses of oat lines grown in three quality grades of environments, namely, low, medium, and high. Lines were grouped into two categories, T (200 lines) and G (480 lines), based on treatment and procedural differences practiced within the original 250-cross composite.

As might be expected the productivities of line groups correlated with the grade of environment: 13.0, 23.2, and 31.3 q/ha for T group, and 11.7, 19.8, and 24.8 q/ha for G group, for low, medium, and high, respectively.

Line selection for yield was made at 10% intensity. Overall actual gains across all environments were 16% for T and 7% for G groups. Gains for the selections partitioned by the low, medium, and high environments were: T group 16%, 13%, and 18%, and G group 7%, 5%, and 8% of population means, respectively. Thus greater gains were found in the T group and in the environmental extremes of low and high in both groups.

Significantly, lines selected only under high productivity gave significantly higher mean yields than the means of all environmental groups. This relationship was not found in lines selected under low and medium environments only.

The best lines from each environment were good high-yielding lines. Nevertheless, the following further deductions were made: low and high were the better environments for selection (nearly equal), but high was judged the best, based on its ability to "differentiate high-yielding lines best." There was a recognition of the protective value of using more than one environmental site. The consistent inferiority of the medium productivity group was puzzling.

The final paragraph has intriguing plant breeding implications: "With the judicious use of disruptive selection (Lu et al., 1967) by manipulating artificial selection successively under stress and nonstress conditions, we might be able to increase the frequency of the lines with that mechanism." [The reference is to Lu, Tsai, and Oka (1967), not reviewed in this book.]

The relationship of the quality of the testing environment to genetic advance is of great interest to the plant breeder and indeed is complex, for one may find guidelines suggesting the use of full resource environments as well as severely stressed ones. Quisenberry, Roark, Fryear, and Kohel (1980) selected for lint yield of cotton in three Texas environments differing as follows:

Location	Heat units	Moisture
Lubbock	Deficient	Adequate (irrigated)
Big Spring	Adequate	Deficient
College Station	Adequate	Adequate

The comparisons were made over a period of two years on F_2-derived F_4 and F_5 lines from a hybrid population. The two generations were assessed for residual dominance effects which were genotype × environment interactions per location and per year (combined data were used for the latter). A scale for heat units was calculated and calibrated with past field data. The definition of "adequacy" was based on 3326 heat units and 85 cm of precipitation. adequacy for precipitation could be attained through irrigation at Lubbock.

The results showed that dominance effects for lint yield either were not present or had dissipated by the F_4. The correlations between F_4 and F_5 yields were good at all locations, the yields being 627 and 624 kg/ha at Lubbock, 506 and 479 kg/ha at Big Spring, and 916 and 906 kg/ha at College Station—these comparisons were all made in the same year. Using data from the F_5 grown for two years, significant interaction differences were found for locations, and years by locations, but not for years. Entries averaged over locations and years produced significant differences; however, the authors showed by partitioning the data that the entries interacted significantly with years but not with locations. This latter finding is interpreted to mean that the highest yielders did relatively well regardless of location.

Generally, second-order interactions are both more difficult to interpret and less likely to be of major importance, but in this study the significant interaction of entry × year × location provided the most telling implications to selection methods. Analysis of the F_5 lint yield data disclosed that:

1. The entry × location interaction was zero.
2. The entry (genotypic) estimates were highly significant only at College Station.
3. The entry × year interactions were highly significant at Lubbock and Big Spring.

The authors' interpretations were these: selection for lint yield should be more effective at College Station because genotypic effects principally were involved. Furthermore, the similar reactions at Lubbock and Big Spring were traceable to different forms of stress. Conclusion: selection would be most effective in the "better," less-stressed, environment. The authors extended this to an expectation that selection for lint yield might be ineffective if done at locations or in season when production is limited by a negative expression of any environmental component.

Stress and nonstress situations interest plant breeders. It is conjectured that if the genotype × environment interactions were understood two consequences would ensue: a selection advantage in one situation might be exploited, and a successful environmentally adapted cultivar could be produced. The assumption that selection under stress is worthwhile has been difficult to either prove or disprove because of the ambiguity surrounding the subject. Rosielle and Hamblin (1981) have gone to the root of this problem in a theoretical examination of the effects of selection for yield under stress and nonstress conditions.

The authors began with two basic definitions: (1) tolerance to stress (Y_3) is the yield difference between stress yield (Y_2) and nonstress yield (Y_1), and (2) average productivity (Y_4) is $Y_1 + Y_2/2$. From this they developed five equations, the results of which produced a number of theoretical considerations about selection for tolerance. Specifically, they concentrated on "selection for tolerance to stress where tolerance is defined by a small difference in productivity between stress and non-stress environment," and on selection for average performance in both environments.

Their description of selection in the two environments follows:

1. *Selection in Stress Environments:* The mean yield in this environment increases, but if the selection is done in a population with relatively lower genetic variance there is a decrease in mean yield in nonstress environments. Only if there is higher genetic variance in the stressed environment during selection can there be higher yields in the nonstressed one.

The impact of this hinges on whether stressed environments have or create lower genetic variances, and past experience suggests that they do. Where this general case persists, selection for tolerance decreases mean yields when grown under favorable environments. Selection for average productivity generally permits an increase in yield under both conditions provided there is a positive genetic correlation between yields and environments.

2. *Effect on Stress-selected Lines when Grown in Nonstress Environments:* generally, their mean yield decreases. For yields to increase, a greater genetic variance in the stressed environment over that of the nonstressed one must have existed.

Practical selection implications must take into consideration other aspects of the breeding and testing program; for example, what is the proportion of stress and nonstress environments in the target area for a commercial cultivar? Can boundary lines be drawn and coordinated with testing sites and the nature of germ plasm entries being tested? The general impression left by Rosielle and Hamblin was that selecting for tolerance, that is, testing in stress situations, carries a risk of lower yields in one or both environments. Selection for stability, low regression coefficient, generally has a negative relationship to selection for high mean yield. The message seems to be, choose testing sites representative of the intended cultivar's production area. If stress situations are a significant part of a region's productivity pattern, two testing programs may be required.

It was shown by Hawkins, Fehr, Hammond, and de Cianzo (1983) that selection for a particular group of soybean fatty acids could be "farmed out" to other world environments. They concluded this from a study of testing results of Iowa-adapted cultivars and lines grown near-concurrently in Iowa and Puerto Rico. Although individual differences were found, the general similarity of performance as shown by phenotypic and correlation coefficients of the lines average performance made it clear that selection for fatty acid composition could be done equally reliably in either climate. The plant breeding value is that it provides additional growing seasons per year for selection.

Breeding crops for yield on saline soils presents a difficult problem because the soils are typically highly variable in salt concentration, and yields are also low and variable. Richards (1983) explored several breeding approaches with wheat, barley, and other cereals with unexpected and fortunate results because he found the best strategy was to select for high yield on nonsaline soils. Some of the explanations for this rest on the soil variability aspect: most of the yield from salt-infested soils is derived from the areas of lowest salinity, which are therefore closest in character to nonsaline soils. Also, the proportion of yield coming from higher-salinity areas is low, approaching zero, and these areas do not represent opportunities for constructive selection. Another way of viewing the problem is that progress in increasing productivity of saline soils is limited to areas of lowest salinity.

Using the basic 'Black Shank Synthetic' as a source population, van Sanford and Matzinger (1983) studied the effect of selection for maximum tobacco seedling weight in an optimal and a stressed environment. The stress was applied by withholding one-half of the optimal level amount of nutrients. The experiment continued for 10 cycles in the two environments.

When the final selections from the two regimes were compared in both environments, it was found that although selection for maximum seedling weight response was effective in both environments the maximum gains were made in the lines selected under the optimal environment. The inference would seem to be that selection under the optimal nutrient environment would be better. The author's discussion, however, makes a subtle distinction whose meaning is not clear to me:

> These results support the concept of Hammond that response to selection will be maximized in the environment in which the trait is most fully expressed. Hammond's other notion, however, that the optimal selection environment is the one in which the testing will be done, is not supported by these data.

This seems to imply that although greater selection response gains were made under optimal conditions, there are other compelling reasons why a breeder might choose instead to select for lesser response under a stressed environment. The answer may lie in the authors' speculation that the same genes may be involved in response to selection and the more favorable environment allows a more favorable response. If this environmentally induced "artifact" were removed it is possible that selection responses would have been similar. Even were this so, the optimal environment still would provide a breeder with the practical advantage of easier selection because of a traits increased phenotypic response and higher visibility in the better environment.

Frey, Simons, Skrdla, Michel, and Patrick (1984) showed that the gain in grain yield of oats from selection was significantly higher (50% more) when selection was made in a high-productivity environment rather than a low-productivity one.

Hayward and Vivero (1984) compared six perennial ryegrass lines, which had been selected under spaced-plant, limited competition conditions, with their performance under increasingly competitive sward conditions. They concluded that there was little correlation between the yield of lines selected under spaced-plant conditions and their yields under competitive sward conditions.

Their interpretations were that spaced-plant selection was equivalent to selecting in a favorable environment and was associated with a high sensitivity to change, all subject to genetic control. In this conceptual framework a competitive sward represents a poor environment for such lines, and in consequence of their sensitivity, performance could be expected to be lower.

A major concern of plant breeders as they approach a release decision on a promising candivar is the level of confidence they place in the accumulated test results of relative yields. The key variable is the testing environment, whose characteristics reflect the interactions of years, locations, weather, and other manifestations. A host of questions arise, chief among which is this one: have the limited data been obtained from representative genotype × environment interactions? Strategies to cope with this problem may range from the use of simple averages to sophisticated assessment of environments.

An illustration of the latter was provided by Brennan and Sheppard (1985), who investigated the characteristics of environments in order to formulate a comprehensive yield testing strategy. Building on earlier work by Boyd, Goodchild, Waterhouse, and Singh (1976), Comstock (1977), and Fox and Rosielle (1982), Brennan and Sheppard proceeded to study larger data sets for cultivar × environment interactions and to correlate these with pattern analysis of environment similarities. A group of 16–18 genotypes constituted the test material, and grain yield was the measured character.

The strategy involved first the processing of information from cultivar behavior obtained from pattern analysis, which grouped genotypes together that performed alike in given environments. The next step, using these same procedures, was to group similar environments. Clustering was based on the relative yield of five common tester genotypes. Relative yield was defined as the percent of each environment average yield.

The unique third and final step involved locating the *centroid*, the center of a population of environments (in this case the population consisted of four yearly environments). The concept is grasped readily by looking at their Figures 1 and 2: Figure 1 showed the points of the individual environments surrounding the calculated centroid, and in Figure 2 the somewhat elliptical yearly environment lines are shown in overlay with one small central area common to all of the four years.

These seems little question but that these procedures would aid breeders in understanding the geographic relationships of the total environment with which they are concerned. A statistic, the squared Euclidean distance (SED), which is the distance between two points in multidimensional space, was found to have special properties. SED values above 150 were found to identify environments that were "aberrant" for specific reasons not typical of the general environment, for example, a nematode-infested soil area. When these high SED value environments were withdrawn, a more accurate relative performance for yield was obtained. The ability to exclude aberrant sites or environments obviously increases the homogeneity of the remainder of the group.

Brennan and Sheppard admitted that the procedures were time-consuming and costly; however, it is not clear that continuous application would be necessary. One suggestion that should be noticed by breeders is that "a series of relevant but diverse environments could be generated in one year at one location."

GENOTYPE × ENVIRONMENT

For a clear understanding of the relationships between the components of testing (locations, years, replications, variance components, costs, entries), Sprague and Federer's (1951) paper on corn yield trials is basic. Assuming budget constraints and fixed entry numbers, the authors provided formulas and solutions for various situations. The design parameters for testing depend somewhat on the researcher's goal. If the objective is to test an hypothesis, a sufficient and considered number of replicates is necessary. However, if the nurseries are to provide information for area recommendations as a goal, fewer replications are necessary (except that there should always be at least two; otherwise the data may be considered insufficient for later use in long-term studies). Even though the most efficient solution might specify no replication, it is a small and inexpensive concession to include a second plot.

Sprague and Federer explained,

> In the simplest possible design involving varieties and locations, the variety by location mean square is made up of a component arising from the failure of plots treated alike to give identical yields and an interaction component arising from the failure of the varieties to give the same relative performance at all locations. It is this interaction component which is of major interest and its magnitude is only slightly influenced by variation in number of replicates per location.

Thus increased replication acts to increase the precision of results at a location, thereby raising the confidence of the researcher in the "rightness" of the data, but it seldom alters the interaction component derived from multiple site testing as described above.

Disregarding costs and the two-replicate caveat, Sprague and Federer described the optimal distribution of a fixed number of plots to be a reduction in replications per location along with an increase in number of locations and years. The increase of the latter components was modest, often involving only a single extra location or year.

White (1958) applied variance component analysis to 97 regional oat variety trials conducted over a 10-year period in New York. The CV estimates, rounded off, were variety 43, location 110, year 200, replication 38, V × L 2, V × Y 23, L × Y 287, V × L × Y 26, and error 63. This study had influence on the conduct of state-wide operations. The small variety × location interaction could be interpreted to mean, as White wrote, that "it may suffice to recommend only one variety for the state." I took it to mean also that an expensive far-flung testing program might be reduced, and that more confidence could be placed in the main station results. On the other hand, the high variety × location × year interaction imparted the information that neither variety × location nor variety × year results were consistent, "nor is the location × year interaction the same for all varieties of oats." I think a high second-order interaction should be taken as a warning signal to the breeder, indicating that the variety × environment confrontations are complex. Perhaps assessment of the locations with an eye to clustering would be a first step toward simplification.

More information on this point has been furnished by Professor Sorrells from the Cornell program (personal communication, 1987). When Sorrells examined performances of wheat entries in local and regional state trials over several years, he found that even with a small variety × location interaction there were startling rank changes between local (Ithaca) and state regional averages, amounting in some cases to complete reversal, that is, the same variety in last place in local trials, and in first place in regional trials. Sorrells commented, "The example I have enclosed shows that certain lines or cultivars interact more (perform differently) than others and their mean performance can be greatly influenced by the location of the trials." Fortunately, a correlation analysis of sites identified one local location, on the home station, that gave results similar to the regional sites in western New York.

Robinson and Moll (1959) concluded, on the basis of five years' yield testing of corn at five sites, that the genotype × environment responses were not unique and tied to repeatable environments at specific locations, nor of course are year environments often repeatable.

Miller, Williams, and Robinson (1959) assessed the variety × environment interactions found in cotton in nine testing locations over a period of three years in North Carolina. The same 15 varieties were used in all tests, and perhaps of significance, one source of variance was eliminated by using the same pre-increased seed lots of the varieties each year. Data were taken on six traits of economic importance. Another source of variance, the presence of two known unadapted cultivars, was eliminated by using only the 13 adapted cultivars.

The authors found variety × year and variety × location effects to be small and nonsignificant for yield, but the second-order interaction, variety × year × location, was large and statistically significant. The interactions for all other traits were small and statistically nonsignificant, and were deemed to be of negligible importance.

The authors concluded on the basis of the low variety × location interaction that it would not be necessary to divide the state into testing subareas. They stressed, however, that this does not mean that breeding and testing a cultivar for a special environment can be ignored; for example, an irrigated situation might require special attention. The implications of the high second-order interaction observed are that varieties must be tested over a number of environments in the general area. The test conditions of nine locations and three years seemed to satisfy this requirement. The replications used perhaps could be reduced; however, this is the least costly element of testing and not much saving would result.

Rasmusson and Lambert (1961) examined variety × environment interactions in barley tests in Minnesota. The tests covered six varieties at eight locations over four years. The effects of both years and locations were small, but significant at the 1% and 5% level, respectively. The variety × location variance component was so small as to be nonsignificant (0.22), and indicated that varieties yielded, relative to each other, similarly at any location. The variety × year interaction was larger (3.99*) and significant. The unexpected finding was the size of the variety × year × location interaction (15.97**), 70 times that of variety × location, and four times that of variety × year. As a result of this study the authors proposed to continue testing under a three-replicate, three-year system, but with locations reduced from eight to six. The highly significant variety × year × location interaction was unexpected in view of the reasonable and low first-order interactions; nevertheless this pattern has been observed before.

Schutz and Bernard (1967) analyzed seven Uniform Soybean Tests over a three-year period. These varied in number of entries, locations, and replications. They found that genotype × year interactions, although smaller than genotype × location, would actually be more important inasmuch as years of testing are more limited than number of locations. It was estimated that a good culling level of low-yielding lines could be obtained in one year at 10–15 locations; advanced lines should receive two or three years of testing.

Baker (1969) studied the genotype × environment interactions involving the performance of six wheat cultivars grown each year for five years at nine stations in Western Canada. Two forms of regression analysis were used: the first model isolated the performance of each cultivar over the 45 total environments; the second model isolated the total environmental effects into three components, namely, genotype × year, genotype × site, and genotype × year-site interactions.

The most revealing part of this study involved the partitioning of sums of squares for genotype × place × year. Two of the cultivars, 'Marquis' and 'Cypress', showed magnitude differences in sum of squares (3× and 2×, respectively, to the next largest) so large that they might well be thought of as not belonging to this set. Baker proposed that cultivar stability "is inversely proportional to the sum of squares for genotype–environment interaction *which is attributable to that cultivar*" (italics mine). Baker showed that when the Finlay–Wilkinson model was applied to the first analysis model data, Marquis and 'Thatcher' were shown to have above average stability, whereas it is difficult to escape the later picture of these two cultivars as being the least stable of the six tested. The high variance may be the consequence of leaf and stem rust susceptibility. Baker's conclusions were that the Finlay–Wilkinson regression model failed to account for a major share of the genotype × environment interaction and would be unsuitable for western Canada. An important point was that his studies showed no significant genotype × year interaction, suggesting that years of testing might be reduced. It was found also that replicates could be reduced with less penalty than would be incurred if the number of sites were curtailed.

Briggs and Shebeski (1971) studied early-generation selection for 13 quality and two yield traits in spring wheat. In the process, three populations from different crosses were used, and F_3–F_5 comparisons of the early generations of each were made, one cross per year, over a three-year period. The confounding of replicates, locations, and generations with years was undoubtedly a factor contributing to a range of results.

Briggs and Shebeski's paper served as a reminder of a nagging problem we plant breeders have with interpreting generation to generation correlations, for example, F_3–F_5 yields. We consider it logical that there should be a positive correlation. Numerous experiments have been done over the years. The results have covered the range of possibilities between negative and positive. The breeder is not given a firm base for either expectations or interpretations. The problem, of course, lies with the interactions of yields, particularly with years and locations. If one finds a positive correlation in a given experiment, chances are that luck played a part. How often do like years follow one another? Indeed, Briggs and Shebeski described the three years of tests as "average," "drought," and "extremely wet and cool," respectively. Their paper compared F_3 results with derived-F_5 results the following year; in only one year was a positive result found. The authors were cognizant of the fact that different hybrid populations were used each year.

A foundation for breeder's expectations might be furnished if information could be obtained on the interaction variance itself. This might be done by a seed buildup of F_3 and its F_5 derivative lines or bulk, large enough so that results from multiple sites and particuarly a few years could be evaluated. For example, a three-year study of interactions of genotype yields with years, and of same locations within and over years, might provide a measure of *average* expectation. Every environment differs in its variations, and fortunate is the breeder who operates in a relatively repeatable environment. If we rule out testing in a controlled artificial environment the dilemma persists: F_3 selections made in a "drought" year must be committed and planted for a subsequent F_5 test before the characteristics of that year's environment can be known. Clustering of similar environments as testing sites might help to improve the year to year correlation.

Freeman and Perkins (1971) examined the important and difficult problem of correctly describing and measuring an environment. There is a consensus that if it were possible, environmental effects should be measured in some way other than using the crop itself. However, this is outweighed by the additional advantages derived from measuring the crop itself. These advantages were discussed by the authors. Proceeding from Finlay and Wilkinson's 1963 regression of genotype yield on the site mean yield representing environment, where the base slope for measurement is unity, $b = 1$ (less than 1 indicates high relative stability, more equals lower stability), they referred to Eberhart and Russell's 1966 improvements relating to deviations from regression and the use of combined E and $G \times E$ terms. But what this discussion led to was a realization that there is no independent way of measuring and expressing what an environment is without filtering it through genotypic responses. They concluded that the organism must be used to measure environmental effects and suggested four ways, all of which would be familiar to a breeder:

1. "Divide the replicates of each genotype into two groups, using one group to measure the interaction and the average of the second group over genotypes to measure the environment." This requires a minimum of three replicates.

2. Use one or more genotypes as standards to measure the environment (that is, checks or control varieties, or parents of crosses).
3. Use like material in the same environment (that is, clustering of genotypes).
4. Use replicates of #3 in similar environments (that is, clustering of environments).

The thrust of the Freeman and Perkins paper, however, was to propose the use of the analysis of variance and component partitioning as a substitute for the generally used linear regression of genotype performance. A general conclusion was that the environment is best evaluated by the responses of simlar genotypes, which suggests some stratification in the choice of entries for a nursery.

A subsequent paper by Freeman (1973) stated, "This is largely a review paper," but it is valuable to know of its existence, especially because it contains almost 100 references to previous work on statistical methods for analyzing interactions, both of the general and genotype × environment kinds.

Tucker and Harding (1974) found that when equivalent F_3 samples of hybrid bulk populations of two lima bean crosses were divided and continuously grown to F_{11} without selection at two separate California locations, tests for yields at one site using seeds produced at both sites the previous year showed that the population evolving at one site was higher yielding than its twin population that had evolved at the other site. Furthermore, the yield difference increased each year, suggesting that natural selection was creating a genetic shift towards local adaptation.

Hill's (1976) review on genotype × environment interactions envelops and unifies the subject to an extent that justifies suggesting it as required reading for aspiring geneticists, plant breeders, and biometricians. The paper gently leads the reader from the landmark views of what historically might be termed a simpler age of discovery to the sophisticated multiple approaches of the present time—even to the views of some who, as Hill put it, "remarked that the development of these techniques had far outstripped their biological usage"!

The vehicle for the review rested heavily on the biometric approach as indeed it must, for measurements are everywhere required. Biological assessments of the environment and genetic priorities were not slighted, however. For example, alluding to the view that genotype × environment interactions themselves might be heritable, Hill cited the routine practice of minimizing genotype × environment interactions under field conditions:

> This in turn reduced the scope of the approach to the point where it was used solely for divorcing genetic effects from GE interaction effects. In so doing it failed to recognize that GE interactions are as much a function of the genotype as they are of the environment and so are partly heritable.

The paper is valuable for its treatment of topics that are the basis of plant breeding activities. Examples are adaptation, heritability, stability, and competition. An extensive list of references was appended.

Wood's (1976) paper is a part of the continuing effort to gain a greater understanding of genotype × environment interactions by partitioning the environment into variables. Independent measures of the environment represent a departure from the earlier and continuing process of expressing the environment in terms of genotype interactions which by definition is flawed because environment is a participating force in the expression. Wood stated the case in this manner: "The problem is to find functions of the environmental variables which can be used to explain the variance in performance of the genotypes from environment to environment."

Certain of the details in Wood's examples are thought-provoking in terms of testing methodology. Bear in mind that breeders have always accepted an environment as given. There are of course choices in selection, for example, sites or dates of planting, but these degrees of freedom are used up early in the life of a nursery and thereafter the environment, whatever it may be for that season, dominates. What Wood and others are doing is asking questions about the role of environmental variables, for example, fertilizer composition in plants and soils, and climate variables. In one example, six elements plus soild pH were studied in plants and in soils, the results targeted phosphate as playing a major role in determining the relative performance of lines of *Nicotiana rustica*.

Campbell and Lafever (1977) used the extensive compiled data from 11 cultivars or lines, nine locations, and three years of the Uniform Eastern Soft Red Winter Wheat Nurseries to explore and describe genotype × environment interactions over the several-state region. Analysis of variance mean squares were significant (5% level) for cultivars, cultivar × year, cultivar × location, and cultivar × year × location. The variance components for cultivar × year, and cultivar by locations, were of the same size, implying equal weight in any consideration of changes. The regression of cultivar yield upon the yield mean of all cultivars brought out the fact that certain varieties had the capacity to respond to favorable environments whereas others did not. They found that sites within the large region differed substantially. Lafayette, Indiana was found to be especially suitable for testing because results here correlated well with all other locations—an important consideration in selecting for wide adaptation.

Beyond the nature and choice of sites, the study provided helpful implications for testing programs. The equivalence of years and locations suggested that they could be substituted for each other; for example, time (years) might be saved by increasing the number of locations. When the expected variance of a cultivar mean was graphed against number of locations for one to four years, it became evident that more than one year was required, but little advantage was gained beyond three years. I still would interpret the third year as worthwhile because the increase in locations in the second year seemed to provide insufficient compensation. Although the authors stated that the number of locations routinely used in this nursery (15–20) were not needed in order to characterize a cultivar and to document its performance, it is also understandable that number of sites has great participatory value to the breeders.

In their follow-up study covering more locations and years of the Uniform Eastern Soft Red Winter Wheat Nurseries, Campbell and Lafever (1980) were more cautious in their interpretation, finding that year-to-year interactions with locations mandated substantial regional testing over years. They suggested that it might not be wise to substitute extra locations for years.

Langer, Frey, and Bailey (1979) examined three multiple-variety sets grown in multiple environments over a period of years to compare the relationship of yield to linear regression response indices and three stability response indices. More than 80% of individual varieties' yield variance could be traced to linear regression response. Mean yields of three sets were correlated ($r = .61$) with regression response indices (b values). However, mean yield showed general independence from three stability indices (coefficients of determination, mean squares deviation from regression, and ecovalence values).

Langer, Frey, and Bailey suggested two simpler methods for practical breeding programs, namely, R_1, a range index to include extreme yields of a variety in all environments, and R_2, a similar index of a variety's yields in the poorest and best environments. They conditionally favored R_1, but the two are confounded because R_2 is a subset of R_1. The authors placed a different interpretation on the meaning of the regression technique used by Finlay and Wilkinson, apropos stability.

Brennan and Byth (1979) reviewed the analysis of interaction approaches taken by different researchers. Their study was based on yield tests of 100 wheats of diverse origin at three locations over three years. Their paper should be read because of the breadth of treatment of genotype × environment interactions; however, certain findings are reported here:

1. Clear separation of genotypes for yield requires multi-environmental testing.
2. A series of selection strategies for lines gave negative ratings to the following approaches:
   a. Use of regression coefficient alone
   b. Use of high mean yield with regression coefficient (similar to the Finlay and wilkinson model)
   c. Selection solely for low $G \times E$ sum of squares.

Mean yield alone over all environments tended to favor the identification of lines that did well in higher-yielding environments. This would be of interest if environments were graded. However, when mean yield appeared in an index weighted for $G \times E$ sum of squares, the 1 : 1 combination was not different from mean yield alone (1 : 0). When the weighting increased (1 : 2, 1 : 3, and 0 : 1) the interpretations were negative toward these indices as being helpful in selection among entries in the yield nursery.

Fowler (1979) found that one of the important sources of winter survival capability and variability was the chance distribution pattern of snow cover on the field. The insulating properties of snow cover were shown by Kinbacher and Jensen (1959), who studied field temperatures, of air and of the ground surface under snow, for two years at Ithaca, New York. They found that the ground temperature, under snow, remained between 30 and 32° F for 83 consecutive days, a period that included air temperatures of −14.2° F.

The foundation for Nor and Cady's (1979) research on measuring stability and wide adaptability in crops was a definition of environment expressed in 1952 by Billings (1952), who pointed out that crops and crop yields in an environment could not be considered as primary forces but were instead results of the interactions between the varieties and the environment. Therefore, crop yields represented biased pictures of an environment, and if they are used it must be in a way that separates their identity from other true measures of the environment.

Nor and Cady's approach accepts the beginning thesis of maintaining the separate identities of crops and environment, but their goal is to finish by linking these to produce an estimate of adaptability: "It is an attempt to link together two quantitative measures of a different nature, crop yield and an environmental index, which are functionally related in the true sense of cause and effect."

DISCUSSION OF GENOTYPE × ENVIRONMENT INTERACTIONS

Genotype × environment interaction poses another of the enduring dilemmas in plant breeding: how to apportion the genetic and nongenetic aspects of cultivar performance. On the one hand, it is important to know the underlying genetic makeup—the genotypic base—especially so that breeding gains in yield can be projected. On the other hand, the genotype never makes it on its own; its performance is always in partnership with the environment. It is not too much to say that the genotype × environment interaction, not the genotype, *is* the cultivar. In researching literature for this book, I was pleased to find a similar statement by my colleague at Cornell University, Professor A. M. Srb (1966), who wrote, "It seems to me that the primary ultimate concern of the plant breeder is with phenotype. He is, of course, concerned with genotype also, because the major part of his job is to get particular phenotypes under predictable and recurrent control."

The problem continues as the breeder faces a series of options in interpreting performance trials. Should one seek adaptation over all environments or the best performance in "best" environments, or should one breed a series of cultivars each adapted to a specific environment?

In the early years at Cornell University, I remember being impressed by a report by Went (1950) that pictured the variation in replicated pea plants grown in environmental growth chambers at the Earhart Plant Research Laboratory at the California Institute of Technology. The different plants were so remarkably alike that one had the impression of seeing the naked genotype. However, performance in a growth chamber is really irrelevant to that in the field environment where the cereal cultivar is subject to interactions from a large number of influences such as years, locations, climate, pathogens, fertility levels, and so forth. It is no wonder that this problem continues to receive the attention of breeders, geneticists, and mathematicians, nor is it surprising that several strategies have been proposed to aid breeders in their nursery testing.

CLUSTERING

Clustering analysis identifies testing sites in geographical areas that have similar variety by location interactions. Application of such analysis results in increased efficiencies of testing. There may be several suggested alterations such as elimination of a surplus site or rearranging of regional boundaries. Sometimes an analysis merely confirms the correctness of a previous arrangement of sites within a region. It is instructive to consider clustering in its expanded connotations, which imply that the same reasoning would be profitable applied, for example, to the variation among genotype entries in nurseries. A simple example would be clustering of entries by maturity groups.

An early paper by Yates and Cochran (1938) is of interest because it distinguished two different goals of agronomic experiments. The first relates to the scientific objective, which might be to determine the value of a treatment or a comparison of varieties. The second goal is technical and concerns the rules of procedures, that is, how the experiments are conducted in order to produce unbiased estimates of average response. There is a good discussion on the meaning of "random" and the interpretation of error estimates. The authors pointed out that it is almost impossible to obtain sites selected at random. Selection is usually influenced by a number of factors such as availability, an attempt to find locations representative of an environment, and in regional tests, a consideration of public relations with growers. One can see in the problems that diverse environments create, the glimmerings of the need for a clustering of environments in local, state, or regional nurseries.

Writing as an agronomist, Salmon (1951) took a statistical look at long-time variety tests of wheat in the Great Plains states. Although the article reviewed many aspects of analyzing long-term data, I would pick out one or two innovative thoughts of Salmon's that have special relevance to plant breeding. One is a recommendation that pair comparisons be made of varieties over years to see if genotype is related to high or low year × variety interactions. Second, he proposed a form of variety stratification that was similar to what is now done at times, for example, in early and mid-season oat tests. If like material is grouped, year × variety interactions and error terms would be expected to be lower, with the result that decision time could be saved by increasing replications and decreasing years. For unlike material, there is no easy solution, and tests over more years are required; in these situations replication numbers can be kept to a minimum.

Plant breeders' geographical responsibilities for variety development and recommendations usually are defined by state boundaries. Such politically determined lines seldom coincide with the biological and environmental parameters within which breeders work. Only in the past few decades have breeders begun to match their breeding objectives and tactics with the reality of the environment. Horner and Frey (1957) undertook such a study for the oat program for Iowa. The data from nine traditional testing sites were examined over a five-year period to see if a more logical subdivision grouping would result in more meaningful variety recommendations. The study involved using yearly variety × location interaction means from the whole state and then trying to find combinations of locations (patterns) whose V × L mean squares would be lower than the whole-region figure. A disconcerting discovery was that homogeneity tests of subareas disclosed there was little similarity. Each site was unique. However, patterns of smallest mean squares were found and worked out with consideration of logical and practical aspects; for example, similar climatic conditions suggested southern Iowa as one region that encompassed three testing sites. The final decision to form four subdivisions, using all sites, lowered the variety × location interaction by 30%. It is may understanding that this regional partition has been in use ever since in Iowa.

For many years Alabama had conducted corn varietal trials with the state divided into three north-to-south regions. The original division was predicated on differences in insect damage and expected adaptive differences arising from latitude change. McCain and Shultz (1959) conducted a reappraisal to assess whether the present divisions were needed, and if so, whether the boundaries were correct. Six years of variety test results were used. Basically, they determined variety × location mean squares and attempted to group sites having similar mean squares. The only meaningful basis for clustering was found to be geographic.

The original division had five, six, and four sites in the northern, central, and southern areas, respectively. The changes dictated by the study altered the boundaries slightly—a bit of gerrymandering necessitated by fixed sites—leaving the above areas with four, six, and five sites, respectively. The characteristics of the northern and southern sections were clear integrally; however, the central region showed such diverse results in large mean squares that it could not be considered a single homogeneous or representative region. Additional sites within the region seemed to be required to restore order to the discordant results.

Guitard (1960) used three years of data from diallel correlation coefficients for all pairs of locations to determine the relative performance of nine barley cultivars grown on 10 regional testing sites. If the correlation coefficients were high it was assumed that the varieties performed similarly because the environments were similar. Several indicators of performance were used, namely, yield, height, weight per bushel, and 1000-kernel weight.

When the correlation coefficients for yield were arranged in a diallel graph of location by location, it was seen that the high coefficients (0.7, exceeding 5% level of significance) lay along the blank center diagonal of same site × same site, and could be arranged in three groups, each overlapping its neighbor(s). Two locations lay outside these groups, indicating different, separate environments. Guitard concluded that one location from each of the three groups would represent the eight locations; thus by adding the two distinct locations, the 10 testing sites could be reduced to five. In general, the clustering of locations for yield resulted in a positive clustering of other traits.

In a study of genotype × environment interactions found in sorghum yield tests in Kansas by Liang and Walter (1966), it was discovered that second-order interactions were significant and so large that the responses could not be damped or controlled by grouping locations and years. This implied first, that clustering of locations would be ineffective, and more generally, that single-year tests were inadequate as a base for variety recommendations. Suggested correctional moves would involve two testing years, increasing locations to 10, and reducing replicates to three.

Liang, Heyne, and Walter (1966) examined the environmental conditions of Kansas wheat, oats, and barley testing sites and came to a conclusion similar to the earlier study in Iowa: the state needed to be divided into testing subareas. The eventual clustering resulted in a four-subarea pattern for winter wheat and three for winter barley. No subdividing for oats seemed necessay inasmuch as the mean square of oat variety × location was not significant, and the five locations in use could be considered a single testing area.

The authors also estimated the effect of different numbers of sites, years, and replicates on the variance of a variety mean. The comparisons showed that the order of importance in lowering variance was location, years, and replicates. Fewer than three replicates was not recommended.

Abou–El-Fittouh, Rawlings, and Miller (1969) examined the genotype × environment interactions found in upland cotton in the Regional Cotton Variety Test results of five years. They used two methods of measuring similarity of locations. One was the "distance coefficient," the meaning of which corresponds to its name, and the other the "product-moment correlation coefficient." The distance coefficient was found to be the more efficient of the two and was used to study the zoning that embraced the southern United States and stretched from coast to coast. A revised zoning plan was presented.

The authors pointed out that any limited-years study depends somewhat on the variety entries used during the period. They recommended a study involving a longer and more extensive variety base before implementing the revised plan.

Pederson's (1974) paper on wheat varietal performance over years is interesting because it proposed an estimation of an environment that is independent of the nursery trials conducted by agronomists. This was done by constructing an over-years profile of environment yield fluctuations obtained from independent, in this case, government reports of commercial acreages and yields of an area. The pattern of seasonal effects linked to probabilities furnished measures of environmental distribution to which varietal trial results may be compared. It is of interest that the environmental patterns are also helpful in clustering, identifying sites with like environments.

In a highly statistically technical paper Mungomery, Shorter, and Byth (1974) proposed pattern analysis (clustering) procedures using a group of 58 soybean genotypes tested in eight environments. Seed yield and protein percentage were the traits studied. The study was an attempt to unify several approaches to genotype × environment interactions, which have included partitioning of variances, reducing variance magnitude, and the use of pattern analysis to cluster similar genotypes and environments. Their approach involved the classification of responses to find patterns or clusters that would simplify interpretation by dealing with clusters without foreclosing the study of individual genotype responses. This specific-general posture was obtained through constructing for each genotype a "*p*-attribute," which was defined as the genotype's mean performance in each environment. The *p*-attributes identified a coordinate position in a graphic dimensional representation of space, such that in these examples, each genotype would accumulate a total of eight *p*-figures over the eight environments. Distances between points in the space graph were related to relative similarities of genotypic responses, and portrayed both genetic similarities and adaptive responses to environments. The authors emphasized that their methods were free of prior assumptions.

The plant breeding implications and potential value of procedures such as Mungomery, Shorter, and Byth espoused are clearly evident, and provide insights into the interrelationships among genotypes. Breeders can use this knowledge to choose parents, evaluate crosses and the responses of their progeny, and look for similarities of response pertaining to adaptation and stability. Even visual examination alone of the clustering patterns in the familiar four-cell quadrants is enlightening. If I were to express any reservation about the working use of these procedures, it would center on the capacity of the average breeding program unilaterally and continuously to apply them to a project. Fortunately, biometry is one of the disciplines where cooperative support traditionally has been available to breeders.

Lin and Thompson (1975) presented a regression approach coupled with a technique of clustering genotypes that was used to analyze genotype × environment interaction situations. I see this in terms of Gardner gridding and moving toward the clustering of both genotypes and environments, that is, subsets of genotype nurseries matching subsets of varying quality environments.

Johnson (1977) discussed clustering of corn hybrids in terms of its relationship to comparative stability as derived from regression coefficient analysis and genotypic similarity (comparative stability plus average performance from weighted means). He suggested also that clustering of environments might be done, pointing out that "If environments vary widely, complete genotypic stability is virtually impossible, and only degree of stability among genotypes is relevant."

A surprising plant breeding implication arose from the research finding that five successful commercial hybrids showed small coefficients of regression on the environmental index but large coefficients of regression on the first residual vector. This combination suggested that these genotypes would have stress resistance to drought, but would not produce the highest yields under favorable conditions. Johnson stated,

> The conformation of five popular commercial hybrids to this pattern of performance suggests that strong demand for a hybrid depends more upon minimization of performance problems in stress or low productivity environments than upon high average yields over all environments.

The effects on six agronomic characters of using 10 different criteria to determine nursery harvest dates were given in a paper by Somerville and Briggs (1979). The conclusions support a conservative, consistent policy of not varying present harvesting procedures. One visualizes the problems that led to the study as being due partly to genotypic extremes that are found in almost all programs. Unless differential harvest dates within a nursery are possible, for example, in order to harvest lines differing in maturity at the same stage of physiological maturity, an efficient alternative is to cluster genotypes having similar important expressions of a trait, in this example, maturity.

Hamblin, Fisher, and Ridings (1980) opened a study of wheat testing site selection by defining the characteristics of an ideal site. Of the four criteria mentioned, only two have general significance: (1) that the site produce performance results representative of the target environment (e.g., for a new cultivar), and (2) that genotypic differences be large enough for selection and ranking purposes. (The other two criteria concerned the adequacy and reliability of seed production.)

Their environment evaluation and selection proposal was based on the correlation of site mean yields at each location with the average yields over all sites. This quickly identified sites that represented overall performance of a whole area. This carried the assumption that the target area had been correctly described: for example, it is quite possible to have two equal-sized regions, one of which is environmentally homogeneous, and the other heterogeneous and needing to be subdivided.

In fact, no single testing site gave consistent predictions, via correlations, of the state mean yields. The region encompassed the large agricultural zone of Western Australia. The authors then tried clustering of data from two or more sites, again with a goal of mirroring the state mean yield. Here, considerable success was found with two-site and three-site combinations showing high correlations with state mean yields. In several of these combinations, the names of the same sites appeared again and again.

The plant breeding implications are clear when it is realized that 26 locations represented the state mean yield. The study isolated four "near ideal" combinations of sites which, when matched against actual performance data over several years, were remarkably consistent in matching their performance against that of the whole state. The authors noted that (1) a large data base must be available for similar examinations, (2) variety changes over years are not important, and (3) caution should be observed that the conclusions drawn are for general rather than specific adaptation.

Ghaderi, Everson, and Cress (1980) used test weight of 41 winter wheats as the responsive trait to characterize testing site environments at eight Michigan sites over two years. Cluster analysis was used to classify locations into similar groups—a homogeneous cluster was one in which the genotype × location interactions were nonsignificant. Genotypes were similarly analyzed according to phenotypic responses and through considering the eight sites × two years as 16 environments.

The analysis of locations showed one major division: one location, St. Joseph's, contributed significant genotype × location interaction, and if removed, left the remaining seven locations in a group with nonsignificant interactions. Within this larger group, however, smaller areas of similarity could be observed, and the authors concluded that six total locations would suffice to provide both adequate sampling and diversity for early-generation testing. The general parameter guidelines were (1) with high genotype × environment interactions, test for specific adaptation to each region of production; and (2) with low genotype × environment interactions, do moderate location sampling to test for broad adaptation.

The cluster analysis of genotypes over environments dealt with test weight means having a range of 74.46–78.60 kg/hectoliter, their coefficients of regression, and variance of deviations from regression; thus measures of stability were included. The results could be divided into three meaningful clusters, the middle group consisting of desirable stable genotypes.

The complex problem of testing environments was addressed by Brown, Sorrells, and Coffman (1983) in a paper that merits attention not only for its orderly review of the historical issues but also for advancing our thinking toward improving cultivar testing. Their proposed method has two components (1) a clustering of selected sites having common environmental variables, and (2) choosing optimal selection environments within clusters. Although the proposal represents a partitioning of overall site environment traits, it draws the picture into a coherent whole by linking these with genotypic performance.

Brown et al. began by trying to identify the most predictive subset of site environmental variables (the trait in question could be yield or any other important character). This was done by regressing mean site response of the trait on environmental variables found at the site. The weighted environmental variables were then clustered for analysis. Clusters chosen for grouping were those showing minimum differences, that is, having like patterns. A background goal in the interests of efficiency was to reduce both group (cluster) size and number of clusters so that the overall number of sites was reduced through elimination. This led to the second stage, the selection of optimal sites within a cluster. The authors proposed a "genotypic index regression" method, which required replacing the site mean performance axis of the Finlay–Wilkinson 1963 regression model with genotype mean performance. The slope of the resulting b_j line carries information as to a site's ability to discriminate between entries. In addition, r^2 is calculated as a measure of relative discrimination accuracy. Both b_j and r^2 are used to evaluate site discrimination characteristics.

The authors added to previous work on grouping homogeneous sites through the approach of identifying the environmental variables and using these to help select sites, and through the use of genotypic index regression to identify optimal selection environments within clusters. They pointed out that clustering is not necessary in a crop growing region where all locations represent simpler environments. This proposal, although complex, has the merit of quickly producing an operating base set of sites for which only periodic updates are needed.

As many breeders have done, Ellison, Latter, and Anttonen (1985) conducted a reappraisal of efficiencies and strategies for the multisite testing of elite lines in the New South Wales wheat program, operated out of Sydney and Narrabri. The objectives and procedures of this sound and comprehensive program were described by Derera, Bhatt, and Ellison (1982). The present study involved site genotype × environments, the function and repeatability of results from large plots, the role of harvest index from hill plots, and the interrelationship of these factors. It appeared that about 10 sites represented district sites that included both dryland and irrigated sites, and accommodated separate early and midseason tests.

A first reasonable objective was the examination of site yields for evidence of any correlation that would allow grouping or elimination of sites. The findings were in agreement with the great diversity of environments encountered in this large region; that is, the sites basically represented dissimilar environments. Therefore, if they represented meaningful production areas they could be neither clustered nor eliminated.

The major findings of the study were that hill plot harvest index determinations appeared to be reliable substitutes for large plot field trials. Coupled with similar results being obtained from dryland and irrigated trials, often duplicated at a site, Ellison, Latter, and Anttonen proposed a reallocation-of-resources strategy that would lower the overall effort while at the same time increasing the probability of greater genetic advance. The correlation between hill plot harvest index and large plot grain yield allowed the reduction of large plot numbers to six or eight sites, with fewer replications, and with no duplication for dryland and irrigation. This was coupled with an expansion of the less costly replicated hill plot tests. Harvest index in hill plots was found also to be more stable than that of large plot harvest index. The authors expected that the use of hill plot harvest index applied earlier in the selection procedures would result in advances in both total biological yield and grain yield. Substantial gains in efficiency were accomplished through the trade-off of unreplicated large plots for replicated hill plots at additional sites.

DATE OF SEEDING

Florell (1929) demonstrated distinct differences in performance of wheat, oats, and barley varieties in nurseries sown at the same site but on different dates, "where the seeding season extends over a wide range in date—the reaction of varieties to each date of seeding seems important."

Florell implied that more than a single date of seeding is required in order to reach average and reliable interpretations. A positive plant breeding derivation from this is that more than one date of seeding at the same or different sites can speed obtaining the necessary data on varieties' average responses to genotype × environment interactions.

Harrington and Horner (1935) obtained conclusive evidence that a variety × date-of-sowing interaction existed. They recommended that variety testing be done with several dates of planting in order to obtain representative results of the varieties' performance.

Harrington (1946) introduced a third element, date of seeding, to the generally accepted testing requirements of years and locations. This new option inherently embodies two favorable elements, namely, expanding the range of environments, which elicits an adaptation response, and neutralizing soil heterogeneity effects when the different rates of seeding are at the same location. The test materials included more than 50 entries each of barley and oats grown under replicated conditions in two dates of seeding between 1939 and 1949 at two Canadian locations.

Harrington examined variety and date interaction for yield on the assumption that a significant interaction also meant that information obtained on yield was different. That left this question for interpretation: which set of yield data correctly characterizes a particular genotype's nature, and what is the effect on relative cultivar performances, that is, their rank positions? Among the total 36 tests conducted, he found 23 that had significant variety × date interactions for yield.

If, in different dates of sowing at the same location, the rank order of yield of varieties is altered, the breeder may wonder which represents the truest comparison of varieties. Harrington arranged tests using a basic 25–variety size and counted how many times differences in rank greater than 6, 9, 12, 15, and 18 positions occurred between sowing dates 1 and 2. The reversals or changes were astounding, for example, differences in rank for yield greater than nine occurred in 25% of all oat and barley tests over years and locations. Of special importance was the finding that interactions often represented reversals in rank—this would not be apparent if only one date were sown. Lower percentages were found for height and earliness.

Harrington said that, unquestionably, two dates gave useful data on adaptation and that they emphasized "the value of good strains and reveal the weakness of others." Planting on two dates at one location reduced the cost of the second test. He would first ensure that a sufficient number of replications were included in each sowing. An intermediate single date of seeding would tend to produce an average, but does not give the information on adaptation to early or late seeding.

A very important implication to plant breeding is in Harrington's statement that the date of sowing method would be equally advantageous for early-generation progeny.

SUMMING UP

The choice of nursery sites for testing was once as simple as planting the same group of entry lines at all available sites, with perhaps a degree of nonrandom choice of sites to satisfy the public obligations of a breeding program. Early on, major differences in product were recognized. These led to separate nurseries for different maturities, and for market classes of grain in the cereals. Later, differences in recognized environments resulted in gridding a total testing area. Sometimes both product and environment combined to separate testing sites, as in the case of winter and spring wheats of the market classes hard red winter and hard red spring. The elements of genotype stability and its measurement over environments entered the picture and raised the question of whether the goal was a widely adapted or narrowly adapted cultivar. Reality testing is of interest to many scientists. For example, one might ask if the obtained yields are consonant with state yields. Recent research has dealt with more precise definitions of both environments and genotypes, accompanied by attempts to segment, examine, and group together homogeneous parts of what would otherwise be an indiscriminate whole, for example, a large environmental area, or the group of all genotypes to be tested. In a sense, current moves are analogous to the application of the Gardner grid principle to both environments and genotypes, and recent and current research show an obvious greater sophistication of attempts to search out the parts of the genotype × environment interaction picture.

There is a philosophic aspect to the question of sites and years. For illustration, give a breeder the figure "15" and tell him he may spend it as he wishes. Some would take five sites over three years, whereas others might take the opposite, three sites over five years. A case could be made for either, or strongly for one, based on genetic, economic, or biometric grounds. The relation of time to plant breeding also should enter into the decisions taken. Given equal choices the decision should veer in the direction of the shorter time. There may be some risk in not testing against all the hazards, but this must be balanced against depriving the public of an improved variety at the earliest opportunity. I remember well the spirited discussion that took place at a scientific meeting in Australia when the speaker, a crop breeder, gave his stated opinion that a new variety must not be introduced until absolutely all possible hazards had been tested for and removed. The obvious consequences seemed to be no release at all or a delay of a decade or more.

In my program at Cornell I used a "rule of three' (three replications at three locations for three years) as a minimum guide for decisions on lines that had passed early screens. Our oat regional locations customarily numbered from 8 to 13 per year and we used eight replications, but in view of Balaam and Hunter's 1962 findings, replications could well be reduced to three.

In the final analysis plant breeders must tailor their responses to their particular environments, but already we can see many of the ways in which they either adapt their procedures to the environment or modify the environment to their requirements. They may arrange or cluster both genotype entries and environments. They may accept the mean yield of all entries at a location as a reliable descriptor of that environment. They must remove all known cultivar × environment interactions, such as responses to insects and pathogens. They must pay attention to planting uniform-sized seeds. They must pay more attention to interactions between check varieties and different environments. Stability responses of genotype entries may be used to determine the homogeneity of sites or environments. Testing efficiency may revolve around the adjustment of testing locations, treatments, years, and replications. Large geographic areas can be subdivided to achieve homogeneity of testing sites. For selection, multiple environments offer opportunities, and Gardner's grid system, log density sowing, and log fertilizer appliation are procedures particularly applicable to pedigree F_2 and bulk F_5 spaced selections.

REFERENCES

Abou-El-Fittouh, H. A., J. O. Rawlings, and P. A Miller. 1969. Classification of environments to control genotype × environment interactions with an application to cotton. Crop Sci. 9:135–140.

Allard, R. W. and A. D. Bradshaw. 1964. Implications of genotype–environmental interactions in applied plant breeding. Crop Sci. 4:503–508.

Allard, R. W. and P. E. Hansche. 1964. Some parameters of population variability and their implications in plant breeding. Adv. Agron. 16:281–325.

Allen, F. L., R. E. Comstock, and D. C. Rasmusson. 1978. Optimal environments for yield testing. Crop Sci. 18:747–751.

Baker, R. J. 1969. Genotype–environment interactions in yield of wheat. Can. J. Plant Sci. 49:743–751.

Balaam, L. N. and R. D. Hunter. 1962. The analysis of a series of wheat varietal trials. Aust. J. Stat. 4:61–70.

Billings, W. D. 1952. The environmental complex in relation to plant growth and distribution. Quart. Rev. Biol. 27:251–265.

Boyd, W. J. R., N. A. Goodchild, W. K. Waterhouse, and B. B. Singh. 1976. An analysis of climatic environments for plant breeding purposes. Aust. J. Agric. Res. 27:19–33.

Brennan, P. S. and D. E. Byth. 1979. Genotype × environmental interactions for wheat yields and selection for widely adapted wheat genotypes. Aust. J. Agric. Res. 30:221–232.

Brennan, P. S. and J. A. Sheppard. 1985. Retrospective assessment of environments in the determination of an objective strategy for the evaluation of the relative yield of wheat cultivars. Euphytica 34:397–408.

Briggs, K. G. and L. H. Shebeski. 1971. Early generation selection for yield and breadmaking quality of hard red spring wheat. Euphytica 20:453–463.

Brown, K. D., M. E. Sorrells, and W. R. Coffman. 1983. A method for classification and evaluation of testing environments. Crop Sci. 23:889–893.

Burton, G. W. 1959. Crop management for improved water use efficiency. Adv. Agron. 11:104–109.

Campbell, L. G. and H. N. Lafever, 1977. Cultivar × environment interactions in soft red winter wheat yield tests. Crop Sci. 17:604–608.

Campbell, L. G. and H. N. Lafever. 1980. Effects of locations and years upon relative yields in the soft red winter wheat region. Crop Sci. 20:23–28.

Comstock, R. E. 1977. Quantitative genetics and the design of breeding programs. *In* Proc. Int. Conf. Quant. Genet. (Ed., E. Pollack, O. Kempthorne, and T. B. Bailey, Jr.), Aug. 16–21, 1976. Iowa State University Press, Ames.

Derera, N. F., G. M. Bhatt, and F. W. Ellison. 1982. Two decades of wheat improvement work at Narrabri. J. Aust. Inst. Agric. Sci. 48:131–140.

Eagles, H. A. and K. J. Frey. 1977. Repeatability of the stability–variance parameter in oats. Crop Sci. 17:253–256.

Eagles, H. A., P. N. Hinz, and K. J. Frey. 1977. Selection of superior cultivars of oats (*Avena sativa* L.) by using regression coefficients. Crop Sci. 17:101–105.

Ellison, F. W., B. D. H. Latter, and T. Anttonen. 1985. Optimal regimes of selection for grain yield and harvest index in spring wheat. Euphytica 34:625–639.

Engledow, F. L. and G. V. Yule. 1930. Principles and practice of yield trials. Empire Cotton Growing Corp., London.

Falconer, D. S. 1952. The problem of environment and selection. Am. Nat. 86:293–298.

Feaster, C. V., E. F. Young, Jr., and E. L. Turcotte. 1980. Comparison of artificial and natural selection in American Pima cotton under different environments. Crop Sci. 20:555–558.

Finlay, K. W. and G. N. Wilkinson. 1963. The analysis of adaptation in a plant-breeding programme. Aust. J. Agric. Res. 14;742–754.

Florell, V. H. 1929. Effect of date of seeding upon yield, lodging, date of maturity, and nitrogen content in cereal variety experiments. J. Am. Soc. Agron. 21:725–731.

Fowler, D. B. 1979. Selection for winterhardiness in wheat. II. Variation within field trials. Crop Sci. 19:773–775.

Fox, P. N. and A. A. Rosielle. 1982. Reference sets of genotypes and selection for yield in unpredictable environments. Crop Sci. 22:1171–1175.

Freeman, G. H. 1973. Statistical methods for the analysis of genotype–environment interactions. Heredity 31:339–354.

Freeman, G. H. and J. M. Perkins. 1971. Environmental and genotype–environmental components of variability. VIII. Relations between genotypes grown in different environments and measures of these environments. Heredity 27:15–23.

Frey, K. J. 1964. Adaptation reaction of oat strains under stress and non-stress environmental conditions. Crop Sci. 4:55–58.

Frey, K. J. 1972. Stability indexes for isolines of oats (*Avena sativa* L.). Crop Sci. 12:809–812.

Frey, K. J. 1977. Protein of oats. Z. Pflanzenzuecht. 78:185–215.

Frey, K. J., M. D. Simons, R. K. Skrdla, L. J. Michel, and G. A. Patrick. 1984. 1983 Oat Newsletter 34:57–58.

Ghaderi, A., E. H. Everson, and C. E. Cress. 1980. Classification of environments and genotypes in wheat. Crop Sci. 20:707–710.

Gotoh, K. and S. Osana. 1959a. Efficiency of selection for yield under different densities in a wheat cross. Jap. J. Breeding 9:7–11.

Gotoh, K. and S. Osana. 1959b. Efficiency of selection for yield under different fertilizer levels in a wheat cross. Jap. J. Breeding 9:173–178.

Guitard, A. A. 1960. The use of diallel correlation for determining the relative locational performance of varieties of barley. Can. J. Plant Sci. 40:645–651.

Hamblin, J., H. M. Fisher. and H. I. Ridings. 1980. The choice of locality for plant breeding when selecting for high yield and general adaptation. Euphytica 29:161–168.

Hammond, J. 1947. Animal breeding in relation to nutrition and environmental conditions. Biol. Rev. 22:195–213.

Hanson, W. D. and C. A. Brim. 1963. Optimum allocation of test material for two-stage testing with an application to evaluation of soybean lines. Crop Sci. 3:43–49.

Harrington, J. B. 1946. The differential response of spring-sown varieties of oats and barley to date of seeding and its breeding significance. J. Am. Soc. Agron. 38:1073–1081.

Harrington, J. B. and W. H. Horner. 1935. The reaction of wheat varieties to different dates of sowing. Sci. Agr. 16:127–134.

Hawkins, S. E., W. R. Fehr, E. G. Hammond, and S. Rodriguez de Cianzio. 1983. Use of tropical environments in breeding for oil composition of soybean genotypes adapted to temperate climates. Crop Sci. 23:897–899.

Hayward, M. D. and J. L. Vivero. 1984. Selection for yield in *Lolium perenne*. II. Performance of spaced plant selections under competitive conditions. Euphytica 33:787–800.

Hill, J. 1976. Genotype–environment interactions—a challenge for plant breeders. J. Agric. Sci. 85:477–493.

Hinz, P. N., R. Shorter, P. A. Dubose, and S. S. Yang. 1977. Probabilities of selecting genotypes when testing at several locations. Crop Sci. 17:325–326.

Horner, T. W. and K. J. Frey. 1957. Methods for determining natural areas for oat varietal recommendations. Agron. J. 49:313–315.

Immer, F. R. 1942. Distribution of yields of single plants of varieties and F_2 crosses of barley. J. Am. Soc. Agron. 34:844–850.

James, J. W. 1961. Selection in two environments. Heredity 16:145–152.

Jensen, N. F. 1976. Floating checks for plant breeding nurseries. Cereal Res. Commun. 4(3):285–295.

Jinks, J. L. and V. Connolly. 1973. Selection for specific and general response to environmental differences. Heredity 30:33–40.

Jinks, J. L. and V. Connolly. 1975. Determination of the environmental sensitivity of selection lines by the selection environment. Heredity 34:401–406.

Johnson, G. R. 1977. Analysis of genotypic similarity in terms of mean yield and stability of environmental response in a set of maize hybrids. Crop Sci. 17:837–842.

Johnson, G. R. and K. J. Frey. 1967. Heritabilities of quantitative attributes of oats (*Avena* sp.) at varying levels of environmental stress. Crop Sci. 7:43–46.

Johnson, V. A., S. L. Shafer, and J. W. Schmidt. 1968. Regression analysis of general adaptation in hard red winter wheat (*Triticum aestivum* L.). Crop Sci. 8:187–191.

Jones, G. L., D. F. Matzinger, and W. K. Collins, 1960. A comparison of flue-cured tobacco varieties, repeated over locations and years, with implications on optimum plot allocation. Agron. J. 52:195–199.

Keim, D. L. and W. E. Kronstad. 1981. Drought response of winter wheat cultivars grown under field stress conditions. Crop Sci. 21:11–15.

Kelker, D. and H. Kelker. 1986. The effect of skewness on selection in a plant breeding program. Euphytica 35:303–309.

Kinbacher, E. J. and N. F. Jensen. 1959. Weather records and winter hardiness. Agron. J. 51:185–186.

Knight, R. 1970. The measurement and interpretation of genotype environment interactions. Euphytica 19:225–235.

Langer, I., K. J. Frey, and T. Bailey. 1979. Associations among productivity, production response, and stability indexes in oat varieties. Euphytica 28:17–24.

Levine, L. and L. Van Valen. 1964. Genetic response to the sequence of two environments. Heredity 19:734–736.

Liang, G. H. L. and T. L. Walter. 1966. Genotype × environment interactions from yield tests and their application to sorghum breeding programs. Can. J. Genet. Cytol. 8:306–311.

Liang, G. H. L., E. G. Heyne, and T. L. Walter. 1966. Estimates of variety × environment interactions in yield tests of three small grains and their significance on the breeding programs. Crop Sci. 6:135–139.

Lin, C. and B. Thompson. 1975. An empirical method of grouping genotypes based on a linear function of the genotype-environment interaction. Heredity 34:255–263.

Lu, Y. C., K. H. Tsai, and H. I. Oka. 1967. Studies on soybean breeding in Taiwan. 2. Breeding experiments with successive hybrid generations grown in different seasons. Bot. Bull. Acad. Sin. (Taipei) 8:80–90.

McCain, F. S. and E. F. Shultz, Jr. 1959. A method for determining areas for corn varietal recommendations. Agron. J. 51:476–478.

McNeill, M. J. and K. J. Frey. 1974. Gains from selection and heritabilities in oat populations tested in environments with varying degrees of productivity levels. Egypt. J. Genet. Cytol. 3:79–86.

McVetty, P. B. E. and L. E. Evans. 1980. Breeding methodology in wheat. I. Determination of characters measured on F_2 spaced plants for yield selection in spring wheat. Crop Sci. 20:583–586.

Mather, K. 1961. Competition and cooperation. Symp. Soc. Exp. Biol. 15:264–281.

Meyers, C. H. 1912. Effect of fertility on variation and correlation in wheat. Ann. Rep. Am. Breeders' Assoc. 7:61–74.

Miller, P. A., J. C. Williams, and H. F. Robinson. 1959. Variety × environment interactions in cotton variety tests and their implications on testing methods. Agron. J. 51:132–134.

Miller, P. A., H. F. Robinson, and O. A. Pope. 1962. Cotton variety testing: additional information on variety × environment interactions. Crop Sci. 2:349–352.

Mungomery, V. E., R. Shorter, and D. E. Byth. 1974. Genotype × environment interactions and environmental adaptation. I. Pattern analysis—application to soybean populations. Aust. J. Agric. Res. 25:59–72.

Nor, K. M. and F. B. Cady. 1979. Methodology for identifying wide adaptability in crops. Agron. J. 71:556–559.

Pedersen, A. R., E. H. Everson, and J. E. Grafius. 1978. The gene pool concept as a basis for cultivar selection and recommendation. Crop Sci. 18:883–886.

Pederson, D. G. 1974. The stability of varietal performance over years. 1. The distribution of seasonal effects for wheat grain yields. Heredity 32:85–94.

Quisenberry, J. E., B. Roark, D. W. Fryear, and R. J. Kohel. 1980. Effectiveness of selection in Upland cotton in stress environments. Crop Sci. 20:450–453.

Rasmusson, D. C. and J. W. Lambert. 1961. Variety × environment interactions in barley variety tests. Crop Sci. 1:261–262.

Richards, R. A. 1978. Genetic analysis of drought stress response in rapeseed (*Brassica campestris* and *napus*). I. Assessment of environments for maximum selection response in grain yield. Euphytica 27:609–615.

Richards, R. A. 1983. Should selection for yield in saline regions be made on saline or non-saline soils? Euphytica 32:431–438.

Roberts, H. F. 1912. Variation and correlation in wheat. Ann. Rep. Am. Breeders' Assoc. 7:80–109.

Robinson, H. F. and R. H. Moll. 1959. Implications of environmental effects on genotypes in relation to breeding. Proc. 14th Hybrid Corn Indust.—Res. Conf., 1959.

Rosielle, A. A. and J. Hamblin. 1981. Theoretical aspects of selection for yield in stress and non-stress environments. Crop Sci. 21:943–946.

Roy, N. N. and B. R. Murty. 1970. A selection procedure in wheat for stress environment. Euphytica 19:509–521.

St. Pierre, C. A., H. R. Klinck, and F. M. Gauthier. 1967. Early generation selection under different environments as it influences adaptation of barley. Can. J. Plant Sci. 47:507–517.

Sakai, K. J. 1955. Competition in plants and its relation to selection. Cold Spring Harbor Symp. Quant. Biol. 20:137–157.

Salmon, S. C. 1951. Analysis of variance and long-time variety tests of wheat. Agron. J. 43:562–570.

Sandfaer, J. 1970. An analysis of the competition between some barley varieties. Danish Atomic Energy Comm., Riso Rep. No. 230:114 pp.

Sandison, A. 1959. Influence of site and season on agricultural variety trials. Nature 184:834.

Schutz, W. M. and R. L. Bernard. 1967. Genotype × environment interactions in the regional testing of soybean strains. Crop Sci. 7:125–130.

Shabana, R., T. Bailey, and K. J. Frey. 1980. Production traits of oats selected under low, medium, and high productivity. Crop Sci. 20:739–744.

Somerville, S. C. and K. G. Briggs. 1979. Evaluation of harvesting methods for research plots. Can. J. Plant Sci. 59:863–867.

Sprague, G. F. and W. T. Federer. 1951. A comparison of variance components in corn yield trials. II. Error, year × variety, location by variety and variety components. Agron. J. 43:535–541.

Srb, A. M. 1966. Genetic control of physiologic processes. Proc. Plant Breeding Symp. Iowa State University Press, Ames, Iowa, pp. 355–389.

Tucker, C. L. and J. Harding. 1974. Effect of the environment on seed yield in bulk populations of lima beans. Euphytica 23:135–139.

Van Sanford, D. A. and D. F. Matzinger. 1983. Mass selection for tobacco seedling weight under two nutrient regimes. Crop Sci. 23:1163–1167.

Vela-Cardenas, M. and K. J. Frey. 1972. Optimum environment for maximizing heritability and genetic gain from selection. Iowa State J. Sci. 46:381–394.

Walker, A. K. and W. R. Fehr. 1978. Yield stability of soybean mixtures and multiple pure stands. Crop Sci. 18:719–723.

Went, F. W. 1950. The response of plants to climate. Science 112:489–494.

White, S. B., Jr. 1958. Variance component analysis for a series of oat yield trials in New York. M.S. dissertation, Cornell University, Ithaca, NY, 76 pp.

Wit, C. T. de. 1961. Space relationships within populations of one or more species. Symp. Soc. Exp. Biol. 15:314–329.

Wood, J. T. 1976. The use of environmental variables in the interpretation of genotype–environment interaction. Heredity 37:1–7.

Yates, F. and W. G. Cochran. 1938. The analysis of groups of experiments. J. Agric. Sci. Cambridge 28:556–580.

37

FIELD ORGANIZATION AND OPERATIONS

This chapter might be called the "nuts and bolts" aspect of running a field program.

NURSERY PLOTS: SIZE, SHAPE, DESIGNS, AND REPLICATIONS

Wiebe (1935) reported on an extensive study of plot size, shape, and relative position in a uniformity trial involving 1500 rod rows of wheat, 1 ft apart. The article must be read for detail, but salient points were these:

1. As distance separating plots expands, the correlation of yields decreases. Wiebe found that statistically significant correlations could not be found beyond 48 ft.
2. Given constant plot size, correlations increase as the shape approaches a square.

Balaam and Hunter (1962) compared lattice and randomized block designs in wheat varietal trials. They found the lattice superior, with gains varying up to 143% percent (this agrees generally with the results of many researchers). They also observed changes in the variance of mean yields as replications and locations were increased. Increases in locations caused a more rapid decline in variance per mean than did increases in replication. They concluded that "the optimum allocation is one replication per location and to increase the number of locations indefinitely within the limits of the research programme."

According to Balaam and Hunter, Skinner (1961), working with sugar cane, "concluded that one replication is sometimes best and that there is little gain in efficiency by using more than two replicates."

Balaam and Hunter's recommendation of one replication per location at as many sites as feasible, although statistically correct, omits the practical consideration of the great disparity between the costs of replication and of additional sites. When one has arrived on site for planting, the costs of sowing (or harvesting) an additional replicate is miniscule compared to the organization, time, and travel required in adding sites. Also overlooked is the plant breeder's desire to see the nursery entries in more than one replicate. I would suggest a modification to add two or three replicates per site.

Ikehashi, Nakane, and Ito (1969) maintained that if there are restrictions on field size, for example, because of land availability, an optimal replication number exists. If plot size is not changed it is inevitable that an increase in replication reduces selection intensity because fewer lines can be entered in the available space. There are, of course, favorable aspects to increased replication, such as effects on heritability, and the authors concluded after statistical treatment of the problem, that smaller plots such as hill plots might be used.

Although Baker and Curnow (1969) discussed maize and the improvement of composite populations as a source of inbreds, their treatment of replicate populations has relevance to population handling and selection methods for the small grain self-pollinated cereals. Recall the experience of Nickell and Grafius (1969), who found a case of negative response to selection—a reversal—for high yield in winter barley. In the year of selection the average yield of the 136 lines was 123.3% of the check cultivar. The next year the mean yield of the same lines was 90.7% of the same check. The use of replicate populations if extended over time would provide some protection against this reversal, because the initial selection taken in each of two years would be made against the background of that year's genotype × environment interaction.

Several years ago I standardized procedures in my breeding programs so that, beginning with the F_2 if there were seed enough, all early-generation plots were the same size, approximately 100 ft^2, the width of the planting drill and of rod row length. Because this represented a rather small population sample, planting rates were doubled, and replicate populations of a given cross were planted to the number justified by an estimate of worth made first from knowledge of the parents, and buttressed later from observation and performance data from the populations. I called these aliquot samples "units." They took the form, for simple, three-way, and double crosses, of one or two extra units planted the same year, with a return to reserve seed for one or two units the following year.

With the development and integration of the diallel selective mating system into the overall project, I adopted a formal guide to the number of replicate units to be planted. This was based on the number of crosses bulked: one field unit to be planted for every five hybrids in the objective base. Thus in my Table 1 (Jensen, 1978), the Hessian fly resistance base bulk had 17 crosses and 350 seeds, and three F_2 units were sown; the high test weight base bulk involved 35 crosses and 877 seeds, and seven F_2 units were planted. Reserve seed from all early-generation production was saved as a matter of policy for at least three years, and longer with certain composites. With especially promising crosses we would return to the reserve F_2 seed for replicate populations in succeeding years. I believe that my understanding of replicate populations is in no way different from Baker and Curnow's discussion on corn, and the same reasons support their use as a sound breeding and selection procedure in the small grain cereals.

Sedcole (1977) considered the problem of how many plants (presumably of a hybrid population) must be grown for a breeder to have reasonable assurance of recovering a particular trait. Four methods were given; all require some estimation of the heritability or probability of occurrence of the trait. Answers to the problem would be helpful in breeding—a breeder does not need to know exact numbers, but the estimates would allow him to provide his own safety margin, that is, $\times$ times the minimum number.

England (1977) discussed the design of field plots for selection purposes, using considerations applicable to genetically homogeneous populations such as pure lines, varieties, F_1 hybrids, clones, and so forth. The formulas and discussion covered the various components the breeder controls, such as number of plants per plot, number of replications, heritability, and intensity of selection. Some of the conclusions represented simple, commonsense reactions to a presented situation; for example, if a project has an area limit of 1600 plots, two replications would permit 800 entries, and four replications only 400. I was impressed with estimates affecting efficiency, which generally seemed to translate into requirements smaller than expected for plants per plot and number of replications. Indeed, the requirements for single-plant testing is favorable also for the hill plot method, which embraces many plants in a plot that takes little more space than a single plant.

Sidwell, Smith, and McNew (1978) made estimates of heritability and genetic advance for several wheat characteristics using two kinds of selection units, single plots (SRS) and multiple replications (MRS). They found that multiple replication selection gave higher heritability estimates and lower standard errors than single replicates. However, the bias in favor of MRS was less when genetic advance was examined, for "as the reference unit changed from MRS to SRS, the predicted genetic advance decreased less than the heritability estimates." Of the traits considered, namely, grain yield, tiller number, kernels/spike, yield/spike, kernel weight, plant height, and heading date, the authors selected increased kernel weight as the character most subject to direct selection and positively related to increases in grain yield.

Briggs and Faris (1979) compared three barley plot conformations with the standard four-row, 5-m plots with the center two rows harvested (CK), for effect on yield, seed size, days to heading and maturity, and height. The plots were (1) SR_5: single row, 5 m, unbordered, flanked by other genotypes, (2) SR_1: single row, 1 m, space-planted at 5 cm, and (3) SP: single-plant plots, each containing 40 plants of a cultivar on a 61-cm^2 grid. Eleven cultivars were used in the tests conducted in Alberta, Canada.

SR_5 and SR_1 were found to show lower correlations with CK for yield. In the first case one has to balance the loss of precision against the 75% saving in nursery space over the CK—the loss was attributed to intergenotypic competition and smaller size. In the second case the loss of precision was considered unacceptable.

The single-plant plots, SP, showed great inconsistency over locations—attributed partly to space planting—and did not inspire confidence in results.

Rather unexpected was the failure of two methods to reduce apparent variability: the introduction of control (check) cultivars did not materially change either the relative performances or correlations between SR_1 and CK, and the moving mean (average of five plots each side of test cultivar) was ineffective and erratic in comparisons of SR_5 and SR_1.

The different plots were equally effective in measuring the cultivars for the other traits, with the exception that SR_5 was not suitable for hectoliter weight.

ROD ROW AND MINIATURE HEAD-HILL PLOTS

Hayes and Arny (1917) established the validity of rod row testing. The method may have been used first as a unit to sample the yields of large plots. Arny and Garber (1919) sampled 0.1-acre plots using nine samples per plot in the development of the rod row method, and also investigated 1 yd^2 units. The rod row length finally became standarized as a row approximating 5 m in length.

Bonnett and Bever (1947) gave a detailed description of the use of head hills as a substitute for head rows.

Ross and Miller (1955) reported on a comparison of hill plots with rod rows and 0.02-acre field drill plots in barley and oats. The putative advantages of hills were small area per plot, low seed requirement, and a simpler design and statistical analyses. They found that the variability was always higher in hill tests, and that measurement data on heading, height, lodging, and test weight were inconsistent and could not be taken with precision. Hill tests correlated better with other methods in oats than in barley (not unexpected because barley in many ways shows greater sensitivity to the environment, perhaps less stability, than oats, for example, a greater negative response to late sowing). The authors concluded that hill plots were not as accurate as row and drill plots and not as reliable. They might be useful for screening tests with large numbers (presumably with a high cutoff point).

Jensen and Robson (1969) outlined a new type of miniature plot, the linear hill plot, for testing for yield. The plots were 1 ft long, with 10 plots laid end to end (separated by 4-in. open areas) within one rod row furrow; a 16-in. border capped each end of the row. Thus a 10-variety × 10-replicate nursery could be planted in the space occupied by 10 rod rows and be placed anywhere on the field within the rod row section arrangement. Seed requirements were 1 g per plot. It was found that rod row precision could be equaled by planting 2.4 times as many replicates; however, slight modifications that were suggested should reduce this figure to 2.

Working with soybeans, Torrie (1962) compared hill nurseries with row nurseries. He found varietal differences in yield to be similar in both nursery types; however, because of greater variance in hill nurseries, more replications would be required.

Jellum, Brown, and Seif (1963) used F_4, F_2-derived lines as a test of hills versus row plots for early-generation material. They used 96 bulk lines (from six hybrids), and four checks in a 10 × 10 lattice square design at two locations. In general the correlation between hill and row plot yields was acceptable (15 of 24 comparisons were significantly correlated). Particularly, a calculation of the number of high-yielding lines, based on saving the top 20%, showed that more lines, common to both kinds of plots, would have been saved using hills than rows. The inconsistencies in yield correlations relegated hill plots to a supplementary testing role.

Frey (1965), using hill plots of 30 seeds per hill, found a correlation of .98 between rod row and hill yields of oats. He concluded that hill plots could be used for early-generation testing, but final evaluation should be done in rod rows (and larger plots). As might be expected, the coefficients of variability for grain yield were from two to five times larger for hills than for rod rows; as a consequence more replicates would be needed for hill plot testing.

Smith, Kleese, and Stuthman (1970) found significant interaction, attributed to competition, in hill plots of five different oat cultivars. The competition experiment used Schutz and Brim's (1967) nine-hill model, where the center hill is the plot measured for effect caused by neighboring hills, which could be the same cultivar or other cultivars. Despite the strong variances observed, the yield results were about as expected under rod row conditions. Nevertheless, hill plots were not viewed with the same confidence as were rod row trials. A strong impression received from this paper supports the need to separate, and test separately in homogeneous groups, lines with striking differences—a familiar example being the separation of oat cultivars into early, midseason, and occasionally, late nurseries.

Khadr, Kassem, and ElKhishen (1970) reported on a comparison of parents and 107 F_4 lines from three wheat crosses when grown in hill and row plots. They found hill plots suitable for heading date, plant height, and seed weight, but not for yield, where rod rows were found to be superior. The latter may be due, as was also pointed out by Frey, to much larger sample sizes generating yield in rod row tests and thereby increasing precision. Matching precision levels could be obtained by increasing plot or replication numbers as suggested by Jensen and Robson (1969) for their miniature yield plots.

Baker and Leisle (1970) compared hill and rod row plots of durum and common wheats over a two-year period. They found that hill tests compared favorably with rod rows in all tests of the same cultivars and might be used for similar genetic and performance testing purposes. The yield range, perhaps as would be expected, was greater in the hills than in the rod rows—this has little meaning unless tied to heritability values. The authors pointed out that the efficiency of hill plots can easily be increased by adding hills. I thought their comments on changes in rank order of yield in different testing procedures were well taken: by themselves rank order changes mean little, and an instance is cited of a change in rank from first in hills, to fifth in rod rows, where it was shown that the top seven rankings in the rod row tests were not significantly different in yield.

Hill and row plots were evaluated for testing suitability in barley by Walsh, Park, and Reinbergs (1976). Doubled haploid lines and commercial cultivars were tested in rows versus hills, with the hills varied by planting methods and rates of seeding. Neither planting methods nor rates showed significant interaction with genotypes. Correlation of hill results with rows was high, and overall, the conclusion supported the view that hill plots were satisfactory for early evaluation of breeding lines. High replication was recommended (10 were used in these tests).

Park, Reinbergs, and Song (1977) explored the use of hill plots to determine the relationship of grain yield and components of doubled haploids of barley relative to comparisons with row plots. Different within and between spacings or densities were used. It was concluded that hill plots could be used to evaluate two of the three components of grain yield, namely, number of seeds per spike and seed weight. The third, spikes per plot, could not be reliably evaluated because the results were so different from those found in row plots—it was, however, highly correlated with grain yield. Differences in seeding rates did not seem of major importance, although heavier seeding rates gave somewhat higher correlations.

Garland and Fehr (1981) compared hill with row plots for selection purposes in soybeans. They found that single unbordered hill plots were sufficient to evaluate height, lodging, maturity, yield, and phenotypic score. Visual discrimination was relied on heavily. When replication was used, a systematic, nonrandom arrangement of entries in replications made for greater efficiency. The authors cited two advantages of hill over row plots: one plant can furnish seed for a replicated hill planting, and they are less expensive to use.

BORDER PLOTS

Hartwig, Johnson, and Carr (1951) found border competition to be present and significant in soybean plot trials. Under conditions of rank growth they found one-row plots unreliable and subject to variety × environment interaction. Their recommendations were for bordered three- or four-row plots, with border rows later discarded.

Rasmusson and Lambert (1961) found no barley variety × plot type interaction in yield comparisons involving the four-row standard plot with the two center rows harvested, a two-row plot bordered by (vegetative) winter wheat, and a two-row plot bordered by random self species, that is, other genotypes under test in the nursery. Although the winter wheat bordered plots yielded higher, they apparently produced the same order ranks. Rasmusson and Lambert preferred them over the self-species bordered plots because of weed control and the imposition of competition acting to reduce border effect. I might add that such plots would also be easier to harvest and would contribute to the maintenance of seed purity.

Rich (1973) found definite differences in wheat variety performance associated with plot row numbers and guard rows in Texas. Reduction or elimination of competition effects was attained either by using six-row plots or four guard rows. With fewer plot rows, changes in yield ranks occurred. Row spacing of 30 cm was recommended.

There is general agreement that nursery yield plots require border rows. I thought it an anomaly, when I first came to Cornell University as a graduate student, that Professor H. H. Love, a distinguished statistician and author of several books, used unshielded plots, actually single-row plots, well replicated, throughout the nursery. Because plots were arranged systematically throughout replicates, any aberrant performance of a cultivar or any localized environmental effect could exert a marked primary effect on three plots, itself and two neighbors. Any number of causes can selectively affect a plot, for example, germination, lodging, or disease, but the most dramatic example I ever saw concerned the oat cultivar 'Vicland', when it succumbed to Victoria blight (*Helminthosporium victoriae*) and left a gaping hole in the nursery at harvest time. In a nursery of systematically arranged entries in replicates, the missing plot would always be flanked by the same neighbors. Crop lodging can create the same effect.

I once measured wheat yields row by row on large nine-row drill plots that were separated by alleys. A graph produced a "saucer" effect, with the greatest yields on the two outside drills (#1 and #9) and lesser yields on the #2 and #8 drills, and the five center rows were quite alike in yield. I was surprised recently to find a report in the literature by Hulbert, Michels, and Burkart (1931), who graphed the row-by-row yield results from a 14-drill plot: "The data on the two plats of wheat show that border effect, due to additional alley space, was only operative through the inside border rows (two and thirteen)."

POPULATION AND GROUP SIZES AND NUMBERS

In "Cereal breeding procedures," designed for world wide use, Harrington (1952) discussed F_2 population sizes at some length. A middle ground of 10,000–20,000 plants was suggested. Depending on the perceived breeding difficulty, and on the relationship of parents, the number could be below 10,000 or above 20,000 plants.

Finney (1958) introduced a term useful for plant breeders, namely, "cohort": ". . . one *cohort* of new varieties, namely all those for which testing was begun in the same year."

I used this concept throughout my breeding programs, labeling each new batch according to the year, and thereafter conducting comparisons both within and among cohorts. This concept of looking first within groups was helpful for selection and discard. This can be extended by examining lines within a cross, lines among crosses within a cohort, and lines among cohorts, in that order. Finney also considered the problem of population and program size.

Bell (1963), believing that selection for yield in barley F_2 populations was ineffective, advocated the use of large F_3 nurseries.

Making comparisons in large nurseries where a check cultivar is the only visible aid is a problem to all plant breeders. A helpful procedure is to break the whole nursery area—which might be four or five acres—into smaller, more coherent groups for which a logical separate identity exists. This is nothing more than an extension of Gardner's grid concept. A first division in my Cornell program was by cohort group—a cohort consisted of the entries that annually were added to the testing program. Within the cohort were the groups of crosses, each containing various numbers of entries. This grouping was particularly helpful at the critical time of visual preharvest selection (as in the large head row or subsequent first rod row nurseries). Instead of viewing a 10,000-head row nursery as one matrix, it may be seen as a series of smaller groups containing a sequence of entries by crosses, for example, groups of crosses having 211, 74, 397, and 16 entries per cross. The problem is reduced to dealing separately with a group of, say, 211 lines, the breeder moving back and forth before this section in the field, concerned only with evaluating that cross. There is time in the second year and later to place the selected genotypes in broader peer competition, as well as to introduce randomization (incidentally, even when the lines from one cross are kept intact as a group, randomization of position within the group can be initiated). To aid in highlighting the position of cross groups in the field I planted marker rows at the start and end of each cross group. For example, in wheat, check cultivars were used to end a cross group, followed by a row of rye or triticale, and another check cultivar to start the next cross group.

CONTROL OR CHECK VARIETIES

Jensen and Federer (1964) found that adjacent nursery row competition in wheat trials favored tall genotypes over short. The effects did not cancel each other, but instead provided a competitive overall advantage (5.0 bu/acre) for the tall vs. a yield decrease of 2.3 bu/acre for the short. The tall variety used, 'Genesee', also happened to be the standard check cultivar in use at the Cornell Agricultural Experiment Station during the four-year period. The size of the introduced error, often greater than the real difference between two wheats, mandated a change to three-row check plots with only the center row harvested.

Baker and McKenzie (1967) tested Yates's (1936) thought that overadjustment of yields through the use of control plots could actually be counterproductive to the efficiency of statistical analysis, and that analysis of covariance would damp overadjustment (Yates introduced pseudo-factorial plot arrangements, for which he claimed greater efficiencies than those obtainable from check varieties). Using two 2-replicate oat nurseries of 20 and 27 entries, with a control plot sown every other plot, the authors found that control plots were of no value in the first nursery, where analysis of covariance failed to reduce the error variance below a b value of 1, but in the second nursery the following year, they gave a 14% gain in efficiency. Considering also that the control plots greatly enlarged the testing area, the authors concluded their use was questionable and should not be substituted for replication.

Shebeski (1967) reported some very revealing experiments with wheat variety check or control plots. These studies began from the premise that "while soils are universally heterogeneous contiguous plots tended to yield alike." In other words, variance expands with distance. He was able to show that the correlation between control yields dropped precipitously as the distance between control plots increased; thus there was a high correlation at 9-ft intervals and essentially no correlation at 50-ft and greater intervals.

Experiments were then organized with a repeating plot sequence of "control–line–line–control," so that every F_3 line, and its subsequent F_5 line, was adjacent to a control plot. The results were remarkable: the agreement between order of yield rankings of F_3 and F_5 lines was significantly correlated with $r = .847$.

The profound statement this research made is that control of environmental variability can expose the orderly and inherent nature of characters controlled by quantitative genes and permit their measurement. Practical considerations raise general questions about how this technique might be used, however. This scheme involved a high proportion of control area to overall land requirement. Shebeski, assuming that F_3 yield predictive values would be positive, also conducted milling and baking tests on the higher-yielding F_3 lines. Subsequently he learned that not only did the F_3 yield data have predictive value, but so did the milling and baking results. This provided the breeder with a two-pronged truncation weapon. One might assume further that the field presentation of the F_3 lines would allow additional on-the-spot discard decisions.

What I am attempting here is to make an obviously successful concept generally useful. If a breeding program involves many crosses per year, the cost of treating all in this way in the F_3 would be large. Without some sort of restriction, the procedures seem too cost-intensive for a general application to very many crosses. On the other hand, they are reasonable for a few elite crosses. Ambiguous as the term may be, especially at the F_2 level where a characterization of a hybrid population would have to be made, I think the breeder must sort crosses across a quality range that has "elite" at the top. This sorting would take its cues from several levels: from knowledge of parents, previous experiences, appearance of F_2, and indeed, I would not exclude the breeder's bias even to the extent that a particular cross is elite because he or she thinks it is. If the number of crosses in a breeding program can be sorted in this way, Shebeski's contiguous control method would find general use among the best crosses where the expenditure of special effort is merited. Shebeski was well aware of the need for early recognition of cross quality; for example, he stated, "We are attempting to determine if we can eliminate whole crosses on the basis of their quality at the F_2 level."

Briggs and Shebeski (1968) showed that control plots more than 50 ft apart are past the point of significant correlation with yield and cease to perform a useful yield comparison for the nursery.

Briggs, Bushuk, and Shebeski (1969) looked at variation in breadmaking quality in wheat controls (field checks). The placement of check varieties to minimize yield variance has received much study, but the variance for breadmaking quality traits has not been looked into relative to spacing of controls. Wheat protein percent ranged from 10.3 to 16.5% among the 354 'Manitou' controls, and it was shown that control plots 9 ft apart were very similar (r = .88), with association decreasing rapidly until at 99 ft apart correlations were nonsignificant. They recommended frequent control plots in F_2 progeny tests for breadmaking quality comparisons, coincidentally discarding lines for negative agronomic traits. They proposed replicating lines at different locations and, because of seed limitations, using hill or miniature plots (cf. Frey, 1965, and Jensen and Robson, 1969, respectively).

DePauw and Shebeski (1973) in a 1973 report narrowed the gap often found between generation means separated by seasons through the use of control plots adjacent to lines under test. F_3 line and F_4 bulk correlation and regression coefficients for yield, as a percentage of adjacent control, were .50** and .39**, respectively, and almost identical, .56* and .39*, respectively, for the same comparison between F_3 line and F_5 family.

Jensen (1976) addressed the problems inherent in the control or check variety concept of measuring entry performances in nursery yield trials. The problems were these:

1. Checks have relatively short life-spans—checks show "time erosion" as breeding improvements turn them into anachronisms.
2. Checks have unique genotype × environment interactions—a check's responses are not specifically matched to any genotype under test.
3. Check performances may be irrelevant in certain conditions, for example, in regional or international tests, where every breeder's checks may not be included because of space limitations.
4. Checks may have specific weaknesses or strengths that invalidate the check's usefulness, for example, in disease situations.

Jensen proposed a universal and perpetual check that would automatically adjust (float) to environmental and breeding improvement changes and thus could be as relevant in the year 2000, for example, as today. This was termed a "floating check" and was defined as the site highest yield (SHY); by definition it was always 100%. Breeders understand the meaning of the highest attainable yield and are always eager to compare entry rankings. However, under the floating check concept genotype identity of SHY is unimportant—it is a statistic only and may be averaged from nurseries where SHY was produced by different genotypes. Jensen stated the case for a floating check in this way:

> A nursery is seen as a battleground for a struggle between a number of genotypes and the environment. SHY is the highest interaction product and is derived from the competitive float of *all* genotypes against each sample of the universal environment. Stability is fostered in this statistic by the fact that the interaction product (yield) is a function of excellence which is not tied to a single genotype across sites as is the case with a check variety but is the mean of all genotypes which, across sites, are able to produce the unique statistic of site highest yield. The positive characteristics of SHY are 1) excellent stability, 2) goal oriented (keyed to highest performance), 3) the "percent SHY" index is always on a scale of one hundred since SHY cannot be exceeded by any entry, and 4) the index shows not only rank but position of each entry.

Examples of floating checks applied to state, regional, and international nurseries were given. The stability of SHY was shown to be outstanding when compared to five standard check varieties used in regional nurseries.

A paper by Brennan and Byth (1979), which discussed the treatment of genotype × environment interactions for wheat yields, explored pathways that might be said to parallel site highest yield, but in the end they used a version of Finlay and Wilkinson's site mean yield. Their best selection strategy for widely adapted wheat genotypes was to express line mean yield at each environment in percent of environment (nursery site) mean yield, average these across environments (sites), and use these composite figures to select for high relative mean yield.

There is another application of the universal check concept, applied to the control or check variety. While working on the multiline concept, I saw the relationship of these ideas to check varieties, which after all, have finite useful lifetimes. A new kind of check composed of an appropriate mixture of cultivars and candivars could be created so that the blend performance would have broader application than a single genotype check. The universal and perpetual aspects of such a control variety are that it would never need to be replaced. Instead, the components would change slightly and slowly over long periods, through add-on drop-off procedures and through adjustment of proportion percentages. To illustrate this for the Cornell program of a decade ago, I might use as components the five Cornell varieties Avon, Arrow, Ticonderoga, Yorkstar, and Houser, with seed proportions adjusted to a sense of each one's age and commercial importance. Perhaps one variety each from Michigan and Canada might be included, as well as a candivar or two. This kind of a check variety should, in my opinion, have great stability, both over years and over locations. When a new cultivar was released it could be added into the blend and proportions adjusted. Occasionally an old cultivar might be dropped. The subtle changes should not invalidate, however, the use of such a check variety for long-time use. The check would have to be prepared anew annually to avoid genetic shift (which might improve check performance, but would also perceptibly change the check's standard of performance over time). If an analogy is needed, we might think of it as the "Dow" check variety: the Dow financial index has been changed many times—the 30-component stocks that make up the Dow today represent several add-on drop-off decisions, yet the Dow is always considered modern and relevant (but it is not irrelevant to say that many financiers favor the more broadly based—more components—indices).

MOVING MEANS OR AVERAGES

Knott (1972) examined three ways of expressing F_3 plot yield data, namely, weight as a percentage of replicate mean yield, weight as a percentage of the closest two check yields, and weight as a percentage of the moving average of the nearest seven hybrid plots. He found that both of the latter two methods resulted in a significant lowering of the error mean square. Knott favored the moving average because it did not increase nursery size (as checks would).

The moving average is an attractive method that seems to have a place in cereal testing programs. A check may be included, but what the moving average does is relate a plot and genotype yield to a slowly changing (add one, drop one) measurement of the environment. The measurement is an interaction one and can handle large (yield) variations whose waves can be damped by choosing longer or shorter distances over which the average moves. In my cereal projects at Cornell University I used the moving average as a supplement to other checks. In Knott's study he used a length of seven hybrid plots, that is, three plots on each side of the target plot; all yields were included, except checks (when they appeared, they were ignored). For the problem that arises at the end of the block, the average ceases to move and the yields of the same seven end plots were used (I used a variation—a serpentine turn to the next block, continuing the moving average).

Townley-Smith and Hurd (1973) reported that moving means were more effective than repeated control cultivars or unadjusted yields. Control plots gave very little improvement over unadjusted yields. When moving means were used, however, error variance was reduced in all cases. Soil heterogeneity and/or choice of control cultivar are offered as reasons for the inconsistency of control influence on the error variance. The use of the analysis of covariance technique gave no better results—lower error variance—than the adjustments for moving means.

The authors also studied the optimal number of plots to use in the moving mean (the yields of bracketing plots excluding the center pivot plot and any included control plots), testing from 6 to 18 plots, and found little effect from using different plot numbers. It was suggested that compensating forces involving soil heterogeneity and block sizes tended to produce equivalent results on reducing error variance. They suggested the use of eight plots as suitable for the moving mean.

Townley-Smith and Hurd cited on the one hand the extra land and labor costs associated with control plots, and on the other hand, the moving mean analysis requirements of complete randomization per replicate and complete harvest of all plots.

Boerma and Cooper (1975) used an ingenious variation of the moving average when they based their yield selection intensity first on a three-environment mean (one F_3 and two F_4), and next on a five-environment mean yield (adding two F_5 environments).

Mak, Harvey, and Berdahl (1978) compared frequent check plots with moving means for efficiency in error control in barley nurseries of single-rod row plots. The study included more than 140 lines with a cultivar check every third row (in the second test year two cultivars were alternated in the every third row position). Although rod rows 1 ft apart were used, the planting arrangement within rows was different in the two years: space planted the first year and crop density the second. Three and two replications, respectively, were grown.

The moving mean was calculated using from 2 to 12 adjacent plots. Measurements were taken on yield and protein content: (1) as is, unadjusted; (2) percentage adjustment, (a) as a percentage of the moving mean or (b) as a percentage of the mean of the two adjacent checks; and (3) covariance adjustment—analysis of covariance using the moving mean and check plot values as the independent covariate. For comparisons the authors used the coefficient of variation. The results were as follows:

Control Plots: The CV for yield was not reduced by controls in 1974 regardless of adjustment, but was reduced in 1975, more so by covariance than by percentage adjustment. For protein content the CVs were reduced in both years by both adjustments to control plots, but more so by covariance than by percentage. The reduction was greater for protein than for yield.

Moving Means: The most effective number of adjacent plots tested was eight. Again, covariance adjustment was more effective in reducing CVs for both yield and protein, but more so with protein.

Overall there was little to choose between control plots and moving means on the basis of error control; however, the advantage would have to lie with the moving mean because of the additional plot space used by the frequent control plots. A point could also be made that the increased size of the nursery occasioned by the numerous control plots also would have increased the error variance.

In the absence of adjustments, the authors found a partially balanced lattice was superior to the randomized complete block design and equal to control plots or moving means with covariance adjustment.

The moving mean, or running average, method uses as a basis for yield comparison the average yields of adjacent plots bracketing an excluded center plot (some prefer to include the center plot). After each determination of a mean all measurements shift one plot forward. I used the moving average as an aid in local nurseries: an 11-plot span, 5 plots before and 5 plots after the excluded center plot; thus the 10-plot moving average was always visible, requiring only the insertion of the decimal.

Mak, Harvey, and Berdahl's use of single-row plots and a running mean reminds me of procedures in use at Cornell University many years ago. I have been critical of Professor Love's use of single-row plots from the standpoint of systematic arrangement and the rigid competitive aspects; however, there is no denying the efficiency in land use and convenience. Dr. Love's greatest interest, aside from plant breeding, was in biometry and statistics, backed by years of extensive field tests. I saw acres of his uniformity trials with oats. In two of his books (Love 1938 and 1943), well worth reading today, he devoted considerable attention to problems of plot technique. He discussed the moving average and several uses of check plots, such as the average of closest checks, the average of all checks (today we would call this the site check mean), a graded method for checks positioned at different distances from each other, and a combination of site check mean and the graded method.

It was the graded, or "theoretical check," method that was in use in the years when I arrived at Cornell as a graduate student. This was a cultivar check every tenth row, which unavoidably required the acceptance of the assumption that soil fertility changes proceeded in a linear manner down the rod row sections. The simple formula to give the calculated or theoretical check for the first row was "$0.9C_1 + 0.1C_2$," proceeding to "$0.1C_1 + 0.9C_2$"

for the ninth row between the checks. The yield of an entry would be expressed, for example, as 56.2 ± 3.8 bu/acre.

Even today, I think this a method hard to surpass for the first screening of large numbers after the head or plant row generation, that is, the first year in rod row yield tests. The plot size is large enough, the statistical control of variance, with check presence, is helpful, and the competitive aspect of adjacent genotypes might be anticipated and allowed for (or, if more than one replicate, damped by random arrangement).

Love's suggestion of using both the mean of all nursery checks and the graded check system [$0.5\ (C_m + 0.8C_1 + 0.2C_2)$ = calculated check for first entry between two checks 10 rows apart] was novel. This formulation split the difference between overall nursery check and local genotype × environment interactions. One wonders whether there might be other possible modifications of control plots that, like an index, would provide a more stable comparison base. A running average applied to, say, the moving group of five checks, might be useful.

COLLECTING AND EXPRESSING DATA

Detail on this subject is, I think, unnecessary. A number of excellent books on practical field designs and statistical treatments are readily available. Basic requirements are assurance of random bases, comparisons of efficiencies, and choices of programs that fit the project operations. Great technological advances have been made in data collecting and handling in the last decade. To cite just one example, Scott, Kronstad, and McCuistion (1978) depicted a fully integrated data processing system for field use in small grain operations. Powered by a portable electric generator, the arrangement permits electronic weighing and data processing through a computer. The system is tied efficiently to field harvesting procedures.

SEED INCREASE AND VARIETY RELEASE

In an interesting discussion of what constitutes varietal purity, Harrington (1939) in effect took a stand opposing extreme homozygosity, except for stocks used for genetic research. His principal thesis was that selection from known varieties for the purpose of purifying seed stocks should be in large enough numbers to be representative.

After pointing out that variability and heterogeneity, not a pure line culture, predominate in natural crop populations, and that these mixed populations suggest advantageous competition effects, Hutchinson (1940) stated,

> If variability, and not uniformity, is the natural characteristic of crop populations, the breeder's ideal must be reconsidered. Selection for uniformity should not process beyond the stage needed to ensure a satisfactory grade in the marketable

product, except where it is possible by further specialization to achieve closer adaptation to local climates.

The preparation of similar lines for a cereal cultivar release, or reselecting an older cultivar for a seed stock rerelease, involves a consideration of many morphological traits, but the procedures seldom extend to yield testing of the final lines. Wheat and Frey's (1961) paper is interesting because it provided a statistical basis for the number of lines needed. They took 400 heads (the method of procurement was not stated) from each of three oat cultivars and sowed them in head rows. Their repurification history was 'Bonham' three times, 'Mo.0–205' two times, and 'Clintland' zero times. Discards of morphological off-types were made in the head rows and the next-year rod rows; then the first 99 rows were harvested. From these, 24 were chosen randomly and used to make composite populations of 3, 6, 12, and 24 lines. In all, including check populations, a total of 54 entries was tested for yield at several Iowa locations.

All lines and entries were morphologically uniform. Using yield as an indicator of physiological differences, the results showed no yield differences except for one line in Bonham. The authors concluded that several hundred heads need not be selected. No specific number was specified, and for a very sound reason: "The exact number of lines that should be composited to reform the repurified breeders seed of an oat variety may vary from one variety to another."

I can vouch for the accuracy of this statement—varieties can be greatly different and in my experience we had difficulty in keeping certain varieties pure and in repurifying their seed stocks. Especially troublesome were the wheat varieties 'Yorkstar' and 'Cornell 595' and the winter barley 'Wong'.

Details of the unusual Cornell 595 case are given in my article (Jensen, 1965) on population variability in small grains (the instability of this variety was traced to a chromosomal defect). This paper is an excellent review treatment of off-types, impurities, instabilities, and the like in the small grains, and includes my experiences with Breeder Seed preparation for 16 varieties and their subsequent maintenance. The systems for seed stock maintenance used by other breeders are described in detail.

One of the most impressive seed increases of cereals was reported in 1969 by Ebeltoft (1969) when, in one calender year, an initial 1 bu of barley seed was translated into 150,000 bu of seed. This was possible by using three sites that effectively used all of the available growing season: August–November in California, from 1 to 50 bu; November–late April in Arizona, from 50 to 3000 bu; and May–August in North Dakota, from 3000 to 150,000 bu.

REFERENCES

Arny, A. C. and R. J. Garber. 1919. Field technic in determining yields of plots of grain by the rod-row method. J. Am. Soc. Agron. 11:33–47.

Baker, L. H. and R. N. Curnow. 1969. Choice of population size and use of variation between replicate populations in plant breeding selection programs. Crop Sci. 9:555–560.

Baker, R. J. and R. I. H. McKenzie. 1967. Use of control plots in yield trials. Crop Sci. 7:335–337.

Baker, R. S. and D. Leisle. 1970. Comparison of hill and row plots in common and durum wheats. Crop Sci. 10:581–583.

Balaam, L. N. and R. D. Hunter. 1962. The analysis of a series of wheat varietal trials. Aust. J. Stat. 4:61–70.

Bell, G. D. H. 1963. Barley Genetics I. Proc. 1st Int. Barley Genet. Symp., Wageningen, pp. 285–302.

Boerma, H. R. and R. L. Cooper. 1975. Performance of pure lines obtained from superior-yielding heterogeneous lines in soybeans. Crop Sci. 15:300–302.

Bonnett, O. T. and W. M. Bever. 1947. Head-hill method of planting head selections of small grains. J. Am. Soc. Agron. 39:442–445.

Brennan, P. S. and D. E. Byth. 1979. Genotype × environmental interactions for wheat yields and selection for widely adapted wheat genotypes. Aust. J. Agric. Res. 30:221–232.

Briggs, K. G. and D. G. Faris. 1979. Comparison of multiple- and single-row plots and spaced plants for testing yield and other variables in barley. Can. J. Plant Sci. 59:493–498.

Briggs, K. G. and L. H. Shebeski. 1968. Implications concerning the frequency of control plots in wheat breeding nurseries. Can. J. Plant Sci. 48:149–153.

Briggs, K. G., W. Bushuk, and L. H. Shebeski. 1969. Variation in breadmaking quality of systematic controls in a wheat breeding nursery and its relation to plant breeding procedures. Can. J. Plant Sci. 49:21–28.

De Pauw, R. M. and L. H. Shebeski. 1973. An evaluation of an early generation yield testing procedure in *Triticum aestivum*. Can. J. Plant Sci. 53:465–470.

Ebeltoft, D. C. 1969. Rapid increase of barley, spring wheat. North Dakota Agric. Exp. Stn. Farm Res. 26:3–6.

England, F. 1977. Response to family selection based on replicated trials. J. Agric. Sci. 88:127–134.

Finney, D. J. 1958. Plant selection for yield improvement. Euphytica 7:83–106.

Frey, K. J. 1965. The utility of hill plots in oat research. Euphytica 14:196–208.

Garland, M. L. and W. R. Fehr. 1981. Selection for agronomic characters in hill and row plots of soybeans. Crop Sci. 21:591–595.

Harrington, J. B. 1939. How should varieties of annual self-fertilized crops be perpetuated? J. Am. Soc. Agron. 31:472–474.

Harrington, J. B. 1952. Cereal breeding procedures. FAO-UN, Rome, Italy.

Hartwig, E. E., H. W. Johnson, and R. B. Carr. 1951. Border effects in soybean test plots. Agron. J. 43:443–445.

Hayes, H. K. and A. C. Arny. 1917. Experiments in field technic in rod row tests. J. Agric. Res. 11(9):399–419.

Hulbert, H. W., C. A. Michels, and F. L. Burkart. 1931. Border effect in variety tests of small grains. Univ. Idaho AES Bull. 9, Moscow, 23 pp.

Hutchinson, J. B. 1940. The application of genetics to plant breeding. 1. The genetic interpretation of plant breeding problems. J. Genet. 40:271–282.

Ikehashi, H., A. Nakane, and R. Ito. 1969. Optimum replication in pedigree selection under restrictions of field size. Jap. J. Breeding 19:317–320.

Jellum, M. D., C. M. Brown, and R. D. Seif. 1963. Hill and row plot comparison for yield of oats. Crop Sci. 3:194–196.

Jensen, N. F. 1965. Population variability in small grains. Agron. J. 57:153–162.

Jensen, N. F. 1976. Floating checks for plant breeding nurseries. Cereal Res. Commun. 4(3):285–295.

Jensen, N. F. 1978. Composite breeding methods and the DSM system in cereals. Crop Sci. 18:622–626.

Jensen, N. F. and W. T. Federer. 1964. Adjacent row competition in wheat. Crop Sci. 4:641–645.

Jensen, N. F. and D. S. Robson. 1969. Miniature plots for cereal testing. Crop Sci. 9:288–289.

Khadr, F. H., A. A. Kassem, and A. A. ElKhishen. 1970. Hill versus row plot for testing wheat lines. Crop Sci. 10:449–450.

Knott, D. R. 1972. Effects of selection for F2 plant yield on subsequent generations in wheat. Can. J. Plant Sci. 52:721–726.

Love, H. H. 1938. Application of statistical methods to agricultural research. The Commercial Press, Changsha, China, 501 pp.

Love, H. H. 1943. Experimental methods in agricultural research. University of Puerto Rico, Rio Piedras, 229 pp.

Mak, C., B. L. Harvey, and J. D. Berdahl. 1978. An evaluation of control plots and moving means for error control in barley nurseries. Crop Sci. 18:870–873.

Nickell, C. D. and J. E. Grafius. 1969. Analysis of a negtive response to selection for high yield in winter barley, *Hordeum vulgare* L. Crop Sci. 9:447–451.

Park, S. J., E. Reinbergs, and L. S. P. Song. 1977. Grain yield and its components in spring barley under row and hill plot conditions. Euphytica 26:521–526.

Rasmusson, D. C. and J. W. Lambert. 1961. Plot types for preliminary yield tests of barley. Crop Sci. 1:419–420.

Rich, P. A. 1973. Influence of cultivar row spacing and number of rows on yield of wheat plots. Agron. J. 65:331–333.

Ross, W. M. and J. D. Miller. 1955. A comparison of hill and conventional yield tests using oats and spring barley. Agron. J. 47:253–255.

Schutz, W. M. and C. A. Brim. 1967. Inter-genotypic competition in soybeans. I. Evaluation of effects and proposed field plot design. Crop Sci. 7:371–376.

Scott, N. H., W. E. Kronstad, and W. L. McCuistion. 1978. An electronic weighing and data processing system for field use in a small grain breeding program. Crop Sci. 18:684–685.

Sedcole, J. R. 1977. Number of plants necessary to recover a trait. Crop Sci. 17:667–668.

Shebeski, L. H. 1967. Wheat and wheat breeding. *In* K. F. Nielson, (Ed.), Proc. Can. Centennial Wheat Symp. Modern Press, Saskatoon, pp. 249–272.

Sidwell, R. J., E. L. Smith, and R. W. McNew. 1978. Heritability and genetic advance of selected agronomic traits in a winter wheat cross. Cereal Res. Commun. 6:103–111.

Skinner, J. C. 1961. Sugar cane selection experiments. 2. Competition between varieties. Bur. Sugar Exp. Stns. Tech. Commun. Year 1961. No. 1, Sugar Exp. Stns. Board, Brisbane, Queensland.

Smith, O. D., R. A. Kleese, and D. D. Stuthman. 1970. Competition among oat varieties grown in hill plots. Crop Sci. 10:381–384.

Torrie, J. H. 1962. Comparison of hills and rows for evaluating soybean strains. Crop Sci. 2:47–49.

Townley-Smith, T. F. and E. A. Hurd. 1973. Use of moving means in wheat yield trials. Can. J. Plant Sci. 53:447–450.

Walsh, E. J., S. J. Park, and E. Reinbergs. 1976. Hill plots for preliminary yield evaluation of doubled haploids in a barley breeding program. Crop Sci. 16:862–866.

Wheat, J. G. and K. J. Frey. 1961. Number of lines needed in oat-variety purification. Agron J. 53:39–41.

Wiebe, G. A. 1935. Variation and correlation in grain yield among 1500 wheat nursery plots. J. Agric. Res. 50:331–357.

Yates, F. 1936. A new method of arranging variety trials involving a large number of varieties. J. Agric. Sci. 26:424–455.

[illegible] of selected agronomic traits in a winter wheat cross. Cereal Res. Comm. [illegible]

[illegible]

[illegible]

[illegible]

[illegible] 1. Plant Sci. [illegible]

[illegible] plots for preliminary yield [illegible]

[illegible]

[illegible]

[illegible]

38

"101 WAYS" TO ENRICH YOUR BREEDING PROGRAM

An ongoing plant breeding program establishes its own momentum and gradually the major planning and operating procedures become routine. There may come a time of complacency and self-satisfaction when one realizes that a certain input of effort is likely to result in a future acceptable level of research performance output. This can be a welcome realization, permitting the breeder to divert time to other activities, but it can also be a danger signal if nothing fills the created time vacuum. There is never a time when a program cannot absorb new innovations, new approaches, and new directions. Where do such new ideas come from? One source that every breeder taps, knowingly or not, is communication. Attendance at a scientific meeting and participation in discussions can be tremendously inspiring and can ignite a breeder to initiate something new in his or her own program. Reading and thinking can do the same thing. The items that follow are guaranteed to stimulate your thinking about your own program:

1. Broaden your breeding program by sharing. Arrange with colleagues at other stations or institutions to exchange hybrid single-cross seeds on a seed-for-seed basis. Then use these as parents to form three-way hybrid populations.
2. Examine the history of crosses made in your project. Are you rigid in your choice of crosses; for example, do you make only single crosses? How about three-way hybrids? Double crosses? Is there a place for a backcross progam?
3. Could you use the many World Collections in your breeding and testing program? Thousands of genotypes reside in these collections and can

be had for the asking. Find out what is available in categories that interest you, for example, disease resistance or dwarf types.

4. Single out a few high-priority projects and give them unremitting attention. An illustration from my own experience of such an item that did not get persistent attention, although the problem was always there and the research always underway, is the case of the dwarf bunt-resistant soft white winter wheat cultivar 'Avon', introduced in 1959. With the passage of time a shorter version was needed in western New York to compete with the new semidwarf cultivars. I had many crosses and a backcross program with this ideal in mind, but failed to isolate the work sufficiently from the overall wheat improvement program. The result is that the research work for a bunt-resistant semidwarf wheat was never completed, the cultivar Avon passed from the list of recommended varieties, and the western soft white winter wheat area in the state is again susceptible to lossses from an important pathogen.

5. Ponder old cultivars that growers continue to use despite the absence of promotion or publicity. Such cultivars often exhibit a valuable trait. For example, Rodney oats persisted a decade past its "normal" life expectancy because of excellent kernel quality—it has an attractive plump kernel.

6. Consider mutation breeding. Passé in North America, it is important in Europe and the Scandinavian countries where sophisticated procedures are in use. In retrospect I can see that mutation breeding might have been a useful first step in reducing height. For example, our too-tall wheats of the 1930s and 1940s did not lack genes for productivity, but their height frequently made them subject to yield losses from lodging.

7. Watch the crop germ plasm registrations, such as those published in Crop Science. Some of these may be valuable additions to your breeding program.

8. Identify and build small breeding projects around a common germ plasm base. For example, the Iowa oat improvement program has emphasized the use of *Avena sterilis* germ plasm. The base can be a single genotype, or it may be several chosen as parents for a central breeding goal.

9. Deliberately create variety in your hybridizing program by inserting exotic parents. How often an outstanding cultivar harbors a parent of unheralded ancestry! The choice of such additions to your crossing program should not be random, but should be guided by a perceived contribution that the genotype could make, for example, earliness, reduced height, rust resistance, or yield.

10. Review the strength and efficiency of your attrition procedures. Strive to keep your program lean. Is there an entire population you can discontinue? Is your selection rate too high (recognizing, of course, that it will vary from population to population depending upon the excellence of the material)? Make your program reflect the stamp of your authority.

11. Use bulk germ plasm collections. For example, for some decades

cereal breeders and geneticists have formed germ plasm collections. Most of these are still in existence, in advanced generations, and available to breeders.

12. Identify key characteristics of the crop you are working on and incorporate these as basic goals for your planning and crossing program. A good example is test weight of grain, which must be given high priority in the eastern soft white winter wheats. High protein in oats is another key trait.

13. Evaluate cross potential. This is a tough one—an area not yet mastered by all plant breeders. As your hybrid populations come on-stream can you realistically evaluate their potential worth, the chances of their containing releasable individuals? Study the pedigrees while you have the F_2 and F_3 population before you in the field. If you can, with confidence and equanimity, discard 5, 10, or even 20% of your hybrid populations, it will do wonders for the efficient movement of your program.

14. Observe the new lines being developed by colleagues elsewhere and bring some of these as parents into your crossing program. You may get a five-year time advantage in using a candivar in this way, as well as ensuring a desirable diversity in your breeding program.

15. Establish a reserve bank of single crosses to use in your hybridization work. These can be used as parents for future three-way crosses (and after they have served as parents the remaining F_2-embryo seeds can still be used as a population for selection). The wheat single-cross reserve bank in my wheat breeding project at Cornell University always contained a few hundred selected single-cross hybrids. Each year when I planned the greenhouse crossing program I would choose numbers of these—usually only one seed from a cross sufficed—to be grown as parents along with the pure line genotypes. The attraction of this bank is the wide latitude of germ plasm choice available at all times to the breeder.

16. Follow the new introductions to the World Collections. In some cases these are listed annually as, for example, in the crop newsletters. Many of these are "unknowns" as to performance. Initially, little more than place of origin may be known; however, there is increased emphasis today on accumulating and cataloging detailed information on each entry.

17. Broaden your breeding plans by choosing as parents some among the current best performers in regional, national and international nurseries. Examine not only yield but other characteristics in which you are interested. For example, I routinely added exotic cultivars such as 'Kavkaz' to my wheat program based on their worldwide performance. Kavkaz, incidentally, was not only excellent in yield but had good cold resistance and lodging resistance.

18. If you are a cereal breeder, visit the World Collection of Small Grains at its annual growing site. These collections are so large now that only a portion can be grown in one year and it may take five or more years

to make a complete increase of one crop collection. Convince your administrators that this can be the most rewarding expenditure of project monies. Note the numbers of genotypes that interest you in the field and request they be sent to you; or if the crop is near maturity you may be permitted to take a few seeds right in the field. Your selections can then be grown in short rows for further observation in your locality. You will be looking for parents, but keep in mind that thousands of genotypes in the Collections are nonproprietary open stock. It is possible to find an outstanding genotype adapted to your region and one could inquire as to its status; perhaps it is available for naming and release as a cultivar.

19. Review the planning and operations of your crossing program. These, and the methods you use—backcross, pedigree, composite—largely determine the size and strength of your program. Could your program benefit from more hybrids per season in order to present to you a wider array of germ plasm and segregation? Do you feel that you are overextended and need to process populations in more detail? If the first is true, perhaps more composite handling is the answer; if the second, you may need to move more to pedigree or backcross methods. The range in workable numbers of annual hybrids made is enormous, and different plant breeders can find success at different points in this range. For example, my predecessor at Cornell averaged fewer than 30 crosses per year yet was a successful wheat breeder with a number of released cultivars, some of which were of world-class quality.

20. Break down your overall project goals to several specific goals. Don't just say that you have a barley breeding program. Instead, speak about and act upon your malting barley project, the powdery mildew division, the height program, and so forth. These are interrelated, part and parcel of each other, yet the separate categories, each with its separate plans, files, and field organizaiton, will aid you in reaching the overall goals of barley breeding.

21. Create crypto single crosses that can substitute for pure lines as parents. Certain single-cross combinations may be better than pure lines in hybrids. They often are single crosses of pure lines that have arrays of traits such that the breeder may think of them as equally desirable when used as a pure line parent. Illustrations are the high-protein oat cultivars 'Wright' × 'Dal' (or substitute high-protein 'Jaycee' for one); crosses of any of the recent soft white wheat cultivars, all of which have a comparable quality base; and combinations of malting quality barley cultivars. Such single crosses have the advantage of introducing significant additional germ plasm breadth to a further cross (three-way), yet this three-way cross may be treated as though a single cross had been made. To illustrate, the crypto single cross of Wright × Dal might be crossed to 'Orbit', creating the three-way cross Wright × Dal 2× Orbit. However, the breeder may consider this as simply a single cross of high protein × Orbit. This permits him to build a further "three-way" cross of, for example, 'Ogle' 2× Orbit × high

protein. The germ plasm breadth advantages of such controlled complex crosses are obvious.

22. Think about strategies of crossing and exploiting certain crosses. Do you spend equal time on all crosses? Try looking over the list of all crosses made during a season and select those that on paper seem worth more effort than the others. Successful or not, this creates real interest in managing the program.

23. Give your program a sense of continuity. See it not as a series of discrete hybridizations with each season an end in itself; rather, think of it as rolling series of crosses that build and cycle and join one season to the next. For example, in the oat high-protein project (Item 21 above) think of ways to extend the crossing sequence, for example, by introducing dwarfness via 'O.T. 184', then crossing back to high protein, followed by crossing dwarf types again to Ogle or Orbit or other cultivars. As soon as high-protein lines are identified from the high-protein project, tie them back to the program through recurrent selection.

24. Become catholic in your reading habits. Expand your journal coverage. Just as important, expand your awareness of research in other crops. Workers in different crop species "reinvent the wheel" time after time and remain unaware that the answers to planned research are already available elsewhere or can be deduced. Researchers in the self-pollinated crops can find a world of information in literature on the cross-pollinated species; indeed, many of the problems are the same, for example, genetic sterility, recurrent selection, genetic advance, pollination, and so forth.

25. This is the computer age. Use the computer to sift the information of decades on breeding, selection, yield, parent contributions, and a host of other unprocessed data. Today, even professional baseball and footabll teams use computers to develop game tactics and strategies, to compute odds and percentages for given moves—even to evaluate players for draft choices.

26. Subscribe to the annual crop newsletters (often free or for a donation).

27. Establish mandatory indoctrination sessions from experienced colleagues to further the training and awareness base of new plant breeders. The purpose should be to save the new breeder the time and effort exploring pathways found by experience to be unprofitable, and to stimulate an historical evaluation of procedures without suggesting rigid operational procedures.

28. Set up high × low crosses, each highlighting a different important plant trait, for example, yield or protein.

29. Explore index selection to see if it will help in your selection program. Try something simple, such as tandem selection, and evaluate it against single character selection.

30. AND, there are at least 71 more moves you can make!

AUTHOR INDEX

SUBJECT INDEX